A COURSE
OF
PURE MATHEMATICS

A COURSE
OF
PURE MATHEMATICS

BY

G. H. HARDY

TENTH EDITION

WITH A FOREWORD
BY T. W. KÖRNER

CAMBRIDGE
UNIVERSITY PRESS

University Printing House, Cambridge CB2 8BS, United Kingdom

Cambridge University Press is part of the University of Cambridge.

It furthers the University's mission by disseminating knowledge in the pursuit of education, learning and research at the highest international levels of excellence.

www.cambridge.org
Information on this title: www.cambridge.org/9780521720557

First published 1908
Tenth edition reissued with Foreword 2008
4th printing 2015

A catalogue record for this publication is available from the British Library

ISBN 978-0-521-72055-7 Paperback

FOREWORD

T. W. Körner

My copy of Hardy's *Pure Mathematics* is the eighth edition, printed in 1941. It must have been one of the first books that my father bought as an almost penniless refugee student in England, and the pencilled notations show that he read most of it. It was the first real mathematics book that I attempted to read and, though much must have passed over my head, I can still feel the thrill of reading the construction of the real numbers by Dedekind cuts. One hundred years after it was first published, CUP is issuing this Centenary edition, not as an act of piety, but because *A Course In Pure Mathematics* remains an excellent seller, bought and read by every new generation of mathematicians.

During most of the nineteenth century, mathematics stood supreme among the subjects studied at Cambridge. Exposure to the absolute truths of mathematics was an essential part of an intellectual education. The most able students could measure themselves against their opponents in mathematical examinations (the Tripos) which tested speed, accuracy and problem-solving abilities to the utmost. However, it was a system directed entirely towards the teaching of undergraduates. In Germany and France there were research schools in centres like Berlin, Göttingen and Paris. In England, major mathematicians like Henry Smith and Cayley remained admired but isolated.

An education that produced Maxwell, Kelvin, Rayleigh and Stokes cannot be dismissed out of hand, but any mathematical school which concentrates on teaching and examining runs the risk of becoming old-fashioned. (Think of the concours for the Grandes Écoles in our day.) It is possible that, even in applied mathematics, the Cambridge approach was falling behind Europe. It is certain that, with a few notable but isolated exceptions, pure mathematical research hardly existed in Britain. Hardy took pleasure in repeating the judgement of an unnamed European colleague that the characteristics of English mathematics had been 'occasional flashes of insight, isolated achievements sufficient to show that the ability is really there, but for the most part, amateurism, ignorance, incompetence and triviality.'

When Hardy arrived as a student at Cambridge, reform was very much in the air.

> I had of course found at school, as every future mathematician does, that I could often do things much better than my teachers; and even at Cambridge I found, though naturally much less frequently, that I could sometimes do things better than the college lecturers. But I was really quite ignorant, even when I took the Tripos, of the subjects on which I have spent my life; and I still thought of mathematics as essentially a 'competitive' subject. My eyes were first opened by Professor Love, who taught me for a few terms and gave me my first serious conception of analysis. But the great debt which I owe to him – he was after all primarily an applied mathematician – was his advice to read Jordan's famous *Cours d'analyse*; and I shall never forget the astonishment with which I read this remarkable work, the first inspiration for so many mathematicians of my generation, and learnt for the first time as I read it what mathematics really meant.
>
> [*A Mathematician's Apology*]

Ever since Newton, mathematicians had struggled with the problem of putting the calculus on as sound a footing as Euclid's geometry. But what were the fundamental axioms on which the calculus was to be founded? How should concepts like a differentiable function be defined? Which theorems were 'obvious' and which 'subtle'? Until these questions were answered, all calculus textbooks would have to mix accurate argument with hand waving. Sometimes the author would be aware of gaps and resort to rhetoric 'Persist and faith will come to you.' More frequently, author and reader would sleepwalk hand in hand through the difficulty – most lecturers will be aware how fatally easy it is to convince an audience of an erroneous proof provided you are convinced of it yourself.

The first edition of Jordan's work (1882–87) belonged to this old tradition but the second edition (1893–96) wove together the work of rigorisers like Weierstrass to produce a complete and satisfactory account of the calculus. The impact on Cambridge of Jordan and of the new 'continental' analysis was immense. Young and Hobson, men who had expected to spend their lifetimes in the comfortable routine of undergraduate teaching, suddenly threw themselves into research and, still more remarkably, became great mathematicians.

This impact can be read in three books which are still in print today. We read them now in various revised editions, but all were first written by young men determined to challenge a century of tradition. The first was Whittaker's *A Course of Modern Analysis* (1902) (later editions by Whittaker and Watson), which showed that the special functions which formed the crown jewels of the old analysis were best treated by modern methods. The second was Hobson's *The Theory of Functions of a real variable* (1907), which set out the new analysis for professional mathematicians. The third is the present text, first published in 1908 and intended for 'first year students... whose abilities reach or approach... scholarship standard'.

The idea of such a text may appear equally absurd to those who deal with the mass university system of today and to those whose view of the old university system is moulded by *Brideshead Revisited* or *Sinister Street.* However, although most of the students in Cambridge came from well-off backgrounds, some came from poorer backgrounds and *needed* to distinguish themselves whilst some of their richer companions *wished* to distinguish themselves. Most of the mathematically able students came from a limited number of schools where they often received an outstanding mathematical education. (Read, for example, Littlewood's account of his mathematical education in *A Mathematician's Miscellany.*)

Hardy's intended audience was small and it is not surprising that CUP made him pay £15 out of his own pocket for corrections. This audience was, however, an audience fully accustomed through the study of Euclidean geometry both to follow and, even more importantly, to construct long chains of reasoning. It was also trained in fast and accurate manipulations, both in algebra and calculus, within a problem-solving context. The modern writer of a first course in analysis must address an audience with much less experience of proof, substantially lower algebraic fluency and little experience of applying calculus to interesting problems in mechanics and geometry. Spivak's *Calculus* is outstanding, but Hardy can illustrate his text with much richer exercises. (The reader should note that questions like Example 1.1, which appear to be simple statements are, in fact, invitations to prove those statements.)

Cambridge and Oxford used Hardy's *Pure Mathematics*, and the two universities dominated the British mathematical scene. (Before World War II, almost every mathematics professor in the British Isles was Cambridge or Oxford trained.) For the next 70 years, Hardy's book defined the first analysis course in Britain. Analysis texts could have borne titles like: 'Hardy made easier', 'An introduction to Hardy' or 'Hardy slimmed down'. Burkill's *First Course in Analysis* represents an outstanding example in the latter class.

In the last 40 years, Hardy's model has been put under strain from two different directions. The expansion of the university system has brought more students into mathematics, but the new students are less well prepared and less willing to study mathematics for its own sake. It is clear that 'Hardy diluted' cannot be appropriate for such students. On the other hand, the frontiers of mathematics have continued to advance and an analysis course for future researchers must prepare them to meet such things as manifolds and infinite dimensional spaces. Dieudonné's *Foundations of Modern Analysis* and Kolmogorov and Fomin's *Introductory Real Analysis* represent two very different but equally inspiring approaches to the problem.

As new topics enter the syllabus, old ones have to be removed. Today's best students get 'Hardy stripped to the bone' followed by a course on metric and topological spaces. They are taught speedily and efficiently, but some things have been lost. The TGV carries you swiftly across France, but isolates you from the land and its people. We claim to give our students the experience of mathematics, but provide plenty of 'routine exercises' and relegate 'the more difficult proofs' to appendices. Perhaps later generations of mathematicians may judge our teaching as harshly as Hardy and his generation judged the teaching of their Cambridge predecessors.

Hardy's book begins with a presentation of the properties of the real number system. In the first edition this is done axiomatically, but, in the second and later editions, Hardy constructs the real numbers starting from the rationals. Bertrand Russell says that there are two methods of presenting mathematics, the postulational and the constructive, with the postulational method having all the advantages of theft over honest toil, but a modern course would either leave the construction until much later or omit it altogether. Hardy allows the reader to skip the construction, but the reader should do at least some of the exercises that conclude the chapter. The next two chapters present material that the modern undergraduate would be expected to have met before embarking on a course of analysis.

The course proper starts with Chapter IV and V which introduce the notion of a limit. The treatment is is more leisurely than would be found in a modern introduction, but the reader who hurries through it is throwing away the advantage of listening to a great analyst talking about the elements of his subject. In one or two places the notation is definitely old-fashioned (as foreshadowed in the footnote in §71 *divergent* now means simply *not convergent*). It is easy, however, to make the transfer to modern notation later. More importantly, the reader should note that the theorems in §101 to §107 lie much deeper than those that precede them. Wherever Hardy makes use of the classes L and R of §17 he is appealing to the basic properties of the real numbers and the reader should pay close attention to the argument.

Chapter VI introduces differentiation and integration. Here, the most subtle arguments are to be found in §122 leading to the mean value theorem in §126. Until §161, Hardy considers integration as the inverse operation to differentiation (though he gives the link to area informally in §148) but in this section he defines the definite integral and completes his presentation of the foundations of the calculus. Once the foundations have been laid, he goes on to develop the standard methods and theorems of the calculus and give a rigorous account of the trigonometric, exponential and logarithmic function, both in the real and complex case.

In his *Mathematical Thought from Ancient to Modern Times* Kline dismisses the 100-year struggle of Bolzano, Cauchy, Abel, Dirichlet, Weierstrass, Cantor, Peano and others to rigorise analysis with the words 'the theorems of analysis only had to be more carefully formulated . . . all that rigour did was to substantiate what mathematicians had always known to be the case'. In fact, the rigorisation process revealed that several things that mathematicians had always known to be the case were, in fact, false. It is not true that every maximisation problem has a solution, it is not true that any continuous function must be differentiable except at a few exceptional points, it is not true that the boundary of a region is a negligible part of it, it is not true that every sufficiently smooth function is equal to its Taylor series, ...

Still more importantly, the process of rigourisation revealed the underlying structure of the real line and produced new tools (such as the Heine–Borel theorem of §106) to exploit that structure. In the decade that Hardy wrote his text, the study of the notion of area by Cantor, Peano, Jordan and Borel reached its apotheosis in the work of Lebesgue. Armed with the new tool of the Lebesgue integral and clear understanding of foundations, analysis entered on a golden century to which Hardy was to contribute such gems as the Hardy spaces and the Hardy–Littlewood maximal theorem.

> I wrote a great deal during . . . [1900–1910], but very little of any importance; there are not more than four or five papers which I can still remember with some satisfaction. The real crises of my career came in 1911, when I began my long collaboration with Littlewood, and in 1913 when I discovered Ramanujan.
>
> [*A Mathematician's Apology*]

For its author and for his audience, *A Course in Pure Mathematics* represented not an end but a beginning.

Hardy published about 350 papers, including nearly 100 with Littlewood, but his contributions to mathematics did not stop there. He taught and inspired generations of research students. As one of them writes about Hardy's lectures, 'Whatever the subject was, he pursued it with an eager single-mindedness which the audience found irresistible. One felt, temporarily at least, that nothing else in the world but the proof of those theorems mattered. There could be no more inspiring director of the work of others. He was always at the head of a team of researchers, both colleagues and students, whom he provided with an inexhaustible stock of ideas on which to work.' Tichmarsh adds, and others confirm, that 'He was an extremely kind-hearted man, who could not bear any of his students to fail in their researches.'

Pólya recalled how Hardy '... valued clarity, yet what he valued most in mathematics was not clarity but power, surmounting great obstacles that others abandoned in despair.' Pólya also recalled how much Hardy loved jokes and told an anecdote which illustrated both aspects of Hardy's character.

> In working with Hardy, I once had an idea of which he approved. But afterwards I did not work sufficiently hard to carry out that idea, and Hardy disapproved. He did not tell me so, of course, yet it came out when he visited a zoological garden in Sweden with Marcel Riesz. In a cage there was a bear. The cage had a gate, and on the gate there was a lock. The bear sniffed at the lock, hit it with his paw, then growled a little, turned around and walked away. 'He is like Pólya', said Hardy. 'He has excellent ideas, but does not carry them out.'

In Hardy's presidential address to the London Mathematical Society in 1928 he was able to boast that he had sat through every word of every lecture of every meeting of every paper since he became secretary in 1917. He oversaw the foundation of the *Journal of the London Mathematical Society* and revived the *Quarterly Journal* at Oxford. The satisfactory state of the London Mathematical Society's finances today is the result of Hardy's bequest of a substantial fortune and the royalties of his books.

Hardy wrote or co-wrote several other classics. Perhaps the most remarkable is *Inequalities* with Littlewood and Pólya. In this book authors magically provide a coherent view of a subject which, though it lies at the heart of analysis, seems impossible to organise.

Hardy's *A Mathematician's Apology* is both a mathematical and a literary triumph and remains unequalled as a meditation on the life of a pure mathematician. It is also a defence of rationality and the free life of the intellect at a time when they were terribly threatened.

However, in my view, the most enchanting of his books is *Number Theory* (written with E. M. Wright). If I had to choose one book to take to a desert island, I would take Zygmund's *Trigonometric Series* if

I thought I might be rescued, but Hardy and Wright's *Number Theory* if I knew that I was never coming back.

To read Hardy is to read a mathematician fully aware of his own abilities but who treats you as a natural equal. May this book give as much pleasure to you as it has given to me.

T. W. Körner

PREFACE TO THE TENTH EDITION

THE changes in the present edition are as follows:

1. An index has been added. Hardy had begun a revision of an index compiled by Professor S. Mitchell; this has been completed, as far as possible on Hardy's lines, by Dr T. M. Flett.

2. The original proof of the Heine-Borel Theorem (pp. 197–199) has been replaced by two alternative proofs due to Professor A. S. Besicovitch.

3. Example 24, p. 394 has been added to.

August, 1950 J. E. LITTLEWOOD

PREFACE TO THE SEVENTH EDITION

THE changes in this edition are more important than in any since the second. The book has been reset, and this has given me the opportunity of altering it freely.

I have cancelled what was Appendix II (on the 'O, o, $\sim$' notation), and incorporated its contents in the appropriate places in the text. I have rewritten the parts of Chs. VI and VII which deal with the elementary properties of differential coefficients. Here I have found de la Vallée-Poussin's *Cours d'analyse* the best guide, and I am sure that this part of the book is much improved. These important changes have naturally involved many minor emendations.

I have inserted a large number of new examples from the papers for the Mathematical Tripos during the last twenty years, which should be useful to Cambridge students. These were collected for me by Mr E. R. Love, who has also read all the proofs and corrected many errors.

The general plan of the book is unchanged. I have often felt tempted, re-reading it in detail for the first time for twenty years, to make much more drastic changes both in substance and in style. It was written when analysis was neglected in Cambridge, and with an emphasis and enthusiasm which seem rather ridiculous now. If I were to rewrite it now I should not write (to use Prof. Littlewood's simile) like 'a missionary talking to cannibals', but with decent terseness and restraint; and, writing more shortly, I should be able to include a great deal more. The book would then be much more like a *Traité d'analyse* of the standard pattern.

It is perhaps fortunate that I have no time for such an undertaking, since I should probably end by writing a much better but much less individual book, and one less useful as an introduction to the books on analysis of which, even in England, there is now no lack.

November, 1937 G. H. H.

EXTRACT FROM THE PREFACE TO THE FIRST EDITION

THIS book has been designed primarily for the use of first year students at the Universities whose abilities reach or approach something like what is usually described as 'scholarship standard'. I hope that it may be useful to other classes of readers, but it is this class whose wants I have considered first. It is in any case a book for mathematicians: I have nowhere made any attempt to meet the needs of students of engineering or indeed any class of students whose interests are not primarily mathematical.

I regard the book as being really elementary. There are plenty of hard examples (mainly at the ends of the chapters): to these I have added, wherever space permitted, an outline of the solution. But I have done my best to avoid the inclusion of anything that involves really difficult ideas.

September, 1908 G. H. H.

CONTENTS

(Entries in small print at the end of the contents of each chapter refer to subjects discussed incidentally in the examples)

CHAPTER I

Real variables

CHAPTER II

Functions of real variables

CHAPTER III

COMPLEX NUMBERS

CHAPTER IV

LIMITS OF FUNCTIONS OF A POSITIVE INTEGRAL VARIABLE

CHAPTER V

Limits of functions of a continuous variable. Continuous and discontinuous functions

CHAPTER VI

Derivatives and integrals

CHAPTER VII

Additional theorems in the differential and integral calculus

CHAPTER VIII

THE CONVERGENCE OF INFINITE SERIES AND INFINITE INTEGRALS

CHAPTER IX

THE LOGARITHMIC, EXPONENTIAL, AND CIRCULAR FUNCTIONS OF A REAL VARIABLE

CHAPTER X

THE GENERAL THEORY OF THE LOGARITHMIC, EXPONENTIAL, AND CIRCULAR FUNCTIONS

CHAPTER I

REAL VARIABLES

1. Rational numbers. A fraction $r = p/q$, where p and q are positive or negative integers, is called a *rational number*. We can suppose (i) that p and q have no common factor, since if they have a common factor we can divide each of them by it, and (ii) that q is positive, since

$$p/(-q) = (-p)/q, \quad (-p)/(-q) = p/q.$$

To the rational numbers thus defined we may add the 'rational number 0' obtained by taking $p = 0$.

We assume that the reader is familiar with the ordinary arithmetical rules for the manipulation of rational numbers. The examples which follow demand no knowledge beyond this.

Examples I. 1. If r and s are rational numbers, then $r+s$, $r-s$, rs, and r/s are rational numbers, unless in the last case $s = 0$ (when r/s is of course meaningless).

2. If λ, m, and n are positive rational numbers, and $m>n$, then $\lambda(m^2-n^2)$, $2\lambda mn$, and $\lambda(m^2+n^2)$ are positive rational numbers. Hence show how to determine any number of right-angled triangles the lengths of all of whose sides are rational.

3. Any terminated decimal represents a rational number whose denominator contains no factors other than 2 or 5. Conversely, any such rational number can be expressed, and in one way only, as a terminated decimal.

[The general theory of decimals will be considered in Ch. IV.]

4. The positive rational numbers may be arranged in the form of a simple series as follows:

$$\tfrac{1}{1}, \tfrac{2}{1}, \tfrac{1}{2}, \tfrac{3}{1}, \tfrac{2}{2}, \tfrac{1}{3}, \tfrac{4}{1}, \tfrac{3}{2}, \tfrac{2}{3}, \tfrac{1}{4}, \ldots.$$

Show that p/q is the $[\frac{1}{2}(p+q-1)(p+q-2)+q]$th term of the series.

[In this series every rational number is repeated indefinitely. Thus 1 occurs as $\frac{1}{1}, \frac{2}{2}, \frac{3}{3}, \ldots$. We can of course avoid this by omitting every number which has already occurred in a simpler form, but then the problem of determining the precise position of p/q becomes more complicated.]

2. The representation of rational numbers by points on a line. It is convenient, in many branches of mathematical analysis, to make a good deal of use of geometrical illustrations.

The use of geometrical illustrations in this way does not, of course, imply that analysis has any sort of dependence upon geometry: they are illustrations and nothing more, and are employed merely for the sake of clearness of exposition. This being so, it is not necessary that we should attempt any logical analysis of the ordinary notions of elementary geometry; we may be content to suppose, however far it may be from the truth, that we know what they mean.

Assuming, then, that we know what is meant by a *straight line*, a *segment* of a line, and the *length* of a segment, let us take a straight line Λ, produced indefinitely in both directions, and a segment A_0A_1 of any length. We call A_0 the *origin*, or *the point* 0, and A_1 *the point* 1, and we regard these points as representing the numbers 0 and 1.

In order to obtain a point which shall represent a positive rational number $r = p/q$, we choose the point A_r such that

$$A_0A_r/A_0A_1 = r,$$

A_0A_r being a stretch of the line extending in the same direction along the line as A_0A_1, a direction which we shall suppose to be from left to right when, as in Fig. 1, the line is drawn horizontally across the paper. In order to obtain a point to represent a

A_{-s} A_{-1} A_0 A_1 A_s

Fig. 1

negative rational number $r = -s$, it is natural to regard length as a magnitude capable of sign, positive if the length is measured in one direction (that of A_0A_1), and negative if measured in the other, so that $AB = -BA$; and to take as the point representing r the point A_{-s} such that

$$A_0A_{-s} = -A_{-s}A_0 = -A_0A_s.$$

We thus obtain a point A_r on the line corresponding to every rational value of r, positive or negative, and such that

$$A_0A_r = r . A_0A_1;$$

and if, as is natural, we take A_0A_1 as our unit of length, and write $A_0A_1 = 1$, then we have

$$A_0A_r = r.$$

We shall call the points A_r the *rational points* of the line.

3. Irrational numbers. If the reader will mark off on the line all the points corresponding to the rational numbers whose denominators are 1, 2, 3, ... in succession, he will readily convince himself that he can cover the line with rational points as closely as he likes. We can state this more precisely as follows: *if we take any segment BC on Λ, we can find as many rational points as we please on BC.*

Suppose, for example, that BC falls within the segment A_1A_2. It is evident that if we choose a positive integer k so that

$$k . BC > 1 \qquad (1)^*,$$

and divide A_1A_2 into k equal parts, then at least one of the points of division (say P) must fall inside BC, without coinciding with either B or C. For if this were not so, BC would be entirely included in one of the k parts into which A_1A_2 has been divided, which contradicts the supposition (1). But P obviously corresponds to a rational number whose denominator is k. Thus at least one rational point P lies between B and C. But then we can find another such point Q between B and P, another between B and Q, and so on indefinitely; i.e., as we asserted above, we can find as many as we please. We may express this by saying that BC includes *infinitely many* rational points.

The meaning of such phrases as '*infinitely many*' or '*an infinity of*', in such sentences as 'BC includes infinitely many rational points' or 'there are an infinity of rational points on BC' or 'there are an infinity of positive integers', will be considered more closely in Ch. IV. The assertion 'there are an infinity of positive integers' means 'given any positive integer n,

* The assumption that this is possible is equivalent to the assumption of what is known as the axiom of Archimedes.

however large, we can find more than n positive integers'. This is plainly true whatever n may be, e.g. for $n = 100{,}000$ or $100{,}000{,}000$. The assertion means exactly the same as 'we can find *as many positive integers as we please*'.

The reader will easily convince himself of the truth of the following assertion, which is substantially equivalent to what was proved in the second paragraph of this section: given any rational number r, and any positive integer n, we can find another rational number lying on either side of r and differing from r by less than $1/n$. It is merely to express this differently to say that we can find a rational number lying on either side of r and differing from r by *as little as we please*. Again, given any two rational numbers r and s, we can interpolate between them a chain of rational numbers in which any two consecutive terms differ by as little as we please, that is to say by less than $1/n$, where n is any positive integer assigned beforehand.

From these considerations the reader might be tempted to infer that an adequate view of the nature of the line could be obtained by imagining it to be formed simply by the rational points which lie on it. And it is certainly the case that if we imagine the line to be made up solely of the rational points, and all other points (if there are any such) to be eliminated, the figure which remained would possess most of the properties which common sense attributes to the straight line, and would, to put the matter roughly, look and behave very much like a line.

A little further consideration, however, shows that this view would involve us in serious difficulties.

Let us look at the matter for a moment with the eye of common sense, and consider some of the properties which we may reasonably expect a straight line to possess if it is to satisfy the idea which we have formed of it in elementary geometry.

The straight line must be composed of points, and any segment of it by all the points which lie between its end points. With any such segment must be associated a certain entity called its *length*, which must be a *quantity* capable of *numerical measurement* in terms of any standard or unit length, and these lengths must be capable of combination with one another, according to the ordinary rules of algebra, by means of addition or multiplication.

Again, it must be possible to construct a line whose length is the sum or product of any two given lengths. If the length PQ, along a given line, is a, and the length QR, along the same straight line, is b, the length PR must be $a+b$. Moreover, if the lengths OP, OQ, along one straight line, are 1 and a, and the length OR along another straight line is b, and if we determine the length OS by Euclid's construction (Euc. VI. 12) for a fourth proportional to the lines OP, OQ, OR, this length must be ab, the algebraical fourth proportional to 1, a, b. And it is hardly necessary to remark that the sums and products thus defined must obey the ordinary 'laws of algebra'; viz.

$$a+b = b+a, \quad a+(b+c) = (a+b)+c,$$

$$ab = ba, \quad a(bc) = (ab)c, \quad a(b+c) = ab+ac.$$

The lengths of our lines must also obey a number of obvious laws concerning inequalities as well as equalities: thus if A, B, C are three points lying along Λ from left to right, we must have $AB < AC$, and so on. Moreover it must be possible, on our fundamental line Λ, to find a point P such that A_0P is equal to any segment whatever taken along Λ or along any other straight line. All these properties of a line, and more, are involved in the presuppositions of our elementary geometry.

Now it is very easy to see that the idea of a straight line as composed of a series of points, each corresponding to a rational number, cannot possibly satisfy all these requirements. There

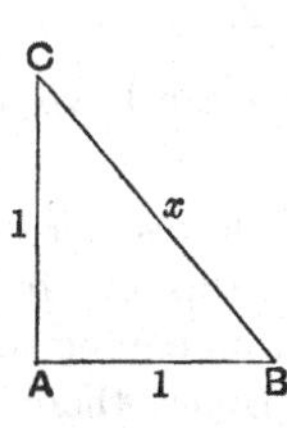

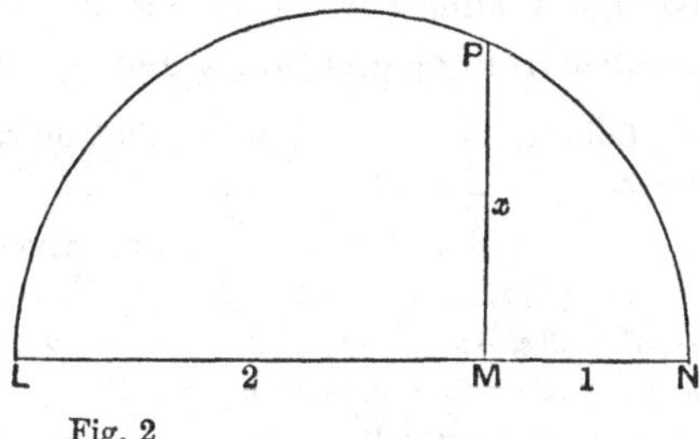

Fig. 2

are various elementary geometrical constructions, for example, which purport to construct a length x such that $x^2 = 2$. For instance, we may construct an isosceles right-angled triangle

ABC such that $AB = AC = 1$. Then if $BC = x$, $x^2 = 2$. Or we may determine the length x by means of Euclid's construction (Euc. VI. 13) for a mean proportional to 1 and 2, as indicated in the figure. Our requirements therefore involve the existence of a length measured by a number x, and a point P on Λ, such that

$$A_0P = x, \quad x^2 = 2.$$

But it is easy to see that *there is no rational number such that its square is* 2. In fact we may go further and say that there is no rational number whose square is m/n, where m/n is any positive fraction in its lowest terms, unless m and n are both perfect squares.

For suppose, if possible, that

$$p^2/q^2 = m/n,$$

p having no factor in common with q, and m no factor in common with n. Then $np^2 = mq^2$. Every factor of q^2 must divide np^2 and, as p and q have no common factor, every factor of q^2 must divide n. Hence $n = \lambda q^2$, where λ is an integer. But this involves $m = \lambda p^2$; and, as m and n have no common factor, λ must be unity. Thus $m = p^2$, $n = q^2$, as was to be proved. In particular it follows, by taking $n = 1$, that an integer cannot be the square of a rational number, unless that rational number is itself integral.

It appears then that our requirements involve the existence of a number x and a point P, not one of the rational points already constructed, such that $A_0P = x$, $x^2 = 2$; and (as the reader will remember from elementary algebra) we write $x = \sqrt{2}$.

The following alternative proof that no rational number can have its square equal to 2 is interesting.

Suppose, if possible, that p/q is a positive fraction, in its lowest terms, such that $(p/q)^2 = 2$ or $p^2 = 2q^2$. It is easy to see that this involves $(2q-p)^2 = 2(p-q)^2$; and so $(2q-p)/(p-q)$ is another fraction having the same property. But clearly $q<p<2q$, and so $p-q<q$. Hence there is another fraction equal to p/q and having a smaller denominator, which contradicts the assumption that p/q is in its lowest terms.

Examples II. 1. Show that no rational number can have its cube equal to 2.

2. Prove generally that a rational fraction p/q in its lowest terms cannot be the cube of a rational number unless p and q are both perfect cubes.

3. A more general proposition, which is due to Gauss and includes those which precede as particular cases, is the following: *an algebraical equation*

$$x^n + p_1 x^{n-1} + p_2 x^{n-2} + \dots + p_n = 0,$$

with integral coefficients, cannot have a rational but non-integral root.

[For suppose that the equation has a root a/b, where a and b are integers without a common factor, and b is positive. Writing a/b for x, and multiplying by b^{n-1}, we obtain

$$-\frac{a^n}{b} = p_1 a^{n-1} + p_2 a^{n-2} b + \dots + p_n b^{n-1},$$

a fraction in its lowest terms equal to an integer, which is absurd. Thus $b = 1$, and the root is a. It is evident that a must be a divisor of p_n. More generally, if a/b is a root of $p_0 x^n + p_1 x^{n-1} + \dots + p_n = 0$, then a is a divisor of p_n and b of p_0.]

4. Show that if $p_n = 1$ and neither of

$$1 + p_1 + p_2 + p_3 + \dots, \quad 1 - p_1 + p_2 - p_3 + \dots$$

is zero, then the equation cannot have a rational root.

5. Find the rational roots (if any) of

$$x^4 - 4x^3 - 8x^2 + 13x + 10 = 0.$$

[The roots can only be integral, and so ± 1, ± 2, ± 5, ± 10 are the only possibilities: whether these are roots can be determined by trial.]

4. Irrational numbers (*continued*). The result of our geometrical representation of the rational numbers is therefore to suggest the desirability of enlarging our conception of 'number' by the introduction of further numbers of a new kind.

The same conclusion might have been reached without the use of geometrical language. One of the central problems of algebra is that of the solution of equations, such as

$$x^2 = 1, \quad x^2 = 2.$$

The first equation has the two rational roots 1 and -1. But, if our conception of number is to be limited to the rational numbers, we can only say that the second equation has no roots; and the same is the case with such equations as $x^3 = 2$, $x^4 = 7$. These facts are plainly sufficient to make some generalisation of our idea of number desirable, if it should prove to be possible.

Let us consider more closely the equation $x^2 = 2$.

We have already seen that there is no rational number x which satisfies this equation. The square of any rational number is either less than or greater than 2. We can therefore divide the positive rational numbers (to which for the present we confine our attention) into two classes, one containing the numbers whose squares are less than 2, and the other those whose squares are greater than 2. We shall call these two classes *the class L*, or *the lower class*, or *the left-hand class*, and *the class R*, or *the upper class*, or *the right-hand class*. It is obvious that every member of R is greater than all the members of L. Moreover it is easy to convince ourselves that we can find a member of the class L whose square, though less than 2, differs from 2 by as little as we please, and a member of R whose square, though greater than 2, also differs from 2 by as little as we please. In fact, if we carry out the ordinary arithmetical process for the extraction of the square root of 2, we obtain a series of rational numbers, viz.

$$1, \quad 1{\cdot}4, \quad 1{\cdot}41, \quad 1{\cdot}414, \quad 1{\cdot}4142, \quad \ldots$$

whose squares

$$1, \quad 1{\cdot}96, \quad 1{\cdot}9881, \quad 1{\cdot}999396, \quad 1{\cdot}99996164, \quad \ldots$$

are all less than 2, but approach nearer and nearer to it; and by taking a sufficient number of the figures given by the process we can obtain as close an approximation as we want. And if we increase the last figure, in each of the approximations given above, by unity, we obtain a series of rational numbers

$$2, \quad 1{\cdot}5, \quad 1{\cdot}42, \quad 1{\cdot}415, \quad 1{\cdot}4143, \quad \ldots$$

whose squares

$$4, \quad 2{\cdot}25, \quad 2{\cdot}0164, \quad 2{\cdot}002225, \quad 2{\cdot}00024449, \quad \ldots$$

are all greater than 2 but approximate to 2 as closely as we please.

The reasoning which precedes, although it will probably convince the reader, is hardly of the precise character required by modern mathematics. We can supply a formal proof as follows. In the first place, we can find a member of L and a member of R, differing by as little as we please. For we saw in § 3 that, given any two rational numbers a and b, we can construct a chain of rational numbers, of which a and b are the first and last, and in which any two consecutive numbers differ by as little as we please. Let us then take a member x of L and a member y of R, and interpolate between them a chain of rational numbers of which x is the first and y the last, and in which any two consecutive numbers differ by less than δ, δ being any positive rational number as small as we please, such as $\cdot 01$ or $\cdot 0001$ or $\cdot 000001$. In this chain there must be a last which belongs to L and a first which belongs to R, and these two numbers differ by less than δ.

We can now prove that *an x can be found in L and a y in R such that $2-x^2$ and y^2-2 are as small as we please*, say less than δ. Substituting $\frac{1}{4}\delta$ for δ in the argument which precedes, we see that we can choose x and y so that $y-x<\frac{1}{4}\delta$; and we may plainly suppose that both x and y are less than 2. Thus

$$y+x<4, \quad y^2-x^2 = (y-x)(y+x)<4(y-x)<\delta;$$

and since $x^2<2$ and $y^2>2$ it follows *a fortiori* that $2-x^2$ and y^2-2 are each less than δ.

It follows also that *there can be no largest member of L or smallest member of R*. For if x is any member of L, then $x^2<2$. Suppose that $x^2 = 2-\delta$. Then we can find a member x_1 of L such that x_1^2 differs from 2 by less than δ, and so $x_1^2>x^2$ or $x_1>x$. Thus there are larger members of L than x; and, since x is *any* member of L, it follows that no member of L can be larger than all the rest. Hence L has no largest member, and similarly R has no smallest.

5. Irrational numbers (*continued*). We have thus divided the positive rational numbers into two classes, L and R, such that (i) every member of R is greater than every member of L, (ii) we can find a member of L and a member of R whose difference is as small as we please, (iii) L has no greatest and R no least member. Our common-sense notion of the attributes of a straight line, the requirements of our elementary geometry and our elementary algebra, alike demand *the existence of a number x greater than all the members of L and less than all the members of R, and of a corresponding point P on Λ such that P divides the points which correspond to members of L from those which correspond to members of R.*

Let us suppose for a moment that there is such a number x, and that it may be operated upon in accordance with the laws of algebra, so that, for example, x^2 has a definite meaning. Then x^2

cannot be either less than or greater than 2. For suppose, for example, that x^2 is less than 2. Then it follows from what precedes that we can find a positive rational number ξ such that ξ^2 lies between x^2 and 2. That is to say, we can find a member of L greater than x; and this contradicts the supposition that x divides the members of L from those of R. Thus x^2 cannot be less than 2, and similarly it cannot be greater than 2. We are therefore driven to the conclusion that $x^2 = 2$, and that x is the number which in algebra we denote by $\sqrt{2}$. And this number $\sqrt{2}$ is not rational, for no rational number has its square equal to 2. It is the simplest example of what is called an *irrational* number.

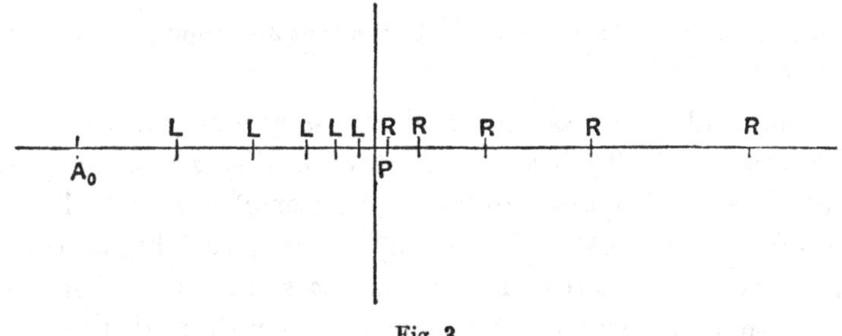

Fig. 3

But the preceding argument may be applied to equations other than $x^2 = 2$, almost word for word; for example to $x^2 = N$, where N is any integer which is not a perfect square, or to

$$x^3 = 3, \quad x^3 = 7, \quad x^4 = 23,$$

or, as we shall see later on, to $x^3 = 3x + 8$. We are thus led to believe in the existence of irrational numbers x and points P on Λ such that x satisfies equations such as these, even when these lengths cannot (as $\sqrt{2}$ can) be constructed by means of elementary geometrical methods.

The reader will no doubt remember that in treatises on elementary algebra the root of such an equation as $x^q = n$ is denoted by $\sqrt[q]{n}$ or $n^{1/q}$, and that a meaning is attached to such symbols as

$$n^{p/q}, \quad n^{-p/q}$$

by means of the equations

$$n^{p/q} = (n^{1/q})^p, \quad n^{p/q}n^{-p/q} = 1.$$

And he will remember how, in virtue of these definitions, the 'laws of indices' such as

$$n^r \times n^s = n^{r+s}, \quad (n^r)^s = n^{rs}$$

are extended so as to cover the case in which r and s are any rational numbers.

The reader may now follow one or other of two alternative courses. He may, if he pleases, be content to assume that 'irrational numbers' such as $\sqrt{2}$, $\sqrt[3]{3}$, ... exist and are amenable to the algebraical laws with which he is familiar*. If he does this he will be able to avoid the more abstract discussions of the next few sections, and may pass on at once to §§ 13 *et seq.*

If, on the other hand, he is not disposed to adopt so naïve an attitude, he will be well advised to pay careful attention to the sections which follow, in which these questions receive fuller consideration.

Examples III. 1. Find the difference between 2 and the squares of the decimals given in § 4 as approximations to $\sqrt{2}$.

2. Find the differences between 2 and the squares of

$$\tfrac{1}{1}, \tfrac{3}{2}, \tfrac{7}{5}, \tfrac{17}{12}, \tfrac{41}{29}, \tfrac{99}{70}.$$

3. Show that if m/n is a good approximation to $\sqrt{2}$, then $(m+2n)/(m+n)$ is a better one, and that the errors in the two cases are in opposite directions. Apply this result to continue the series of approximations in the preceding example.

4. If x and y are approximations to $\sqrt{2}$, by defect and by excess respectively, and $2-x^2<\delta$, $y^2-2<\delta$, then $y-x<\delta$.

5. The equation $x^2 = 4$ is satisfied by $x = 2$. Examine how far the argument of the preceding sections applies to this equation (writing 4 for 2 throughout). [If we define the classes L, R as before, they do not include *all* rational numbers. The rational number 2 is an exception, since 2^2 is neither less than nor greater than 4.]

6. Irrational numbers (*continued*). In § 4 we discussed a special mode of division of the positive rational numbers x into two classes, such that $x^2 < 2$ for the members of one class and $x^2 > 2$ for those of the other. Such a mode of division is called a *section* of the numbers in question. It is plain that we could

* This is the point of view which was adopted in the first edition of this book.

equally well construct a section in which the numbers of the two classes were characterised by the inequalities $x^3 < 2$ and $x^3 > 2$, or $x^4 < 7$ and $x^4 > 7$. Let us now attempt to state the principles of the construction of such sections of the positive rational numbers in quite general terms.

Suppose that P and Q stand for two properties which are mutually exclusive and one of which must be possessed by every positive rational number. Further, suppose that every such number which possesses P is less than any such number which possesses Q. Thus P might be the property '$x^2 < 2$' and Q the property '$x^2 > 2$'. Then we call the numbers which possess P the lower or left-hand class L and those which possess Q the upper or right-hand class R. In general both classes will exist; but it may happen in special cases that one does not, every number belonging to the other. This would obviously happen, for example, if P (or Q) were the property of being rational, or of being positive. For the present, however, we shall confine ourselves to cases in which both classes exist; and then it follows, as in § 4, that we can find a member of L and a member of R whose difference is as small as we please.

In the particular case which we considered in § 4, L had no greatest member and R no least; but one or other of the classes may have a greatest or least member, and it is important to distinguish the different possibilities. It is not possible that L should have a greatest member *and* R a least. For if l were the greatest member of L, and r the least of R, so that $l < r$, then $\frac{1}{2}(l+r)$ would be a positive rational number lying between l and r, and so could belong neither to L nor to R; and this contradicts our assumption that every such number belongs to one class or to the other. This being so, there are but three possibilities, which are mutually exclusive. Either (i) L has a greatest member l, or (ii) R has a least member r, or (iii) L has no greatest member and R no least.

The section of § 4 gives an example of the last possibility. An example of the first is obtained by taking P to be '$x^2 \leqq 1$' and Q to be '$x^2 > 1$'; here

$l = 1$. If P is '$x^2 < 1$' and Q is '$x^2 \geqq 1$', we have an example of the second possibility, with $r = 1$. It should be observed that we do not obtain a section at all by taking P to be '$x^2 < 1$' and Q to be '$x^2 > 1$'; for the special number 1 escapes classification (cf. Ex. III. 5).

7. Irrational numbers (*continued*). In the first two cases we say that the section *corresponds to* a positive rational number a, which is l in the one case and r in the other. Conversely, it is clear that to any such number a corresponds a section which we shall denote by α*. For we might take P and Q to be the properties expressed by

$$x \leqq a, \quad x > a$$

respectively, or by $x < a$ and $x \geqq a$. In the first case a would be the greatest member of L, and in the second case the least member of R. There are in fact just two sections corresponding to any positive rational number. In order to avoid ambiguity we select one of them; let us select that in which the number itself belongs to the *upper* class. In other words, let us agree that we will consider only sections in which the lower class L has no greatest number.

There being this correspondence between the positive rational numbers and the sections defined by means of them, it would be perfectly legitimate, for mathematical purposes, to replace the numbers by the sections, and to regard the symbols which occur in our formulae as standing for the sections instead of for the numbers. Thus, for example, $\alpha > \alpha'$ would mean the same as $a > a'$, if α and α' are the sections which correspond to a and a'.

But when we have in this way substituted sections of rational numbers for the rational numbers themselves, we are almost forced to a generalisation of our number system. For there are sections (such as that of § 4) which do *not* correspond to any rational number. The aggregate of sections is a larger aggregate than that of the positive rational numbers; it includes sections corresponding to all these numbers, and more besides. It is this fact which we make the basis of our generalisation of the idea of

* It will be convenient to denote a section, corresponding to a rational number denoted by an English letter, by the corresponding Greek letter.

number. We accordingly frame the following definitions, which will however be modified in the next section, and must therefore be regarded as temporary and provisional.

A section of the positive rational numbers, in which both classes exist and the lower class has no greatest member, is called a **positive real number.**

A positive real number which does not correspond to a positive rational number is called a positive **irrational** *number.*

8. Real numbers. We have confined ourselves so far to certain sections of the positive rational numbers, which we have agreed provisionally to call 'positive real numbers'. Before we frame our final definitions, we must alter our point of view a little. We shall consider sections, or divisions into two classes, not merely of the positive rational numbers, but of all rational numbers, including zero. We may then repeat all that we have said about sections of the positive rational numbers in §§ 6, 7, merely omitting the word positive occasionally.

DEFINITIONS. *A section of the rational numbers, in which both classes exist and the lower class has no greatest member, is called a* **real number**, *or simply a* **number**.

A real number which does not correspond to a rational number is called an **irrational** *number.*

If the real number does correspond to a rational number, we shall use the term 'rational' as applying to the real number also.

The term 'rational number' will, as a result of our definitions, be ambiguous; it may mean the rational number of § 1, or the corresponding real number. If we say that $\frac{1}{2} > \frac{1}{3}$, we may be asserting either of two different propositions, one a proposition of elementary arithmetic, the other a proposition concerning sections of the rational numbers. Ambiguities of this kind are common in mathematics, and are perfectly harmless, since the relations between different propositions are exactly the same whichever interpretation is attached to the propositions themselves. From $\frac{1}{2} > \frac{1}{3}$ and $\frac{1}{3} > \frac{1}{4}$ we can infer $\frac{1}{2} > \frac{1}{4}$; the inference is in no way affected by any doubt whether $\frac{1}{2}$, $\frac{1}{3}$, and $\frac{1}{4}$ are arithmetical fractions or real numbers. Sometimes, of course, the context in which (e.g.) '$\frac{1}{2}$'

occurs is sufficient to fix its interpretation. When we say (see § 9) that $\frac{1}{2} < \sqrt{(\frac{1}{3})}$, we *must* mean by '$\frac{1}{2}$' the real number $\frac{1}{2}$.

The reader should observe, moreover, that no particular logical importance is to be attached to the precise form of definition of a 'real number' that we have adopted. We defined a 'real number' as being a section, i.e. a pair of classes. We might equally well have defined it as being the lower, or the upper, class; indeed it would be easy to define an infinity of classes of entities each of which would possess the properties of the class of real numbers. What is essential in mathematics is that its symbols should be capable of *some* interpretation; generally they are capable of *many*, and then, so far as mathematics is concerned, it does not matter which we adopt. Bertrand Russell has said that 'mathematics is the science in which we do not know what we are talking about, and do not care whether what we say about it is true', a remark which is expressed in the form of a paradox but which in reality embodies a number of important truths. It would take too long to analyse the meaning of Russell's epigram in detail, but one at any rate of its implications is this, that the symbols of mathematics are capable of varying interpretations, and that we are in general at liberty to adopt whichever we prefer.

There are now three cases to distinguish. It may happen that all negative rational numbers belong to the lower class, and zero and all positive rational numbers to the upper. We describe this section as the *real number zero*. Or again it may happen that the lower class includes some positive numbers. Such a section we describe as a *positive real number*. Finally it may happen that some negative numbers belong to the upper class. Such a section we describe as a *negative real number**.

The difference between our present definition of a positive real number α and that of § 7 amounts to the addition to the lower class of zero and all the negative rational numbers. An example of a negative real number is given by taking the property P of § 6 to be $x+1<0$ and Q to be $x+1 \geqslant 0$.

* There are also sections in which every number belongs to the lower or to the upper class. The reader may be tempted to ask why we do not regard these sections also as defining numbers, which we might call the *real numbers positive and negative infinity*.

There is no logical objection to such a procedure, but it proves to be inconvenient in practice. The most natural definitions of addition and multiplication do not work in a satisfactory way. Moreover, for a beginner, the chief difficulty in the elements of analysis is that of learning to attach precise senses to phrases containing the word 'infinity'; and experience seems to show that he is likely to be confused by any addition to their number.

This section plainly corresponds to the negative rational number -1. It we took P to be $x^3 < -2$ and Q to be $x^3 > -2$, we should obtain a negative real number which is not rational.

9. Relations of magnitude between real numbers. It is plain that, now that we have extended our conception of number, we are bound to make corresponding extensions of our conceptions of equality, inequality, addition, multiplication, and so on. We have to show that these ideas can be applied to the new numbers, and that, when this extension of them is made, all the ordinary laws of algebra retain their validity, so that we can operate with real numbers in general in exactly the same way as with the rational numbers of § 1. To do all this systematically would occupy a considerable space, and we shall be content to indicate summarily how a more systematic discussion would proceed.

We denote a real number by a Greek letter such as α, β, γ, ...; the rational numbers of its lower and upper classes by the corresponding English letters a, A; b, B; c, C; .... The classes themselves we denote by (a), (A), ...

If α and β are two real numbers, there are three possibilities:

(i) every a is a b and every A a B; in this case (a) is identical with (b) and (A) with (B);

(ii) every a is a b, but not all A's are B's; in this case (a) is a proper part of (b)*, and (B) a proper part of (A);

(iii) every A is a B, but not all a's are b's.

These three cases may be indicated graphically as in Fig. 4.

In case (i) we write $\alpha = \beta$, in case (ii) $\alpha < \beta$, and in case (iii) $\alpha > \beta$. It is clear that, when α and β are both rational, these definitions agree with the ideas of equality and inequality between rational numbers which we began by taking for granted; and that any positive number is greater than any negative number.

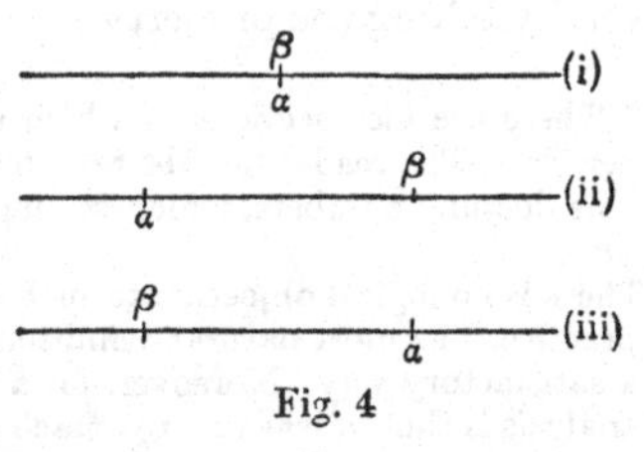

Fig. 4

* I.e. is included in but not identical with (b).

It will be convenient to define at this stage the negative $-\alpha$ of a positive number α. We suppose first that α is irrational. If (a), (A) are the classes which constitute α, we can define another section of the rational numbers by putting all numbers $-A$ in the lower class and all numbers $-a$ in the upper. The real number thus defined, which is clearly negative, we denote by $-\alpha$. Similarly we can define $-\alpha$ when α is negative; if α is negative, $-\alpha$ is positive. It is plain also that $-(-\alpha) = \alpha$. Of the two numbers α and $-\alpha$ one is always positive. The one which is positive we denote by $|\alpha|$ and call the *modulus* of α.

There is a complication if α is rational. In this case α belongs to (A), and the classes $(-A)$, $(-a)$ do not define a real number in the sense of § 8, since $-\alpha$ belongs to the lower class instead of to the upper. We must therefore modify our definition of $-\alpha$ by agreeing that, when α is rational, the rational $-\alpha$ is to be included in the upper class.

Examples IV. 1. Prove that $0 = -0$.

2. Prove that $\beta = \alpha$, $\beta < \alpha$, or $\beta > \alpha$ according as $\alpha = \beta$, $\alpha > \beta$, or $\alpha < \beta$.

3. If $\alpha = \beta$ and $\beta = \gamma$, then $\alpha = \gamma$.

4. If $\alpha \leqq \beta$ and $\beta < \gamma$, then $\alpha < \gamma$.

5. Prove that $-\beta < -\alpha$ if $\alpha < \beta$.

6. Prove that $\alpha > 0$ if α is positive, and $\alpha < 0$ if α is negative.

7. Prove that $\alpha \leqq |\alpha|$.

8. Prove that $1 < \sqrt{2} < \sqrt{3} < 2$.

[All these results are immediate consequences of our definitions.]

10. Algebraical operations with real numbers. We now proceed to define the meaning of the elementary algebraical operations such as addition, as applied to real numbers in general.

(i) *Addition.* In order to define the sum of two numbers α and β, we consider the following two classes: (i) the class (c) formed by all sums $c = a + b$, (ii) the class (C) formed by all sums $C = A + B$. Plainly $c < C$ in all cases.

Again, there cannot be more than one rational number which does not belong either to (c) or to (C). For suppose there were two, say r and s, and let s be the greater. Then both r and s must be greater than every c and less than every C; and so $C-c$ cannot be less than $s-r$. But

$$C-c = (A-a)+(B-b);$$

and we can choose a, b, A, B so that both $A-a$ and $B-b$ are as small as we like; and this plainly contradicts our hypothesis.

If every rational number belongs to (c) or to (C), the classes (c), (C) form a section of the rational numbers, that is to say, a number γ. If there is one which does not, we add it to (C). We have now a section or real number γ, which must clearly be rational, since it corresponds to the least member of (C). *In any case we call γ the sum of α and β, and write*

$$\gamma = \alpha+\beta.$$

If both α and β are rational, they are the least members of the upper classes (A) and (B). In this case it is clear that $\alpha+\beta$ is the least member of (C), so that our definition agrees with our previous ideas of addition.

(ii) *Subtraction.* We define $\alpha-\beta$ by the equation

$$\alpha-\beta = \alpha+(-\beta).$$

The idea of subtraction accordingly presents no fresh difficulties.

Examples V. 1. Prove that $\alpha+(-\alpha) = 0$.

2. Prove that $\alpha+0 = 0+\alpha = \alpha$.

3. Prove that $\alpha+\beta = \beta+\alpha$. [This follows at once from the fact that the classes $(a+b)$ and $(b+a)$, or $(A+B)$ and $(B+A)$, are the same, since, e.g., $a+b = b+a$ when a and b are rational.]

4. Prove that $\alpha+(\beta+\gamma) = (\alpha+\beta)+\gamma$.

5. Prove that $\alpha-\alpha = 0$.

6. Prove that $\alpha-\beta = -(\beta-\alpha)$.

7. From the definition of subtraction, and Exs. 4, 1, and 2 above, it follows that

$$(\alpha-\beta)+\beta = \{\alpha+(-\beta)\}+\beta = \alpha+\{(-\beta)+\beta\} = \alpha+0 = \alpha.$$

We might therefore define the difference $\alpha-\beta = \gamma$ by the equation $\gamma+\beta = \alpha$.

8. Prove that $\alpha-(\beta-\gamma) = \alpha-\beta+\gamma$.

9. Give a definition of subtraction which does not depend upon a previous definition of addition. [To define $\gamma = \alpha-\beta$, form the classes (c),

(C) for which $c = a - B$, $C = A - b$. It is easy to show that this definition is equivalent to that which we adopted in the text.]

10. Prove that

$$||\alpha| - |\beta|| \leqq |\alpha \pm \beta| \leqq |\alpha| + |\beta|.$$

11. Algebraical operations with real numbers *(continued).* (iii) *Multiplication.* When we come to multiplication, it is most convenient to begin with *positive* numbers, and to go back for a moment to the sections of positive rational numbers only which we considered in §§ 4–7. We may then follow practically the same road as in the case of addition, taking (c) to be (ab) and (C) to be (AB). The argument is the same, except when we are proving that all rational numbers with at most one exception must belong to (c) or (C). This depends, as in the case of addition, on showing that we can choose a, A, b, and B so that $C - c$ is as small as we please. Here we use the identity

$$C - c = AB - ab = (A - a)B + a(B - b).$$

We include negative numbers within the scope of our definition by agreeing that, if α and β are positive, then

$$(-\alpha)\beta = -\alpha\beta, \quad \alpha(-\beta) = -\alpha\beta, \quad (-\alpha)(-\beta) = \alpha\beta.$$

Finally we agree that $(0)\alpha = \alpha(0) = 0$ for all α.

(iv) *Division.* In order to define division, we begin by defining the reciprocal $1/\alpha$ of a number α (other than zero). Confining ourselves in the first instance to positive numbers and sections of positive rational numbers, we define the reciprocal of a positive number α by means of the lower class $(1/A)$ and the upper class $(1/a)$. We then define the reciprocal of a negative number $-\alpha$ by the equation $1/(-\alpha) = -(1/\alpha)$. Finally we define α/β by the equation

$$\alpha/\beta = \alpha \times (1/\beta).$$

We are then in a position to apply to all real numbers, rational or irrational, the whole of the ideas and methods of elementary algebra. Naturally we do not propose to carry out this task in detail. It will be more profitable and more interesting to turn our attention to some special, but particularly important, classes of irrational numbers.

Examples VI. Prove the theorems expressed by the following formulae:

1. $\alpha \times 1 = 1 \times \alpha = \alpha$. 2. $\alpha \times (1/\alpha) = 1$. 3. $\alpha\beta = \beta\alpha$.
4. $\alpha(\beta\gamma) = (\alpha\beta)\gamma$. 5. $\alpha(\beta+\gamma) = \alpha\beta+\alpha\gamma$. 6. $(\alpha+\beta)\gamma = \alpha\gamma+\beta\gamma$.
7. $|\alpha\beta| = |\alpha||\beta|$.

12. The number $\sqrt{2}$. Let us now return for a moment to the particular irrational number which we discussed in §§ 4–5. We there constructed a section by means of the inequalities $x^2 < 2$, $x^2 > 2$. This was a section of the positive rational numbers only; but we replace it (as was explained in § 8) by a section of all the rational numbers. We denote the section or number thus defined by the symbol $\sqrt{2}$.

The classes by means of which the product of $\sqrt{2}$ by itself is defined are (i) (aa'), where a and a' are positive rational numbers whose squares are less than 2, (ii) (AA'), where A and A' are positive rational numbers whose squares are greater than 2. These classes exhaust all positive rational numbers save one, which can only be 2 itself. Thus

$$(\sqrt{2})^2 = \sqrt{2}\sqrt{2} = 2.$$

Again

$$(-\sqrt{2})^2 = (-\sqrt{2})(-\sqrt{2}) = \sqrt{2}\sqrt{2} = (\sqrt{2})^2 = 2.$$

Thus *the equation $x^2 = 2$ has the two roots $\sqrt{2}$ and $-\sqrt{2}$*. Similarly we could discuss the equations $x^2 = 3$, $x^3 = 7$, ... and the corresponding irrational numbers $\sqrt{3}$, $-\sqrt{3}$, $\sqrt[3]{7}$, ...

13. Quadratic surds. A number of the form $\pm\sqrt{a}$, where a is a positive rational number which is not the square of another rational number, is called a *pure quadratic surd*. A number of the form $a \pm \sqrt{b}$, where a is rational, and $\sqrt{b}$ is a pure quadratic surd, is sometimes called a mixed quadratic surd.

The two numbers $a \pm \sqrt{b}$ are the roots of the quadratic equation

$$x^2 - 2ax + a^2 - b = 0.$$

Conversely, the equation $x^2 + 2px + q = 0$, where p and q are rational, and $p^2 - q > 0$, has as its roots the two quadratic surds $-p \pm \sqrt{(p^2 - q)}$.

The only kind of irrational numbers whose existence was suggested by the geometrical considerations of § 3 are these quadratic surds, pure and mixed, and the more complicated irrationals which may be expressed in a form involving the repeated extraction of square roots, such as

$$\sqrt{2}+\sqrt{(2+\sqrt{2})}+\sqrt{\{2+\sqrt{(2+\sqrt{2})}\}}.$$

It is easy to construct geometrically a line whose length is equal to any number of this form, as the reader will easily see for himself. That irrational numbers of these kinds *only* can be constructed by Euclidean methods (i.e. by geometrical constructions with ruler and compasses) is a point the proof of which must be deferred for the present*. This property of quadratic surds makes them especially interesting.

Examples VII. 1. Give geometrical constructions for

$$\sqrt{2}, \quad \sqrt{(2+\sqrt{2})}, \quad \sqrt{\{2+\sqrt{(2+\sqrt{2})}\}}.$$

2. The quadratic equation $ax^2+2bx+c=0$ has two real roots† if $b^2-ac>0$. Suppose a, b, c rational. Nothing is lost by taking all three to be integers, for we can multiply the equations by the least common multiple of their denominators.

The reader will remember that the roots are $\{-b\pm\sqrt{(b^2-ac)}\}/a$. It is easy to construct these lengths geometrically, first constructing $\sqrt{(b^2-ac)}$. A more elegant, though less straightforward, construction is the following.

Draw a circle of unit radius, a diameter PQ, and the tangents at the ends of the diameters.

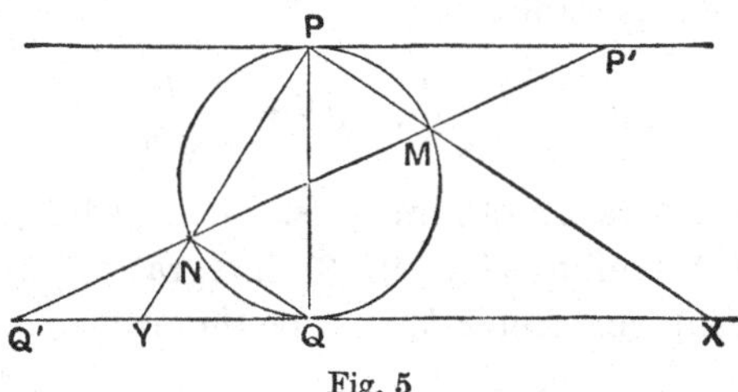

Fig. 5

Take $PP'=-2a/b$ *and* $QQ'=-c/2b$, *having regard to sign*‡. *Join* $P'Q'$,

* See Ch. II, Misc. Exs. 22.

† I.e. there are two values of x for which $ax^2+2bx+c=0$. If $b^2-ac<0$, there are no such values of x. The reader will remember that in books on elementary algebra the equation is said to have two 'complex' roots. The meaning to be attached to this statement will be explained in Ch. III.

When $b^2=ac$ the equation has only one root. For the sake of uniformity it is generally said in this case to have 'two equal' roots, but this is a mere convention.

‡ The figure is drawn to suit the case in which b and c have the same and a the opposite sign. The reader should draw figures for other cases.

*cutting the circle in M and N. Draw PM and PN, cutting QQ′ in X and Y. Then QX and QY are the roots of the equation with their proper signs**.

The proof is simple and we leave it as an exercise to the reader. Another, perhaps even simpler, construction is the following. *Take a line AB of unit length. Draw $BC = -2b/a$ perpendicular to AB, and $CD = c/a$ perpendicular to BC and in the same direction as BA. On AD as diameter describe a circle cutting BC in X and Y. Then BX and BY are the roots.*

3. If ac is positive, PP' and QQ' will be drawn in the same direction. Verify that $P'Q'$ will not meet the circle if $b^2 < ac$, while if $b^2 = ac$ it will be a tangent. Verify also that if $b^2 = ac$ the circle in the second construction will touch BC.

4. Prove that
$$\sqrt{(pq)} = \sqrt{p} \times \sqrt{q}, \quad \sqrt{(p^2q)} = p\sqrt{q}.$$

14. Some theorems concerning quadratic surds. Two pure quadratic surds are said to be *similar* if they can be expressed as rational multiples of the same surd, and otherwise to be *dissimilar*. Thus
$$\sqrt{8} = 2\sqrt{2}, \quad \sqrt{\tfrac{25}{2}} = \tfrac{5}{2}\sqrt{2},$$
and so $\sqrt{8}$, $\sqrt{\tfrac{25}{2}}$ are similar surds. On the other hand, if M and N are integers which have no common factor, and neither of which is a perfect square, then $\sqrt{M}$ and $\sqrt{N}$ are dissimilar surds.

For suppose, if possible,
$$\sqrt{M} = \frac{p}{q}\sqrt{\frac{t}{u}}, \quad \sqrt{N} = \frac{r}{s}\sqrt{\frac{t}{u}},$$
where all the letters denote integers. Then $\sqrt{(MN)}$ is evidently rational, and therefore (Ex. II. 3) integral. Thus $MN = P^2$, where P is an integer. Let $a, b, c, \ldots$ be the prime factors of P, so that
$$MN = a^{2\alpha}b^{2\beta}c^{2\gamma}\ldots,$$
where $\alpha, \beta, \gamma, \ldots$ are positive integers. Then MN is divisible by $a^{2\alpha}$, and therefore either (1) M is divisible by $a^{2\alpha}$, or (2) N is divisible by $a^{2\alpha}$, or (3) M and N are both divisible by a. The last case may be ruled out, since M and N have no common factor. This argument may be applied to each of the factors $a^{2\alpha}, b^{2\beta}, c^{2\gamma}, \ldots$, so

* I have taken this construction from Klein's *Vorträge über ausgewählte Fragen der Elementargeometrie* (Leipzig, 1895).

that M must be divisible by some of these factors and N by the remainder. Thus

$$M = P_1^2, \quad N = P_2^2,$$

where P_1^2 denotes the product of some of the factors $a^{2\alpha}, b^{2\beta}, c^{2\gamma}, \ldots$ and P_2^2 the product of the rest. Hence M and N are both perfect squares, which is contrary to our hypothesis.

THEOREM. *If A, B, C, D are rational and*

$$A + \sqrt{B} = C + \sqrt{D},$$

then either (i) $A = C$, $B = D$ *or* (ii) *B and D are both squares of rational numbers.*

For $B - D$ is rational, and so is

$$\sqrt{B} - \sqrt{D} = C - A.$$

If B is not equal to D (in which case it is obvious that A is also not equal to C), it follows that

$$\sqrt{B} + \sqrt{D} = (B - D)/(\sqrt{B} - \sqrt{D})$$

is also rational. Hence $\sqrt{B}$ and $\sqrt{D}$ are rational.

COROLLARY. *If $A + \sqrt{B} = C + \sqrt{D}$, then $A - \sqrt{B} = C - \sqrt{D}$ (unless $\sqrt{B}$ and $\sqrt{D}$ are both rational).*

Examples VIII. 1. Prove *ab initio* that $\sqrt{2}$ and $\sqrt{3}$ are not similar surds.

2. Prove that $\sqrt{a}$ and $\sqrt{(1/a)}$, where a is rational, are similar surds (unless both are rational).

3. If a and b are rational, then $\sqrt{a} + \sqrt{b}$ cannot be rational unless $\sqrt{a}$ and $\sqrt{b}$ are rational. The same is true of $\sqrt{a} - \sqrt{b}$, unless $a = b$.

4. If $$\sqrt{A} + \sqrt{B} = \sqrt{C} + \sqrt{D},$$

then either (a) $A = C$ and $B = D$, or (b) $A = D$ and $B = C$, or (c) $\sqrt{A}$, $\sqrt{B}$, $\sqrt{C}$, $\sqrt{D}$ are all rational or all similar surds. [Square the given equation and apply the theorem above.]

5. Neither $(a + \sqrt{b})^3$ nor $(a - \sqrt{b})^3$ can be rational unless $\sqrt{b}$ is rational.

6. Prove that if $x = p + \sqrt{q}$, where p and q are rational, then x^m, where m is any integer, can be expressed in the form $P + Q\sqrt{q}$, where P and Q are rational. For example,

$$(p + \sqrt{q})^2 = p^2 + q + 2p\sqrt{q}, \quad (p + \sqrt{q})^3 = p^3 + 3pq + (3p^2 + q)\sqrt{q}.$$

Deduce that any polynomial in x with rational coefficients (i.e. any expression of the form
$$a_0x^n + a_1x^{n-1} + \dots + a_n,$$
where $a_0, \dots, a_n$ are rational numbers) can be expressed in the form $P + Q\sqrt{q}$.

7. If $a + \sqrt{b}$, where b is not a perfect square, is the root of an algebraical equation with rational coefficients, then $a - \sqrt{b}$ is another root of the same equation.

8. Express $1/(p + \sqrt{q})$ in the form prescribed in Ex. 6. [Multiply numerator and denominator by $p - \sqrt{q}$.]

9. Deduce from Exs. 6 and 8 that any expression of the form $G(x)/H(x)$, where $G(x)$ and $H(x)$ are polynomials in x with rational coefficients, can be expressed in the form $P + Q\sqrt{q}$, where P and Q are rational.

10. If p, q, and $p^2 - q$ are positive, we can express $\sqrt{(p + \sqrt{q})}$ in the form $\sqrt{x} + \sqrt{y}$, where
$$x = \tfrac{1}{2}\{p + \sqrt{(p^2 - q)}\}, \quad y = \tfrac{1}{2}\{p - \sqrt{(p^2 - q)}\}.$$

11. Determine the conditions that it may be possible to express $\sqrt{(p + \sqrt{q})}$, where p and q are rational, in the form $\sqrt{x} + \sqrt{y}$, where x and y are rational.

12. If $a^2 - b$ is positive, then a necessary and sufficient condition that
$$\sqrt{(a + \sqrt{b})} + \sqrt{(a - \sqrt{b})}$$
should be rational is that $a^2 - b$ and $\tfrac{1}{2}\{a + \sqrt{(a^2 - b)}\}$ should both be squares of rational numbers.

15. The continuum. The aggregate of all real numbers, rational and irrational, is called the *arithmetical continuum*.

It is convenient to suppose that the straight line Λ of § 2 is composed of points corresponding to all the numbers of the arithmetical continuum, and of no others*. The points of the line, the aggregate of which may be said to constitute the *linear continuum*, then supply us with a convenient image of the arithmetical continuum.

We have considered in some detail the chief properties of a few classes of real numbers, such, for example, as rational

* This supposition is merely a hypothesis adopted (i) because it suffices for the purposes of our geometry and (ii) becauses it provides us with convenient geometrical illustrations of analytical processes. As we use geometrical language only for purposes of illustration, it is not part of our business to study the foundations of geometry.

numbers or quadratic surds. We add a few further examples to show how very special these particular classes of numbers are, and how, to put it roughly, they comprise only a minute fraction of the infinite variety of numbers which constitute the continuum.

(i) Let us consider a more complicated surd expression such as

$$z = \sqrt[3]{(4+\sqrt{15})}+\sqrt[3]{(4-\sqrt{15})}.$$

Our argument for supposing that the expression for z has a meaning might be as follows. We first show, as in § 12, that there is a number $y = \sqrt{15}$ such that $y^2 = 15$, and we can then, as in § 10, define the numbers $4+\sqrt{15}$, $4-\sqrt{15}$. Now consider the equation in z_1,

$$z_1^3 = 4+\sqrt{15}.$$

The right-hand side of this equation is not rational: but the same reasoning which leads us to suppose that there is a real number x such that $x^3 = 2$ (or any other rational number) also leads us to the conclusion that there is a number z_1 such that $z_1^3 = 4+\sqrt{15}$. We thus define $z_1 = \sqrt[3]{(4+\sqrt{15})}$, and similarly we can define $z_2 = \sqrt[3]{(4-\sqrt{15})}$; and then, as in § 10, we define $z = z_1+z_2$.

It is easy to verify that

$$z^3 = 3z+8;$$

and it is not difficult to give a direct proof of the existence of a unique number satisfying this equation.

In the first place, z (if it exists) must be positive. For $z = -\zeta$ gives $\zeta^3-3\zeta+8 = 0$ or $3-\zeta^2 = 8/\zeta$. But this is impossible if ζ is positive, for then $\zeta^2<3$, $\zeta<2$, and $8/\zeta>4$, whereas $3-\zeta^2<3$.

Next, the equation cannot be satisfied by two different numbers z_1 and z_2. For suppose, if possible, that

$$z_1^3 = 3z_1+8, \quad z_2^3 = 3z_2+8.$$

Then z_1 and z_2 are positive, and $z_1^3>8$, $z_2^3>8$, or $z_1>2$, $z_2>2$; and this is impossible because, when we subtract and divide by z_1-z_2, we obtain

$$z_1^2+z_1z_2+z_2^2 = 3.$$

Hence there is at most one z for which $z^3 = 3z+8$; and it cannot be rational. For any rational root of the equation must be integral and a divisor of 8 (Ex. II. 3), and no one of 1, 2, 4, 8 is a root.

We can now divide the positive rationals x into two classes L, R according as $x^3<3x+8$ or $x^3>3x+8$. If x belongs to R, and $y>x$, then y also belongs to R, since $y>x>2$ and

$$y^3-3y-(x^3-3x) = (y-x)(y^2+xy+x^2-3)>0.$$

Similarly we can show that if x belongs to L, and $y<x$, then y belongs to L.

Finally, it is evident that the classes L and R both exist; and they define a section of the positive rational numbers, or positive real number z, which satisfies the equation.

The reader who knows how to solve cubic equations by Cardan's method will be able to obtain the explicit expression of z directly from the equation.

(ii) The direct argument applied above to the equation $x^3 = 3x + 8$ could be applied (though the application would be a little more difficult) to the equation

$$x^5 = x + 16,$$

and would lead us to the conclusion that there is a unique positive real number which satisfies this equation. In this case, however, it is not possible to obtain a simple explicit expression for x composed of any combination of surds. It is indeed known (though the proof is difficult) that it is *generally* impossible to find such an expression for the root of an equation of higher degree than 4. Thus, besides irrational numbers which can be expressed as pure or mixed quadratic or other surds, or combinations of such surds, there are others which are roots of algebraical equations but cannot be so expressed. It is only in very special cases that such expressions can be found.

(iii) But even when we have added to our list of irrational numbers roots of equations (such as $x^5 = x + 16$) which cannot be explicitly expressed as surds, we have not exhausted the different kinds of irrational numbers contained in the continuum. Let us draw a circle whose diameter is equal to A_0A_1, i.e. to unity. It is natural to suppose* that the circumference of such a circle has a length capable of numerical measurement. This length is usually denoted by π; and it has been shown† (though the proof is again difficult) that this number π is not the root of any algebraical equation with integral coefficients, such, for example, as

$$\pi^2 = n, \quad \pi^3 = n, \quad \pi^5 = \pi + n,$$

where n is an integer. In this way it is possible to define a number which is not rational and does not belong to any of the classes of

* See Hobson's *Plane trigonometry* (5th edition), pp. 7 *et seq.*

† See Hobson, *loc. cit.*, pp. 305 *et seq.*, or the same writer's *Squaring the circle* (Cambridge, 1913).

irrational numbers which we have so far considered. This number π is not an isolated or exceptional case. Only special classes of irrational numbers are roots of such algebraical equations; and only still more special classes are expressible by means of surds.

16. The continuous real variable. The 'real numbers' may be regarded from two points of view. We may think of them *as an aggregate*, the 'arithmetical continuum' defined in the preceding section, or *individually*. And when we think of them individually, we may think either of a particular *specified* number (such as 1, $-\frac{1}{2}$, $\sqrt{2}$, or π) or we may think of *any* number, *an unspecified* number, *the number x*. This last is our point of view when we make such assertions as 'x is a number', 'x is the measure of a length', 'x may be rational or irrational'. The x which occurs in propositions such as these is called *the continuous real variable*: and the individual numbers are called the *values* of the variable.

A 'variable', however, need not necessarily be continuous. Instead of considering the aggregate of *all* real numbers, we might consider some partial aggregate contained in the former aggregate, such as the aggregate of rational numbers, or the aggregate of positive integers. Let us take the last case. Then in statements about *any* positive integer, or *an unspecified* positive integer, such as 'n is either odd or even', n is called the variable, a *positive integral variable*, and the individual positive integers are its values.

Naturally 'x' and 'n' are only examples of variables, the variable whose 'field of variation' is formed by all the real numbers, and that whose field is formed by the positive integers. These are the most important examples, but we have often to consider other cases. In the theory of decimals, for instance, we may denote by x any figure in the expression of any number as a decimal. Then x is a variable, but a variable which has only ten different values, viz. 0, 1, 2, 3, 4, 5, 6, 7, 8, 9. We may say, shortly, that the variables with which we shall be concerned are classes of integers or of real numbers, and that the values of the variables are the members of these classes.

17. Sections of the real numbers. In §§ 4–7 we considered 'sections' of the rational numbers, i.e. modes of division of the rational numbers (or of the positive rational numbers only) into two classes L and R possessing the following characteristic properties:

(i) that every number of the type considered belongs to one and only one of the two classes;

(ii) that both classes exist;

(iii) that any member of L is less than any member of R.

It is plainly possible to apply the same idea to the aggregate of all real numbers, and the process is, as the reader will find in later chapters, of very great importance.

Let us then suppose* that P and Q are two properties which are mutually exclusive, and one of which is possessed by every real number. Further let us suppose that any number which possesses P is less than any which possesses Q. We call the numbers which possess P the *lower* or *left-hand class* L, and those which possess Q the *upper* or *right-hand class* R.

Thus P might be $x \leqq \sqrt{2}$ and Q be $x > \sqrt{2}$. It is important to observe that a pair of properties which suffice to define a section of the rational numbers may not suffice to define one of the real numbers. This is so, for example, with the pair '$x < \sqrt{2}$' and '$x > \sqrt{2}$' or (if we confine ourselves to positive numbers) with '$x^2 < 2$' and '$x^2 > 2$'. Every rational number possesses one or other of the properties, but not every real number, since in either case $\sqrt{2}$ escapes classification.

There are now two possibilities†. Either L has a greatest member l, or R has a least member r. *Both* of these events cannot

* The discussion which follows is in many ways similar to that of § 6. We have not attempted to avoid a certain amount of repetition. The idea of a 'section', first brought into prominence in Dedekind's famous pamphlet *Stetigkeit und irrationale Zahlen*, is one which must be grasped by every reader of this book, even if he be one of those who prefer to omit the discussion of the notion of an irrational number contained in §§ 6–12.

† There were three in § 6.

occur. For if L had a greatest member l, and R a least member r, the number $\frac{1}{2}(l+r)$ would be greater than all members of L and less than all members of R, and so could not belong to either class. On the other hand *one* event must occur*.

For let L_1 and R_1 denote the classes formed from L and R by taking only the rational members of L and R. Then the classes L_1 and R_1 form a section of the rational numbers. There are now two cases to distinguish.

It may happen that L_1 has a greatest member α. In this case α must be also the greatest member of L. For if not we can find a greater, say β. There are rational numbers lying between α and β, and these, being less than β, belong to L, and therefore to L_1; and this is plainly a contradiction. Hence α is the greatest member of L.

On the other hand it may happen that L_1 has no greatest member. In this case the section of the rational numbers formed by L_1 and R_1 is a real number α. This number α must belong to L or to R. If it belongs to L we can show, precisely as before, that it is the greatest member of L; and similarly that, if it belongs to R, it is the least member of R.

Thus in any case either L has a greatest member or R a least. Any section of the real numbers therefore 'corresponds' to a real number in the sense in which a section of the rational numbers sometimes, but not always, corresponds to a rational number. This conclusion is of very great importance; for it shows that the consideration of sections of all the real numbers does not lead to any further generalisation of our idea of number. Starting from the rational numbers, we found that the idea of a section of the rational numbers led us to a new conception of a number, that of a real number, more general than that of a rational number; and it might have been expected that the idea of a section of the real numbers would have led us to a conception more general still. The discussion which precedes shows that this is not the case, and that the aggregate of real numbers, or the continuum

* This was not the case in § 6.

has a kind of completeness which the aggregate of the rational numbers lacked, a completeness which is expressed in technical language by saying that the continuum is closed.

The result which we have just proved may be stated as follows:

Dedekind's theorem. *If the real numbers are divided into two classes L and R in such a way that*

(i) *every number belongs to one or other of the two classes,*

(ii) *each class contains at least one number,*

(iii) *any member of L is less than any member of R,*

then there is a number α, which has the property that all the numbers less than it belong to L and all the numbers greater than it to R. The number α itself may belong to either class.

In applications we have often to consider sections not of *all* numbers but of all those contained in an *interval* (β, γ), that is to say of all numbers x such that $\beta \leqq x \leqq \gamma$. A 'section' of such numbers is of course a division of them into two classes possessing the properties (i), (ii), and (iii). Such a section may be converted into a section of *all* numbers by adding to L all numbers less than β and to R all numbers greater than γ. It is clear that the conclusion stated in Dedekind's theorem still holds if we substitute 'the real numbers of the interval (β, γ)' for 'the real numbers', and that the number α in this case satisfies the inequalities $\beta \leqq \alpha \leqq \gamma$.

18. Points of accumulation. A system of real numbers, or of the points on a straight line corresponding to them, defined in any way whatever, is called an *aggregate* or *set* of numbers or points. The set might consist, for example, of all the positive integers, or of all the rational points.

It is most convenient here to use the language of geometry*. Suppose then that we are given a set of points, which we will denote by S. Take any point ξ, which may or may not belong to S. Then there are two possibilities. Either (i) it is possible to choose a positive number δ so that the interval $(\xi - \delta, \xi + \delta)$ does not contain any point of S, other than ξ itself†, or (ii) this is not possible.

* The reader will hardly require to be reminded that this course is adopted solely for reasons of linguistic convenience.

† This clause is of course unnecessary if ξ does not itself belong to S.

Suppose, for example, that S consists of the points corresponding to all the positive integers. If ξ is itself a positive integer, we can take δ to be any number less than 1, and (i) will be true; or, if ξ is halfway between two positive integers, we can take δ to be any number less than $\frac{1}{2}$. On the other hand, if S consists of all the rational points, then, whatever the value of ξ, (ii) is true, for any interval whatever contains an infinity of rational points.

Let us suppose that (ii) is true. Then any interval $(\xi-\delta, \xi+\delta)$, however small its length, contains at least one point ξ_1 which belongs to S and does not coincide with ξ; and this whether ξ itself be a member of S or not. In this case we shall say that ξ is a *point of accumulation* of S. It is easy to see that the interval $(\xi-\delta, \xi+\delta)$ must contain, not merely one, but infinitely many points of S. For, when we have determined ξ_1, we can take an interval $(\xi-\delta_1, \xi+\delta_1)$ surrounding ξ but not reaching as far as ξ_1. But this interval also must contain a point, say ξ_2, which is a member of S and does not coincide with ξ. Obviously we may repeat this argument, with ξ_2 in the place of ξ_1; and so on indefinitely. In this way we can determine as many points

$$\xi_1, \xi_2, \xi_3, \dots$$

as we please, all belonging to S, and all lying inside the interval $(\xi-\delta, \xi+\delta)$.

A point of accumulation of S may or may not be itself a point of S. The examples which follow illustrate the various possibilities.

Examples IX. 1. If S consists of the points corresponding to the positive integers, or all the integers, there are no points of accumulation.

2. If S consists of all the rational points, every point of the line is a point of accumulation.

3. If S consists of the points $1, \frac{1}{2}, \frac{1}{3}, \dots$, there is one point of accumulation, viz. the origin.

4. If S consists of all the positive rational points, the points of accumulation are the origin and all positive points of the line.

19. Weierstrass's theorem. The general theory of sets of points is of the utmost interest and importance in the higher branches of analysis; but it is for the most part too difficult to be included in a book such as this. There is however one fundamental theorem which is easily deduced from Dedekind's theorem and which we shall require later.

THEOREM. *If a set S contains infinitely many points, and is entirely situated in an interval (α, β), then at least one point of the interval is a point of accumulation of S.*

We divide the points of the line Λ into two classes in the following manner. The point P belongs to L if there are an infinity of points of S to the right of P, and to P in the contrary case. Then it is evident that conditions (i) and (iii) of Dedekind's theorem are satisfied; and since α belongs to L and β to R, condition (ii) is satisfied also.

Hence there is a point ξ such that, however small be δ, $\xi-\delta$ belongs to L and $\xi+\delta$ to R, so that the interval $(\xi-\delta, \xi+\delta)$ contains an infinity of points of S. Hence ξ is a point of accumulation of S.

This point may of course coincide with α or β, as for instance when $\alpha = 0$, $\beta = 1$, and S consists of the points $1, \frac{1}{2}, \frac{1}{3}, \ldots$. In this case 0 is the sole point of accumulation. An alternative proof is given in §71, p. 139.

MISCELLANEOUS EXAMPLES ON CHAPTER I

1. What are the conditions that $ax+by+cz = 0$, (1) for all values of x, y, z; (2) for all values of x, y, z subject to $\alpha x+\beta y+\gamma z = 0$; (3) for all values of x, y, z subject to both $\alpha x+\beta y+\gamma z = 0$ and $Ax+By+Cz = 0$?

2. Any positive rational number can be expressed in one and only one way in the form

$$a_1+\frac{a_2}{1.2}+\frac{a_3}{1.2.3}+\ldots+\frac{a_k}{1.2.3\ldots k},$$

where $a_1, a_2, \ldots, a_k$ are integers, and

$$0\leqq a_1,\quad 0\leqq a_2<2,\quad 0\leqq a_3<3,\quad \ldots,\quad 0<a_k<k.$$

3. Any positive rational number can be expressed in one and only one way as a simple continued fraction

$$a_1+\frac{1}{a_2+}\,\frac{1}{a_3+}\ldots\frac{1}{+a_n},$$

where $a_1, a_2, \ldots, a_n$ are integers, and

$$a_1\geqq 0,\quad a_2>0,\quad \ldots,\quad a_{n-1}>0,\quad a_n>1.$$

[Accounts of the theory of continued fractions will be found in text-books of algebra, or in Hardy and Wright, *An introduction to the theory of numbers*, ch. 10.]

4. Find the rational roots (if any) of $9x^3 - 6x^2 + 15x - 10 = 0$.

5. A line AB is divided at C *in aurea sectione* (Euc. II. 11), i.e. so that $AB.AC = BC^2$. Show that the ratio AC/AB is irrational.

[A direct geometrical proof will be found in Bromwich's *Infinite series*, 2nd edition, § 136, p. 400.]

6. A is irrational. In what circumstances can $\dfrac{aA+b}{cA+d}$, where a, b, c, d are rational, be rational?

7. **Some elementary inequalities.** In what follows $a_1, a_2, \ldots$ denote positive numbers (including zero) and $p, q, \ldots$ positive integers. Since $a_1^p - a_2^p$ and $a_1^q - a_2^q$ have the same sign, we have $(a_1^p - a_2^p)(a_1^q - a_2^q) \geqq 0$, or

$$a_1^{p+q} + a_2^{p+q} \geqq a_1^p a_2^q + a_1^q a_2^p \quad \text{.....................(1)},$$

an inequality which may also be written in the form

$$\frac{a_1^{p+q} + a_2^{p+q}}{2} \geqq \left(\frac{a_1^p + a_2^p}{2}\right)\left(\frac{a_1^q + a_2^q}{2}\right) \quad \text{..................(2)}.$$

By repeated application of this formula we obtain

$$\frac{a_1^{p+q+r+\cdots} + a_2^{p+q+r+\cdots}}{2} \geqq \left(\frac{a_1^p + a_2^p}{2}\right)\left(\frac{a_1^q + a_2^q}{2}\right)\left(\frac{a_1^r + a_2^r}{2}\right)\cdots \quad \text{...(3)},$$

and in particular

$$\frac{a_1^p + a_2^p}{2} \geqq \left(\frac{a_1 + a_2}{2}\right)^p \quad \text{..........................(4)}.$$

When $p = q = 1$ in (1), or $p = 2$ in (4), the inequalities are merely different forms of the inequality $a_1^2 + a_2^2 \geqq 2a_1 a_2$, which expresses the fact that the arithmetic mean of two positive numbers is not less than their geometric mean.

8. **Generalisations for n numbers.** If we write down the $\frac{1}{2}n(n-1)$ inequalities of the type (1) which can be formed with n numbers $a_1, a_2, \ldots, a_n$, and add the results, we obtain the inequality

$$n\,\Sigma a^{p+q} \geqq \Sigma a^p\, \Sigma a^q \quad \text{..........................(5)},$$

or

$$\frac{1}{n}\Sigma a^{p+q} \geqq \left(\frac{1}{n}\Sigma a^p\right)\left(\frac{1}{n}\Sigma a^q\right) \quad \text{.....................(6)}.$$

Hence we can deduce an obvious extension of (3) which the reader may formulate for himself, and in particular the inequality

$$\frac{1}{n}\Sigma a^p \geqq \left(\frac{1}{n}\Sigma a\right)^p \quad \text{..........................(7)}.$$

9. **The general form of the theorem concerning the arithmetic and geometric means.** An inequality of a slightly different character is that which asserts that the arithmetic mean of $a_1, a_2, \dots, a_n$ is not less than their geometric mean. Suppose that a_r and a_s are the greatest and least of the a's (if there are several greatest or least a's we may choose any of them indifferently), and let G be their geometric mean. We may suppose $G > 0$, as the truth of the proposition is obvious when $G = 0$. If now we replace a_r and a_s by

$$a_r' = G, \quad a_s' = a_r a_s/G,$$

we do not alter the value of the geometric mean; and, since

$$a_r' + a_s' - a_r - a_s = (a_r - G)(a_s - G)/G \leqq 0,$$

we certainly do not increase the arithmetic mean.

It is clear that we may repeat this argument until we have replaced each of $a_1, a_2, \dots, a_n$ by G; at most n repetitions will be necessary. Since the final value of the arithmetic mean is G, the initial value cannot have been less.

10. **Cauchy's inequality.** Suppose that $a_1, a_2, \dots, a_n$ and $b_1, b_2, \dots, b_n$ are any two sets of numbers positive or negative. It is easy to verify the identity

$$(\Sigma a_r b_r)^2 = \Sigma a_r^2 \, \Sigma b_s^2 - \Sigma (a_r b_s - a_s b_r)^2,$$

where r and s assume the values $1, 2, \dots, n$. It follows that

$$(\Sigma a_r b_r)^2 \leqq \Sigma a_r^2 \, \Sigma b_r^2.$$

11. If $a_1, a_2, \dots, a_n$ are positive, then

$$\Sigma a_r \, \Sigma \frac{1}{a_r} \geqq n^2.$$

12. If a, b, c are positive, and $a + b + c = 1$, then

$$\left(\frac{1}{a} - 1\right)\left(\frac{1}{b} - 1\right)\left(\frac{1}{c} - 1\right) \geqq 8. \qquad \textit{(Math. Trip. 1932)}$$

13. If a and b are positive, and $a + b = 1$, then

$$\left(a + \frac{1}{a}\right)^2 + \left(b + \frac{1}{b}\right)^2 \geqq \frac{25}{2}. \qquad \textit{(Math. Trip. 1926)}$$

14. If $a_1, a_2, \dots, a_n$ are all positive, and $s_n = a_1 + a_2 + \dots + a_n$, then

$$(1 + a_1)(1 + a_2)\dots(1 + a_n) \leqq 1 + s_n + \frac{s_n^2}{2!} + \dots + \frac{s_n^n}{n!}.$$

(*Math. Trip.* 1909)

15. If $a_1, a_2, \dots, a_n$ and $b_1, b_2, \dots, b_n$ are two sets of positive numbers, arranged in descending order of magnitude, then

$$(a_1 + a_2 + \dots + a_n)(b_1 + b_2 + \dots + b_n) \leqq n(a_1 b_1 + a_2 b_2 + \dots + a_n b_n).$$

16. If $a, b, c, \ldots, k$ and $A, B, C, \ldots, K$ are two sets of numbers, and all of the first set are positive, then

$$\frac{aA+bB+\ldots+kK}{a+b+\ldots+k}$$

lies between the algebraically least and greatest of $A, B, \ldots, K$.

[Examples 7–16 are, for the most part, very special cases of well-known general theorems, which are discussed systematically in Hardy, Littlewood, and Pólya, *Inequalities* (Cambridge, 1934). See also § 74 of Ch. IV, and Appendix I.]

17. If $\sqrt{p}$, $\sqrt{q}$ are dissimilar surds, and $a+b\sqrt{p}+c\sqrt{q}+d\sqrt{(pq)} = 0$, where a, b, c, d are rational, then $a = 0, b = 0, c = 0, d = 0$.

[Express $\sqrt{p}$ in the form $M+N\sqrt{q}$, where M and N are rational, and apply the theorem of § 14.]

18. Show that if $a\sqrt{2}+b\sqrt{3}+c\sqrt{5} = 0$, where a, b, c are rational numbers, then $a = 0, b = 0, c = 0$.

19. Any polynomial in $\sqrt{p}$ and $\sqrt{q}$, with rational coefficients (i.e. any sum of a finite number of terms of the form $A(\sqrt{p})^m(\sqrt{q})^n$, where m and n are integers, and A rational), can be expressed in the form

$$a+b\sqrt{p}+c\sqrt{q}+d\sqrt{(pq)},$$

where a, b, c, d are rational.

20. Express $\dfrac{a+b\sqrt{p}+c\sqrt{q}}{d+e\sqrt{p}+f\sqrt{q}}$, where a, b, etc. are rational, in the form

$$A+B\sqrt{p}+C\sqrt{q}+D\sqrt{(pq)},$$

where A, B, C, D are rational.

[Evidently

$$\frac{a+b\sqrt{p}+c\sqrt{q}}{d+e\sqrt{p}+f\sqrt{q}} = \frac{(a+b\sqrt{p}+c\sqrt{q})(d+e\sqrt{p}-f\sqrt{q})}{(d+e\sqrt{p})^2-f^2q} = \frac{\alpha+\beta\sqrt{p}+\gamma\sqrt{q}+\delta\sqrt{(pq)}}{\epsilon+\zeta\sqrt{p}},$$

where α, β, etc. are rational numbers which can easily be found. The required reduction may now be easily completed by multiplication of numerator and denominator by $\epsilon-\zeta\sqrt{p}$. For example, prove that

$$\frac{1}{1+\sqrt{2}+\sqrt{3}} = \frac{1}{2}+\frac{1}{4}\sqrt{2}-\frac{1}{4}\sqrt{6}.]$$

21. If a, b, x, y are rational numbers such that

$$(ay-bx)^2+4(a-x)(b-y) = 0,$$

then either (i) $x = a, y = b$ or (ii) $1-ab$ and $1-xy$ are squares of rational numbers. (*Math. Trip.* 1903)

22. If all the values of x and y given by

$$ax^2+2hxy+by^2 = 1, \quad a'x^2+2h'xy+b'y^2 = 1$$

(where a, h, b, a', h', b' are rational) are rational, then

$$(h-h')^2-(a-a')(b-b'), \quad (ab'-a'b)^2+4(ah'-a'h)(bh'-b'h)$$

are both squares of rational numbers. (*Math. Trip.* 1899)

23. Show that $\sqrt{2}$ and $\sqrt{3}$ are cubic functions of $\sqrt{2}+\sqrt{3}$, with rational coefficients, and that $\sqrt{2}-\sqrt{6}+3$ is the ratio of two linear functions of $\sqrt{2}+\sqrt{3}$. (*Math. Trip.* 1905)

24. Show that

$$\sqrt{\{a+2m\sqrt{(a-m^2)}\}}+\sqrt{\{a-2m\sqrt{(a-m^2)}\}}$$

is equal to $2m$ if $2m^2>a>m^2$, and to $2\sqrt{(a-m^2)}$ if $a>2m^2$.

25. Show that any polynomial in $\sqrt[3]{2}$, with rational coefficients, can be expressed in the form

$$a+b\sqrt[3]{2}+c\sqrt[3]{4},$$

where a, b, c are rational.

More generally, if p is any rational number, any polynomial in $\sqrt[m]{p}$ with rational coefficients can be expressed in the form

$$a_0+a_1\alpha+a_2\alpha^2+\ldots+a_{m-1}\alpha^{m-1},$$

where a_0, a_1, ... are rational and $\alpha = \sqrt[m]{p}$. For any such polynomial is of the form

$$b_0+b_1\alpha+b_2\alpha^2+\ldots+b_k\alpha^k,$$

where the b's are rational. If $k \leqq m-1$, this is already of the form required. If $k>m-1$, let α^r be any power of α higher than the $(m-1)$th. Then $r = \lambda m+s$, where λ is an integer and $0 \leqq s \leqq m-1$; and $\alpha^r = \alpha^{\lambda m+s} = p^\lambda \alpha^s$. Hence we can get rid of all powers of α higher than the $(m-1)$th.

26. Express $(\sqrt[3]{2}-1)^5$ and $(\sqrt[3]{2}-1)/(\sqrt[3]{2}+1)$ in the form $a+b\sqrt[3]{2}+c\sqrt[3]{4}$, where a, b, c are rational. [Multiply numerator and denominator of the second expression by $\sqrt[3]{4}-\sqrt[3]{2}+1$.]

27. If

$$a+b\sqrt[3]{2}+c\sqrt[3]{4} = 0,$$

where a, b, c are rational, then $a = 0$, $b = 0$, $c = 0$.

[Let $y = \sqrt[3]{2}$. Then $y^3 = 2$ and

$$cy^2+by+a = 0.$$

Hence $2cy^2+2by+ay^3 = 0$ or

$$ay^2+2cy+2b = 0.$$

Multiplying these two quadratic equations by a and c and subtracting, we obtain $(ab-2c^2)y+a^2-2bc = 0$ or $y = -(a^2-2bc)/(ab-2c^2)$, a rational number, which is impossible. The only alternative is that $ab-2c^2 = 0$, $a^2-2bc = 0$.

Hence $ab = 2c^2$, $a^4 = 4b^2c^2$. If neither a nor b is zero, we can divide the second equation by the first, which gives $a^3 = 2b^3$: and this is impossible, since $\sqrt[3]{2}$ cannot be equal to the rational number a/b. Hence $ab = 0$, $c = 0$, and it follows from the original equation that a, b, and c are all zero.

As a corollary, if $a+b\sqrt[3]{2}+c\sqrt[3]{4} = d+e\sqrt[3]{2}+f\sqrt[3]{4}$, then $a = d$, $b = e$, $c = f$.

It may be proved, more generally, that if

$$a_0+a_1p^{1/m}+\ldots+a_{m-1}p^{(m-1)/m} = 0,$$

p not being a perfect mth power, then $a_0 = a_1 = \ldots = a_{m-1} = 0$; but the proof is less simple.]

28. If $A+\sqrt[3]{B} = C+\sqrt[3]{D}$, then either $A = C$, $B = D$, or B and D are both cubes of rational numbers.

29. If $\sqrt[3]{A}+\sqrt[3]{B}+\sqrt[3]{C} = 0$, then either one of A, B, C is zero, and the other two equal and opposite, or $\sqrt[3]{A}$, $\sqrt[3]{B}$, $\sqrt[3]{C}$ are rational multiples of the same surd $\sqrt[3]{X}$.

30. Find rational numbers α, β such that

$$\sqrt[3]{(7+5\sqrt{2})} = \alpha+\beta\sqrt{2}.$$

31. If $(a-b^3)\,b>0$, then

$$\sqrt[3]{\left\{a+\frac{9b^3+a}{3b}\sqrt{\left(\frac{a-b^3}{3b}\right)}\right\}}+\sqrt[3]{\left\{a-\frac{9b^3+a}{3b}\sqrt{\left(\frac{a-b^3}{3b}\right)}\right\}}$$

is rational. [Each of the numbers under a cube root is of the form

$$\left\{\alpha+\beta\sqrt{\left(\frac{a-b^3}{3b}\right)}\right\}^3,$$

where α and β are rational.]

32. Prove that

$$\sqrt{(\sqrt[3]{5}-\sqrt[3]{4})} = \tfrac{1}{3}(\sqrt[3]{2}+\sqrt[3]{20}-\sqrt[3]{25}),$$

$$\sqrt[3]{(\sqrt[3]{2}-1)} = \sqrt[3]{(\tfrac{1}{9})}-\sqrt[3]{(\tfrac{2}{9})}+\sqrt[3]{(\tfrac{4}{9})},$$

$$\sqrt[4]{\left(\frac{3+2\sqrt[4]{5}}{3-2\sqrt[4]{5}}\right)} = \frac{\sqrt[4]{5}+1}{\sqrt[4]{5}-1}.$$

33. If $\alpha = \sqrt[n]{p}$, then any polynomial in α is the root of an equation of degree n, with rational coefficients.

[We can express the polynomial (x say) in the form

$$x = l_1+m_1\alpha+\ldots+r_1\alpha^{(n-1)},$$

where l_1, m_1, ... are rational, as in Ex. 25.

Similarly

$$x^2 = l_2+m_2\alpha+\ldots+r_2\alpha^{(n-1)},$$

$$\cdots\cdots\cdots\cdots\cdots\cdots$$

$$x^n = l_n+m_n\alpha+\ldots+r_n\alpha^{(n-1)}.$$

Hence

$$L_1x+L_2x^2+\ldots+L_nx^n = \Delta,$$

where Δ is the determinant

$$\begin{vmatrix} l_1 & m_1 & \dots & r_1 \\ l_2 & m_2 & \dots & r_2 \\ \dots & \dots & \dots & \dots \\ l_n & m_n & \dots & r_n \end{vmatrix}$$

and L_1, L_2, ... the minors of l_1, l_2, ...]

34. Apply this process to $x = p+\sqrt{q}$, and deduce the theorem of § 14.

35. Show that $y = a+bp^{1/3}+cp^{2/3}$ satisfies the equation

$$y^3-3ay^2+3y(a^2-bcp)-a^3-b^3p-c^3p^2+3abcp = 0.$$

36. **Algebraic numbers.** We have seen that some irrational numbers (such as $\sqrt{2}$) are roots of equations of the type

$$a_0x^n+a_1x^{n-1}+\dots+a_n = 0,$$

where $a_0, a_1, \dots, a_n$ are integers. Such irrational numbers are called *algebraic* numbers: all other irrational numbers, such as π (§ 15), are called *transcendental* numbers.

37. If x and y are algebraic numbers, then so are $x+y$, $x-y$ and xy; and x/y is algebraic if $y \neq 0$.

[A little knowledge of algebra is required. We must use the theorems that the elementary symmetric functions Σx_r, $\Sigma x_r x_s, \dots$ of the roots of an equation

$$(1) \qquad x^m-p_1x^{m-1}+p_2x^{m-2}-\dots\pm p_m = 0$$

are $p_1, p_2, \dots$, and that any symmetric polynomial (see §§ 23, 31) in $x_1, x_2, \dots$, with integral coefficients, is a polynomial in $p_1, p_2, \dots$ with integral coefficients.

We can write the equations satisfied by x and y in the forms (1) and

$$(2) \qquad y^n-q_1y^{n-1}+q_2y^{n-2}-\dots\pm q_n = 0,$$

$p_1, p_2, \dots$ and $q_1, q_2, \dots$ being rational. We suppose that the roots of (1) and (2) are $x_1, x_2, \dots$ and $y_1, y_2, \dots$, x and y being x_1 and y_1, and form the product

$$P(z) = \prod_{h=1}^{m}\prod_{k=1}^{n}(z-x_h-y_k)$$

extended over the mn pairs of values of h and k. Then $P(z)$ is a polynomial of degree mn in z, and its coefficients are symmetric polynomials in the x_h's and the y_k's, with integral coefficients. It follows that the coefficients are polynomials in $p_1, p_2, \dots, q_1, q_2, \dots$ with integral coefficients. Thus $P(z) = 0$ is an equation of degree mn, with rational coefficients, one of whose roots is $x+y$.

The proof for $x-y$ and xy is similar. If $y \neq 0$ and we suppose, as we then may, that $q_n \neq 0$, then $z = 1/y$ satisfies

$$z^n - r_1 z^{n-1} + r_2 z^{n-2} - \ldots \pm r_n = 0,$$

where $r_1 = q_{n-1}/q_n$, $r_2 = q_{n-2}/q_n$, ... Hence z is algebraic, and therefore $x/y = xz$ is algebraic.

In particular $x+k$ and kx are algebraic if k is rational.]

38. If $$x^m + \alpha_1 x^{m-1} + \alpha_2 x^{m-2} + \ldots + \alpha_m = 0,$$

where $\alpha_1, \alpha_2, \ldots, \alpha_m$ are algebraic, then x is algebraic.

[This may be proved similarly. Each α_r satisfies an equation

$$\alpha_r^{n_r} - p_{r,1}\alpha_r^{n_r-1} + \ldots \pm p_{r,n_r} = 0$$

with rational coefficients. We suppose that the roots of this equation are $\alpha_{r,1}, \alpha_{r,2}, \ldots, \alpha_{r,n_r}$ (α_r being $\alpha_{r,1}$), and form the product

$$P(x) = \Pi\,(x^m + \alpha_{1,s_1} x^{m-1} + \alpha_{2,s_2} x^{m-2} + \ldots + \alpha_{m,s_m})$$

extended over the $N = n_1 n_2 \ldots n_m$ combinations of the suffixes $s_1, s_2, \ldots, s_m$, thus obtaining a polynomial in x, of degree mN, with rational coefficients.

In particular $x^{m/n}$ is algebraic if x is algebraic and m and n integral.]

39. If $$x^2 - 2x\sqrt{2} + \sqrt{3} = 0$$

then $$x^8 - 16x^6 + 58x^4 - 48x^2 + 9 = 0.$$

40. Find equations, with rational coefficients, satisfied by

$$1+\sqrt{2}+\sqrt{3}, \quad \frac{\sqrt{3}+\sqrt{2}}{\sqrt{3}-\sqrt{2}}, \quad \sqrt{\{\sqrt{3}+\sqrt{2}\}}+\sqrt{\{\sqrt{3}-\sqrt{2}\}}, \quad \sqrt[3]{2}+\sqrt[3]{3}.$$

41. If $x^3 = x+1$, then $x^{3n} = a_n x + b_n + c_n x^{-1}$, where

$$a_{n+1} = a_n + b_n, \quad b_{n+1} = a_n + b_n + c_n, \quad c_{n+1} = a_n + c_n.$$

42. If $x^6 + x^5 - 2x^4 - x^3 + x^2 + 1 = 0$ and $y = x^4 - x^2 + x - 1$, then y satisfies a quadratic equation with rational coefficients.

(*Math. Trip.* 1903)

[It will be found that $y^2 + y + 1 = 0$.]

CHAPTER II

FUNCTIONS OF REAL VARIABLES

20. The idea of a function. Suppose that x and y are two continuous real variables, which we may suppose to be represented geometrically by distances $A_0P = x$, $B_0Q = y$ measured from fixed points A_0, B_0 along two straight lines Λ, M. And let us suppose that the positions of the points P and Q are not independent, but are connected by a relation which we can imagine expressed as a relation between x and y; so that, when P and x are known, Q and y are also known. We might, for example, suppose that $y = x$, or $2x$, or $\frac{1}{2}x$, or $x^2 + 1$. In all of these cases the value of x determines that of y. Or again we might suppose that the relation between x and y is given, not by means of an explicit formula for y in terms of x, but by means of a geometrical construction which enables us to determine Q when P is known.

In these circumstances y is said to be a *function* of x. This notion of functional dependence of one variable upon another is perhaps the most important in the whole range of higher mathematics. In order to enable the reader to be certain that he understands it clearly, we shall, in this chapter, illustrate it by means of a large number of examples.

But before we proceed to do this, we must point out that the simple examples of functions mentioned above possess three characteristics which are by no means involved in the general idea of a function, viz.:

(1) y is determined *for every value of* x;

(2) to each value of x for which y is given corresponds *one and only one value of* y;

(3) the relation between x and y is expressed by means of *an analytical formula*, from which the value of y corresponding to a

given value of x can be calculated by direct substitution of the latter.

It is indeed the case that these particular characteristics are possessed by many of the most important functions. But the consideration of the following examples will make it clear that they are by no means essential to a function. All that is essential is that there should be some relation between x and y such that to some values of x at any rate correspond values of y.

Examples X. 1. Let $y = x$ or $2x$ or $\frac{1}{2}x$ or $x^2 + 1$. Nothing further need be said at present about cases such as these.

2. Let $y = 0$ whatever be the value of x. Then y is a function of x, for we can give x any value, and the corresponding value of y (viz. 0) is known. In this case the functional relation makes the same value of y correspond to all values of x. The same would be true were y equal to 1 or $-\frac{1}{2}$ or $\sqrt{2}$ instead of 0. Such a function of x is called *a constant.*

3. Let $y^2 = x$. Then if x is positive this equation defines *two* values of y corresponding to each value of x, viz. $\pm\sqrt{x}$. If $x = 0$, $y = 0$. Hence to the particular value 0 of x corresponds *one* and only one value of y. But if x is negative there is *no* value of y which satisfies the equation. That is to say, the function y is not defined for negative values of x. This function therefore possesses the characteristic (3), but neither (1) nor (2).

4. Consider a volume of gas maintained at a constant temperature and contained in a cylinder closed by a sliding piston*.

Let A be the area of the cross-section of the piston and W its weight. The gas, held in a state of compression by the piston, exerts a certain pressure p_0 per unit of area on the piston, which balances the weight W, so that

$$W = Ap_0.$$

Let v_0 be the volume of the gas when the system is thus in equilibrium. If additional weight is placed upon the piston the latter is forced downwards. The volume (v) of the gas diminishes; the pressure (p) which it exerts upon unit area of the piston increases. Boyle's experimental law asserts that the product of p and v is very nearly constant, a correspondence which, if exact, would be represented by an equation of the type

$$pv = a \quad \text{..............................(i)},$$

where a is a number which can be determined approximately by experiment.

* I borrow this instructive example from Prof. H. S. Carslaw's *Introduction to the calculus.*

Boyle's law, however, only gives a reasonable approximation to the facts provided the gas is not compressed too much. When v is decreased and p increased beyond a certain point, the relation between them is no longer expressed with tolerable exactness by the equation (i). It is known that a much better approximation to the true relation can then be found by means of what is known as 'van der Waals' law', expressed by the equation

$$\left(p+\frac{\alpha}{v^2}\right)(v-\beta) = \gamma \quad \text{......................(ii)},$$

where α, β, γ are numbers which can also be determined approximately by experiment.

Of course the two equations, even taken together, do not give anything like a complete account of the relation between p and v. This relation is no doubt in reality much more complicated, and its form changes, as v varies, from a form nearly equivalent to (i) to a form nearly equivalent to (ii). But, from a mathematical point of view, there is nothing to prevent us from contemplating an ideal state of things in which, for all values of v not less than a certain value V, (i) would be exactly true, and (ii) exactly true for all values of v less than V. And then we might regard the two equations as together defining p as a function of v. It is an example of a function which for some values of v is defined by one formula and for other values of v is defined by another.

This function possesses the characteristic (2): to any value of v only one value of p corresponds: but it does not possess (1). For p is not defined as a function of v for negative values of v; a 'negative volume' means nothing, and negative values of v are irrelevant.

5. Suppose that a perfectly elastic ball is dropped (without rotation) from a height $\frac{1}{2}g\tau^2$ on to a fixed horizontal plane, and rebounds continually.

The ordinary formulae of elementary dynamics, with which the reader is probably familiar, show that $h = \frac{1}{2}gt^2$ if $0 \leqq t \leqq \tau$, $h = \frac{1}{2}g(2\tau - t)^2$ if $\tau \leqq t \leqq 3\tau$, and generally

$$h = \tfrac{1}{2}g(2n\tau - t)^2$$

if $(2n-1)\tau \leqq t \leqq (2n+1)\tau$, h being the depth of the ball, at time t, below its original position. Here also h is a function of t which is only defined for positive values of t.

6. Suppose that y is defined as being *the largest prime factor of* x. This is an instance of a definition which only applies to a particular class of values of x, viz. *integral* values. 'The largest prime factor of $\frac{11}{3}$ or of $\sqrt{2}$ or of π' means nothing, and so our defining relation fails to define for such values of x as these. Thus this function does not possess the characteristic

(1). It possesses (2), but not (3), since there is no simple formula which expresses y in terms of x.

7. Let y be defined as *the denominator of x when x is expressed in its lowest terms*. This is an example of a function which is defined if and only if x is *rational*. Thus $y = 7$ if $x = -11/7$, but y is not defined for $x = \sqrt{2}$.

21. The graphical representation of functions. Suppose that the variable y is a function of the variable x. It will generally be open to us also to regard x as a function of y, in virtue of the functional relation between x and y. But for the present we shall look at this relation from the first point of view. We shall then call x the *independent variable* and y the *dependent variable*; and, when the particular form of the functional relation is not specified, we shall express it by writing

$$y = f(x)$$

(or $F(x)$, $\phi(x)$, $\psi(x)$, ..., as the case may be).

The nature of particular functions may, in very many cases, be illustrated and made easily intelligible as follows. Draw two lines OX, OY at right angles to one another and produced indefinitely in both directions. We can represent values of x and y by distances measured from O along the lines OX, OY respectively, regard being paid, of course, to sign, and the positive directions of measurement being those indicated by arrows in Fig. 6.

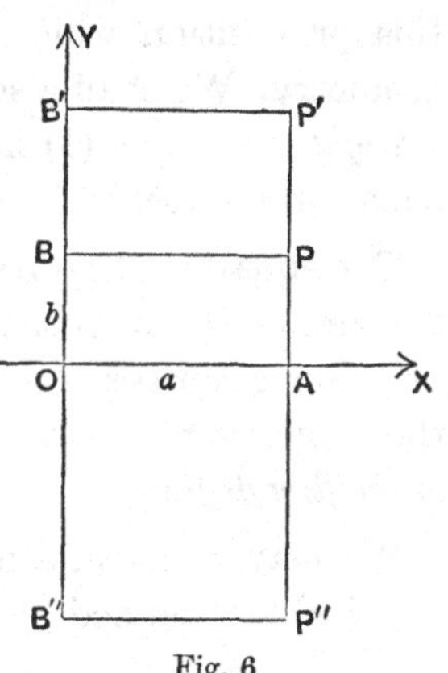

Fig. 6

Let a be any value of x for which y is defined and has (let us suppose) the single value b. Take $OA = a$, $OB = b$, and complete the rectangle $OAPB$. Imagine the point P marked on the diagram. This marking of the point P may be regarded as showing that the value of y for $x = a$ is b.

If to the value a of x correspond several values of y (say b, b', b''), we have, instead of the single point P, a number of points P, P', P'', ...

We shall call P the *point* (a, b); a and b the *coordinates of* P *referred to the axes* OX, OY; a the *abscissa*, b the *ordinate of* P; OX and OY the *axis of* x and the *axis of* y, or together the *axes of coordinates*, and O the *origin of coordinates*, or simply the *origin*.

Let us now suppose that for all values a of x for which y is defined, the value b (or values $b, b', b'', \ldots$) of y, and the corresponding point P (or points $P, P', P'', \ldots$), have been determined. We call the aggregate of all these points the *graph* of the function y.

To take a very simple example, suppose that y is defined as a function of x by the equation

$$Ax + By + C = 0 \qquad \ldots\ldots\ldots\ldots\ldots\ldots\ldots(1),$$

where A, B, C are any fixed numbers*. Then y is a function of x which possesses all the characteristics (1), (2), (3) of § 20. It is easy to show that *the graph of* y *is a straight line*. The reader is in all probability familiar with one or other of the various proofs of this proposition which are given in text-books of analytical geometry. We shall also say that *the locus of the point* (x, y) *is a straight line*, that (1) is *the equation of the locus*, and that the equation *represents* the locus.

The equation $Ax + By + C = 0$ is the most general equation of the first degree in both x and y. Hence *the general equation of the first degree represents a straight line*. It is equally easy to prove the converse proposition that *the equation of any straight line is of the first degree*.

We may mention a few further examples of interesting geometrical loci defined by equations. An equation of the form

$$(x - \alpha)^2 + (y - \beta)^2 = \rho^2,$$

or

$$x^2 + y^2 + 2Gx + 2Fy + C = 0,$$

where $G^2 + F^2 - C > 0$, represents a circle. The equation

$$Ax^2 + 2Hxy + By^2 + 2Gx + 2Fy + C = 0$$

* If $B = 0$, y does not occur in the equation. We must then regard y as a function of x defined for one value only of x, viz. $x = -C/A$, and then having *all* values.

(the general equation of the second degree) represents, assuming that the coefficients satisfy certain inequalities, a conic section, i.e. an ellipse, parabola, or hyperbola. For further discussion of these loci we must refer to books on analytical geometry.

22. Polar coordinates. In what precedes we have determined the position of P by the lengths of its coordinates $OM = x$, $MP = y$. If $OP = r$ and $MOP = \theta$, θ being an angle between 0 and 2π (measured in the positive direction), it is evident that

$$x = r\cos\theta, \quad y = r\sin\theta,$$

$$r = \sqrt{(x^2+y^2)}, \quad \cos\theta : \sin\theta : 1 :: x : y : r,$$

and that the position of P is equally well determined by a knowledge of r and θ. We call r and θ the *polar coordinates* of P. The former, it should be observed, is essentially positive*.

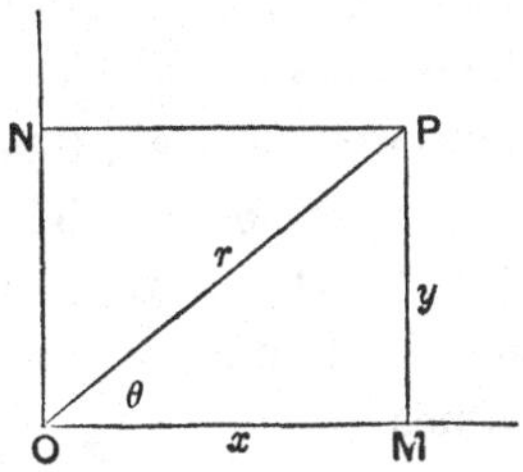

Fig. 7

If P moves on a locus there will be some relation between r and θ, say $r = f(\theta)$ or $\theta = F(r)$. This we call the *polar equation* of the locus. The polar equation may be deduced from the (x, y) equation (or *vice versa*) by means of the formulae above.

Thus the polar equation of a straight line is of the form

$$r\cos(\theta - \alpha) = p,$$

where p and α are constants. The equation $r = 2a\cos\theta$ represents a circle passing through the origin; and the general equation of a circle is of the form

$$r^2 + c^2 - 2rc\cos(\theta - \alpha) = A^2,$$

where A, c, and α are constants.

* Polar coordinates are sometimes defined so that r may be positive or negative. In this case two pairs of coordinates—e.g. $(1, 0)$ and $(-1, \pi)$—correspond to the same point. The distinction between the two systems may be illustrated by means of the equation $l/r = 1 - e\cos\theta$, where $l > 0$, $e > 1$. According to our definitions r must be positive and therefore $\cos\theta < 1/e$: the equation represents one branch only of a hyperbola, the other having the equation $-l/r = 1 - e\cos\theta$. With the system of coordinates which admits negative values of r, the equation represents the whole hyperbola.

23. Further examples of functions and their graphical representation. The examples which follow will give the reader a better notion of the infinite variety of possible types of functions.

A. **Polynomials.** A *polynomial in* x is a function of the form

$$a_0x^m + a_1x^{m-1} + \dots + a_m,$$

where $a_0, a_1, \dots, a_m$ are constants. The simplest polynomials are the powers $y = x, x^2, x^3, \dots, x^m, \dots$. The graph of the function x^m is of two distinct types, according as m is even or odd.

First let $m = 2$. Then three points on the graph are $(0, 0)$, $(1, 1)$, $(-1, 1)$. Any number of additional points on the graph may be found by assigning other special values to x: thus the values

$$x = \tfrac{1}{2},\ 2,\ 3,\ -\tfrac{1}{2},\ -2,\ -3$$

give

$$y = \tfrac{1}{4},\ 4,\ 9,\ \ \tfrac{1}{4},\ \ 4,\ \ 9.$$

If the reader will plot off a fair number of points on the graph, he will be led to conjecture that the form of the graph is something like that shown in Fig. 8. If he draws a curve through the special points which he has proved to lie on the graph and then tests its accuracy by giving x new values, and calculating the corresponding values of y, he will find that they lie as near to the curve as it is reasonable to expect, when the inevitable inaccuracies of drawing are considered. The curve is of course a parabola.

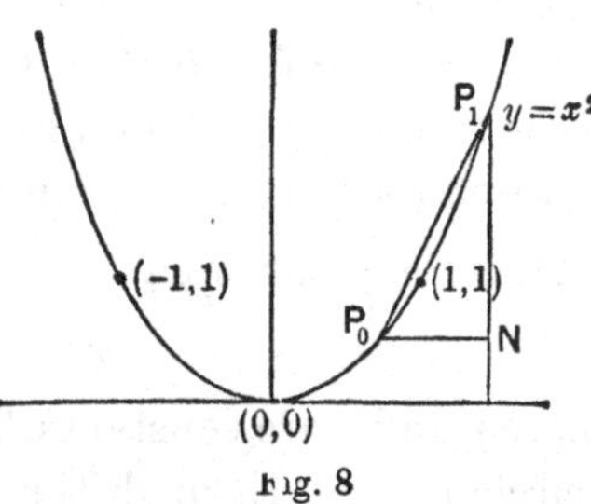

Fig. 8

There is, however, one fundamental question which we cannot answer adequately at present. The reader has no doubt some notion as to what is meant by a *continuous* curve, a curve without breaks or jumps; such a curve, in fact, as is roughly represented in Fig. 8. The question is whether the graph of the function $y = x^2$ is in fact such a curve. This cannot be *proved* by merely constructing any number of isolated points on the curve, although the more such points we construct the more probable it will appear.

This question cannot be discussed properly until Ch. V. In that chapter we shall consider in detail what our common-sense idea of continuity really means, and how we can prove that such graphs as the one now considered, and others which we shall consider later on in this chapter, are really continuous curves. For the present the reader may be content to draw his curves as common sense dictates.

It is easy to see that the curve $y = x^2$ is everywhere convex to the axis of x. Let P_0, P_1 (Fig. 8) be the points (x_0, x_0^2), (x_1, x_1^2). Then the coordinates of a point on the chord P_0P_1 are $x = \lambda x_0 + \mu x_1$, $y = \lambda x_0^2 + \mu x_1^2$, where λ and μ are positive numbers whose sum is 1. And

$$y - x^2 = (\lambda + \mu)(\lambda x_0^2 + \mu x_1^2) - (\lambda x_0 + \mu x_1)^2 = \lambda\mu(x_1 - x_0)^2 \geqq 0,$$

so that the chord lies entirely above the curve.

The curve $y = x^4$ is similar to $y = x^2$ in general appearance, but flatter near O, and steeper beyond the points A, A' (Fig. 9), and $y = x^m$, where m is even and greater than 4, is still more so. As m gets larger and larger the flatness and steepness grow more and more pronounced, until the curve is practically indistinguishable from the thick line in the figure.

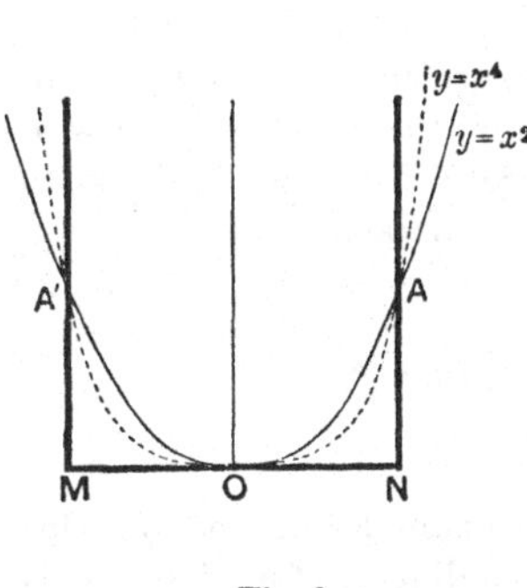

Fig. 9

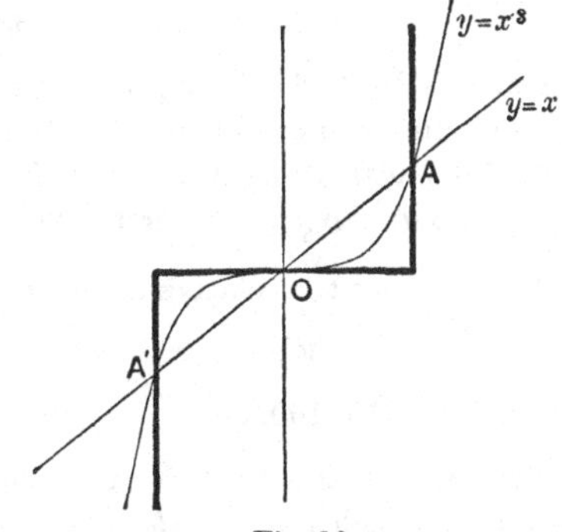

Fig. 10

The reader should consider next the curves given by $y = x^m$, when m is odd. The fundamental difference between the two cases is that whereas when m is even $(-x)^m = x^m$, so that the curve is symmetrical about OY, when m is odd $(-x)^m = -x^m$, so that y is negative when x is negative. Fig. 10 shows the curves $y = x$, $y = x^3$, and the form to which $y = x^m$ approximates for larger odd values of m.

It is now easy to see how (theoretically at any rate) the graph of any polynomial may be constructed. In the first place, from the graph of $y = x^m$ we can at once derive that of Cx^m, where C is a constant, by multiplying the ordinate of every point of the curve by C. And if we know the graphs of $f(x)$ and $F(x)$, we can find that of $f(x) + F(x)$ by taking the ordinate of every point to be the sum of the ordinates of the corresponding points on the two original curves.

The drawing of graphs of polynomials is however so much facilitated by the use of more advanced methods, which will be explained later on, that we shall not pursue the subject further here.

Examples XI. 1. Trace the curves $y = 7x^4$, $y = 3x^5$, $y = x^{10}$.

[The reader should draw the curves carefully, and all three should be drawn in one figure*. He will then realise how rapidly the higher powers of x increase, as x gets larger and larger, and will see that, in such a polynomial as

$$x^{10} + 3x^5 + 7x^4$$

(or even $x^{10} + 30x^5 + 700x^4$), it is the *first* term which is of really preponderant importance when x is fairly large. Thus even when $x = 4$, $x^{10} > 1{,}000{,}000$, while $30x^5 < 35{,}000$ and $700x^4 < 180{,}000$; while if $x = 10$ the preponderance of the first term is still more marked.]

2. Compare the relative magnitudes of

$$x^{12}, \quad 1{,}000{,}000x^6, \quad 1{,}000{,}000{,}000{,}000x$$

when $x = 1, 10, 100$, etc.

[The reader should make up a number of examples of this type for himself. This idea of the *relative rate of growth* of different functions of x is one with which we shall often be concerned in the following chapters.]

* It will be found convenient to take the scale of measurement along the axis of y a good deal smaller than that along the axis of x, in order to prevent the figure becoming of an awkward size.

3. Draw the graph of $ax^2+2bx+c$.

[Here
$$y-\frac{ac-b^2}{a}=a\left(x+\frac{b}{a}\right)^2.$$

If we take new axes parallel to the old and passing through the point $x=-b/a$, $y=(ac-b^2)/a$, the new equation is $y'=ax'^2$. The curve is a parabola.]

4. Trace the curves $y=x^3-3x+1$, $y=x^2(x-1)$, $y=x(x-1)^2$.

24. B. Rational functions. The class of functions which ranks next to that of polynomials in simplicity and importance is that of *rational functions*. A rational function is the quotient of one polynomial by another: thus if $P(x)$, $Q(x)$ are polynomials, we may denote the general rational function by

$$R(x)=\frac{P(x)}{Q(x)}.$$

In the particular case when $Q(x)$ is constant, $R(x)$ reduces to a polynomial: thus the class of rational functions includes that of polynomials as a sub-class. The following points concerning the definition should be noticed.

(1) We usually suppose that $P(x)$ and $Q(x)$ have no common factor $x+a$ or $x^p+ax^{p-1}+bx^{p-2}+\dots+k$, all such factors being removed by division.

(2) It should however be observed that this removal of common factors *does as a rule change the function*. Consider for example the function x/x, which is a rational function. On removing the common factor x we obtain $1/1=1$. But the original function is not *always* equal to 1: it is equal to 1 only so long as $x\neq 0$. If $x=0$ it takes the form $0/0$, which is meaningless. Thus the function x/x is equal to 1 if $x\neq 0$ and is undefined when $x=0$. It therefore differs from the function 1, which is *always* equal to 1.

(3) Such a function as

$$\left(\frac{1}{x+1}+\frac{1}{x-1}\right)\Big/\left(\frac{1}{x}+\frac{1}{x-2}\right)$$

may be reduced, by the ordinary rules of algebra, to the form

$$\frac{x^2(x-2)}{(x-1)^2(x+1)},$$

which is a rational function of the standard form. But here again it must be noticed that the reduction is not *always* legitimate. In order to calculate the value of a function for a given value of x we must substitute the value for x in the function *in the form in which it is given*. In this case the formula is meaningless for the values $x = -1, 1, 0, 2$, and so the function is not defined for these values. The same is true of the reduced form, so far as the values -1 and 1 are concerned. But $x = 0$ and $x = 2$ give the value 0. Thus once more the two functions are not the same.

(4) But, as appears from the particular example considered under (3), there will generally be a certain number of values of x for which the function is not defined even when it has been reduced to a rational function of the standard form. These are the values of x (if any) for which the denominator vanishes.

(5) Generally we agree, in dealing with expressions such as those considered in (2) and (3), to disregard the exceptional values of x for which such processes of simplification as were used there are illegitimate, and to reduce our function to the standard form of rational function. The reader will easily verify that (on this understanding) the sum, product, or quotient of two rational functions may themselves be reduced to rational functions of the standard type. And generally *a rational function of a rational function is itself a rational function*: i.e. if in $z = P(y)/Q(y)$, where P and Q are polynomials, we substitute $y = P_1(x)/Q_1(x)$, we obtain on simplification an equation of the form $z = P_2(x)/Q_2(x)$.

(6) It is in no way presupposed in the definition of a rational function that the constants which occur as coefficients should be rational *numbers*. The word rational has reference solely to the way in which the variable x appears in the function. Thus

$$\frac{x^2+x+\sqrt{3}}{x\sqrt[3]{2}-\pi}$$

is a rational function.

The use of the word rational arises as follows. The rational function $P(x)/Q(x)$ may be generated from x by a finite number of operations upon x, including only multiplication of x by itself or a constant, addition of terms thus obtained, and division of one function, obtained by such multiplications and additions, by another. In so far as the variable x is concerned, this procedure is very much like that by which all rational numbers can be obtained from unity, a procedure exemplified in the equation

$$\frac{5}{3} = \frac{1+1+1+1+1}{1+1+1}.$$

Again, *any* function which can be deduced from x by the elementary operations mentioned above, using at each stage of the process functions which have already been obtained from x in the same way, can be reduced to the standard type of rational function. Thus

$$\left(\frac{x}{x^2+1}+\frac{2x+7}{x^2+\dfrac{11x-3\sqrt{2}}{9x+1}}\right)\Bigg/\left(17+\frac{2}{x^3}\right)$$

can be reduced to the standard type of rational function.

25. The graphical study of rational functions depends, even more than that of polynomials, on the methods of the differential calculus. We shall therefore content ourselves at present with a very few examples.

Examples XII. 1. Draw the graphs of $y = 1/x$, $y = 1/x^2$, $y = 1/x^3, \ldots$.

[The figures show the graphs of the first two curves. It should be observed that the functions are not defined for $x = 0$.]

2. Trace

$$y = x+\frac{1}{x},\quad x-\frac{1}{x},\quad x^2+\frac{1}{x^2},\quad x^2-\frac{1}{x^2},\quad ax+\frac{b}{x},$$

taking various values, positive and negative, for a and b.

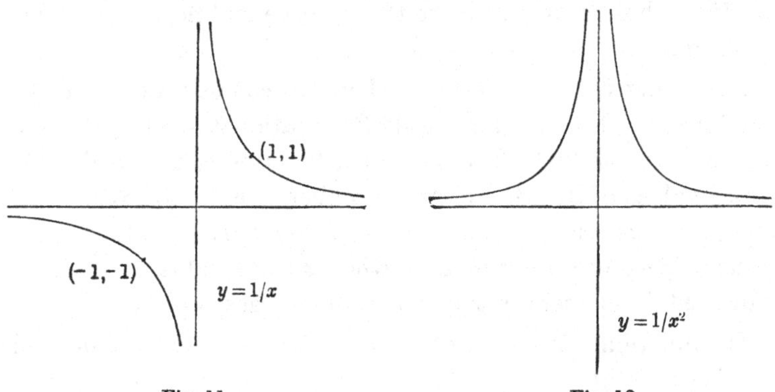

Fig. 11 Fig. 12

3. Trace

$$y = \frac{x+1}{x-1},\quad \left(\frac{x+1}{x-1}\right)^2,\quad \frac{1}{(x-1)^2},\quad \frac{x^2+1}{x^2-1}.$$

4. Trace $y = 1/(x-a)(x-b)$, $1/(x-a)(x-b)(x-c)$, where $a<b<c$.

5. Sketch the general form assumed by the curves $y = 1/x^m$ as m becomes larger and larger, considering separately the cases in which m is odd or even.

26. C. Explicit algebraical functions. The next important class of functions is that of *explicit algebraical functions.* These are functions which can be generated from x by a finite number of operations such as those used in generating rational functions, together with a finite number of operations of root extraction. Thus

$$\frac{\sqrt{(1+x)}-\sqrt[3]{(1-x)}}{\sqrt{(1+x)}+\sqrt[3]{(1-x)}},\quad \sqrt{x}+\sqrt{(x+\sqrt{x})},\quad \left(\frac{x^2+x+\sqrt{3}}{x\sqrt[3]{2}-\pi}\right)^{\frac{2}{3}}$$

are explicit algebraical functions, and so is $x^{m/n}$ (i.e. $\sqrt[n]{x^m}$), where m and n are any integers.

It should be noticed that there is an ambiguity of notation involved in such an equation as $y=\sqrt{x}$. We have up to the present regarded $\sqrt{2}$, for example, as denoting the *positive* square root of 2, and it would be natural to denote by $\sqrt{x}$, where x is any positive number, the positive square root of x, in which case $y=\sqrt{x}$ would be a one-valued function of x. It is however often more convenient to regard $\sqrt{x}$ as standing for the two-valued function whose two values are the positive and negative square roots of x.

The reader will observe that, when this course is adopted, the function $\sqrt{x}$ differs fundamentally from rational functions in two respects. In the first place a rational function is always defined for all values of x with a certain number of isolated exceptions. But $\sqrt{x}$ is undefined for *a whole range* of values of x (i.e. all negative values). Secondly the function, when x has a value for which it is defined, has generally two values of opposite signs.

The functions $\sqrt[3]{x}$, on the other hand, is one-valued and defined for all values of x.

Examples XIII. 1. $\sqrt{\{(x-a)(b-x)\}}$, where $a<b$, is defined only for $a\leqq x\leqq b$. If $a<x<b$, it has two values; if $x=a$ or b, only one, viz. 0.

2. Consider similarly

$$\sqrt{\{(x-a)(x-b)(x-c)\}}\quad (a<b<c),$$

$$\sqrt{\{x(x^2-a^2)\}},\quad \sqrt[3]{\{(x-a)^2(b-x)\}}\quad (a<b),$$

$$\frac{\sqrt{(1+x)}-\sqrt{(1-x)}}{\sqrt{(1+x)}+\sqrt{(1-x)}},\quad \sqrt{(x+\sqrt{x})}.$$

3. Trace the curves $y^2 = x$, $y^3 = x$, $y^2 = x^3$.

4. Draw the graphs of the functions $y = \sqrt{(a^2 - x^2)}$, $y = b\sqrt{\left(1 - \frac{x^2}{a^2}\right)}$.

27. D. Implicit algebraical functions. It is easy to verify that if

$$y = \frac{\sqrt{(1+x)} - \sqrt[3]{(1-x)}}{\sqrt{(1+x)} + \sqrt[3]{(1-x)}}$$

then
$$\left(\frac{1+y}{1-y}\right)^6 = \frac{(1+x)^3}{(1-x)^2},$$

and if
$$y = \sqrt{x} + \sqrt{(x + \sqrt{x})}$$

then
$$y^4 - (4y^2 + 4y + 1)x = 0.$$

Each of these equations may be expressed in the form

$$y^m + R_1 y^{m-1} + \dots + R_m = 0 \qquad \text{(1)},$$

where $R_1, R_2, \dots, R_m$ are rational functions of x; and the reader will easily verify that, if y is any one of the functions considered in the last set of examples, then y satisfies an equation of this form. It is naturally suggested that the same is true of any explicit algebraic function. And this is in fact true, and indeed not difficult to prove, though we shall not delay to write out a formal proof here. An example should make clear to the reader the lines on which such a proof would proceed. Let

$$y = \frac{x + \sqrt{x} + \sqrt{(x+\sqrt{x})} + \sqrt[3]{(1+x)}}{x - \sqrt{x} + \sqrt{(x+\sqrt{x})} - \sqrt[3]{(1+x)}}.$$

Then
$$y = \frac{x+u+v+w}{x-u+v-w},$$

$$u^2 = x, \quad v^2 = x + u, \quad w^3 = 1 + x;$$

and we have only to eliminate u, v, w between these equations in order to obtain an equation of the form desired.

We are therefore led to give the following definition: *y is an algebraical function of x, of degree m, if it is a root of an equation of degree m in y whose coefficients are rational functions of x.* There is no real loss of generality in supposing that, as in equation (1), the first coefficient is unity.

This class of functions includes all the explicit algebraical functions considered in § 26. But it also includes other functions which cannot be expressed as explicit algebraical functions. For it is known that in general such an equation as (1) cannot be solved explicitly for y in terms of x, when m is greater than 4, though such a solution is always possible if $m = 1, 2, 3$, or 4 and in special cases for higher values of m.

The definition of an algebraical function should be compared with that of an algebraical number given in the preceding chapter (Misc. Ex. 36).

Examples XIV. 1. If $m = 1$, y is a rational function.

2. If $m = 2$, the equation is $y^2 + R_1 y + R_2 = 0$, so that

$$y = \tfrac{1}{2}\{-R_1 \pm \sqrt{(R_1^2 - 4R_2)}\}.$$

This function is defined for all values of x for which $R_1^2 \geqq 4R_2$. It has two values if $R_1^2 > 4R_2$ and one if $R_1^2 = 4R_2$.

If $m = 3$ or 4, we can use the methods explained in treatises on algebra for the solution of cubic and biquadratic equations. But as a rule the process is complicated and the results inconvenient in form, and we can study the properties of the function better by means of the original equation.

3. Consider the functions defined by the equations

$$y^2 - 2y - x^2 = 0, \quad y^2 - 2y + x^2 = 0, \quad y^4 - 2y^2 + x^2 = 0,$$

in each case obtaining y as an explicit function of x, and stating for what values of x it is defined.

4. Find algebraical equations, with coefficients rational in x, satisfied by each of the functions

$$\sqrt{x} + \sqrt{(1/x)}, \quad \sqrt[3]{x} + \sqrt[3]{(1/x)}, \quad \sqrt{(x + \sqrt{x})}, \quad \sqrt{\{x + \sqrt{(x + \sqrt{x})}\}}.$$

5. Consider the equation $\quad y^4 = x^2$.

[Here $y^2 = \pm x$. If x is positive, $y = \sqrt{x}$; if negative, $y = \sqrt{(-x)}$. Thus the function has two values for all values of x save $x = 0$.]

6. An algebraical function of an algebraical function of x is itself an algebraical function of x.

[This may be proved on the general lines of Exs. 37 and 38, p. 38. We start from equations

$$y^m + R_1(z)y^{m-1} + \dots + R_m(z) = 0, \quad z^n + S_1(x)z^{n-1} + \dots + S_n(x) = 0,$$

with rational coefficients, and form the product

$$\Pi\{y^m + R_1(z_h)y^{m-1} + \dots + R_m(z_h)\}$$

extended over the n roots z_h of the second equation.]

7. An example should perhaps be given of an algebraical function which cannot be expressed in an explicit algebraical form. Such an example is the function y defined by the equation

$$y^5 - y - x = 0.$$

But the proof that we cannot express y explicitly in terms of x is difficult, and cannot be attempted here.

28. Transcendental functions. All functions of x which are not algebraical are called *transcendental* functions. The definition is negative. We do not attempt any systematic classification of the transcendental functions, but we can pick out one or two sub-classes of particular importance.

E. **The direct and inverse trigonometrical or circular functions.** These are the sine and cosine functions of elementary trigonometry, their inverses, and the functions derived from them. We may assume provisionally that the reader is familiar with their most important properties*.

Examples XV. 1. Draw the graphs of

$$\cos x, \quad \sin x, \quad a\cos x + b\sin x.$$

[Since $a\cos x + b\sin x = \beta\cos(x-\alpha)$, where $\beta = \sqrt{(a^2+b^2)}$ and α is an angle whose cosine and sine are $a/\sqrt{(a^2+b^2)}$ and $b/\sqrt{(a^2+b^2)}$, the graphs of these three functions are similar in character.]

2. Draw the graphs of $\cos^2 x$, $\sin^2 x$, $a\cos^2 x + b\sin^2 x$.

3. Suppose the graphs of $f(x)$ and $F(x)$ drawn. Then the graph of

$$f(x)\cos^2 x + F(x)\sin^2 x$$

is a wavy curve which oscillates between the curves $y = f(x)$, $y = F(x)$. Draw the graph when $f(x) = x$, $F(x) = x^2$.

4. Show that the graph of $\cos px + \cos qx$ lies between those of $2\cos\frac{1}{2}(p-q)x$ and $-2\cos\frac{1}{2}(p+q)x$, touching each in turn. Sketch the graph when $(p-q)/(p+q)$ is small. (*Math. Trip.* 1908)

5. Draw the graphs of $x + \sin x$, $(1/x) + \sin x$, $x\sin x$, $(\sin x)/x$.

6. Draw the graph of $\sin(1/x)$.

[If $y = \sin(1/x)$, then $y = 0$ when $x = 1/m\pi$, where m is any integer. Similarly $y = 1$ when $x = 1/(2m+\frac{1}{2})\pi$ and $y = -1$ when $x = 1/(2m-\frac{1}{2})\pi$.

* The definitions of the circular functions given in elementary trigonometry presuppose that any sector of a circle has associated with it a definite number called its *area*. How this assumption is justified will appear in Chs. VII and IX.

The curve is entirely comprised between the lines $y = -1$ and $y = 1$ (Fig. 13). It oscillates up and down, the rapidity of the oscillations becoming greater and greater as x approaches 0. For $x = 0$ the function is undefined. When x is large y is small*. The negative half of the curve is similar in character to the positive half.]

7. Draw the graph of $x \sin(1/x)$.

[This curve is comprised between the lines $y = -x$ and $y = x$ just as the curve of Ex. 6 is comprised between the lines $y = -1$ and $y = 1$ (Fig. 14).]

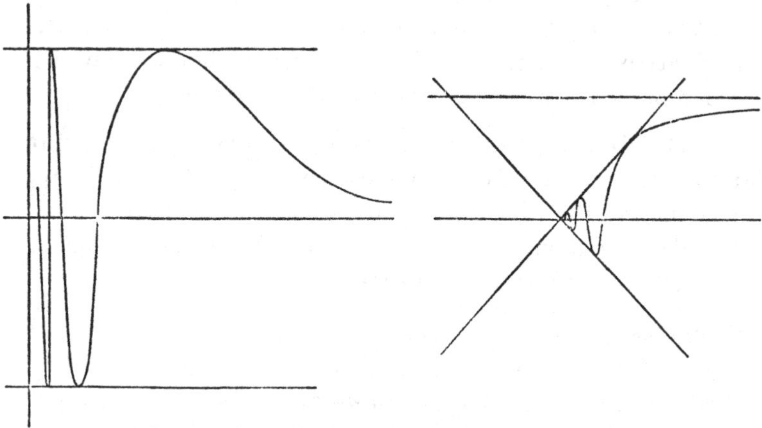

Fig. 13　　　　Fig. 14

8. Draw the graphs of

$$x^2 \sin\frac{1}{x},\quad \frac{1}{x}\sin\frac{1}{x},\quad \left(x\sin\frac{1}{x}\right)^2,\quad \sin x + \sin\frac{1}{x},\quad \sin x \sin\frac{1}{x}.$$

9. Draw the graphs of $\cos x^2$, $\sin x^2$, $a\cos x^2 + b\sin x^2$.

10. Draw the graphs of $\arccos x$ and $\arcsin x$ (the inverse cosine and sine, sometimes written $\cos^{-1}x$ and $\sin^{-1}x$).

[If $y = \arccos x$, $x = \cos y$. This enables us to draw the graph of x, considered as a function of y, and the same curve shows y as a function of x. It is clear that y is only defined for $-1 \leqq x \leqq 1$, and is infinitely many-valued for these values of x. As the reader no doubt remembers, there is, when $-1 < x < 1$, a value of y between 0 and π, say α, and the other values of y are given by the formula $2n\pi \pm \alpha$, where n is any integer.]

11. Draw the graphs of

$$\tan x,\quad \cot x,\quad \sec x,\quad \operatorname{cosec} x,\quad \tan^2 x,\quad \cot^2 x,\quad \sec^2 x,\quad \operatorname{cosec}^2 x.$$

* See Chs. IV and V for explanations of the precise meaning of this phrase.

12. Draw the graphs of $\operatorname{arc} \tan x$, $\operatorname{arc} \cot x$, $\operatorname{arc} \sec x$, $\operatorname{arc} \operatorname{cosec} x$. Give formulae (as in Ex. 10) expressing all the values of each of these functions in terms of any particular value.

13. Draw the graphs of $\tan(1/x)$, $\cot(1/x)$, $\sec(1/x)$, $\operatorname{cosec}(1/x)$.

14. Show that $\cos x$ and $\sin x$ are not rational functions of x.

[A function is said to be *periodic*, with period a, if $f(x) = f(x+a)$ for all values of x for which $f(x)$ is defined. Thus $\cos x$ and $\sin x$ have the period 2π. It is easy to see that no periodic function can be a rational function, unless it is a constant. For suppose that

$$f(x) = P(x)/Q(x),$$

where P and Q are polynomials, and that $f(x) = f(x+a)$, each of these equations holding for all values of x. Let $f(0) = k$. Then the equation $P(x) - kQ(x) = 0$ is satisfied by an infinite number of values of x, viz. $x = 0$, a, $2a$, etc., and therefore for all values of x. Thus $f(x) = k$ for all values of x, i.e. $f(x)$ is a constant.]

15. Show, more generally, that no function with a period can be an algebraical function of x.

[Let the equation which defines the algebraical function be

$$y^m + R_1 y^{m-1} + \dots + R_m = 0 \quad \text{.....................(1)},$$

where $R_1, \dots, R_m$ are rational functions of x. This may be put in the form

$$P_0 y^m + P_1 y^{m-1} + \dots + P_m = 0,$$

where $P_0, P_1, \dots, P_m$ are polynomials in x. Arguing as above, we see that

$$P_0 k^m + P_1 k^{m-1} + \dots + P_m = 0$$

for all values of x. Hence $y = k$ satisfies the equation (1) for all values of x, and one set of values of our algebraical function reduces to a constant.

Now divide (1) by $y - k$ and repeat the argument. Our final conclusion is that our algebraical function has, for any value of x, the same set of values $k, k', \dots$; i.e. it is composed of a certain number of constants.]

16. The inverse sine and inverse cosine are not rational or algebraical functions. [This follows from the fact that, for any value of x between -1 and $+1$, $\operatorname{arc} \sin x$ and $\operatorname{arc} \cos x$ have infinitely many values.]

29. F. Other classes of transcendental functions. Next in importance to the trigonometrical functions come the exponential and logarithmic functions, which will be discussed in Chs. IX and X. But these functions are beyond our range at present. And most of the other classes of transcendental functions whose properties have been studied, such as the elliptic functions,

Bessel's and Legendre's functions, gamma-functions, and so forth, lie altogether beyond the scope of this book. There are however some elementary types of functions which, though of much less importance theoretically than the rational, algebraical, or trigonometrical functions, are particularly instructive as illustrations of the possible varieties of the functional relation.

Examples XVI. 1. Let $y = [x]$, where $[x]$ denotes the greatest integer not greater than x. The graph is shown in Fig. 15*a*. The left-hand end points of the thick lines, but not the right-hand ones, belong to the graph.

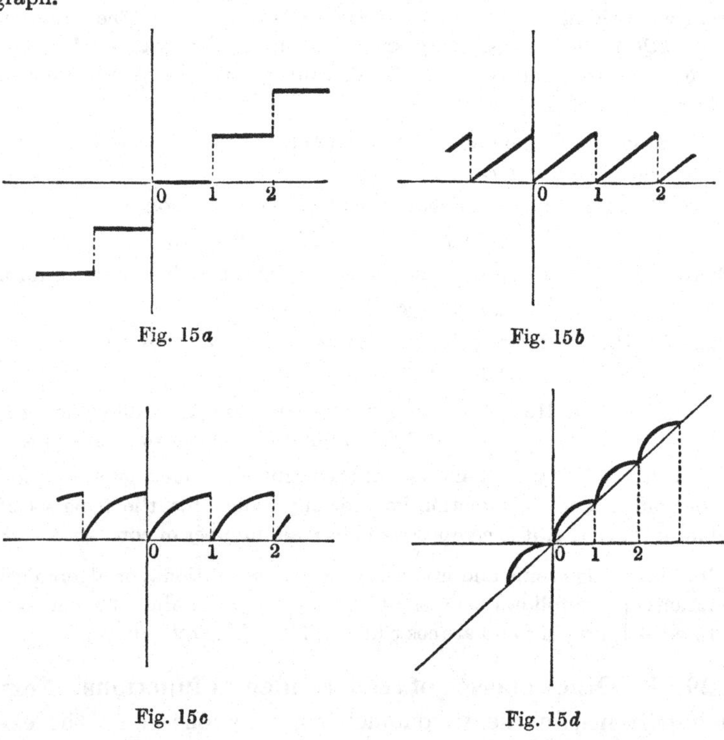

Fig. 15*a*

Fig. 15*b*

Fig. 15*c*

Fig. 15*d*

2. $y = x-[x]$. (Fig. 15*b*.)

3. $y = \sqrt{\{x-[x]\}}$. (Fig. 15*c*.)

4. $y = [x]+\sqrt{\{x-[x]\}}$. (Fig. 15*d*.)

5. $y = (x-[x])^2$, $[x]+(x-[x])^2$.

6. $y = [\sqrt{x}]$, $[x^2]$, $\sqrt{x}-[\sqrt{x}]$, $x^2-[x^2]$, $[1-x^2]$.

7. Let y be defined as *the largest prime factor of* x (cf. Ex. x. 6). Then y is defined only for integral values of x. If

$$x = 1, 2, 3, 4, 5, 6, 7, 8, 9, 10, 11, 12, 13, \ldots,$$

then $$y = 1, 2, 3, 2, 5, 3, 7, 2, 3, \quad 5, 11, \quad 3, 13, \ldots.$$

The graph consists of a number of isolated points.

8. Let y be *the denominator of* x (Ex. x. 7). In this case y is defined only for rational values of x. We can mark off as many points on the graph as we please, but the result is not in any ordinary sense of the word a curve, and there are no points corresponding to any irrational values of x.

Draw the straight line joining the points $(N-1, N)$, (N, N), where N is a positive integer. Show that the number of points of the locus which lie on this line is equal to the number of positive integers less than and prime to N.

9. Let $y = 0$ when x is an integer, $y = x$ when x is not an integer. The graph is derived from the straight line $y = x$ by taking out the points

$$\ldots(-1, -1), \quad (0, 0), \quad (1, 1), \quad (2, 2),$$

and adding the points $(-1, 0)$, $(0, 0)$, $(1, 0)$, ... on the axis of x.

10. Let $y = 1$ when x is rational, but $y = 0$ when x is irrational. The graph consists of two series of points arranged upon the lines $y = 1$ and $y = 0$. To the eye it is not distinguishable from two continuous straight lines, but in reality an infinite number of points are missing from each line.

11. Let $y = x$ when x is irrational and $y = \sqrt{\{(1+p^2)/(1+q^2)\}}$ when x is a rational fraction p/q.

The irrational values of x contribute to the graph a curve in reality discontinuous, but apparently not to be distinguished from the straight line $y = x$.

Now consider the rational values of x. First let x be positive. Then $\sqrt{\{(1+p^2)/(1+q^2)\}}$ cannot be equal to p/q unless $p = q$, i.e. $x = 1$. Thus all the points which correspond to rational values of x lie off the line, except the one point $(1, 1)$. Again, if $p<q$, $\sqrt{\{(1+p^2)/(1+q^2)\}}>p/q$; if $p>q$, $\sqrt{\{(1+p^2)/(1+q^2)\}}<p/q$. Thus the points lie above the line $y = x$ if $0<x<1$, below if $x>1$. If p and q are large, $\sqrt{\{(1+p^2)/(1+q^2)\}}$ is nearly equal to p/q. Near any value of x we can find any number of rational fractions with large numerators and denominators. Hence the graph contains a large number of points which crowd round the line $y = x$. Its general appearance (for positive values of x) is that of a line surrounded by a swarm of isolated points which gets denser and denser as the points approach the line.

The part of the graph which corresponds to negative values of x consists of the rest of the discontinuous line together with the reflections of all these isolated points in the axis of y. Thus to the left of the axis of y the swarm of points is not round $y = x$ but round $y = -x$, which is not itself part of the graph.

30. Graphical solution of equations containing a single unknown number. Many equations can be expressed in the form

$$f(x) = \phi(x) \quad \text{..........................(1)},$$

where $f(x)$ and $\phi(x)$ are functions whose graphs are easy to draw. And if the curves

$$y = f(x), \quad y = \phi(x)$$

intersect in a point P whose abscissa is ξ, then ξ is a root of the equation (1).

Examples XVII. 1. **The quadratic equation** $ax^2 + 2bx + c = 0$. This may be solved graphically in a variety of ways. For instance we may draw the graphs of

$$y = ax + 2b, \quad y = -c/x,$$

whose intersections, if any, give the roots. Or we may take

$$y = x^2, \quad y = -(2bx + c)/a.$$

See also Ex. VII. 2.

2. Solve by any of these methods

$$x^2 + 2x - 3 = 0, \quad x^2 - 7x + 4 = 0, \quad 3x^2 + 2x - 2 = 0.$$

3. **The equation** $x^m + ax + b = 0$. This may be solved by constructing the curves $y = x^m$, $y = -ax - b$. Verify the following table for the number of roots of

$$x^m + ax + b = 0:$$

(a) *m even* $\begin{cases} b \text{ positive, } \textit{two or none,} \\ b \text{ negative, } \textit{two}. \end{cases}$

(b) *m odd* $\begin{cases} a \text{ positive, } \textit{one,} \\ a \text{ negative, } \textit{three or one}. \end{cases}$

Construct numerical examples to illustrate all possible cases.

4. Show that the equation $\tan x = ax + b$ has always an infinite number of roots.

5. Determine the number of roots of

$$\sin x = x, \quad \sin x = \tfrac{1}{3}x, \quad \sin x = \tfrac{1}{8}x, \quad \sin x = \tfrac{1}{120}x.$$

6. Show that if a is small and positive (e.g. $a = \cdot 01$), the equation

$$x - a = \tfrac{1}{2}\pi \sin^2 x$$

has three roots. Consider also the case in which a is small and negative. Explain how the number of roots varies as a varies.

31. Functions of two variables and their graphical representation. In § 20 we considered two variables connected by a relation. We may similarly consider *three* variables (x, y, and z) connected by a relation such that when the values of x and y are both given, the value or values of z are known. In this case we call z a *function of the two variables* x and y; x and y the *independent* variables, z the *dependent* variable; and we express this dependence of z upon x and y by writing

$$z = f(x, y).$$

The remarks of § 20 may all be applied, *mutatis mutandis*, to this more complicated case.

The method of representing such functions of two variables graphically is the same in principle as for functions of a single variable. We take three axes OX, OY, OZ in space of three dimensions, each axis being perpendicular to the other two. The point (a, b, c) is the point whose distances from the planes YOZ, ZOX, XOY, measured parallel to OX, OY, OZ, are a, b, and c. Regard must of course be paid to sign, lengths measured in the directions OX, OY, OZ being regarded as positive. The definitions of *coordinates*, *axes*, and *origin* are the same as before.

Now let
$$z = f(x, y).$$
As x and y vary, the point (x, y, z) will move in space. The aggregate of all the positions it assumes is called the *locus* of the point (x, y, z) or the *graph* of the function $z = f(x, y)$. When the relation between x, y, and z which defines z can be expressed in an analytical formula, this formula is called the *equation* of the locus. It is easy to show, for example, that the equation

$$Ax + By + Cz + D = 0$$

(the general equation of the first degree) represents a *plane*, and that the equation of any plane is of this form. The equation

$$(x-\alpha)^2 + (y-\beta)^2 + (z-\gamma)^2 = \rho^2,$$

or
$$x^2 + y^2 + z^2 + 2Fx + 2Gy + 2Hz + C = 0,$$

where $F^2 + G^2 + H^2 - C > 0$, represents a *sphere*; and so on. For proofs of these propositions we must again refer to text-books of analytical geometry.

32. Curves in a plane. We have hitherto used the notation

$$y = f(x) \quad \text{.............................(1)}$$

to express functional dependence of y upon x. It is evident that this notation is most appropriate in the case in which y is defined by an explicit formula in x.

We have however very often to deal with functional relations which it is impossible or inconvenient to express in this form. If, for example, $y^5 - y - x = 0$ or $x^5 + y^5 - ay = 0$, it is known to be impossible to express y explicitly as an algebraical function of x. If

$$x^2 + y^2 + 2Gx + 2Fy + C = 0,$$

then

$$y = -F + \sqrt{(F^2 - x^2 - 2Gx - C)};$$

but the functional dependence of y upon x is more simply expressed by the original equation.

In all these cases the functional relation is expressed *by equating a function of the two variables x and y to zero*, i.e. by means of an equation

$$f(x, y) = 0 \quad \text{..........................(2).}$$

We shall adopt this equation as the standard method of expressing the functional relation. It includes the equation (1) as a special case, since $y - f(x)$ is a special form of a function of x and y. We can then speak of the locus of the point (x, y) subject to $f(x, y) = 0$, the graph of the function y defined by $f(x, y) = 0$, the curve or locus $f(x, y) = 0$, and the equation of this curve or locus.

There is another method of representing curves which is often useful. Suppose that x and y are both functions of a third variable t, which may or may not have some particular geometrical significance. We may write

$$x = f(t), \quad y = F(t) \quad \text{....................(3).}$$

If a particular value is assigned to t, the corresponding values of x and of y are known. Each pair of such values defines a point (x, y). If we construct all the points which correspond in this way to different values of t, we obtain *the graph of the locus defined by the equations* (3). Suppose for example

$$x = a\cos t, \quad y = a\sin t.$$

Let t vary from 0 to 2π. Then it is easy to see that the point (x, y) describes the circle whose centre is the origin and whose radius is a. If t varies beyond these limits, (x, y) describes the circle over and over again.

Elimination of t gives $x^2+y^2 = a^2$, the ordinary equation of the circle.

Examples XVIII. 1. The points of intersection of the two curves whose equations are $f(x, y) = 0$, $\phi(x, y) = 0$, where f and ϕ are polynomials, can be determined if these equations can be solved as a pair of simultaneous equations in x and y. The solution generally consists of a finite number of pairs of values of x and y. The two equations therefore generally represent a finite number of isolated points.

2. Trace the curves $(x+y)^2 = 1$, $xy = 1$, $x^2-y^2 = 1$.

3. The curve $f(x, y)+\lambda\phi(x, y) = 0$ represents a curve passing through the points of intersection of $f = 0$ and $\phi = 0$.

4. What loci are represented by

$$(\alpha)\ x = at+b,\quad y = ct+d, \qquad (\beta)\ \frac{x}{a} = \frac{2t}{1+t^2},\quad \frac{y}{b} = \frac{1-t^2}{1+t^2},$$

when t varies through all real values?

33. Loci in space. In space of three dimensions there are two fundamentally different kinds of loci, of which the simplest examples are the plane and the straight line.

A particle which moves along a straight line has only *one degree of freedom*. Its direction of motion is fixed; its position can be completely fixed by one measurement of position, e.g. by its distance from a fixed point on the line. It we take the line as our fundamental line Λ of Ch. I, the position of any of its points is determined by a single coordinate x. A particle which moves in a plane, on the other hand, has *two* degrees of freedom; to fix its position requires the determination of two coordinates.

A locus represented by a single equation

$$z = f(x, y)$$

plainly belongs to the second of these two classes of loci, and is called a *surface*. It may or may not satisfy our common-sense notion of what a surface should be.

The considerations of § 31 may evidently be generalised so as to give definitions of a function $f(x, y, z)$ of *three* variables (or of functions of any number of variables). And as in § 32 we agreed to adopt $f(x, y) = 0$ as the standard form of the equation of a plane curve, so now we shall agree to adopt

$$f(x, y, z) = 0$$

as the standard form of equation of a surface.

The locus represented by *two* equations of the form $z = f(x, y)$ or $f(x, y, z) = 0$ belongs to the first class of loci, and is called a *curve*. Thus a *straight line* may be represented by two equations of the type $Ax + By + Cz + D = 0$. A *circle* in space may be regarded as the intersection of a sphere and a plane; it may therefore be represented by two equations of the forms

$$(x-\alpha)^2 + (y-\beta)^2 + (z-\gamma)^2 = \rho^2, \quad Ax + By + Cz + D = 0.$$

Examples XIX. 1. What is represented by *three* equations of the type $f(x, y, z) = 0$?

2. Three linear equations in general represent a single point. What are the exceptional cases?

3. What are the equations of a plane curve $f(x, y) = 0$ in the plane XOY, when regarded as a curve in space? [$f(x, y) = 0$, $z = 0$.]

4. **Cylinders.** What is the meaning of a single equation $f(x, y) = 0$, considered as a locus in space of three dimensions?

[All points on the surface satisfy $f(x, y) = 0$, whatever be the value of z. The curve $f(x, y) = 0$, $z = 0$ is the curve in which the locus cuts the plane XOY. The locus is the surface formed by drawing lines parallel to OZ through all points of this curve. Such a surface is called a *cylinder*.]

5. **Graphical representation of a surface on a plane. Contour maps.** It might seem to be impossible to represent a surface adequately by a drawing on a plane; but a very fair notion of the nature of the surface may often be obtained as follows. Let the equation of the surface be $z = f(x, y)$.

If we give z a particular value a, we have an equation $f(x, y) = a$, which we may regard as determining a plane curve on the paper. We trace this curve and mark it (a). Actually the curve (a) is the projection on the plane XOY of the section of the surface by the plane $z = a$. We do this for all values of a (practically, of course, for a selection of values of a). We obtain some such figure as is shown in Fig. 16. It will at once suggest a contoured Ordnance Survey map: and in fact this is the principle on which such maps are constructed. The contour line 1000 is the projection, on the plane of the sea level, of the section of the surface of the land by the plane parallel to the plane of the sea level and 1000 ft. above it*.

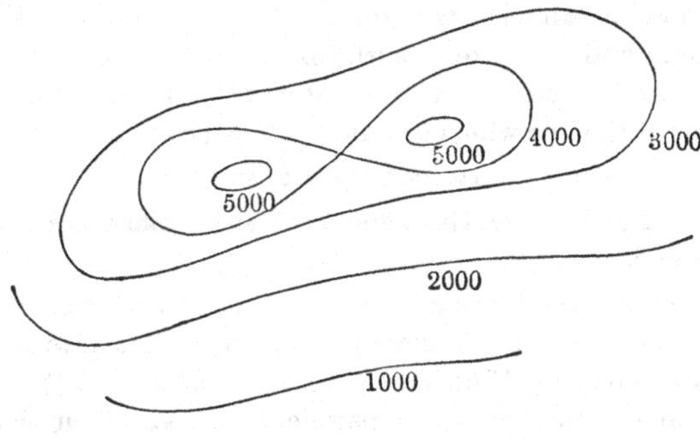

Fig. 16

6. Draw a series of contour lines to illustrate the form of the surface $2z = 3xy$.

7. **Right circular cones.** Take the origin of coordinates at the vertex of the cone and the axis of z along the axis of the cone; and let α be the semi-vertical angle of the cone. The equation of the cone (which must be regarded as extending both ways from its vertex) is

$$x^2+y^2-z^2\tan^2\alpha = 0.$$

8. **Surfaces of revolution in general.** The cone of Ex. 7 cuts ZOX in two lines whose equations may be combined in the equation $x^2 = z^2\tan^2\alpha$. That is to say, the equation of the surface generated by the revolution of the curve $y = 0$, $x^2 = z^2\tan^2\alpha$ round the axis of z is derived from the second of these equations by changing x^2 into x^2+y^2. Show generally that the equation of the surface generated by the revolution of the curve $y = 0$, $x = f(z)$, round the axis of z, is

$$\sqrt{(x^2+y^2)} = f(z).$$

9. **Cones in general.** A surface formed by straight lines passing through a fixed point is called a *cone*: the point is called the *vertex*. A particular case is given by the right circular cone of Ex. 7. Show that the equation of a cone whose vertex is O is of the form $f(z/x, z/y) = 0$, and that any equation of this form represents a cone. [If (x, y, z) lies on the cone, so must $(\lambda x, \lambda y, \lambda z)$, for any value of λ.]

* We assume that the effects of the earth's curvature may be neglected.

10. **Ruled surfaces.** Cylinders and cones are special cases of *surfaces composed of straight lines*. Such surfaces are called *ruled surfaces*.

The two equations
$$x = az + b, \quad y = cz + d \quad \text{.......................(1)}$$
represent the intersection of two planes, i.e. a straight line. Now suppose that a, b, c, d instead of being fixed are functions of an auxiliary variable t. For any particular value of t the equations (1) define a line. As t varies, this line moves and generates a surface, whose equation may be found by eliminating t between the two equations (1). For instance, in Ex. 7 the equations of the line which generates the cone are
$$x = z \tan\alpha \cos t, \quad y = z \tan\alpha \sin t,$$
where t is the angle between the plane XOZ and a plane through the line and the axis of z.

Another simple example of a ruled surface may be constructed as follows. Take two sections of a right circular cylinder perpendicular to the axis and at a distance l apart (Fig. 17 a). We can imagine the surface of the cylinder to be made up of a number of thin parallel rigid rods of length l, such as PQ, the ends of the rods being fastened to two circular rods of radius a.

Now let us take a third circular rod of the same radius and place it round the surface of the cylinder at a distance h from one of the first two rods (see Fig. 17 a, where $Pq = h$). Unfasten the end Q of the rod PQ and turn PQ about P until Q can be fastened to the third circular rod in the position Q'. The angle $qOQ' = \alpha$ in the figure is given by
$$l^2 - h^2 = qQ'^2 = (2a \sin \tfrac{1}{2}\alpha)^2.$$
Let all the other rods of which the cylinder was composed be treated in the same way. We obtain a ruled surface whose form is indicated in Fig. 17 b. It is entirely built up of straight lines; but the surface is curved everywhere, and is in general shape not unlike certain forms of table-napkin rings (Fig. 17 c).

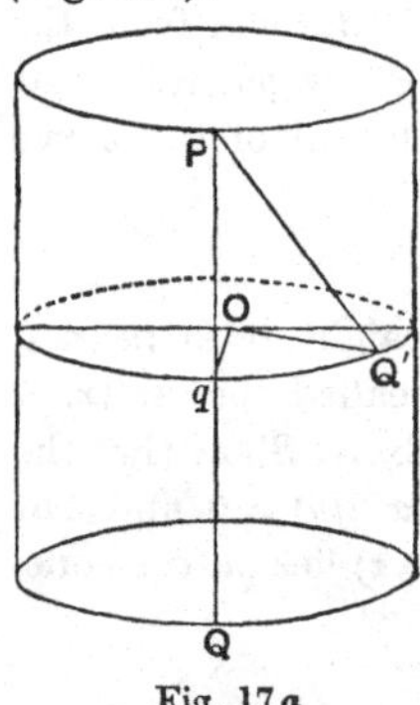

Fig. 17 a

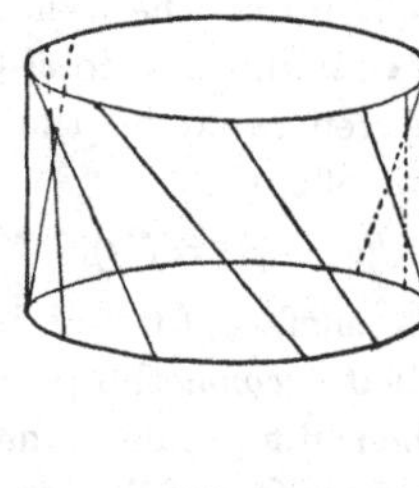
Fig. 17 b

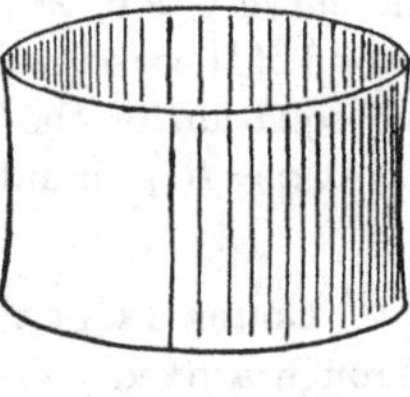
Fig. 17 c

MISCELLANEOUS EXAMPLES ON CHAPTER II

1. Show that if $y = f(x) = (ax+b)/(cx-a)$ then $x = f(y)$.

2. If $f(x) = f(-x)$ for all values of x, $f(x)$ is called an *even* function. If $f(x) = -f(-x)$, it is called an *odd* function. Show that any function of x, defined for all values of x, is the sum of an even and an odd function of x.

[Use the identity $f(x) = \frac{1}{2}\{f(x)+f(-x)\}+\frac{1}{2}\{f(x)-f(-x)\}$.]

3. Draw the graphs of the functions

$$3\sin x+4\cos x, \quad \sin\left(\frac{\pi}{\sqrt{2}}\sin x\right). \quad (\textit{Math. Trip.}\ 1896)$$

4. Draw the graphs of the functions

$$\sin x\,(a\cos^2 x+b\sin^2 x), \quad \frac{\sin x}{x}(a\cos^2 x+b\sin^2 x), \quad \left(\frac{\sin x}{x}\right)^2.$$

5. Draw the graphs of the functions $x[1/x]$, $[x]/x$.

6. Draw the graphs of the functions

$$\text{(i)}\ \arccos(2x^2-1)-2\arccos x,$$

$$\text{(ii)}\ \arctan\frac{a+x}{1-ax}-\arctan a-\arctan x,$$

where the symbols $\arccos\alpha$, $\arctan\alpha$ denote, for any value of α, the least positive (or zero) angle, whose cosine or tangent is α.

7. Verify the following method of constructing the graph of $f\{\phi(x)\}$ by means of the line $y = x$ and the graphs of $f(x)$ and $\phi(x)$: take $OA = x$ along OX, draw AB parallel to OY to meet $y = \phi(x)$ in B, BC parallel to OX to meet $y = x$ in C, CD parallel to OY to meet $y = f(x)$ in D, and DP parallel to OX to meet AB in P; then P is a point on the graph required.

8. Show that the roots of $x^3+px+q = 0$ are the abscissae of the points of intersection (other than the origin) of the parabola $y = x^2$ and the circle

$$x^2+y^2+(p-1)y+qx = 0.$$

9. The roots of $x^4+nx^3+px^2+qx+r = 0$ are the abscissae of the points of intersection of the parabola $x^2 = y-\frac{1}{2}nx$ and the circle

$$x^2+y^2+(\tfrac{1}{8}n^3-\tfrac{1}{2}pn+\tfrac{1}{2}n+q)x+(p-1-\tfrac{1}{4}n^2)y+r = 0.$$

10. Discuss the graphical solution of the equation

$$x^m+ax^2+bx+c = 0$$

by means of the curves $y = x^m$, $y = -ax^2-bx-c$. Draw up a table of the various possible numbers of roots.

11. Solve the equation $\sec\theta + \operatorname{cosec}\theta = 2\sqrt{2}$; and show that the equation $\sec\theta + \operatorname{cosec}\theta = c$ has two roots between 0 and 2π if $c^2 < 8$ and four if $c^2 > 8$.

12. Show that the equation

$$2x = (2n+1)\pi(1-\cos x),$$

where n is a positive integer, has $2n+3$ roots and no more, indicating their localities roughly. (*Math. Trip.* 1896)

13. Show that the equation $\frac{2}{3}x\sin x = 1$ has four roots between $-\pi$ and π.

14. Discuss the number and values of the roots of the equations

(1) $\cot x + x - \frac{3}{2}\pi = 0$, (2) $x^2 + \sin^2 x = 1$, (3) $(1+x^2)\tan x = 2x$,

(4) $\sin x - x + \frac{1}{6}x^3 = 0$, (5) $(1-\cos x)\tan\alpha - x + \sin x = 0$.

15. The polynomial of the second degree which assumes, when $x = a$, b, c, the values α, β, γ is

$$\alpha\frac{(x-b)(x-c)}{(a-b)(a-c)} + \beta\frac{(x-c)(x-a)}{(b-c)(b-a)} + \gamma\frac{(x-a)(x-b)}{(c-a)(c-b)}.$$

Give a similar formula for the polynomial of the $(n-1)$th degree which assumes, when $x = a_1, a_2, \dots, a_n$, the values $\alpha_1, \alpha_2, \dots, \alpha_n$.

16. Find a polynomial in x of the second degree which for the values 0, 1, 2 of x takes the values $1/c$, $1/(c+1)$, $1/(c+2)$; and show that when $x = c+2$ its value is $1/(c+1)$. (*Math. Trip.* 1911)

17. Show that if x is a rational function of y, and y is a rational function of x, then $Axy + Bx + Cy + D = 0$.

18. If y is an algebraical function of x, then x is an algebraical function of y.

19. Verify that the equation

$$\cos\tfrac{1}{2}\pi x = 1 - \frac{x^2}{x + (1-x)\sqrt{\left(\frac{2-x}{3}\right)}}$$

is approximately true for all values of x between 0 and 1. [Take $x = 0, \frac{1}{6}, \frac{1}{3}, \frac{1}{2}, \frac{2}{3}, \frac{5}{6}, 1$, and use tables. For which of these values is the formula exact?]

20. What is the form of the graph of the functions

$$z = [x] + [y], \quad z = x + y - [x] - [y]?$$

21. What is the form of the graph of the functions $z = \sin x + \sin y$, $z = \sin x \sin y$, $z = \sin xy$, $z = \sin(x^2+y^2)$?

22. **Geometrical constructions for irrational numbers.** In Ch. I we indicated one or two simple geometrical constructions for a length

equal to $\sqrt{2}$, starting from a given unit length. We also showed how to construct the roots of any quadratic equation $ax^2+2bx+c=0$, it being supposed that we can construct lines whose lengths are equal to any of the ratios of the coefficients a, b, c, as is certainly the case if a, b, c are rational. All these constructions were what may be called Euclidean constructions; they depended on the ruler and compasses only.

It is fairly obvious that we can construct by these methods the length measured by any irrational number which is defined by any combination of square roots, however complicated. Thus

$$\sqrt[4]{\left\{\sqrt{\left(\frac{17+3\sqrt{11}}{17-3\sqrt{11}}\right)}-\sqrt{\left(\frac{17-3\sqrt{11}}{17+3\sqrt{11}}\right)}\right\}}$$

is a case in point. This expression contains a fourth root, but this is of course the square root of a square root. We should begin by constructing $\sqrt{11}$, e.g. as the mean between 1 and 11: then $17+3\sqrt{11}$ and $17-3\sqrt{11}$, and so on. Or these two mixed surds might be constructed directly as the roots of $x^2-34x+190=0$.

Conversely, *only* irrationals of this kind can be constructed by Euclidean methods. Starting from a unit length we can construct any *rational* length. And hence we can construct the line $Ax+By+C=0$, provided that the ratios of A, B, C are rational, and the circle

$$(x-\alpha)^2+(y-\beta)^2=\rho^2$$

(or $x^2+y^2+2gx+2fy+c=0$), provided that α, β, ρ are rational, a condition which implies that g, f, c are rational.

Now in any Euclidean construction each new point introduced into the figure is determined as the intersection of two lines or circles, or a line and a circle. But if the coefficients are rational, such a pair of equations as

$$Ax+By+C=0, \quad x^2+y^2+2gx+2fy+c=0$$

give, on solution, values of x and y of the form $m+n\sqrt{p}$, where m, n, p are rational: for if we substitute for x in terms of y in the second equation we obtain a quadratic in y with rational coefficients. Hence the coordinates of all points obtained by means of lines and circles with rational coefficients are expressible by rational numbers and quadratic surds. And so the same is true of the distance $\sqrt{\{(x_1-x_2)^2+(y_1-y_2)^2\}}$ between any two points so obtained.

With the irrational distances thus constructed we may proceed to construct a number of lines and circles whose coefficients may now themselves involve quadratic surds. It is evident, however, that all the lengths which we can construct by the use of such lines and circles are still expressible by square roots only, though our surd expressions may now be of a more complicated form. And this remains true however often our constructions are repeated. Hence *Euclidean methods will construct any surd expression involving square roots only, and no others.*

One of the famous problems of antiquity was that of the duplication of the cube, that is to say of the construction by Euclidean methods of a length measured by $\sqrt[3]{2}$. It can be shown that $\sqrt[3]{2}$ cannot be expressed by means of any finite combination of rational numbers and square roots, so that no solution is possible. See Hobson, *Squaring the circle*, pp. 47 *et seq.*; the first stage of the proof, viz. the proof that $\sqrt[3]{2}$ cannot be a root of a quadratic equation $ax^2+2bx+c=0$ with rational coefficients, was given in Ch. I (Misc. Exs. 27).

23. Show that the only lengths which can be constructed with the ruler only, starting from a given unit length, are rational lengths.

24. **Approximate quadrature of the circle.** Let O be the centre of a circle of radius R. On the tangent at A take $AP = \frac{11}{5}R$ and $AQ = \frac{13}{5}R$, in the same direction. On AO take $AN = OP$ and draw NM parallel to OQ and cutting AP in M. Show that

$$AM/R = \tfrac{13}{25}\sqrt{146},$$

and that to take AM as being equal to the circumference of the circle would lead to a value of π correct to five places of decimals. If R is the earth's radius, the error in supposing AM to be its circumference is less than 11 yards.

[We stated in § 15 that π is transcendental; but we cannot prove in this book even that it is irrational. This was proved first by Lambert in 1761 by use of continued fractions.

The most familiar approximations to π are $\frac{22}{7}$ and $\frac{355}{113}$, the last correct to 6 places of decimals. The Indians used the approximation $\sqrt{10}$ (in error in the second place). A large number of curious approximations will be found in Ramanujan's *Collected papers*, pp. 23–39. Among the simplest are

$$\frac{19}{16}\sqrt{7},\quad \frac{7}{3}\left(1+\frac{\sqrt{3}}{5}\right),\quad \left(9^2+\frac{19^2}{22}\right)^{\frac{1}{4}},\quad \frac{63}{25}\left(\frac{17+15\sqrt{5}}{7+15\sqrt{5}}\right);$$

these are correct to 3, 3, 8, and 9 places respectively.]

25. **Constructions for $\sqrt[3]{2}$.** O is the vertex and S the focus of the parabola $y^2 = 4x$, and P is one of its points of intersection with the parabola $x^2 = 2y$. Show that OP meets the latus rectum of the first parabola in a point Q such that $SQ = \sqrt[3]{2}$.

26. Take a circle of unit diameter, a diameter OA and the tangent at A. Draw a chord OBC cutting the circle at B and the tangent at C. On this line take $OM = BC$. Taking O as origin and OA as axis of x, show that the locus of M is the curve

$$(x^2+y^2)\,x-y^2 = 0$$

(the *Cissoid of Diocles*). Sketch the curve. Take along the axis of y a length $OD = 2$. Let AD cut the curve in P and OP cut the tangent to the circle at A in Q. Show that $AQ = \sqrt[3]{2}$.

CHAPTER III

COMPLEX NUMBERS

34. Displacements along a line and in a plane. The 'real number' x, with which we have been concerned in the two preceding chapters, may be regarded from many different points of view. It may be regarded as a pure number, destitute of geometrical significance, or a geometrical significance may be attached to it in at least three different ways. It may be regarded as *the measure of a length*, viz. the length A_0P along the line Λ of Ch. I. It may be regarded as *the mark of a point*, viz. the point P whose distance from A_0 is x. Or it may be regarded as *the measure of a displacement* or *change of position* on the line Λ. It is on this last point of view that we shall now concentrate our attention.

Imagine a small particle placed at P on the line Λ and then displaced to Q. We shall call the displacement or change of position which is needed to transfer the particle from P to Q *the displacement* $\overline{PQ}$. To specify a displacement completely three things are needed: its *magnitude*, its *sense* forwards or backwards along the line, and what may be called its *point of application*, i.e. the original position P of the particle. But, when we are thinking merely of the change of position produced by the displacement, it is natural to disregard the point of application and to consider all displacements as equivalent whose lengths and senses are the same. Then the displacement is completely specified by the length $PQ = x$, the sense of the displacement being fixed by the sign of x. We may therefore, without ambiguity, speak of *the displacement* $[x]$*, and we may write $\overline{PQ} = [x]$.

* It is hardly necessary to caution the reader against confusing this use of the symbol $[x]$ and that of Ch. II (Exs. XVI and Misc. Ex. 20).

We use the square bracket to distinguish the displacement $[x]$ from the length or number x*. If the coordinate of P is a, that of Q will be $a+x$; the displacement $[x]$ therefore transfers a particle from the point a to the point $a+x$.

We pass on to consider *displacements in a plane*. We may define the displacement $\overline{PQ}$ as before. But now more data are required in order to specify it completely. We require to know: (i) the *magnitude* of the displacement, i.e. the length of the straight line PQ; (ii) the *direction* of the displacement, which is determined by the angle which PQ makes with some fixed line in the plane; (iii) the *sense* of the displacement; and (iv) its *point of application*. Of these requirements we may disregard the fourth, if we consider two displacements as equivalent if they are the same in magnitude, direction, and sense. In other words, if PQ and RS are equal and parallel, and the sense of motion from P to Q is the same as that of motion from R to S, we regard the displacements $\overline{PQ}$ and $\overline{RS}$ as equivalent, and write

$$\overline{PQ} = \overline{RS}.$$

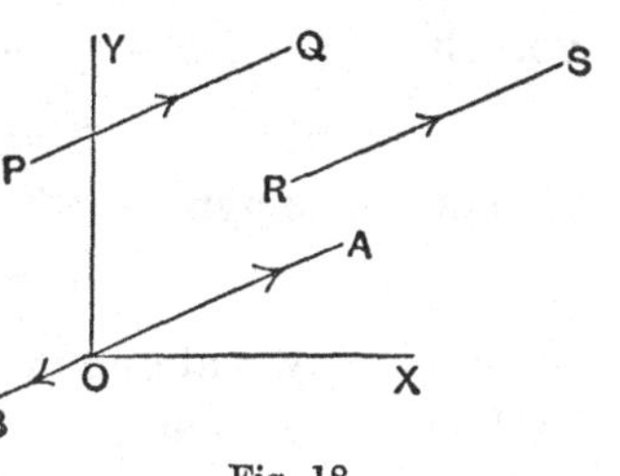

Fig. 18

Now let us take any pair of coordinate axes in the plane (such as OX, OY in Fig. 18). Draw a line OA equal and parallel to PQ, the sense of motion from O to A being the same as that from P to Q. Then $\overline{PQ}$ and $\overline{OA}$ are equivalent displacements. Let x and y be the coordinates of A. Then it is evident that $\overline{OA}$ is completely specified if x and y are given. We call $\overline{OA}$ *the displacement* $[x, y]$ and write

$$\overline{OA} = \overline{PQ} = \overline{RS} = [x, y].$$

* Strictly speaking we ought, by some similar difference of notation, to distinguish the actual length x from the number x which measures it. The reader will perhaps be inclined to consider such distinctions pedantic. But increasing experience of mathematics will reveal to him the great importance of distinguishing clearly between things which, however intimately connected, are not the same.

35. Equivalence of displacements. Multiplication of displacements by numbers. If ξ and η are the coordinates of P, and ξ' and η' those of Q, it is evident that

$$x = \xi' - \xi, \quad y = \eta' - \eta.$$

The displacement from (ξ, η) to (ξ', η') is therefore

$$[\xi' - \xi, \eta' - \eta].$$

It is clear that two displacements $[x, y]$, $[x', y']$ are equivalent if, and only if, $x = x'$, $y = y'$. Thus $[x, y] = [x', y']$ if and only if

$$x = x', \quad y = y' \quad \text{.....................(1)}.$$

The reverse displacement $\overline{QP}$ would be $[\xi - \xi', \eta - \eta']$, and it is natural to agree that

$$[\xi - \xi', \eta - \eta'] = -[\xi' - \xi, \eta' - \eta],$$

$$\overline{QP} = -\overline{PQ},$$

these equations being really definitions of the meaning of the symbols $-[\xi' - \xi, \eta' - \eta]$, $-\overline{PQ}$. Having thus agreed that

$$-[x, y] = [-x, -y],$$

it is natural to agree further that

$$\alpha[x, y] = [\alpha x, \alpha y] \quad \text{.....................(2)},$$

where α is any real number, positive or negative. Thus (Fig. 18) if $OB = -\frac{1}{2}OA$ then

$$\overline{OB} = -\tfrac{1}{2}\overline{OA} = -\tfrac{1}{2}[x, y] = [-\tfrac{1}{2}x, -\tfrac{1}{2}y].$$

The equations (1) and (2) define the first two important ideas connected with displacements, viz. *equivalence* of displacements, and *multiplication of displacements by numbers.*

36. Addition of displacements. We have not yet given any definition which enables us to attach a meaning to the expressions

$$\overline{PQ} + \overline{P'Q'}, \quad [x, y] + [x', y'].$$

Common sense at once suggests that we should define the sum of two displacements as the displacement which is the result of the successive application of the two given displacements. In other words, it suggests that if QQ_1 be drawn equal and parallel to $P'Q'$, so that the result of successive displacements $\overline{PQ}$, $\overline{P'Q'}$ on a particle at P is to transfer it first to Q and then to Q_1, we should define the sum of $\overline{PQ}$ and $\overline{P'Q'}$ as being $\overline{PQ_1}$. If then we draw OA equal and parallel to PQ, and OB equal and parallel to $P'Q'$, and complete the parallelogram $OACB$, we have

$$\overline{PQ}+\overline{P'Q'} = \overline{PQ_1} = \overline{OA}+\overline{OB} = \overline{OC}.$$

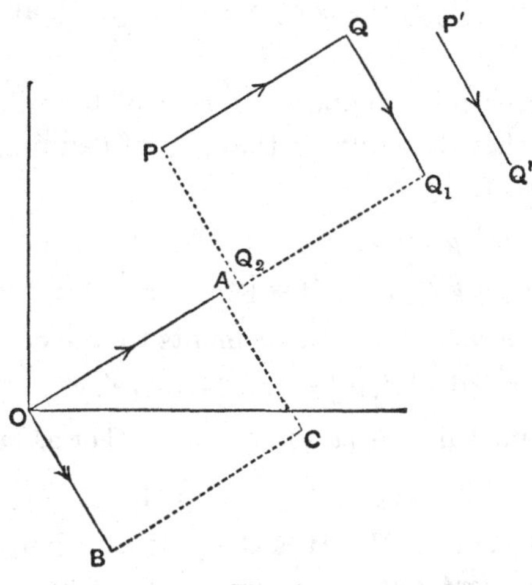

Fig. 19

Let us consider the consequences of adopting this definition. If the coordinates of B are x', y', then those of the middle point of AB are $\frac{1}{2}(x+x')$, $\frac{1}{2}(y+y')$, and those of C are $x+x'$, $y+y'$. Hence

$$[x, y]+[x', y'] = [x+x', y+y'] \quad \text{...........(3)},$$

which may be regarded as the symbolic definition of addition of displacements. We observe that

$$\begin{aligned}[x', y']+[x, y] &= [x'+x, y'+y] \\ &= [x+x', y+y'] = [x, y]+[x', y'].\end{aligned}$$

In other words, *addition of displacements obeys the commutative law* expressed in ordinary algebra by the equation $a+b = b+a$. This law expresses the obvious geometrical fact that if we move from P first through a distance PQ_2 equal and parallel to $P'Q'$, and then through a distance equal and parallel to PQ, we shall arrive at the same point Q_1 as before.

In particular $$[x, y] = [x, 0] + [0, y] \quad \text{.....................(4)}.$$

Here $[x, 0]$ denotes a displacement through a distance x in a direction parallel to OX. It is in fact what we previously denoted by $[x]$, when we were considering only displacements along a line. We call $[x, 0]$ and $[0, y]$ the *components* of $[x, y]$, and $[x, y]$ their *resultant*.

When we have once defined addition of two displacements, there is no further difficulty in the way of defining addition of any number. Thus, by definition,

$$[x, y] + [x', y'] + [x'', y''] = ([x, y] + [x', y']) + [x'', y'']$$
$$= [x + x', y + y'] + [x'', y''] = [x + x' + x'', y + y' + y''].$$

We define *subtraction* of displacements by the equation

$$[x, y] - [x', y'] = [x, y] + (-[x', y']) \quad \text{.........(5)},$$

which is the same thing as $[x, y] + [-x', -y']$ or as $[x - x', y - y']$. In particular $$[x, y] - [x, y] = [0, 0].$$

The displacement $[0, 0]$ leaves the particle where it was; it is the *zero displacement*, and we agree to write $[0, 0] = 0$.

Examples XX. 1. Prove that

(i) $\alpha[\beta x, \beta y] = \beta[\alpha x, \alpha y] = [\alpha\beta x, \alpha\beta y]$,

(ii) $([x, y] + [x', y']) + [x'', y''] = [x, y] + ([x', y'] + [x'', y''])$,

(iii) $[x, y] + [x', y'] = [x', y'] + [x, y]$,

(iv) $(\alpha + \beta)[x, y] = \alpha[x, y] + \beta[x, y]$,

(v) $\alpha\{[x, y] + [x', y']\} = \alpha[x, y] + \alpha[x', y']$.

[We have already proved (iii). The remaining equations follow with equal ease from the definitions. The reader should in each case consider the geometrical significance of the equation, as we did above in the case of (iii).]

2. If M is the middle point of PQ, then $\overline{OM} = \frac{1}{2}(\overline{OP} + \overline{OQ})$. More generally, if M divides PQ in the ratio $\mu : \lambda$, then

$$\overline{OM} = \frac{\lambda}{\lambda + \mu}\overline{OP} + \frac{\mu}{\lambda + \mu}\overline{OQ}.$$

3. If G is the centre of mass of equal particles at $P_1, P_2, \ldots, P_n$, then

$$\overline{OG} = \frac{1}{n}(\overline{OP}_1 + \overline{OP}_2 + \ldots + \overline{OP}_n).$$

4. If P, Q, R are collinear points in the plane, then it is possible to find real numbers α, β, γ, not all zero, and such that

$$\alpha.\overline{OP} + \beta.\overline{OQ} + \gamma.\overline{OR} = 0;$$

and conversely. [This is really only another way of stating Ex. 2.]

5. If $\overline{AB}$ and $\overline{AC}$ are two displacements not in the same straight line, and

$$\alpha.\overline{AB} + \beta.\overline{AC} = \gamma.\overline{AB} + \delta.\overline{AC},$$

then $\alpha = \gamma$ and $\beta = \delta$.

[Take $\overline{AB_1} = \alpha.\overline{AB}$, $\overline{AC_1} = \beta.\overline{AC}$. Complete the parallelogram $AB_1P_1C_1$. Then $\overline{AP_1} = \alpha.\overline{AB} + \beta.\overline{AC}$. It is evident that $\overline{AP_1}$ can be expressed in this form in one way only, whence the theorem follows.]

6. $ABCD$ is a parallelogram. Through Q, a point inside the parallelogram, RQS and TQU are drawn parallel to the sides. Show that RU, TS intersect on AC.

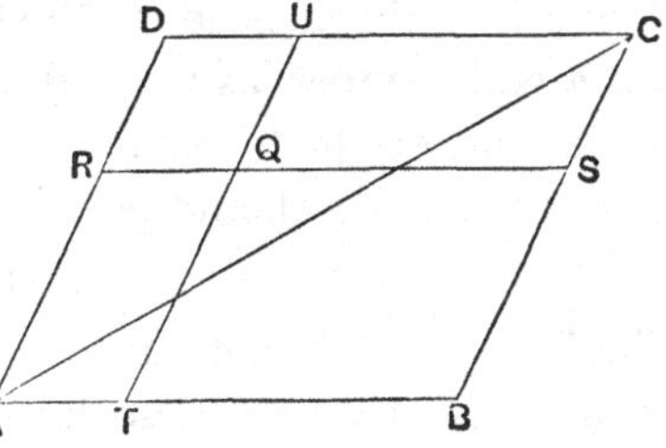

Fig. 20

[Let the ratios $AT:AB$, $AR:AD$ be denoted by α, β. Then

$$\overline{AT} = \alpha.\overline{AB}, \quad \overline{AR} = \beta.\overline{AD},$$
$$\overline{AU} = \alpha.\overline{AB} + \overline{AD},$$
$$\overline{AS} = \overline{AB} + \beta.\overline{AD}.$$

Let RU meet AC in P. Then, since R, U, P are collinear,

$$\overline{AP} = \frac{\lambda}{\lambda+\mu}\overline{AR} + \frac{\mu}{\lambda+\mu}\overline{AU},$$

where μ/λ is the ratio in which P divides RU. That is to say

$$\overline{AP} = \frac{\alpha\mu}{\lambda+\mu}\overline{AB} + \frac{\beta\lambda+\mu}{\lambda+\mu}\overline{AD}.$$

But since P lies on AC, $\overline{AP}$ is a numerical multiple of $\overline{AC}$; say

$$\overline{AP} = k.\overline{AC} = k.\overline{AB} + k.\overline{AD}.$$

Hence (Ex. 5) $\alpha\mu = \beta\lambda + \mu = (\lambda+\mu)k$, from which we deduce

$$k = \frac{\alpha\beta}{\alpha+\beta-1}.$$

The symmetry of this result shows that a similar argument would also give

$$\overline{AP'} = \frac{\alpha\beta}{\alpha+\beta-1}\overline{AC},$$

if P' is the point where TS meets AC. Hence P and P' are the same point.]

7. $ABCD$ is a parallelogram, and M the middle point of AB. Show that DM trisects and is trisected by AC*.

37. Multiplication of displacements. So far we have made no attempt to attach any meaning to the notion of the *product* of two displacements. The only kind of multiplication which we have considered is that in which a displacement is multiplied by a number. The expression

$$[x, y]\,[x', y']$$

so far means nothing, and we are at liberty to define it as we please.

Our choice of a definition is decided by the following principles. It is clear (1) that the product of two displacements should be a displacement. Next we have defined $\alpha[x, y]$, where α is a real number, as $[\alpha x, \alpha y]$; and α may be regarded as a displacement, viz. $[\alpha, 0]$. Hence, changing our notation, we see that (2) our definition should make

$$[x, 0]\,[x', y'] = [xx', xy'].$$

Finally (3) that the definition should obey the ordinary commutative, distributive, and associative laws of multiplication, so that

$$[x, y]\,[x', y'] = [x', y']\,[x, y],$$
$$([x, y] + [x', y'])\,[x'', y''] = [x, y]\,[x'', y''] + [x', y']\,[x'', y''],$$
$$[x, y]\,([x', y'] + [x'', y'']) = [x, y]\,[x', y'] + [x, y]\,[x'', y''],$$

and $$[x, y]\,([x', y']\,[x'', y'']) = ([x, y]\,[x', y'])\,[x'', y''].$$

Thus

$$[x, y]\,[x', y'] = [xx', yy']$$

would not be a suitable definition, since it would give

$$[x, 0]\,[x', y'] = [xx', 0],$$

and contradict (2).

38. The right definition to take is suggested as follows. We know that, if OAB, OCD are two similar triangles, the angles corresponding in the order in which they are written, then

$$OB/OA = OD/OC,$$

* The two last examples are taken from Willard Gibbs's *Vector analysis*.

or $OB.OC = OA.OD$. This suggests that we should try to define multiplication and division of displacements in such a way that

$$\overline{OB}/\overline{OA} = \overline{OD}/\overline{OC}, \quad \overline{OB}.\overline{OC} = \overline{OA}.\overline{OD}.$$

Now let

$$\overline{OB} = [x, y], \quad \overline{OC} = [x', y'], \quad \overline{OD} = [X, Y],$$

and suppose that A is the point $(1, 0)$, so that $\overline{OA} = [1, 0]$. Then

$$\overline{OA}.\overline{OD} = [1, 0][X, Y] = [X, Y],$$

and so

$$[x, y][x', y'] = [X, Y].$$

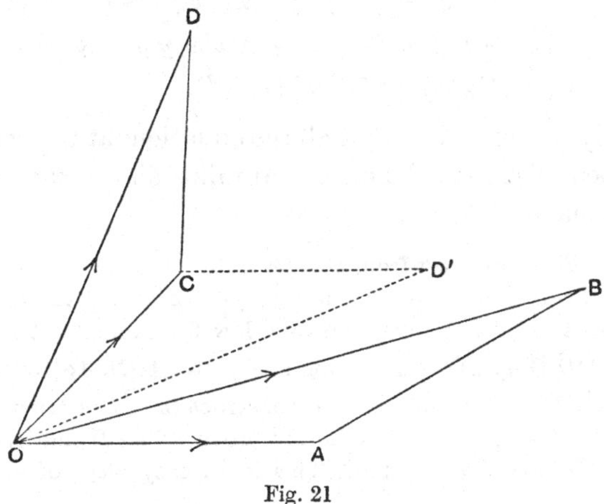

Fig. 21

The product $\overline{OB}.\overline{OC}$ is therefore to be defined as $\overline{OD}$, D being obtained by constructing on OC a triangle similar to OAB. In order to free this definition from ambiguity, it should be observed that on OC we can describe *two* such triangles, OCD and OCD'. We choose that for which the angle COD is equal to AOB in sign as well as in magnitude. We say that the two triangles are then *similar in the same sense*.

If the polar coordinates of B and C are (ρ, θ) and (σ, ϕ), so that

$$x = \rho\cos\theta, \quad y = \rho\sin\theta, \quad x' = \sigma\cos\phi, \quad y' = \sigma\sin\phi,$$

then the polar coordinates of D are $(\rho\sigma, \theta+\phi)$. Hence

$$X = \rho\sigma\cos(\theta+\phi) = xx' - yy',$$
$$Y = \rho\sigma\sin(\theta+\phi) = xy' + yx'.$$

The required definition is therefore

$$[x, y][x', y'] = [xx' - yy', xy' + yx'] \quad \text{.........(6)}.$$

We observe (1) that if $y = 0$, then $X = xx'$, $Y = xy'$, as we desired; (2) that the right-hand side is not altered if we interchange x and x', and y and y', so that

and (3) that $$[x, y][x', y'] = [x', y'][x, y];$$

$$\begin{aligned}\{[x, y] + [x', y']\}[x'', y''] &= [x + x', y + y'][x'', y''] \\ &= [(x + x')x'' - (y + y')y'', (x + x')y'' + (y + y')x''] \\ &= [xx'' - yy'', xy'' + yx''] + [x'x'' - y'y'', x'y'' + y'x''] \\ &= [x, y][x'', y''] + [x', y'][x'', y''].\end{aligned}$$

Similarly we can verify that all the equations at the end of § 37 are satisfied. Thus the definition (6) fulfils all the requirements which we made of it in § 37.

Example. Show directly from the geometrical definition given above that multiplication of displacements obeys the commutative and distributive laws. [Take the commutative law for example. The product $\overline{OB}.\overline{OC}$ is $\overline{OD}$ (Fig. 21), COD being similar to AOB. To construct the product $\overline{OC}.\overline{OB}$ we should have to construct on OB a triangle BOD_1 similar to AOC; and so what we want to prove is that D and D_1 coincide, or that BOD is similar to AOC. This is an easy piece of elementary geometry.]

39. Complex numbers. Just as to a displacement $[x]$ along OX correspond a point (x) and a real number x, so to a displacement $[x, y]$ in the plane correspond a point (x, y) and *a pair of real numbers* x, y.

We shall find it convenient to denote this pair of real numbers x, y by the symbol

$$x + yi.$$

The reason for the choice of this notation will appear later. For the present the reader must regard $x + yi$ as *simply another way of writing* $[x, y]$. The symbol $x + yi$ is called a *complex number*.

We proceed next to define *equivalence, addition,* and *multiplication* of complex numbers. To every complex number corresponds a displacement. Two complex numbers are equivalent if the corresponding displacements are equivalent. The sum or product of two complex numbers is the complex number which corresponds to the sum or product of the two corresponding displacements. Thus

$$x+yi = x'+y'i \quad \text{..................(1)},$$

if and only if $x = x'$, $y = y'$;

$$(x+yi)+(x'+y'i) = (x+x')+(y+y')i \quad \text{.........(2)};$$

$$(x+yi)(x'+y'i) = xx'-yy'+(xy'+yx')i \quad \text{......(3)}.$$

In particular we have, as special cases of (2) and (3),

$$x+yi = (x+0i)+(0+yi),$$

$$(x+0i)(x'+y'i) = xx'+xy'i;$$

and these equations suggest that there will be no danger of confusion if, when dealing with complex numbers, we write x for $x+0i$ and yi for $0+yi$, as we shall henceforth.

The reader will easily verify for himself that addition and multiplication of complex numbers obey the laws of algebra expressed by the equations

$$(x+yi)+(x'+y'i) = (x'+y'i)+(x+yi),$$

$$\{(x+yi)+(x'+y'i)\}+(x''+y''i) = (x+yi)+\{(x'+y'i)+(x''+y''i)\},$$

$$(x+yi)(x'+y'i) = (x'+y'i)(x+yi),$$

$$(x+yi)\{(x'+y'i)+(x''+y''i)\} = (x+yi)(x'+y'i)+(x+yi)(x''+y''i),$$

$$\{(x+yi)+(x'+y'i)\}(x''+y''i) = (x+yi)(x''+y''i)+(x'+y'i)(x''+y''i),$$

$$(x+yi)\{(x'+y'i)(x''+y''i)\} = \{(x+yi)(x'+y'i)\}(x''+y''i),$$

the proofs of these equations being practically the same as those of the corresponding equations for the corresponding displacements.

Subtraction and division of complex numbers are defined as in ordinary algebra. Thus we may define $(x+yi)-(x'+y'i)$ as

$$(x+yi)+\{-(x'+y'i)\} = x+yi+(-x'-y'i) = (x-x')+(y-y')i;$$

or, what is the same thing, as the number $\xi+\eta i$ such that

$$(x'+y'i)+(\xi+\eta i) = x+yi.$$

And $(x+yi)/(x'+y'i)$ is defined as the complex number $\xi+\eta i$ such that

$$(x'+y'i)(\xi+\eta i) = x+yi,$$

or

$$x'\xi-y'\eta+(x'\eta+y'\xi)i = x+yi,$$

or

$$x'\xi-y'\eta = x, \quad x'\eta+y'\xi = y \quad \ldots\ldots\ldots\ldots\ldots(4).$$

Solving these equations for ξ and η, we obtain

$$\xi = \frac{xx'+yy'}{x'^2+y'^2}, \quad \eta = \frac{yx'-xy'}{x'^2+y'^2}.$$

This solution fails if x' and y' are both zero, i.e. if $x'+y'i = 0$. Thus subtraction is always possible; division is always possible unless the divisor is zero.

We may now define positive integral powers of $x+yi$, polynomials in $x+yi$, and rational functions of $x+yi$, as in ordinary algebra.

Examples. (1) From a geometrical point of view, the problem of the division of the displacement $\overline{OD}$ by $\overline{OC}$ is that of finding B so that the triangles COD, AOB are similar, and this is evidently possible (and the solution unique) unless C coincides with O, or $\overline{OC} = 0$.

(2) The numbers $x+yi$, $x-yi$ are said to be *conjugate*. Verify that

$$(x+yi)(x-yi) = x^2+y^2,$$

so that the product of two conjugate numbers is real, and that

$$\frac{x+yi}{x'+y'i} = \frac{(x+yi)(x'-y'i)}{(x'+y'i)(x'-y'i)} = \frac{xx'+yy'+(x'y-xy')i}{x'^2+y'^2}.$$

40. One most important property of real numbers is that know as *the factor theorem*, which asserts that *the product of two numbers cannot be zero unless one of the two is itself zero*. To prove that this is also true of complex numbers we put $x = 0$, $y = 0$ in the equations (4) of the preceding section. Then

$$x'\xi-y'\eta = 0, \quad x'\eta+y'\xi = 0.$$

These equations give $\xi = 0$, $\eta = 0$, i.e.

$$\xi+\eta i = 0,$$

unless $x' = 0$ and $y' = 0$, or $x'+y'i = 0$. Thus $x+yi$ cannot vanish unless either $x'+y'i$ or $\xi+\eta i$ vanishes.

41. The equation $i^2 = -1$. We agreed to simplify our notation by writing x instead of $x+0i$ and yi instead of $0+yi$. The particular complex number $1i$ we shall denote simply by i. It is the number which corresponds to a unit displacement along OY. Also

$$i^2 = ii = (0+1i)(0+1i) = (0.0-1.1)+(0.1+1.0)i = -1.$$

Similarly $(-i)^2 = -1$. Thus the complex numbers i and $-i$ satisfy the equation $x^2 = -1$.

The reader will now easily satisfy himself that the upshot of the rules for addition and multiplication of complex numbers is this, that *we operate with complex numbers in exactly the same way as with real numbers, treating the symbol i as itself a number, but replacing the product $ii = i^2$ by -1 whenever it occurs.* Thus, for example,

$$\begin{aligned}(x+yi)(x'+y'i) &= xx'+xy'i+yx'i+yy'i^2\\ &= (xx'-yy')+(xy'+yx')i.\end{aligned}$$

42. The geometrical interpretation of multiplication by i. Since

$$(x+yi)i = -y+xi,$$

it follows that if $x+yi$ corresponds to $\overline{OP}$, and OQ is drawn equal to OP and so that POQ is a positive right angle, then $(x+yi)i$ corresponds to $\overline{OQ}$. In other words, *multiplication of a complex number by i turns the corresponding displacement through a right angle.*

We might have developed the whole theory of complex numbers from this point of view. Starting with the ideas of x as representing a displacement along OX, and of i as a symbol of operation equivalent to turning x through a right angle, we should have been led to regard yi as a displacement of magnitude y along OY. It would then have been natural to define $x+yi$ as in §§ 36 and 39, and $(x+yi)i$ would have represented the displacement obtained by turning $x+yi$ through a right angle, i.e. $-y+xi$. Finally, we should naturally have defined $(x+yi)x'$ as $xx'+yx'i$, $(x+yi)y'i$ as $-yy'+xy'i$, and $(x+yi)(x'+y'i)$ as the sum of these displacements, i.e. as

$$xx'-yy'+(xy'+yx')i.$$

43. The equations $z^2+1=0$, $az^2+2bz+c=0$. There is no real number z such that $z^2+1=0$; this is expressed by saying that the equation has *no real roots*. But, as we have just seen, the two complex numbers i and $-i$ satisfy this equation. We express this by saying that the equation has *the two complex roots* i and $-i$. Since i satisfies $z^2=-1$, it is sometimes written in the form $\sqrt{(-1)}$.

Complex numbers are sometimes called *imaginary**. The expression is not a happy one, but it is firmly established and must be accepted. But an 'imaginary number' is no more 'imaginary', in any ordinary sense of the word, than a 'real' number or any other mathematical object.

A real number is not a number in the same sense as a rational number, and a complex number is not a number in the same sense as a real number. It is, as should be clear from the preceding discussion, *a pair of numbers* (x, y), united symbolically, for purposes of technical convenience, in the form $x+yi$. Thus

$$i = 0+1i$$

stands for the pair of numbers $(0, 1)$, and may be represented geometrically by a point or by the displacement $[0, 1]$. And when we say that i is a root of the equation $z^2+1=0$, what we mean is simply that we have defined a method of combining such pairs of numbers (or displacements) which we call 'multiplication', and which, when we so combine $(0, 1)$ with itself, gives the result $(-1, 0)$.

Now let us consider the more general equation

$$az^2+2bz+c=0,$$

where a, b, c are real numbers. If $b^2>ac$, the ordinary method of solution gives two real roots $\{-b\pm\sqrt{(b^2-ac)}\}/a$. If $b^2<ac$, the equation has no real roots. It may be written in the form

$$\left(z+\frac{b}{a}\right)^2 = -\frac{ac-b^2}{a^2},$$

* The phrase 'real number' was introduced as an antithesis to 'imaginary number'.

which is true if $z+(b/a)$ is either of the complex numbers $\pm i\sqrt{(ac-b^2)}/a$*. We express this by saying that the equation has *the two complex roots*

$$-\frac{b}{a}\pm\frac{i\sqrt{(ac-b^2)}}{a}.$$

If we agree as a matter of convention to say that when $b^2 = ac$ (in which case the equation is satisfied by *one* value of x only, viz. $-b/a$), the equation has *two equal roots*, then *a quadratic equation with real coefficients has two roots in all cases, either two distinct real roots, or two equal real roots, or two distinct complex roots.*

The question is naturally suggested whether a quadratic equation may not, when complex roots are once admitted, have more than two roots. It is easy to see that this is not possible. Its impossibility may in fact be proved by precisely the same chain of reasoning as is used in elementary algebra to prove that an equation of the nth degree cannot have more than n real roots. Let us denote the complex number $x+yi$ by the single letter z, a convention which we may express by writing $z = x+yi$. Let $f(z)$ denote any polynomial in z, with real or complex coefficients. Then we prove in succession:

(1) that the remainder, when $f(z)$ is divided by $z-a$, a being any real or complex number, is $f(a)$;

(2) that if a is a root of the equation $f(z) = 0$, then $f(z)$ is divisible by $z-a$;

(3) that if $f(z)$ is of the nth degree, and $f(z) = 0$ has the n roots $a_1, a_2, \ldots, a_n$, then

$$f(z) = A(z-a_1)(z-a_2)\ldots(z-a_n),$$

where A is a constant, real or complex, in fact the coefficient of z^n in $f(z)$. From the last result, and the theorem of § 40, it follows that $f(z)$ cannot have more than n roots.

We conclude that a quadratic equation with real coefficients has exactly two roots. We shall see later on that a similar theorem is true for an equation of any degree and with either real or complex coefficients: *an equation of the nth degree has exactly n*

* We shall sometimes write $x+iy$ instead of $x+yi$ for convenience in printing.

roots. The only point in the proof which presents any difficulty is the first, viz. the proof that any equation must have *at least one* root. This we must postpone for the present*. We may, however, at once call attention to one very interesting result of this theorem. In the theory of number we start from the positive integers and from the ideas of addition and multiplication and the converse operations of subtraction and division. We find that these operations are not always possible unless we admit new kinds of numbers. We can only attach a meaning to $3-7$ if we admit *negative* numbers, or to $\frac{3}{7}$ if we admit *rational fractions*. When we extend our list of arithmetical operations so as to include root extraction and the solution of equations, we find that some of them, such as that of the extraction of the square root of a number which (like 2) is not a perfect square, are not possible unless we widen our conception of a number, and admit the *irrational* numbers of Ch. I.

Others, such as the extraction of the square root of -1, are not possible unless we go still further, and admit the *complex* numbers of this chapter. And it would not be unnatural to suppose that, when we come to consider equations of higher degree, some might prove to be insoluble even by the aid of complex numbers, and that thus we might be led to the consideration of numbers of still more types. The fact that the roots of any algebraical equation whatever are ordinary complex numbers shows that this is not the case.

All theorems of elementary algebra which are proved merely by the application of the rules of addition and multiplication are true *whether the numbers which occur in them are real or complex*, since the rules referred to apply to complex as well as real numbers. For example, we know that, if α and β are the roots of

$$az^2+2bz+c=0,$$

then
$$\alpha+\beta=-(2b/a), \quad \alpha\beta=(c/a).$$

Similarly, if α, β, γ are the roots of

$$az^3+3bz^2+3cz+d=0,$$

* See Appendix I.

then

$$\alpha+\beta+\gamma = -(3b/a), \quad \beta\gamma+\gamma\alpha+\alpha\beta = (3c/a), \quad \alpha\beta\gamma = -(d/a).$$

All such theorems as these are true whether $a, b, \ldots, \alpha, \beta, \ldots$ are real or complex.

44. Argand's diagram. Let P (Fig. 22) be the point (x, y), r the length OP, and θ the angle XOP, so that

$$x = r\cos\theta, \quad y = r\sin\theta, \quad r = \sqrt{(x^2+y^2)}, \quad \cos\theta : \sin\theta : 1 :: x : y : r.$$

We denote the complex number $x+yi$ by z, as in § 43, and we call z the *complex variable*. We call P *the point* z, or the point corresponding to z; z the *argument of* P, x the *real part*, y the *imaginary part*, r the *modulus*, and θ the *amplitude* of z; and we write

$$x = \mathbf{R}(z), \quad y = \mathbf{I}(z),$$

$$r = |z|, \quad \theta = \operatorname{am} z.$$

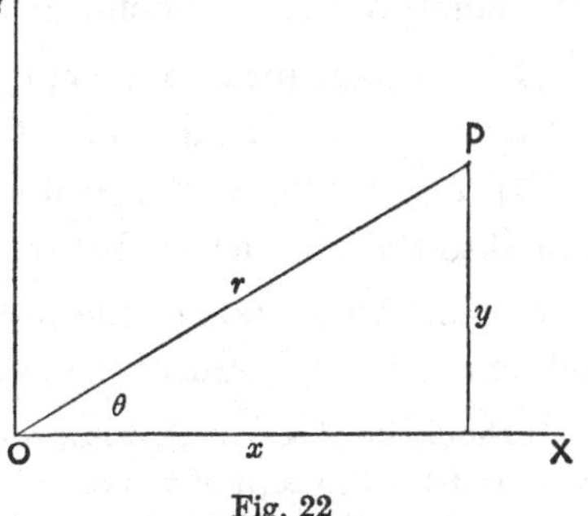

Fig. 22

When $y = 0$ we say that *z is real*, when $x = 0$ that *z is purely imaginary*. Two numbers $x+yi$, $x-yi$ which differ only in the signs of their imaginary parts, we call *conjugate*. It will be observed that the sum $2x$ of two conjugate numbers and their product x^2+y^2 are both real, that they have the same modulus $\sqrt{(x^2+y^2)}$, and that their product is equal to the square of the modulus of either. The roots of a quadratic with real coefficients, for example, are conjugate, when not real.

It must be observed that θ or $\operatorname{am} z$ is a many-valued function of x and y, having an infinity of values, which are angles differing by multiples of 2π*. A line originally lying along OX will, if turned through any of these angles, come to lie along OP. We shall describe that one of these angles which lies between $-\pi$ and

* It is evident that $|z|$ is identical with the polar coordinate r of P, and that the other polar coordinate θ is one value of $\operatorname{am} z$. This value is not necessarily the *principal* value, as defined below, for the polar coordinate of § 22 lies between 0 and 2π, and the principal value between $-\pi$ and π.

π as the *principal value* of the amplitude of z. This definition is unambiguous except when one of the values is π, in which case $-\pi$ is also a value. In this case we must make some special provision as to which value is to be regarded as the principal value. In general, when we speak of the amplitude of z we shall, unless the contrary is stated, mean the principal value of the amplitude.

Fig. 22 is usually known as Argand's diagram.

45. De Moivre's theorem. The following propositions follow immediately from the definitions of addition and multiplication.

(1) The real (or imaginary) part of the sum of two complex numbers is equal to the sum of their real (or imaginary) parts.

(2) The modulus of the product of two complex numbers is equal to the product of their moduli.

(3) The amplitude of the product of two complex numbers is either equal to the sum of their amplitudes, or differs from it by 2π.

It should be observed that it is not always true that the principal value of $\mathrm{am}(zz')$ is the sum of the principal values of $\mathrm{am}\,z$ and $\mathrm{am}\,z'$. For example, if $z = z' = -1+i$, then the principal values of the amplitudes of z and z' are each $\frac{3}{4}\pi$. But $zz' = -2i$, and the principal value of $\mathrm{am}(zz')$ is $-\frac{1}{2}\pi$ and not $\frac{3}{2}\pi$.

The last two theorems may be expressed in the equation

$$r(\cos\theta + i\sin\theta) \times \rho(\cos\phi + i\sin\phi)$$
$$= r\rho\{\cos(\theta+\phi) + i\sin(\theta+\phi)\},$$

which may be proved at once by multiplying out and using the ordinary trigonometrical formulae for $\cos(\theta+\phi)$ and $\sin(\theta+\phi)$. More generally

$$r_1(\cos\theta_1 + i\sin\theta_1) \times r_2(\cos\theta_2 + i\sin\theta_2) \times \dots \times r_n(\cos\theta_n + i\sin\theta_n)$$
$$= r_1 r_2 \dots r_n\{\cos(\theta_1+\theta_2+\dots+\theta_n) + i\sin(\theta_1+\theta_2+\dots+\theta_n)\}.$$

A particularly interesting case is that in which

$$r_1 = r_2 = \dots = r_n = 1, \quad \theta_1 = \theta_2 = \dots = \theta_n = \theta.$$

We then obtain the equation

$$(\cos\theta + i\sin\theta)^n = \cos n\theta + i\sin n\theta,$$

where n is any positive integer; a result known as *de Moivre's theorem**.

Again, if $$z = r(\cos\theta + i\sin\theta),$$

then $$1/z = (\cos\theta - i\sin\theta)/r.$$

Thus the modulus of the reciprocal of z is the reciprocal of the modulus of z, and the amplitude of the reciprocal is the negative of the amplitude of z. We can now state the theorems for quotients which correspond to (2) and (3).

(4) The modulus of the quotient of two complex numbers is equal to the quotient of their moduli.

(5) The amplitude of the quotient of two complex numbers is either equal to the difference of their amplitudes, or differs from it by 2π.

Again
$$\begin{aligned}(\cos\theta + i\sin\theta)^{-n} &= (\cos\theta - i\sin\theta)^n \\ &= \{\cos(-\theta) + i\sin(-\theta)\}^n \\ &= \cos(-n\theta) + i\sin(-n\theta).\end{aligned}$$

Hence *de Moivre's theorem holds for all integral values of n, positive or negative.*

To the theorems (1)–(5) we may add the following theorem, which is also of great importance.

(6) The modulus of the sum of any number of complex numbers is not greater than the sum of their moduli.

Let $\overline{OP}$, $\overline{OP'}$, ... be the displacements corresponding to the various complex numbers. Draw PQ equal and parallel to OP', QR equal and parallel to OP'', and so on. Finally we reach a point U, such that

$$\overline{OU} = \overline{OP} + \overline{OP'} + \overline{OP''} + \dots.$$

The length OU is the modulus of the sum of the complex numbers,

* It will sometimes be convenient, for the sake of brevity, to denote $\cos\theta + i\sin\theta$ by $\text{Cis}\,\theta$: in this notation, suggested by Profs. Harkness and Morley, de Moivre's theorem is expressed by the equation $(\text{Cis}\,\theta)^n = \text{Cis}\,n\theta$.

whereas the sum of their moduli is the total length of the broken line $OPQR\ldots U$, which is not less than OU.

A purely arithmetical proof of this theorem is outlined in Exs. XXI. 1.

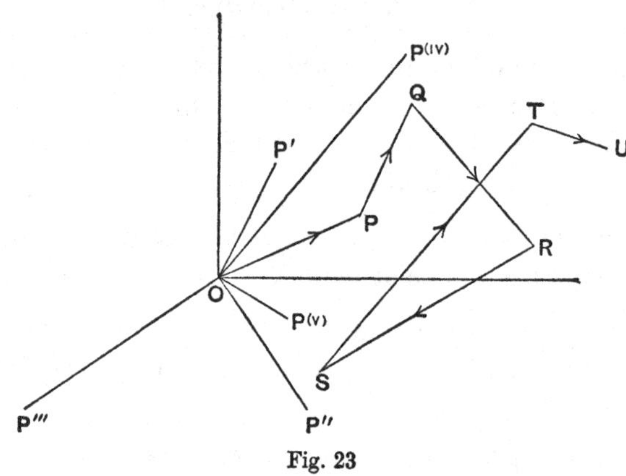

Fig. 23

46. We add some theorems concerning rational functions of complex numbers. A *rational function* of the complex variable z is defined as is a rational function of a real variable x, viz. as the quotient of two polynomials in z.

THEOREM 1. *Any rational function $R(z)$ can be reduced to the form $X + Yi$, where X and Y are rational functions of x and y with real coefficients.*

In the first place it is evident that any polynomial $P(x+yi)$ can be reduced, in virtue of the definitions of addition and multiplication, to the form $A+Bi$, where A and B are polynomials in x and y with real coefficients. Similarly $Q(x+yi)$ can be reduced to the form $C+Di$. Hence

$$R(x+yi) = P(x+yi)/Q(x+yi)$$

can be expressed in the form

$$(A+Bi)/(C+Di) = (A+Bi)(C-Di)/(C+Di)(C-Di)$$
$$= \frac{AC+BD}{C^2+D^2} + \frac{BC-AD}{C^2+D^2}i,$$

which proves the theorem.

THEOREM 2. *If $R(x+yi) = X+Yi$, R denoting a rational function as before, but with* real *coefficients, then $R(x-yi) = X-Yi$.*

In the first place this is easily verified for a power $(x+yi)^n$ by actual expansion. It follows by addition that the theorem is true for any polynomial with real coefficients. Hence, in the notation used above,

$$R(x-yi) = \frac{A-Bi}{C-Di} = \frac{AC+BD}{C^2+D^2} - \frac{BC-AD}{C^2+D^2}i,$$

the reduction being the same as before except that the sign of i is changed throughout. It is evident that results similar to those of Theorems 1 and 2 hold for functions of any number of complex variables.

THEOREM 3. *The roots of an equation*

$$a_0 z^n + a_1 z^{n-1} + \dots + a_n = 0,$$

whose coefficients are real, may, in so far as they are not themselves real, be arranged in conjugate pairs.

For it follows from Theorem 2 that if $x+yi$ is a root then so is $x-yi$. A particular case of this theorem is the result (§ 43) that the roots of a quadratic equation with real coefficients are either real or conjugate.

This theorem is sometimes stated as follows: *in an equation with real coefficients complex roots occur in conjugate pairs.* It should be compared with the result of Exs. VIII. 7, which may be stated as follows: *in an equation with rational coefficients irrational roots occur in conjugate pairs**.

Examples XXI. 1. Prove theorem (6) of § 45 directly from the definitions and without the aid of geometrical considerations.

[First, to prove that $|z+z'| \leqq |z| + |z'|$ is to prove that

$$\sqrt{\{(x+x')^2+(y+y')^2\}} \leqq \sqrt{(x^2+y^2)} + \sqrt{(x'^2+y'^2)}.$$

The theorem is then easily extended to the general case. The theorem is a special case of 'Minkowski's inequality'; see Hardy, Littlewood, and Pólya, *Inequalities*, pp. 30–39.]

* The numbers $a+\sqrt{b}$, $a-\sqrt{b}$, where a, b are rational, are sometimes said to be 'conjugate'.

2. The only case in which

$$|z|+|z'|+\ldots = |z+z'+\ldots|$$

is that in which the numbers $z, z', \ldots$ have all the same amplitude. Prove this both geometrically and analytically.

3. Prove that $$|z-z'| \geqq ||z|-|z'||.$$

4. If the sum and product of two complex numbers are both real, then the two numbers must either be real or conjugate.

5. If $$a+b\sqrt{2}+(c+d\sqrt{2})i = A+B\sqrt{2}+(C+D\sqrt{2})i,$$

where a, b, c, d, A, B, C, D are real rational numbers, then

$$a = A, \quad b = B, \quad c = C, \quad d = D.$$

6. Express the following numbers in the form $A+Bi$, where A and B are real numbers:

$$(1+i)^2, \quad \left(\frac{1+i}{1-i}\right)^2, \quad \left(\frac{1-i}{1+i}\right)^2, \quad \frac{\lambda+\mu i}{\lambda-\mu i}, \quad \left(\frac{\lambda+\mu i}{\lambda-\mu i}\right)^2 - \left(\frac{\lambda-\mu i}{\lambda+\mu i}\right)^2,$$

where λ and μ are real numbers.

7. Express the following functions of $z = x+yi$ in the form $X+Yi$, where X and Y are real functions of x and y: z^2, z^3, z^n, $1/z$, $z+(1/z)$, $(\alpha+\beta z)/(\gamma+\delta z)$, where $\alpha, \beta, \gamma, \delta$ are real numbers.

8. Find the moduli of the numbers and functions in the two preceding examples.

9. The two lines joining the points $z = a$, $z = b$ and $z = c$, $z = d$ will be perpendicular if

$$\operatorname{am}\left(\frac{a-b}{c-d}\right) = \pm\tfrac{1}{2}\pi;$$

i.e. if $(a-b)/(c-d)$ is purely imaginary. What is the condition that the lines should be parallel?

10. The three angular points of a triangle are given by $z = \alpha$, $z = \beta$, $z = \gamma$, where α, β, γ are complex numbers. Establish the following propositions:

(i) *the centre of gravity is given by* $z = \frac{1}{3}(\alpha+\beta+\gamma)$;

(ii) *the circum-centre is given by* $|z-\alpha| = |z-\beta| = |z-\gamma|$;

(iii) *the three perpendiculars from the angular points on the opposite sides meet in a point given by*

$$\mathbf{R}\left(\frac{z-\alpha}{\beta-\gamma}\right) = \mathbf{R}\left(\frac{z-\beta}{\gamma-\alpha}\right) = \mathbf{R}\left(\frac{z-\gamma}{\alpha-\beta}\right) = 0;$$

(iv) *there is a point P inside the triangle such that*

$$CBP = ACP = BAP = \omega,$$

and $$\cot\omega = \cot A + \cot B + \cot C.$$

[To prove (iii) we observe that if A, B, C are the vertices, and P any point z, then the condition that AP should be perpendicular to BC is (Ex. 9) that $(z-\alpha)/(\beta-\gamma)$ should be purely imaginary, or that

$$\mathbf{R}(z-\alpha)\,\mathbf{R}(\beta-\gamma)+\mathbf{I}(z-\alpha)\,\mathbf{I}(\beta-\gamma)=0.$$

This equation, and the two similar equations obtained by permuting α, β, γ cyclically, are satisfied by the same value of z, as appears from the fact that the sum of the three left-hand sides is zero.

To prove (iv), take BC parallel to the positive direction of the axis of x. Then*

$$\gamma-\beta=a,\quad \alpha-\gamma=-b\operatorname{Cis}(-C),\quad \beta-\alpha=-c\operatorname{Cis}B.$$

We have to determine z and ω from the equations

$$\frac{(z-\alpha)(\beta_0-\alpha_0)}{(z_0-\alpha_0)(\beta-\alpha)}=\frac{(z-\beta)(\gamma_0-\beta_0)}{(z_0-\beta_0)(\gamma-\beta)}=\frac{(z-\gamma)(\alpha_0-\gamma_0)}{(z_0-\gamma_0)(\alpha-\gamma)}=\operatorname{Cis}2\omega,$$

where z_0, α_0, β_0, γ_0 denote the conjugates of z, α, β, γ.

Adding the numerators and denominators of the three equal fractions, and using the equation

$$i\cot\omega=(1+\operatorname{Cis}2\omega)/(1-\operatorname{Cis}2\omega),$$

we find that

$$i\cot\omega=\frac{(\beta-\gamma)(\beta_0-\gamma_0)+(\gamma-\alpha)(\gamma_0-\alpha_0)+(\alpha-\beta)(\alpha_0-\beta_0)}{\beta\gamma_0-\beta_0\gamma+\gamma\alpha_0-\gamma_0\alpha+\alpha\beta_0-\alpha_0\beta}.$$

From this it is easily deduced that the value of $\cot\omega$ is $(a^2+b^2+c^2)/4\Delta$, where Δ is the area of the triangle; and this is equivalent to the result given.

To determine z, we multiply the numerators and denominators of the equal fractions by $(\gamma_0-\beta_0)/(\beta-\alpha)$, $(\alpha_0-\gamma_0)/(\gamma-\beta)$, $(\beta_0-\alpha_0)/(\alpha-\gamma)$, and add to form a new fraction. It will be found that

$$z=\frac{a\alpha\operatorname{Cis}A+b\beta\operatorname{Cis}B+c\gamma\operatorname{Cis}C}{a\operatorname{Cis}A+b\operatorname{Cis}B+c\operatorname{Cis}C}.]$$

11. The two triangles whose vertices are the points a, b, c and x, y, z respectively will be similar if

$$\begin{vmatrix}1 & 1 & 1\\ a & b & c\\ x & y & z\end{vmatrix}=0.$$

[The condition required is that $\overline{AB}/\overline{AC}=\overline{XY}/\overline{XZ}$ (large letters denoting the points whose arguments are the corresponding small letters), or $(b-a)/(c-a)=(y-x)/(z-x)$, which is the same as the condition given.]

* We suppose that as we go round the triangle in the direction ABC we leave it on our left.

12. Deduce from the last example that if the points x, y, z are collinear then we can find *real* numbers α, β, γ such that $\alpha+\beta+\gamma=0$ and $\alpha x+\beta y+\gamma z=0$, and conversely (cf. Exs. xx. 4). [Use the fact that in this case the triangle formed by x, y, z is similar to a certain line-triangle on the axis OX, and apply the result of the last example.]

13. **The general linear equation with complex coefficients.** The equation $\alpha z+\beta=0$ has the one solution $z=-\beta/\alpha$, unless $\alpha=0$. If we put

$$\alpha=a+Ai,\quad \beta=b+Bi,\quad z=x+yi,$$

and equate real and imaginary parts, we obtain two equations to determine the two real numbers x and y. The equation will have a real root if $y=0$, which gives $ax+b=0$, $Ax+B=0$, and the condition that these equations should be consistent is $aB-bA=0$.

14. **The general quadratic equation with complex coefficients.** This equation is

$$(a+Ai)z^2+2(b+Bi)z+(c+Ci)=0.$$

Unless a and A are both zero we can divide through by $a+iA$. Hence we may consider

$$z^2+2(b+Bi)z+(c+Ci)=0 \quad \text{.....................(1)}$$

as the standard form of our equation. Putting $z=x+yi$, and equating real and imaginary parts, we obtain a pair of simultaneous equations for x and y, viz.

$$x^2-y^2+2(bx-By)+c=0,\quad 2xy+2(by+Bx)+C=0.$$

If we put

$$x+b=\xi,\quad y+B=\eta,\quad b^2-B^2-c=h,\quad 2bB-C=k,$$

these equations become $\xi^2-\eta^2=h$, $2\xi\eta=k$.

Squaring and adding we obtain

$$\xi^2+\eta^2=\sqrt{(h^2+k^2)},\quad \xi=\pm\sqrt{[\tfrac{1}{2}\{\sqrt{(h^2+k^2)}+h\}]},\quad \eta=\pm\sqrt{[\tfrac{1}{2}\{\sqrt{(h^2+k^2)}-h\}]}.$$

We must choose the signs so that $\xi\eta$ has the sign of k: i.e. if k is positive we must take like signs, if k is negative unlike signs.

Conditions for equal roots. The two roots can only be equal if both the square roots above vanish, i.e. if $h=0$, $k=0$, or if $c=b^2-B^2$, $C=2bB$. These conditions are equivalent to the single condition $c+Ci=(b+Bi)^2$, which expresses the fact that the left-hand side of (1) is a perfect square.

Condition for a real root. If $x^2+2(b+Bi)x+(c+Ci)=0$, where x is real, then $x^2+2bx+c=0$, $2Bx+C=0$. Eliminating x we find that the required condition is

$$C^2-4bBC+4cB^2=0.$$

Condition for a purely imaginary root. This is easily found to be

$$C^2-4bBC-4b^2c=0.$$

Conditions for a pair of conjugate complex roots. Since the sum and the product of two conjugate complex numbers are both real, $b+Bi$ and $c+Ci$ must both be real, i.e. $B = 0$, $C = 0$. Thus the equation (1) can have a pair of conjugate complex roots only if its coefficients are real. The reader should verify this conclusion by means of the explicit expressions of the roots. Moreover, if $b^2 \geqq c$, the roots will be real even in this case. Hence for a pair of conjugate roots we must have $B = 0$, $C = 0$, $b^2 < c$.

15. **The cubic equation.** Consider the cubic equation

$$z^3 + 3Hz + G = 0,$$

where G and H are complex numbers, it being given that the equation has (*a*) a real root, (*b*) a purely imaginary root, (*c*) a pair of conjugate roots. If $H = \lambda + \mu i$, $G = \rho + \sigma i$, we arrive at the following conclusions.

(*a*) *Conditions for a real root.* If μ is not zero, then the real root is $-\sigma/3\mu$, and $\sigma^3 + 27\lambda\mu^2\sigma - 27\mu^3\rho = 0$. On the other hand, if $\mu = 0$ then we must also have $\sigma = 0$, so that the coefficients of the equation are real. In this case there may be three real roots.

(*b*) *Conditions for a purely imaginary root.* If μ is not zero then the purely imaginary root is $\rho i/3\mu$, and $\rho^3 - 27\lambda\mu^2\rho - 27\mu^3\sigma = 0$. If $\mu = 0$ then also $\rho = 0$, and the root is yi, where y is given by the equation $y^3 - 3\lambda y - \sigma = 0$, which has real coefficients. In this case there may be three purely imaginary roots.

(*c*) *Conditions for a pair of conjugate complex roots.* Let these be $x+yi$ and $x-yi$. Then since the sum of the three roots is zero the third root must be $-2x$. From the relations between the coefficients and the roots of an equation we deduce

$$y^2 - 3x^2 = 3H, \quad 2x(x^2+y^2) = G.$$

Hence G and H must both be real.

In each case we can either find a root (in which case the equation can be reduced to a quadratic by dividing by a known factor) or we can reduce the solution of the equation to the solution of a cubic equation with real coefficients.

16. The cubic equation $x^3 + a_1x^2 + a_2x + a_3 = 0$, where $a_1 = A_1 + A_1'i, \ldots$ has a pair of conjugate complex roots. Prove that the remaining root is $-A_1'a_3/A_3'$, unless $A_3' = 0$. Examine the case in which $A_3' = 0$.

17. Prove that, if $z^3 + 3Hz + G = 0$ has two conjugate complex roots, the equation

$$8\alpha^3 + 6\alpha H - G = 0$$

has one real root which is the real part α of the complex roots of the original equation; and that α has the same sign as G.

18. An equation of any order with complex coefficients will in general have no real roots nor pairs of conjugate complex roots. How many conditions must be satisfied by the coefficients in order that the equation should have (*a*) a real root, (*b*) a pair of conjugate roots?

19. **Coaxal circles.** In Fig. 24, let a, b, z be the arguments of A, B, P. Then

$$\operatorname{am}\frac{z-b}{z-a} = APB,$$

if the principal value of the amplitude is chosen. If the two circles shown in the figure are equal, z', z_1, z_1' are the arguments of P', P_1, P_1', and $APB = \theta$, then it is easy to see that

$$\operatorname{am}\frac{z'-b}{z'-a} = \pi-\theta, \quad \operatorname{am}\frac{z_1-b}{z_1-a} = -\theta,$$

and

$$\operatorname{am}\frac{z_1'-b}{z_1'-a} = -\pi+\theta.$$

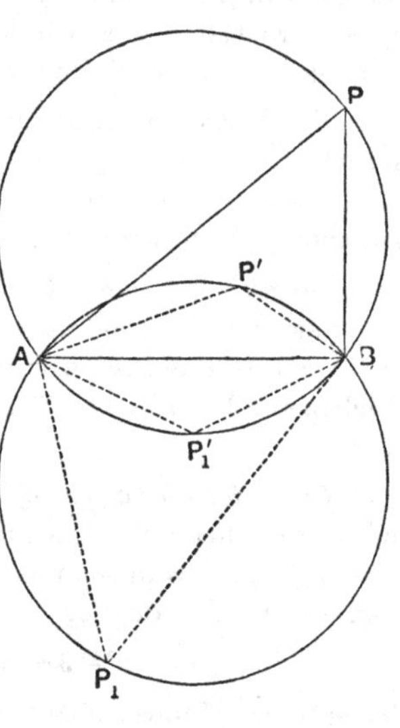

Fig. 24

The locus defined by the equation

$$\operatorname{am}\frac{z-b}{z-a} = \theta,$$

where θ is constant, is the arc APB. By writing $\pi-\theta$, $-\theta$, $-\pi+\theta$ for θ, we obtain the other three arcs shown.

The system of equations obtained by supposing that θ is a parameter, varying from $-\pi$ to $+\pi$, represents *the system of circles which can be drawn through the points A, B*. It should however be observed that each circle has to be divided into two parts to which correspond different values of θ.

20. Now let us consider the equation

$$\left|\frac{z-b}{z-a}\right| = \lambda \quad \text{..............................(1)},$$

where λ is a constant, not **1**.

Let K be the point in which the tangent to the circle ABP at P meets AB. Then the triangles KPA, KBP are similar, and so

$$AP/PB = PK/BK = KA/KP = 1/\lambda.$$

Hence $KA/KB = 1/\lambda^2$, and therefore K is a fixed point for all positions of P which satisfy the equation (1). Also $KP^2 = KA.KB$, and so is constant. Hence *the locus of P is a circle whose centre is K.*

The system of equations obtained by varying λ represents a system of circles, and every circle of this system cuts at right angles every circle of the system of Ex. 19. The circle becomes a straight line when $\lambda = 1$.

The system of Ex. 19 is called *a system of coaxal circles of the common point kind.* The system of Ex. 20 is called *a system of coaxal circles of the limiting point kind, A* and *B* being the *limiting points* of the system. If λ is very large or very small, then the circle is a very small circle containing A or B in its interior.

21. **Bilinear transformations.** Consider the equation

$$z = Z + a \qquad (1),$$

where $z = x + yi$ and $Z = X + Yi$ are two complex variables which we may suppose to be represented in two planes xoy, XOY. To every value of z corresponds one of Z, and conversely. If $a = \alpha + \beta i$, then

$$x = X + \alpha, \quad y = Y + \beta,$$

and to the point (x, y) corresponds the point (X, Y). If (x, y) describes a curve of any kind in its plane, (X, Y) describes a curve in its plane. Thus to any figure in one plane corresponds a figure in the other. A passage of this kind from a figure in the plane xoy to a figure in the plane XOY by means of a relation such as (1) between z and Z is called a *transformation.* In this particular case the relation between corresponding figures is very easily defined. The (X, Y) figure is the same in size, shape, and orientation as the (x, y) figure, but is shifted a distance α to the left, and a distance β downwards. Such a transformation is called a *translation.*

Now consider the equation

$$z = \rho Z \qquad (2),$$

where ρ is positive. This gives $x = \rho X$, $y = \rho Y$. The two figures are similar and similarly situated about their respective origins, but the scale of the (x, y) figure is ρ times that of the (X, Y) figure. Such a transformation is called a *magnification.*

Next consider the equation

$$z = (\cos\phi + i\sin\phi)\, Z \qquad (3).$$

It is clear that $|z| = |Z|$ and that one value of $\operatorname{am} z$ is $\operatorname{am} Z + \phi$, and that the two figures differ only in that the (x, y) figure is the (X, Y) figure turned about the origin through an angle ϕ in the positive direction. Such a transformation is called a *rotation.*

The general linear transformation

$$z = aZ + b \qquad (4)$$

is a combination of the three transformations (1), (2), (3). For, if $|a| = \rho$ and $\operatorname{am} a = \phi$, we can replace (4) by the three equations

$$z = z' + b, \quad z' = \rho Z', \quad Z' = (\cos\phi + i\sin\phi)\, Z.$$

Thus *the general linear transformation is equivalent to the combination of a translation, a magnification, and a rotation.*

Next consider the transformation

$$z = 1/Z \quad \text{.................................(5).}$$

If $|Z| = R$ and $\text{am}\, Z = \Theta$, then $|z| = 1/R$ and $\text{am}\, z = -\Theta$, and to pass from the (x, y) figure to the (X, Y) figure we invert the former with respect to o, with unit radius of inversion, and then construct the image of the new figure in the axis ox (i.e. the symmetrical figure on the other side of ox).

Finally consider the transformation

$$z = \frac{aZ+b}{cZ+d} \quad \text{..............................(6).}$$

This is equivalent to the combination of the transformations

$$z = (a/c) + (bc - ad)(z'/c), \quad z' = 1/Z', \quad Z' = cZ + d,$$

i.e. to a certain combination of transformations of the types already considered.

The transformation (6) is called the *general bilinear transformation.* Solving for Z we obtain

$$Z = -\frac{dz-b}{cz-a}.$$

The general bilinear transformation is the most general type of transformation for which one and only one value of z corresponds to each value of Z, and conversely.

22. *The general bilinear transformation transforms circles into circles.* This may be proved in a variety of ways. We may assume the well-known theorem in pure geometry, that inversion transforms circles into circles (which may of course in particular cases be straight lines). Or we may use the results of Exs. 19 and 20. If, e.g., the (x, y) circle is

$$|(z-\sigma)/(z-\rho)| = \lambda,$$

and we substitute for z in terms of Z, we obtain

$$|(Z-\sigma')/(Z-\rho')| = \lambda',$$

where $$\sigma' = -\frac{b-\sigma d}{a-\sigma c}, \quad \rho' = -\frac{b-\rho d}{a-\rho c}, \quad \lambda' = \left|\frac{a-\rho c}{a-\sigma c}\right|\lambda.$$

23. Consider the transformations $z = 1/Z$, $z = (1+Z)/(1-Z)$, and draw the (X, Y) curves which correspond to (1) circles whose centre is the origin, (2) straight lines through the origin.

24. The condition that the transformation $z = (aZ+b)/(cZ+d)$ should make the circle $x^2 + y^2 = 1$ correspond to a straight line in the (X, Y) plane is $|a| = |c|$.

25. **Cross ratios.** The cross ratio $(z_1 z_2, z_3 z_4)$ is defined to be

$$\frac{(z_1-z_3)(z_2-z_4)}{(z_1-z_4)(z_2-z_3)}.$$

If the four points z_1, z_2, z_3, z_4 are on the same line, this definition agrees with that adopted in elementary geometry. There are 24 cross ratios which can be formed from z_1, z_2, z_3, z_4 by permuting the suffixes. These consist of six groups of four equal cross ratios. If one ratio is λ, then the six distinct cross ratios are λ, $1-\lambda$, $1/\lambda$, $1/(1-\lambda)$, $(\lambda-1)/\lambda$, $\lambda/(\lambda-1)$. The four points are said to be *harmonic* or *harmonically related* if any one of these is equal to -1. In this case the six ratios are $-1, 2, -1, \frac{1}{2}, 2, \frac{1}{2}$.

If any cross ratio is real then all are real and the four points lie on a circle. For in this case

$$\operatorname{am}\frac{(z_1-z_3)(z_2-z_4)}{(z_1-z_4)(z_2-z_3)}$$

must have one of the three values $-\pi, 0, \pi$, so that $\operatorname{am}\{(z_1-z_3)/(z_1-z_4)\}$ and $\operatorname{am}\{(z_2-z_3)/(z_2-z_4)\}$ must either be equal or differ by π (cf. Ex. 19).

If $(z_1 z_2, z_3 z_4) = -1$, we have the two equations

$$\operatorname{am}\frac{z_1-z_3}{z_1-z_4} = \pm\pi + \operatorname{am}\frac{z_2-z_3}{z_2-z_4}, \quad \left|\frac{z_1-z_3}{z_1-z_4}\right| = \left|\frac{z_2-z_3}{z_2-z_4}\right|.$$

The four points A_1, A_2, A_3, A_4 lie on a circle, A_1 and A_2 being separated by A_3 and A_4. Also $A_1A_3/A_1A_4 = A_2A_3/A_2A_4$. Let O be the middle point of A_3A_4. The equation

$$\frac{(z_1-z_3)(z_2-z_4)}{(z_1-z_4)(z_2-z_3)} = -1$$

may be put in the form

$$(z_1+z_2)(z_3+z_4) = 2(z_1z_2+z_3z_4),$$

or, what is the same thing,

$$\{z_1-\tfrac{1}{2}(z_3+z_4)\}\{z_2-\tfrac{1}{2}(z_3+z_4)\} = \{\tfrac{1}{2}(z_3-z_4)\}^2.$$

But this is equivalent to $\overline{OA_1}.\overline{OA_2} = \overline{OA_3}^2 = \overline{OA_4}^2$. Hence OA_1 and OA_2 make equal angles with A_3A_4, and $OA_1.OA_2 = OA_3^2 = OA_4^2$. It will be observed that the relation between the pairs A_1, A_2 and A_3, A_4 is symmetrical. Hence, if O' is the middle point of A_1A_2, $O'A_3$ and $O'A_4$ are equally inclined to A_1A_2, and $O'A_3.O'A_4 = O'A_1^2 = O'A_2^2$.

26. If the points A_1, A_2 are given by $az^2+2bz+c=0$, the points A_3, A_4 by $a'z^2+2b'z+c'=0$, O is the middle point of A_3A_4, and $ac'+a'c-2bb'=0$, then OA_1, OA_2 are equally inclined to A_3A_4 and $OA_1.OA_2 = OA_3^2 = OA_4^2$. (*Math. Trip.* 1901)

27. AB, CD are two intersecting lines in Argand's diagram, and P, Q their middle points. Prove that, if AB bisects the angle CPD and $PA^2 = PB^2 = PC.PD$, then CD bisects the angle AQB and

$$QC^2 = QD^2 = QA.QB. \qquad (\textit{Math. Trip.}\ 1909)$$

28. The condition that four points should lie on a circle. A sufficient condition is that one (and therefore all) of the cross ratios should be real (Ex. 25); this condition is also necessary. Another form of the condition is that it should be possible to choose real numbers α, β, γ such that

$$\begin{vmatrix} 1 & 1 & 1 \\ \alpha & \beta & \gamma \\ z_1z_4+z_2z_3 & z_2z_4+z_3z_1 & z_3z_4+z_1z_2 \end{vmatrix} = 0.$$

[To prove this we observe that the transformation $Z = 1/(z-z_4)$ is equivalent to an inversion with respect to the point z_4, coupled with a certain reflexion (Ex. 21). If z_1, z_2, z_3 lie on a circle through z_4, the corresponding points $Z_1 = 1/(z_1-z_4)$, $Z_2 = 1/(z_2-z_4)$, $Z_3 = 1/(z_3-z_4)$ lie on a straight line. Hence (Ex. 12) we can find real numbers α', β', γ' such that $\alpha'+\beta'+\gamma' = 0$ and $\alpha'/(z_1-z_4)+\beta'/(z_2-z_4)+\gamma'/(z_3-z_4) = 0$, and it is easy to prove that this is equivalent to the given condition.]

29. Prove the following analogue of de Moivre's theorem for real numbers: if ϕ_1, ϕ_2, ϕ_3, ... is a series of positive acute angles such that

$$\tan\phi_{m+1} = \tan\phi_m \sec\phi_1 + \sec\phi_m \tan\phi_1,$$

then

$$\tan\phi_{m+n} = \tan\phi_m \sec\phi_n + \sec\phi_m \tan\phi_n,$$
$$\sec\phi_{m+n} = \sec\phi_m \sec\phi_n + \tan\phi_m \tan\phi_n,$$

and

$$\tan\phi_m + \sec\phi_m = (\tan\phi_1 + \sec\phi_1)^m.$$

[Use the method of mathematical induction.]

30. The transformation $z = Z^m$. In this case $r = R^m$, and θ and $m\Theta$ differ by a multiple of 2π. If Z describes a circle round the origin then z describes a circle round the origin m times.

The whole (x, y) plane corresponds to any one of m sectors in the (X, Y) plane, each of angle $2\pi/m$. To each point in the (x, y) plane correspond m points in the (X, Y) plane.

31. Complex functions of a real variable. If $f(t)$, $\phi(t)$ are two real functions of a real variable t defined for a certain range of values of t, we call

$$z = f(t) + i\phi(t) \qquad (1)$$

a complex function of t. We can represent it graphically by drawing the curve

$$x = f(t), \quad y = \phi(t).$$

If z is a polynomial in t, or rational function of t, with complex coefficients, we can express it in the form (1) and so determine the curve represented by the function.

(i) Let

$$z = a + (b-a)t,$$

where a and b are complex numbers. If $a = \alpha + \alpha' i$, $b = \beta + \beta' i$, then

$$x = \alpha + (\beta-\alpha)t, \quad y = \alpha' + (\beta'-\alpha')t.$$

The curve is the straight line joining the points $z = a$ and $z = b$. The segment between the points corresponds to the range of values of t from 0 to 1. Find the values of t which correspond to the two produced segments of the line.

(ii) If
$$z = c+\rho\left(\frac{1+ti}{1-ti}\right),$$
where ρ is positive, then the curve is the circle of centre c and radius ρ. As t varies through all real values z describes the circle once.

(iii) In general the equation $z = (a+bt)/(c+dt)$ represents a circle. This can be proved by calculating x and y and eliminating: but this process is rather cumbrous. A simpler method is obtained by using the result of Ex. 22. Let $z = (a+bZ)/(c+dZ)$, $Z = t$. As t varies Z describes a straight line, viz. the axis of X. Hence z describes a circle.

(iv) The equation
$$z = a+2bt+ct^2$$
represents a parabola generally, a straight line if b/c is real.

(v) The equation $z = (a+2bt+ct^2)/(\alpha+2\beta t+\gamma t^2)$, where α, β, γ are real, represents a conic section.

[Eliminate t from
$$x = (A+2Bt+Ct^2)/(\alpha+2\beta t+\gamma t^2), \quad y = (A'+2B't+C't^2)/(\alpha+2\beta t+\gamma t^2),$$
where $A+A'i = a$, $B+B'i = b$, $C+C'i = c$.]

47. Roots of complex numbers. We have not, up to the present, attributed any meaning to symbols such as $\sqrt[n]{a}$, $a^{m/n}$, when a is a complex number, and m and n integers. It is, however, natural to adopt the definitions which are given in elementary algebra for real values of a. Thus we define $\sqrt[n]{a}$ or $a^{1/n}$, where n is a positive integer, as a number z which satisfies the equation $z^n = a$; and $a^{m/n}$, where m is an integer, as $(a^{1/n})^m$. These definitions do not prejudge the question whether there are or are not roots of the equation.

48. Solution of the equation $z^n = a$. Let
$$a = \rho(\cos\phi+i\sin\phi),$$
where ρ is positive and ϕ is an angle such that $-\pi<\phi\leqq\pi$. If we put $z = r(\cos\theta+i\sin\theta)$, the equation takes the form
$$r^n(\cos n\theta+i\sin n\theta) = \rho(\cos\phi+i\sin\phi);$$
so that
$$r^n = \rho, \quad \cos n\theta = \cos\phi, \quad \sin n\theta = \sin\phi \quad ......(1).$$

The only possible value of r is $\sqrt[n]{\rho}$, the ordinary arithmetical nth root of ρ; and in order that the last two equations should be satisfied it is necessary and sufficient that $n\theta = \phi + 2k\pi$, where k is an integer, or

$$\theta = (\phi + 2k\pi)/n.$$

If $k = pn + q$, where p and q are integers, and $0 \leqq q < n$, then the value of θ is $2p\pi + (\phi + 2q\pi)/n$, and what value we choose for p here is indifferent. Hence *the equation*

$$z^n = a = \rho(\cos\phi + i\sin\phi)$$

has just n roots, given by $z = r(\cos\theta + i\sin\theta)$, *where*

$$r = \sqrt[n]{\rho}, \quad \theta = (\phi + 2q\pi)/n, \quad (q = 0, 1, 2, \dots, n-1).$$

That the n roots are distinct is easily seen by plotting them on Argand's diagram. The particular root

$$\sqrt[n]{\rho}\,\{\cos(\phi/n) + i\sin(\phi/n)\}$$

is called the *principal value* of $\sqrt[n]{a}$.

The case in which $a = 1$, $\rho = 1$, $\phi = 0$ is of particular interest. The n roots of the equation $x^n = 1$ are

$$\cos(2q\pi/n) + i\sin(2q\pi/n), \qquad (q = 0, 1, \dots, n-1).$$

These numbers are called the nth roots of unity; the principal value is unity itself. If we write ω_n for $\cos(2\pi/n) + i\sin(2\pi/n)$, we see that the nth roots of unity are

$$1, \omega_n, \omega_n^2, \dots, \omega_n^{n-1}.$$

Examples XXII. 1. The two square roots of 1 are $1, -1$; the three cube roots are 1, $\frac{1}{2}(-1 + i\sqrt{3})$, $\frac{1}{2}(-1 - i\sqrt{3})$; the four fourth roots are $1, i, -1, -i$; and the five fifth roots are

$$1, \quad \tfrac{1}{4}[\sqrt{5} - 1 + i\sqrt{\{10 + 2\sqrt{5}\}}], \qquad \tfrac{1}{4}[-\sqrt{5} - 1 + i\sqrt{\{10 - 2\sqrt{5}\}}],$$
$$\tfrac{1}{4}[-\sqrt{5} - 1 - i\sqrt{\{10 - 2\sqrt{5}\}}], \qquad \tfrac{1}{4}[\sqrt{5} - 1 - i\sqrt{\{10 + 2\sqrt{5}\}}].$$

2. Prove that $\quad 1 + \omega_n + \omega_n^2 + \dots + \omega_n^{n-1} = 0.$

3. Prove that

$$(x + y\omega_3 + z\omega_3^2)(x + y\omega_3^2 + z\omega_3) = x^2 + y^2 + z^2 - yz - zx - xy.$$

4. The nth roots of a are the products of the nth roots of unity by the principal value of $\sqrt[n]{a}$.

5. It follows from Exs. XXI. 14 that the roots of

$$z^2 = \alpha + \beta i$$

are $$\pm\sqrt{[\tfrac{1}{2}\{\sqrt{(\alpha^2+\beta^2)}+\alpha\}]} \pm i\sqrt{[\tfrac{1}{2}\{\sqrt{(\alpha^2+\beta^2)}-\alpha\}]},$$

like or unlike signs being chosen according as β is positive or negative. Show that this result agrees with the result of § 48.

6. Show that $(x^{2m}-a^{2m})/(x^2-a^2)$ is equal to

$$\left(x^2 - 2ax\cos\frac{\pi}{m} + a^2\right)\left(x^2 - 2ax\cos\frac{2\pi}{m} + a^2\right)\ldots\left(x^2 - 2ax\cos\frac{(m-1)\pi}{m} + a^2\right).$$

[The factors of $x^{2m}-a^{2m}$ are

$$(x-a),\quad (x-a\omega_{2m}),\quad (x-a\omega_{2m}^2),\quad \ldots\quad (x-a\omega_{2m}^{2m-1}).$$

The factor $x-a\omega_{2m}^m$ is $x+a$. The factors $(x-a\omega_{2m}^s)$, $(x-a\omega_{2m}^{2m-s})$ taken together give a factor $x^2-2ax\cos(s\pi/m)+a^2$.]

7. Resolve $x^{2m+1}-a^{2m+1}$, $x^{2m}+a^{2m}$, and $x^{2m+1}+a^{2m+1}$ into factors in a similar way.

8. Show that $x^{2n}-2x^n a^n\cos\theta+a^{2n}$ is equal to

$$\left(x^2 - 2xa\cos\frac{\theta}{n} + a^2\right)\left(x^2 - 2xa\cos\frac{\theta+2\pi}{n} + a^2\right)\ldots$$
$$\ldots\left(x^2 - 2xa\cos\frac{\theta+2(n-1)\pi}{n} + a^2\right).$$

[Use the formula

$$x^{2n}-2x^n a^n\cos\theta+a^{2n} = \{x^n - a^n(\cos\theta+i\sin\theta)\}\{x^n - a^n(\cos\theta - i\sin\theta)\},$$

and split up each of the last two expressions into n factors.]

9. Find all the roots of the equation $x^6-2x^3+2=0$.

(*Math. Trip.* 1910)

10. The problem of finding the value of ω_n in a form involving square roots only, as in the formula $\omega_3 = \frac{1}{2}(-1+i\sqrt{3})$, is the algebraical equivalent of the geometrical problem of inscribing a regular polygon of n sides in a circle of unit radius by Euclidean methods, i.e. by ruler and compasses. For this construction will be possible if and only if we can construct lengths measured by $\cos(2\pi/n)$ and $\sin(2\pi/n)$; and this is possible (Ch. II, Misc. Ex. 22), if and only if these numbers are expressible in a form involving square roots only.

Euclid gives constructions for $n = 3$, 4, 5, 6, 8, 10, 12, and 15. It is evident that the construction is possible for any value of n which can be found from these by multiplication by any power of 2. There are other special values of n for which such constructions are possible, the most interesting being $n = 17$.

Gauss proved the construction possible when n is of the form

$$2^{2^k}+1$$

and is prime. The numbers 3, 5, 17, 257, and 65537, corresponding to $k = 0, 1, 2, 3, 4$ are prime, and the construction is then possible. But $k = 5, 6, 7$, and 8 give composite values of n, and it is not known whether there are more prime values.

The simplest construction of the 17-agon, due to Richmond, will be found in H. P. Hudson's *Ruler and compasses*, p. 34; in F. and F. V. Morley's *Inversive geometry*, p. 167; and in Klein's book referred to on p. 22.

49. The general form of de Moivre's theorem. It follows from the results of the last section that if q is a positive integer then one of the values of $(\cos\theta+i\sin\theta)^{1/q}$ is

$$\cos(\theta/q)+i\sin(\theta/q).$$

Raising each of these expressions to the power p (where p is any integer positive or negative), we obtain the theorem that one of the values of $(\cos\theta+i\sin\theta)^{p/q}$ is $\cos(p\theta/q)+i\sin(p\theta/q)$, or that *if α is any rational number then one of the values of* $(\cos\theta+i\sin\theta)^{\alpha}$ *is*

$$\cos\alpha\theta+i\sin\alpha\theta.$$

This is a generalized form of de Moivre's theorem (§ 45).

MISCELLANEOUS EXAMPLES ON CHAPTER III

1. The condition that a triangle (xyz) should be equilateral is that

$$x^2+y^2+z^2-yz-zx-xy = 0.$$

[Let XYZ be the triangle. The displacement $\overline{ZX}$ is $\overline{YZ}$ turned through an angle $\frac{2}{3}\pi$ in the positive or negative direction. Since $\operatorname{Cis}\frac{2}{3}\pi = \omega_3$, $\operatorname{Cis}(-\frac{2}{3}\pi) = 1/\omega_3 = \omega_3^2$, we have $x-z = (z-y)\,\omega_3$ or $x-z = (z-y)\,\omega_3^2$. Hence $x+y\omega_3+z\omega_3^2 = 0$ or $x+y\omega_3^2+z\omega_3 = 0$. The result follows from Exs. XXII. 3.]

2. If XYZ, $X'Y'Z'$ are two triangles, and

$$\overline{YZ}\,.\,\overline{Y'Z'} = \overline{ZX}\,.\,\overline{Z'X'} = \overline{XY}\,.\,\overline{X'Y'},$$

then both triangles are equilateral. [From the equations

$$(y-z)(y'-z') = (z-x)(z'-x') = (x-y)(x'-y') = \kappa^2,$$

say, we deduce $\Sigma 1/(y'-z') = 0$, or $\Sigma x'^2-\Sigma y'z' = 0$. Now apply the result of the last example.]

3. Similar triangles BCX, CAY, ABZ are described on the sides of a triangle ABC. Show that the centres of gravity of ABC, XYZ are coincident.

[We have $(x-c)/(b-c) = (y-a)/(c-a) = (z-b)/(a-b) = \lambda$, say. Express $\frac{1}{3}(x+y+z)$ in terms of a, b, c.]

4. If X, Y, Z are points on the sides of the triangle ABC, such that

$$BX/XC = CY/YA = AZ/ZB = r,$$

and if ABC, XYZ are similar, then either $r = 1$ or both triangles are equilateral.

5. If A, B, C, D are four points in a plane, then

$$AD.BC \leqq BD.CA + CD.AB.$$

[Let z_1, z_2, z_3, z_4 be the complex numbers corresponding to A, B, C, D. Then we have identically

$$(z_1-z_4)(z_2-z_3)+(z_2-z_4)(z_3-z_1)+(z_3-z_4)(z_1-z_2) = 0.$$

Hence

$$|(z_1-z_4)(z_2-z_3)| = |(z_2-z_4)(z_3-z_1)+(z_3-z_4)(z_1-z_2)|$$
$$\leqq |(z_2-z_4)(z_3-z_1)|+|(z_3-z_4)(z_1-z_2)|.]$$

6. Deduce Ptolemy's theorem concerning cyclic quadrilaterals from the fact that the cross ratios of four concyclic points are real. [Use the same identity as in the last example.]

7. If $z^2+z'^2 = 1$, then the points z, z' are ends of conjugate diameters of an ellipse whose foci are the points 1, -1. [If CP, CD are conjugate semi-diameters of an ellipse and S, H its foci, then CD is parallel to the external bisector of the angle SPH, and $SP.HP = CD^2$.]

8. Prove that $|a+b|^2+|a-b|^2 = 2\{|a|^2+|b|^2\}$. [This is the analytical equivalent of the geometrical theorem that, if M is the middle point of PQ, then $OP^2+OQ^2 = 2OM^2+2MP^2$.]

9. Deduce from Ex. 8 that

$$|a+\sqrt{(a^2-b^2)}|+|a-\sqrt{(a^2-b^2)}| = |a+b|+|a-b|.$$

[If $a+\sqrt{(a^2-b^2)} = z_1$, $a-\sqrt{(a^2-b^2)} = z_2$, we have

$$|z_1|^2+|z_2|^2 = \tfrac{1}{2}|z_1+z_2|^2+\tfrac{1}{2}|z_1-z_2|^2 = 2|a|^2+2|a^2-b^2|,$$

and so

$$(|z_1|+|z_2|)^2 = 2\{|a|^2+|a^2-b^2|+|b|^2\} = |a+b|^2+|a-b|^2+2|a^2-b^2|.$$

Another way of stating the result is: if z_1 and z_2 are the roots of

$$\alpha z^2+2\beta z+\gamma = 0,$$

then
$$|\alpha|(|z_1|+|z_2|) = |\beta+\sqrt{(\alpha\gamma)}|+|\beta-\sqrt{(\alpha\gamma)}|.]$$

10. Show that the necessary and sufficient conditions that both the roots of the equation $z^2+az+b=0$ should be of unit modulus are

$$|a|\leqq 2, \quad |b|=1, \quad \operatorname{am} b = 2\operatorname{am} a.$$

[The amplitudes have not necessarily their principal values.]

11. If $x^4+4a_1x^3+6a_2x^2+4a_3x+a_4=0$ is an equation with real coefficients and has two real and two complex roots, concyclic in the Argand diagram, then

$$a_3^2+a_1^2a_4+a_2^3-a_2a_4-2a_1a_2a_3=0.$$

12. The four roots of $a_0x^4+4a_1x^3+6a_2x^2+4a_3x+a_4=0$ will be harmonically related if

$$a_0a_3^2+a_1^2a_4+a_2^3-a_0a_2a_4-2a_1a_2a_3=0.$$

[Express $Z_{23,14}Z_{31,24}Z_{12,34}$, where

$$Z_{23,14}=(z_1-z_2)(z_3-z_4)+(z_1-z_3)(z_2-z_4)$$

and z_1, z_2, z_3, z_4 are the roots of the equation, in terms of the coefficients.]

13. **Imaginary points and straight lines.** Let $ax+by+c=0$ be an equation with complex coefficients. If we give x any particular real or complex value, we can find the corresponding value of y. The aggregate of pairs of real or complex values of x and y which satisfy the equation is called an *imaginary straight line*; the pairs of values are called *imaginary points*, and are said *to lie on the line*. The values of x and y are called the *coordinates* of the point (x, y). When x and y are real, the point is called a *real point*: when a, b, c are all real (or can be made all real by division by a common factor), the line is called a *real line*. The points $x=\alpha+\beta i$, $y=\gamma+\delta i$ and $x=\alpha-\beta i$, $y=\gamma-\delta i$ are said to be *conjugate*; and so are the lines

$$(A+A'i)x+(B+B'i)y+C+C'i=0,$$
$$(A-A'i)x+(B-B'i)y+C-C'i=0.$$

Verify the following assertions: every real line contains infinitely many pairs of conjugate imaginary points; an imaginary line in general contains one and only one real point; an imaginary line cannot contain a pair of conjugate imaginary points: and find the conditions (a) that the line joining two given imaginary points should be real, and (b) that the point of intersection of two imaginary lines should be real.

14. Prove the identities

$$(x+y+z)(x+y\omega_3+z\omega_3^2)(x+y\omega_3^2+z\omega_3)=x^3+y^3+z^3-3xyz,$$
$$(x+y+z)(x+y\omega_5+z\omega_5^4)(x+y\omega_5^2+z\omega_5^3)(x+y\omega_5^3+z\omega_5^2)(x+y\omega_5^4+z\omega_5)$$
$$=x^5+y^5+z^5-5x^3yz+5xy^2z^2.$$

15. Solve the equations

$$x^3-3ax+(a^3+1)=0, \quad x^5-5ax^3+5a^2x+(a^5+1)=0.$$

16. If $f(x) = a_0+a_1x+\ldots+a_kx^k$, then

$$\frac{1}{n}\{f(x)+f(\omega x)+\ldots+f(\omega^{n-1}x)\} = a_0+a_nx^n+a_{2n}x^{2n}+\ldots+a_{\lambda n}x^{\lambda n},$$

ω being any root of $x^n = 1$ (except $x=1$), and λn the greatest multiple of n contained in k. Find a similar formula for $a_\mu+a_{\mu+n}x^n+a_{\mu+2n}x^{2n}+\ldots$, where $0<\mu<n$.

17. If $(1+x)^n = p_0+p_1x+p_2x^2+\ldots$, n being a positive integer, then

$$p_0-p_2+p_4-\ldots = 2^{\frac{1}{2}n}\cos\tfrac{1}{4}n\pi, \quad p_1-p_3+p_5-\ldots = 2^{\frac{1}{2}n}\sin\tfrac{1}{4}n\pi.$$

18. Sum the series

$$\frac{x}{2!\,n-2!}+\frac{x^2}{5!\,n-5!}+\frac{x^3}{8!\,n-8!}+\ldots+\frac{x^{\frac{1}{3}n}}{n-1!},$$

n being a multiple of 3. (*Math. Trip.* 1899)

19. If t is a complex number such that $|t| = 1$, then the point $x = (at+b)/(t-c)$ describes a circle as t varies, unless $|c| = 1$, when it describes a straight line.

20. If t varies as in the last example then the point $x = \frac{1}{2}\{at+(b/t)\}$ in general describes an ellipse whose foci are given by $x^2 = ab$, and whose axes are $|a|+|b|$ and $|a|-|b|$. But if $|a| = |b|$ then x describes the finite straight line joining the points $-\sqrt{(ab)}$, $\sqrt{(ab)}$.

21. Prove that if t is real and $z = t^2-1+\sqrt{(t^4-t^2)}$, then, when $t^2<1$, z is represented by a point which lies on the circle $x^2+y^2+x = 0$. Assuming that, when $t^2>1$, $\sqrt{(t^4-t^2)}$ denotes the positive square root of t^4-t^2, discuss the motion of the point which represents z, as t diminishes from a large positive value to a large negative value. (*Math. Trip.* 1912)

22. The coefficients of the transformation $z = (aZ+b)/(cZ+d)$ are subject to the condition $ad-bc = 1$. Show that if $c\neq 0$ there are two *fixed points* α, β, i.e. points unaltered by the transformation, except when $(a+d)^2 = 4$, when there is only one fixed point α; and that in these two cases the transformation may be expressed in the forms

$$\frac{z-\alpha}{z-\beta} = K\frac{Z-\alpha}{Z-\beta}, \quad \frac{1}{z-\alpha} = \frac{1}{Z-\alpha}+K.$$

Show further that if $c = 0$ there will be one fixed point α unless $a = d$, and that in these two cases the transformation may be expressed in the forms

$$z-\alpha = K(Z-\alpha), \quad z = Z+K.$$

Finally, if a, b, c, d are further restricted to positive integral values (including zero), show that the only transformations with less than two fixed points are of the forms

$$\frac{1}{z} = \frac{1}{Z}+K, \quad z = Z+K. \qquad (\textit{Math. Trip.}\ 1911)$$

23. Prove that the relation $z = (1+Zi)/(Z+i)$ transforms the part of the axis of x between the points $z = 1$ and $z = -1$ into a semi-circle passing through the points $Z = 1$ and $Z = -1$. Find all the figures that can be obtained from the originally selected part of the axis of x by successive applications of the transformation. (*Math. Trip.* 1912)

24. Prove that the transformation

$$z = (\cos\theta + i\sin\theta)\frac{Z-a}{1-\bar{a}Z},$$

where a is any complex number whose modulus is not 1, $\bar{a}$ is the conjugate of a, and θ is real, transforms the inside of the unit circle in the z-plane into the inside or outside of the unit circle in the Z-plane; and distinguish the two cases. (*Math. Trip.* 1933)

25. If $z = 2Z+Z^2$ then the circle $|Z| = 1$ corresponds to a cardioid in the plane of z.

26. Discuss the transformation $z = \frac{1}{2}\{Z+(1/Z)\}$, showing in particular that to the circles $X^2+Y^2 = \alpha^2$ correspond the confocal ellipses

$$\frac{x^2}{\left\{\frac{1}{2}\left(\alpha+\frac{1}{\alpha}\right)\right\}^2}+\frac{y^2}{\left\{\frac{1}{2}\left(\alpha-\frac{1}{\alpha}\right)\right\}^2} = 1.$$

27. If $(z+1)^2 = 4/Z$ then the unit circle in the z-plane corresponds to the parabola $R\cos^2\frac{1}{2}\Theta = 1$ in the Z-plane, and the inside of the circle to the outside of the parabola.

28. Show that the transformation $z = (Z+a)^2/(Z-a)^2$, where a is real, transforms the upper half of the z-plane into the interior of a semicircle in the Z-plane. (*Math. Trip.* 1919)

29. If $z = Z^2-1$, then as z describes the circle $|z| = \kappa$, the two corresponding positions of Z each describe the Cassinian oval $\rho_1\rho_2 = \kappa$, where ρ_1, ρ_2 are the distances of Z from the points -1, 1. Trace the ovals for different values of κ.

30. Consider the relation $az^2+2hzZ+bZ^2+2gz+2fZ+c = 0$. Show that there are two values of Z for which the corresponding values of z are equal, and *vice versa*. We call these the *branch points* in the Z and z-planes respectively. Show that, if z describes an ellipse whose foci are the branch points, then so does Z.

[We can without loss of generality, take the given relation in the form

$$z^2+2zZ\cos\omega+Z^2 = 1:$$

the reader should satisfy himself that this is so. The branch points in either plane are then $\operatorname{cosec}\omega$ and $-\operatorname{cosec}\omega$. An ellipse of the form specified is given by

$$|z+\operatorname{cosec}\omega|+|z-\operatorname{cosec}\omega| = C,$$

where C is a constant. This is equivalent (Ex. 9) to

$$|z+\sqrt{(z^2-\operatorname{cosec}^2\omega)}|+|z-\sqrt{(z^2-\operatorname{cosec}^2\omega)}| = C.$$

Express this in terms of Z.]

31. If $z = aZ^m+bZ^n$, where m, n are positive integers and a, b real, then as Z describes the unit circle z describes a hypo- or epi-cycloid.

32. Show that the transformation

$$z = \frac{(a+di)\,\overline{Z}+b}{c\overline{Z}-(a-di)},$$

where a, b, c, d are real and $a^2+d^2+bc>0$, and $\overline{Z}$ denotes the conjugate of Z, is equivalent to an inversion with respect to the circle

$$c(x^2+y^2)-2ax-2dy-b = 0.$$

What is the geometrical interpretation of the transformation when

$$a^2+d^2+bc<0?$$

33. The transformation

$$\frac{1-z}{1+z} = \left(\frac{1-Z}{1+Z}\right)^c,$$

where c is rational and $0<c<1$, transforms the circle $|z| = 1$ into the boundary of a circular lune of angle π/c.

34. Prove that the transformation

$$\frac{z(z-\alpha)}{\alpha z-1} = Z,$$

where α is real and $0<\alpha<1$, transforms the inside of the unit circle in the z-plane into the inside, taken twice, of the unit circle in the Z-plane.

(*Math. Trip.* 1933)

CHAPTER IV

LIMITS OF FUNCTIONS OF A POSITIVE INTEGRAL VARIABLE

50. Functions of a positive integral variable. In Ch. II we discussed the notion of a function of a real variable x, and illustrated the discussion by a large number of examples of such functions. And the reader will remember that there was one important particular with regard to which the functions which we took as illustrations differed very widely. Some were defined for *all* values of x, some for *rational* values only, some for *integral* values only, and so on.

Consider, for example, the following functions: (i) x, (ii) $\sqrt{x}$, (iii) the denominator of x, (iv) the square root of the product of the numerator and the denominator of x, (v) the largest prime factor of x, (vi) the product of $\sqrt{x}$ and the largest prime factor of x, (vii) the xth prime number, (viii) the height measured in inches of convict x in Dartmoor prison.

Then the aggregates of values of x for which these functions are defined or, as we may say, the *fields of definition* of the functions, consist of (i) *all* values of x, (ii) *all positive* values of x, (iii) *all rational* values of x, (iv) *all positive rational* values of x, (v) *all integral* values of x, (vi), (vii) *all positive integral* values of x, (viii) a certain number of positive integral values of x, viz., 1, 2, ..., N, where N is the total number of convicts at Dartmoor at a given moment of time*.

Now let us consider a function, such as (vii) above, which is defined for all positive integral values of x and no others. This function may be regarded from two slightly different points of view. We may consider it, as has so far been our custom, as a function of the real variable x defined for some only of the values

* In the last case N depends on the time, and convict x, where x has a definite value, is a different individual at different moments of time. Thus if we take different moments of time into consideration we have a simple example of a function $y = F(x, t)$ of two variables, defined for a certain range of values of t, viz. from the time of the establishment of Dartmoor prison to the time of its abandonment, and for a certain number of positive integral values of x, this number varying with t.

of x, viz. positive integral values, and say that for all other values of x the definition fails. Or we may leave values of x other than positive integral values entirely out of account, and regard our function as a function of the *positive integral variable n*, whose values are the positive integers

$$1, 2, 3, 4, \ldots.$$

In this case we may write

$$y = \phi(n)$$

and regard y now as a function of n defined for all values of n.

It is obvious that any function of x defined for all values of x gives rise to a function of n defined for all values of n. Thus from the function $y = x^2$ we deduce the function $y = n^2$ by merely omitting from consideration all values of x other than positive integers, and the corresponding values of y. On the other hand from any function of n we can deduce any number of functions of x by merely assigning values to y, corresponding to values of x other than positive integral values, in any way we please.

51. Interpolation. The problem of determining a function of x which shall assume, for all positive integral values of x, values agreeing with those of a given function of n, is of great importance in higher mathematics. It is called the *problem of functional interpolation.*

Were the problem however merely that of finding *some* function of x to fulfil the condition stated, it would of course present no difficulty whatever. We could, as explained above, simply fill in the missing values as we pleased: we might indeed simply regard the given values of the function of n as *all* the values of the function of x and say that the definition of the latter function failed for all other values of x. But such solutions are obviously not what is usually wanted. What is usually wanted is some *formula* involving x (of as simple a kind as possible) which assumes the given values for $x = 1, 2, \ldots$.

In some cases, especially when the function of n is itself defined by a formula, there is an obvious solution. If for example $y = \phi(n)$, where $\phi(n)$ is a function of n, such as n^2 or $\cos n\pi$, which would have a meaning even were n not a positive integer, we naturally take our function of x to be $y = \phi(x)$. But even in this very simple case it is easy to write down other almost equally obvious solutions of the problem. For example

$$y = \phi(x) + \sin x\pi$$

assumes the value $\phi(n)$ for $x = n$, since $\sin n\pi = 0$.

In other cases $\phi(n)$ may be defined by a formula, such as $(-1)^n$, which ceases to define for some values of x (as here in the case of fractional values of x with even denominators, or irrational values). But it may be possible to transform the formula in such a way that it does define for all values of x. In this case, for example,

$$(-1)^n = \cos n\pi,$$

if n is an integer, and the problem of interpolation is solved by the function $\cos x\pi$.

In other cases $\phi(x)$ may be defined for some values of x other than positive integers, but not for all. Thus from $y = n^n$ we are led to $y = x^x$. This expression has a meaning for some only of the remaining values of x. If for simplicity we confine ourselves to positive values of x, then x^x has a meaning for all rational values of x, in virtue of the definitions of fractional powers adopted in elementary algebra. But when x is *irrational* x^x has (so far as we are in a position to say at the present moment) no meaning at all. We are thus led to consider the question of extending our definitions in such a way that x^x shall have a meaning even when x is irrational. We shall see later on how the extension desired may be effected.

Again, consider the case in which

$$y = 1.2 \ldots n = n!.$$

In this case there is no obvious formula in x which reduces to $n!$ for $x = n$, since $x!$ means nothing for values of x other than the positive integers. This is a case in which attempts to solve the problem of interpolation have led to important advances in mathematics. For mathematicians have succeeded in discovering a function (the gamma-function) which possesses the desired property and many other interesting and important properties besides.

52. Finite and infinite classes. Before we proceed further it is necessary to make a few remarks about certain ideas of an abstract nature which are of constant occurrence in pure mathematics.

In the first place, the reader is probably familiar with the notion of *a class*. It is unnecessary to discuss here any logical difficulties which may be involved in the notion of a class: roughly speaking we may say that a class is the aggregate or collection of all the entities or objects which possess a certain property, simple or complex. Thus we have the classes of British subjects, or members of Parliament, or positive integers, or real numbers.

Moreover, the reader has probably an idea of what is meant by a *finite* or *infinite* class. Thus the class of *British subjects* is a finite class: the aggregate of all British subjects, past, present, and future, has a finite number n, though of course we cannot tell at present the actual value of n. The class of *present British subjects*, on the other hand, has a number n which could be ascertained by counting, were the methods of the census effective enough.

On the other hand the class of positive integers is not finite but infinite. This may be expressed more precisely as follows. If n is any positive integer, such as 1000, 1,000,000 or any number we like to think of, then there are more than n positive integers. Thus, if the number we think of is 1,000,000, there are obviously at least 1,000,001 positive integers. Similarly the class of rational numbers, or of real numbers, is infinite. It is convenient to express this by saying that there are *an infinite number* of positive integers, or rational numbers, or real numbers. But the reader must be careful always to remember that by saying this we mean *simply* that the class in question has not a finite number of members such as 1000, or 1,000,000.

53. Properties possessed by a function of n for large values of n. We may now return to the 'functions of n' which we were discussing in §§ 50–51. They have many points of difference from the functions of x which we discussed in Ch. II. But there is one fundamental characteristic which the two classes of functions have in common: *the values of the variable for which they are defined form an infinite class*. It is this fact which forms the basis of all the considerations which follow and which, as we shall see in the next chapter, apply, *mutatis mutandis*, to functions of x as well.

Suppose that $\phi(n)$ is any function of n, and that P is any property which $\phi(n)$ may or may not have, such as that of being a positive integer or of being greater than 1. Consider, for each of the values $n = 1, 2, 3, \ldots$, whether $\phi(n)$ has the property P or not. Then there are three possibilities:

(*a*) $\phi(n)$ may have the property P for *all* values of n, or for all values of n except a finite number N of such values;

(*b*) $\phi(n)$ may have the property for *no* values of n, or only for a finite number N of such values;

(*c*) neither (*a*) nor (*b*) may be true.

If (*b*) is true, the values of n for which $\phi(n)$ has the property form a finite class. If (*a*) is true, the values of n for which $\phi(n)$ has not the property form a finite class. In the third case neither class is finite. Let us consider some particular cases.

(1) Let $\phi(n) = n$, and let P be the property of being a positive integer. Then $\phi(n)$ has the property P for all values of n.

If on the other hand P denotes the property of being a positive integer greater than or equal to 1000, then $\phi(n)$ has the property for all values of n except a finite number of values of n, viz. 1, 2, 3, ..., 999. In either of these cases (*a*) is true.

(2) If $\phi(n) = n$, and P is the property of being less than 1000, then (*b*) is true.

(3) If $\phi(n) = n$, and P is the property of being odd, then (*c*) is true. For $\phi(n)$ is odd if n is odd and even if n is even, and both the odd and the even values of n form an infinite class.

Example. Consider, in each of the following cases, whether (*a*), (*b*), or (*c*) is true:

(i) $\phi(n) = n$, P being the property of being a perfect square,

(ii) $\phi(n) = p_n$, where p_n denotes the nth prime number, P being the property of being odd,

(iii) $\phi(n) = p_n$, P being the property of being even,

(iv) $\phi(n) = p_n$, P being the property $\phi(n) > n$,

(v) $\phi(n) = 1-(-1)^n(1/n)$, P being the property $\phi(n) < 1$,

(vi) $\phi(n) = 1-(-1)^n(1/n)$, P being the property $\phi(n) < 2$,

(vii) $\phi(n) = 1000\{1+(-1)^n\}/n$, P being the property $\phi(n) < 1$,

(viii) $\phi(n) = 1/n$, P being the property $\phi(n) < \cdot 001$,

(ix) $\phi(n) = (-1)^n/n$, P being the property $|\phi(n)| < \cdot 001$,

(x) $\phi(n) = 10000/n$, or $(-1)^n 10000/n$, P being either of the properties $\phi(n) < \cdot 001$ or $|\phi(n)| < \cdot 001$,

(xi) $\phi(n) = (n-1)/(n+1)$, P being the property $1-\phi(n) < \cdot 0001$.

54. Let us now suppose that the assertion (a) is true for the $\phi(n)$ and P in question, i.e. that $\phi(n)$ has the property P, if not for all values of n, at any rate for all values of n except a finite number N of values. We may denote these exceptional values by

$$n_1,\ n_2,\ \ldots,\ n_N.$$

There is of course no reason why these N values should be the *first* N values $1, 2, \ldots, N$, though, as the preceding examples show, this is frequently the case in practice. But whether this is so or not, we know that $\phi(n)$ has the property P if $n > n_N$. Thus the nth prime is odd if $n > 2$, $n = 2$ being the only exception to the statement; and $1/n < \cdot 001$ if $n > 1000$, the first 1000 values of n being the exceptions; and

$$1000\{1+(-1)^n\}/n < 1$$

if $n > 2000$, the exceptional values being $2, 4, 6, \ldots, 2000$. That is to say, in each of these cases the property is possessed *for all values of n from a definite value onwards.*

We shall frequently express this by saying that $\phi(n)$ has the property for *large* or *very large* or *all sufficiently large* values of n. Thus when we say that $\phi(n)$ *has the property P* (which will as a rule be a property expressed by some relation of inequality) *for large values of n*, what we mean is that we can determine some definite number, n_0 say, such that $\phi(n)$ has the property for all values of n greater than or equal to n_0. This number n_0, in the examples considered above, may be taken to be any number greater than n_N, the greatest of the exceptional numbers: it is most natural to take it to be $n_N + 1$.

Thus we may say that 'all large primes are odd', or that '$1/n$ is less than $\cdot 001$ for large values of n'. And the reader must make himself familiar with the use of the word *large* in statements of this kind. *Large* is in fact a word which, standing by itself, has no more absolute meaning in mathematics than in the language of common life. It is a truism that in common life a number which is large in one connection is small in another; 6 goals is a large score in a football match, but 6 runs is not a large score in a cricket match; and 400 runs is a large score, but £400 is not a large income: and so of course in mathematics *large* generally means *large enough*, and what is large enough for one purpose may not be large enough for another.

We know now what is meant by the assertion '$\phi(n)$ has the property P for large values of n'. It is with assertions of this kind that we shall be concerned throughout this chapter.

55. The phrase 'n tends to infinity'. There is a somewhat different way of looking at the matter which it is natural to adopt. Suppose that n assumes successively the values 1, 2, 3, ... The word 'successively' naturally suggests succession in time, and we may suppose n, if we like, to assume these values at successive moments of time (e.g. at the beginnings of successive seconds). Then as the seconds pass n gets larger and larger and there is no limit to the extent of its increase. However large a number we may think of, a time will come when n has become larger than this number.

It is convenient to have a short phrase to express this unending growth of n, and we shall say that n *tends to infinity*, or $n \to \infty$, this last symbol being usually employed as an abbreviation for 'infinity'. The phrase 'tends to' like the word 'successively' naturally suggests the idea of change in time, and it is sometimes convenient to think of the variation of n as accomplished in time in the manner described above. This however is a mere matter of convenience, the variation of n having usually nothing to do with time.

The reader cannot impress upon himself too strongly that when we say that n 'tends to ∞' we mean simply that n is supposed to assume a series of values which increase beyond all limit. *There is no number 'infinity'*: such an equation as

$$n = \infty$$

is as it stands *meaningless*: a number n cannot be equal to ∞, because 'equal to ∞' means nothing. So far in fact the symbol ∞ means nothing at all except in the one phrase 'tends to ∞', the meaning of which we have explained above. Later on we shall learn how to attach a meaning to other phrases involving the symbol ∞, but the reader will always have to bear in mind

(1) that ∞ *by itself* means nothing, although *phrases containing it* sometimes mean something,

(2) that in every case in which a phrase containing the symbol ∞ means something it will do so simply because we have previously attached a meaning to this particular phrase by means of a special definition.

Now it is clear that if $\phi(n)$ has the property P for large values of n, and if n 'tends to ∞', in the sense which we have just explained, then n will ultimately assume values large enough to ensure that $\phi(n)$ has the property P. And so another way of putting the question 'what properties has $\phi(n)$ for sufficiently large values of n?' is 'how does $\phi(n)$ behave as n tends to ∞?'

56. The behaviour of a function of n as n tends to infinity. We shall now proceed, in the light of the remarks made in the preceding sections, to consider the meaning of some kinds of statements which recur continually in higher mathematics. Let us consider, for example, the two following statements: (*a*) $1/n$ *is small for large values of* n, (*b*) $1-(1/n)$ *is nearly equal to* 1 *for large values of* n. Obvious as they may seem, there is a good deal in them which will repay the reader's attention. Let us take (*a*) first, as being slightly the simpler.

We have already considered the statement '$1/n$ *is less than* $\cdot 001$ *for large values of* n'. This, we saw, means that the inequality $1/n < \cdot 001$ is true for all values of n greater than some definite value, in fact greater than 1000. Similarly it is true that '$1/n$ *is less than* $\cdot 0001$ *for large values of* n': in fact $1/n < \cdot 0001$ if $n > 10000$. And instead of $\cdot 001$ or $\cdot 0001$ we might take $\cdot 00001$ or $\cdot 000001$, or any positive number we like.

It is obviously convenient to have some way of expressing the fact that *any* such statement as '$1/n$ *is less than* $\cdot 001$ *for large values of* n' is true, when we substitute for $\cdot 001$ any smaller number, such as ·0001 or ·00001 or any other number we care to choose. And clearly we can do this by saying that '*however small* δ *may be* (provided of course it is positive), *then* $1/n < \delta$ *for sufficiently large values of* n'. That this is true is obvious. For $1/n < \delta$ if $n > 1/\delta$, so that our 'sufficiently large' values of n need only all be greater than $1/\delta$. The assertion is however a complex one, in that it really stands for the whole class of assertions which we obtain by giving to δ special values such as ·001. And of course the smaller δ is, and the larger $1/\delta$, the larger must be the least of the 'sufficiently large' values of n, values which are sufficiently large when δ has one value being inadequate when it has a smaller.

The last statement italicised is what is really meant by the statement (*a*), that $1/n$ is small when n is large. Similarly (*b*) really means '*if* $\phi(n) = 1-(1/n)$, *then the statement* "$1-\phi(n) < \delta$ *for sufficiently large values of* n" *is true whatever positive value* (*such as* ·001 *or* ·0001) *we attribute to* δ'. That the statement (*b*) is true is obvious from the fact that $1-\phi(n) = 1/n$.

There is another way in which it is common to state the facts expressed by the assertions (*a*) and (*b*). This is suggested at once by § 55. Instead of saying '$1/n$ is small for large values of n' we say '$1/n$ tends to 0 as (or when) n tends to ∞'. Similarly we say that '$1-(1/n)$ tends to 1 as n tends to ∞'; and these statements are to be regarded as strictly equivalent to (*a*) and (*b*). Thus the statements

'$1/n$ is small when n is large',

'$1/n$ tends to 0 as n tends to ∞',

are equivalent to one another and to the more formal statement

> 'if δ is any positive number, however small, then $1/n < \delta$ for sufficiently large values of n',

or to the still more formal statement

> 'if δ is any positive number, however small, then we can find a number n_0 such that $1/n < \delta$ for all values of n greater than or equal to n_0'.

The number n_0 which occurs in the last statement is of course a function of δ. We shall sometimes emphasise this fact by writing n_0 in the form $n_0(\delta)$.

The reader should imagine himself confronted by an opponent who questions the truth of the statement. He would name a series of numbers growing smaller and smaller. He might begin with $\cdot 001$. The reader would reply that $1/n < \cdot 001$ as soon as $n > 1000$. The opponent would be bound to admit this, but would try again with some smaller number, such as $\cdot 0000001$. The reader would reply that $1/n < \cdot 0000001$ as soon as $n > 10000000$: and so on. In this simple case it is evident that the reader would always have the better of the argument.

We shall now introduce yet another way of expressing this property of the function $1/n$. We shall say that '*the* **limit** *of* $1/n$ *as (or when) n tends to ∞ is* 0', a statement which we may express symbolically in the form

$$\lim_{n\to\infty} \frac{1}{n} = 0,$$

or simply $\lim (1/n) = 0$. We shall also sometimes write

$$\text{'}1/n \to 0 \text{ as } n \to \infty\text{'}$$

which may be read '$1/n$ tends to 0 as n tends to ∞'; or simply '$1/n \to 0$'. In the same way we shall write

$$\lim_{n\to\infty}\left(1-\frac{1}{n}\right) = 1, \quad \lim\left(1-\frac{1}{n}\right) = 1,$$

or $1-(1/n) \to 1$.

57. Now let us consider a different example: let $\phi(n) = n^2$. Then '*n^2 is large when n is large*'. This statement is equivalent to the more formal statements.

'if Δ is any positive number, however large, then $n^2 > \Delta$ for sufficiently large values of n',

'we can find a number $n_0(\Delta)$ such that $n^2 > \Delta$ for all values of n greater than or equal to $n_0(\Delta)$'.

And it is natural in this case to say that 'n^2 tends to ∞ as n tends to ∞', or 'n^2 tends to ∞ with n', and to write

$$n^2 \to \infty.$$

Finally consider the function $\phi(n) = -n^2$. In this case $\phi(n)$ is large, but negative, when n is large, and we naturally say that

'$-n^2$ tends to $-\infty$ as n tends to ∞' and write

$$-n^2 \to -\infty.$$

And the use of the symbol $-\infty$ in this sense suggests that it will sometimes be convenient to write $n^2 \to +\infty$ for $n^2 \to \infty$, and generally to use $+\infty$ instead of ∞, in order to secure greater uniformity of notation.

But we must once more repeat that in all these statements the symbols ∞, $+\infty$, $-\infty$ mean nothing whatever by themselves, and only acquire a meaning when they occur in certain special connections in virtue of the explanations which we have just given.

58. Definition of a limit. After the discussion which precedes the reader should be in a position to appreciate the general notion of a *limit*. Roughly we may say that *$\phi(n)$ tends to a limit l as n tends to ∞ if $\phi(n)$ is nearly equal to l when n is large.* But although the meaning of this statement should be clear enough after the preceding explanations, it is not, as it stands, precise enough to serve as a strict mathematical definition. It is, in fact, equivalent to a whole class of statements of the type '*for sufficiently large values of n, $\phi(n)$ differs from l by less than δ*'. This statement has to be true for $\delta = \cdot 01$ or $\cdot 0001$ or *any* positive number; and for any such value of δ it has to be true for *any* value of n after a certain definite value $n_0(\delta)$, though the smaller δ is the larger, as a rule, will be this value $n_0(\delta)$.

We accordingly frame the following formal definition:

DEFINITION I. *The function $\phi(n)$ is said to tend to the limit l as n tends to ∞, if, however small be the positive number δ, $\phi(n)$ differs from l by less than δ for sufficiently large values of n; that is to say if, however small be the positive number δ, we can determine a number $n_0(\delta)$ corresponding to δ, such that $\phi(n)$ differs from l by less than δ for all values of n greater than or equal to $n_0(\delta)$.*

It is usual to denote the difference between $\phi(n)$ and l, taken positively, by $|\phi(n)-l|$. It is equal to $\phi(n)-l$ or to $l-\phi(n)$, whichever is positive, and agrees with the definition of the *modulus* of $\phi(n)-l$, as given in Ch. III, though at present we are only considering real values, positive or negative.

With this notation the definition may be stated more shortly as follows: '*if, given any positive number, δ, however small, we can find $n_0(\delta)$ so that $|\phi(n)-l|<\delta$ when $n \geqq n_0(\delta)$, then we say that $\phi(n)$ tends to the limit l as n tends to ∞, and write*

$$\lim_{n\to\infty} \phi(n) = l\text{'}.$$

Sometimes we may omit the '$n\to\infty$'; and sometimes it is convenient, for brevity, to write $\phi(n)\to l$.

The reader will find it instructive to work out, in a few simple cases, the explicit expression of n_0 as a function of δ. Thus if $\phi(x) = 1/n$ then $l = 0$, and the condition reduces to $1/n<\delta$ for $n \geqq n_0$. which is satisfied if $n_0 = 1+[1/\delta]$*. There is one and only one case in which *the same* n_0 will do for *all* values of δ. If, from a certain value N of n onwards, $\phi(n)$ is constant, say equal to C, then it is evident that $\phi(n)-C = 0$ for $n \geqq N$, so that the inequality $|\phi(n)-C|<\delta$ is satisfied for $n \geqq N$ and all positive values of δ. And if $|\phi(n)-l|<\delta$ for $n \geqq N$ and all positive values of δ, then it is evident that $\phi(n) = l$ when $n \geqq N$, so that $\phi(n)$ is constant for all such values of n.

59. The definition of a limit may be illustrated geometrically as follows. The graph of $\phi(n)$ consists of a number of points corresponding to the values $n = 1, 2, 3, \ldots$.

Draw the line $y = l$, and the parallel lines $y = l-\delta$, $y = l+\delta$ at distance δ from it. Then

$$\lim_{n\to\infty} \phi(n) = l$$

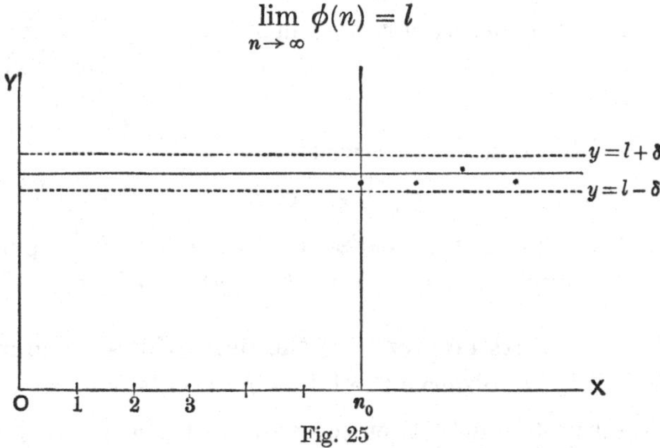

Fig. 25

* Here and henceforward we shall use $[x]$ in the sense of Ch. II, i.e. as the greatest integer not greater than x.

if, when once these lines have been drawn, no matter how close they may be together, we can always draw a line $x = n_0$, as in the figure, in such a way that the point of the graph on this line, and all points to the right of it, lie between them. We shall find this geometrical way of looking at our definition particularly useful when we come to deal with functions defined for all values of a real variable and not merely for positive integral values.

60. So much for functions of n which tend to a limit as n tends to ∞. We must now frame corresponding definitions for functions which, like the functions n^2 or $-n^2$, tend to positive or negative infinity. The reader should by now find no difficulty in appreciating the point of

DEFINITION II. *The function $\phi(n)$ is said to tend to $+\infty$ (positive infinity) with n if, when any number Δ, however large, is assigned, we can determine $n_0(\Delta)$ so that $\phi(n) > \Delta$ when $n \geqq n_0(\Delta)$; that is to say if, however large Δ may be, $\phi(n) > \Delta$ for sufficiently large values of n.*

Another, less precise, form of statement is '*if we can make $\phi(n)$ as large as we please by sufficiently increasing n*'. This is open to the objection that it obscures a fundamental point, viz. that $\phi(n)$ must be greater than Δ for *all* values of n such that $n \geqq n_0(\Delta)$, and not merely for *some* such values. But there is no harm in using this form of expression if we are clear what it means.

When $\phi(n)$ tends to $+\infty$ we write

$$\phi(n) \to +\infty.$$

We may leave it to the reader to frame the corresponding definition for functions which tend to negative infinity.

61. Some points concerning the definitions. The reader should be careful to observe the following points.

(1) We may obviously alter the values of $\phi(n)$ for any finite number of values of n, in any way we please, without in any way affecting the behaviour of $\phi(n)$ as n tends to ∞. For example

$1/n$ tends to 0 as n tends to ∞. We may deduce any number of new functions from $1/n$ by altering a finite number of its values. For instance we may consider the function $\phi(n)$ which is equal to 3 for $n = 1, 2, 7, 11, 101, 107, 109, 237$ and equal to $1/n$ for all other values of n. For this function, just as for the original function $1/n$, $\lim \phi(n) = 0$. Similarly, for the function $\phi(n)$ which is equal to 3 if $n = 1, 2, 7, 11, 101, 107, 109, 237$, and to n^2 otherwise, it is true that $\phi(n) \to +\infty$.

(2) On the other hand we cannot as a rule alter an *infinite* number of the values of $\phi(n)$ without affecting fundamentally its behaviour as n tends to ∞. If for example we altered the function $1/n$ by changing its value to 1 whenever n is a multiple of 100, it would no longer be true that $\lim \phi(n) = 0$. So long as a finite number of values only were affected, we could always choose the number n_0 of the definition so as to be greater than the greatest of the values of n for which $\phi(n)$ was altered. In the examples above, for instance, we could always take $n_0 > 237$, and indeed we should be compelled to do so as soon as our imaginary opponent of § 56 had assigned a value of δ as small as 3 (in the first example) or a value of Δ as great as 3 (in the second). But now *however* large n_0 may be there will be greater values of n for which $\phi(n)$ has been altered.

(3) In applying the test of Definition I it is essential that the inequality $|\phi(n) - l| < \delta$ should be satisfied not merely when $n = n_0$ but when $n \geqq n_0$, i.e. *for* n_0 *and for all larger values of* n. It is obvious, for example, that, if $\phi(n)$ is the function last considered, then given δ we can choose n_0 so that $|\phi(n)| < \delta$ when $n = n_0$: we have only to choose a sufficiently large value of n which is not a multiple of 100. But, when n_0 is thus chosen, it is not true that $|\phi(n)| < \delta$ when $n \geqq n_0$: all the multiples of 100 which are greater than n_0 are exceptions to this statement when $\delta \leqq 1$.

(4) If $\phi(n)$ is always greater than l, we can replace $|\phi(n) - l|$ by $\phi(n) - l$. Thus the test whether $1/n$ tends to the limit 0 as n tends to ∞ is simply whether $1/n < \delta$ when $n \geqq n_0$. If however $\phi(n) = (-1)^n/n$, then l is again 0, but $\phi(n) - l$ is sometimes

positive and sometimes negative. In such a case we must state the condition in the form $|\phi(n)-l|<\delta$, for example, in this particular case, in the form $|\phi(n)|<\delta$.

(5) *The limit l may itself be one of the actual values of $\phi(n)$.* Thus if $\phi(n) = 0$ for all values of n, it is obvious that $\lim \phi(n) = 0$. Again, if we had, in (2) and (3) above, altered the value of the function, when n is a multiple of 100, to 0 instead of to 1, we should have obtained a function $\phi(n)$ which is equal to 0 when n is a multiple of 100 and to $1/n$ otherwise. The limit of this function as n tends to ∞ is still zero. This limit is itself the value of the function for an infinite number of values of n, viz. all multiples of 100.

On the other hand *the limit itself need not (and in general will not) be the value of the function for any value of n.* This is sufficiently obvious in the case $\phi(n) = 1/n$. The limit is zero; but the function is never equal to zero for any value of n.

The reader cannot impress these facts too strongly on his mind. *A limit is not a value of the function*: it is something quite distinct from these values, though it is defined by its relations to them and may possibly be equal to some of them. For the functions

$$\phi(n) = 0,\ 1,$$

the limit is equal to *all* the values of $\phi(n)$: for

$$\phi(n) = 1/n, \quad (-1)^n/n, \quad 1+(1/n), \quad 1+\{(-1)^n/n\}$$

it is not equal to *any* value of $\phi(n)$: for

$$\phi(n) = (\sin \tfrac{1}{2}n\pi)/n, \quad 1+\{(\sin \tfrac{1}{2}n\pi)/n\}$$

(whose limits as n tends to ∞ are easily seen to be 0 and 1, since $\sin \frac{1}{2}n\pi$ is never numerically greater than 1) the limit is equal to the value which $\phi(n)$ assumes for all even values of n, but the values assumed for odd values of n are all different from the limit and from one another.

(6) A function may be always numerically very large when n is very large without tending either to $+\infty$ or to $-\infty$. A sufficient illustration of this is given by $\phi(n) = (-1)^n n$. A function can

only tend to $+\infty$ or to $-\infty$ if, after a certain value of n, it maintains a constant sign.

Examples XXIII. Consider the behaviour of the following functions of n as n tends to ∞:

1. $\phi(n) = n^k$, where k is a positive or negative integer or rational fraction. If k is positive, then n^k tends to $+\infty$ with n. If k is negative, then $\lim n^k = 0$. If $k = 0$, then $n^k = 1$ for all values of n. Hence $\lim n^k = 1$.

The reader will find it instructive, even in so simple a case as this, to write down a formal proof that the conditions of our definitions are satisfied. Take for instance the case of $k>0$. Let Δ be any assigned positive number. We wish to choose n_0 so that $n^k>\Delta$ when $n \geqq n_0$. We have in fact only to take for n_0 any number greater than $\sqrt[k]{\Delta}$. If e.g. $k = 4$, then $n^4 > 10000$ when $n \geqq 11$, $n^4 > 100000000$ when $n \geqq 101$, and so on.

2. $\phi(n) = p_n$, where p_n is the nth prime number. If there were only a finite number of primes then $\phi(n)$ would be defined only for a finite number of values of n. There are however, as was first shown by Euclid, infinitely many primes. Euclid's proof is as follows. Let $2, 3, 5, \dots, p_N$ be all the primes up to p , and let $P = (2.3.5\dots p_N)+1$. Then P is not divisible by any of $2, 3, 5, \dots, p_N$. Hence either P is prime, or P is divisible by a prime p between p_N and P. In either case there is a prime greater than p_N, and so an infinity of primes.

Since $\phi(n)>n$, $\phi(n)\to\infty$.

3. Let $\phi(n)$ be the number of primes less than n. Here again $\phi(n) \to +\infty$.

4. $\phi(n) = [\alpha n]$, where α is any positive number. Here

$$\phi(n) = 0 \quad (0 \leqq n < 1/\alpha), \qquad \phi(n) = 1 \quad (1/\alpha \leqq n < 2/\alpha),$$

and so on; and $\phi(n) \to +\infty$.

5. If $\phi(n) = 1000000/n$, then $\lim \phi(n) = 0$: and if $\psi(n) = n/1000000$, then $\psi(n) \to +\infty$. These conclusions are in no way affected by the fact that at first $\phi(n)$ is much larger than $\psi(n)$, being in fact larger until $n = 1000000$.

6. $\phi(n) = 1/\{n-(-1)^n\}$, $n-(-1)^n$, $n\{1-(-1)^n\}$. The first function tends to 0, the second to $+\infty$, the third does not tend either to a limit or to $+\infty$.

7. $\phi(n) = (\sin n\theta\pi)/n$, where θ is any real number. Here $|\phi(n)|<1/n$, since $|\sin n\theta\pi| \leqq 1$, and $\lim \phi(n) = 0$.

8. $\phi(n) = (\sin n\theta\pi)/\sqrt{n}$, $(a\cos^2 n\theta + b\sin^2 n\theta)/n$, where a and b are any real numbers.

9. $\phi(n) = \sin n\theta\pi$. If θ is integral then $\phi(n) = 0$ for all values of n, and therefore $\lim \phi(n) = 0$.

Next let θ be rational, e.g. $\theta = p/q$, where p and q are positive integers. Let $n = aq+b$, where a is the quotient and b the remainder when n is divided by q. Then $\sin(np\pi/q) = (-1)^{ap}\sin(bp\pi/q)$. Suppose, for example, p even; then, as n increases from 0 to $q-1$, $\phi(n)$ takes the values

$$0,\quad \sin\frac{p\pi}{q},\quad \sin\frac{2p\pi}{q},\quad \dots,\quad \sin\frac{(q-1)p\pi}{q}.$$

When n increases from q to $2q-1$ these values are repeated; and so also as n goes from $2q$ to $3q-1$, $3q$ to $4q-1$, and so on. Thus the values of $\phi(n)$ form a cyclic repetition of a finite series of different values. It is evident that when this is the case $\phi(n)$ cannot tend to a limit, nor to $+\infty$, nor to $-\infty$, as n tends to infinity.

The case in which θ is irrational is a little more difficult. It is discussed in the next set of examples.

62. Oscillating functions. DEFINITION. *When $\phi(n)$ does not tend to a limit, nor to $+\infty$, nor to $-\infty$, as n tends to ∞, we say that $\phi(n)$* **oscillates** *as n tends to ∞.*

A function $\phi(n)$ certainly oscillates if its values form, as in the case considered in the last example above, a continual repetition of a cycle of different values; but of course it may oscillate without possessing this peculiarity. Oscillation is defined in a negative manner: a function oscillates when it does not do certain other things.

The simplest example of an oscillatory function is given by

$$\phi(n) = (-1)^n,$$

which is equal to $+1$ when n is even and to -1 when n is odd. In this case the values recur cyclically. But consider

$$\phi(n) = (-1)^n + n^{-1},$$

the values of which are

$$-1+1,\quad 1+\tfrac{1}{2},\quad -1+\tfrac{1}{3},\quad 1+\tfrac{1}{4},\quad -1+\tfrac{1}{5},\quad \dots.$$

When n is large every value is nearly equal to $+1$ or -1, and obviously $\phi(n)$ does not tend to a limit or to $+\infty$ or to $-\infty$, and therefore it oscillates: but the values do not recur. It is to be observed that in this case every value of $\phi(n)$ is numerically less than or equal to $\frac{3}{2}$. Similarly

$$\phi(n) = (-1)^n\, 100 + 1000\, n^{-1}$$

oscillates. When n is large, every value is nearly equal to 100 or to -100. The numerically greatest value is 900 (for $n=1$). But now consider $\phi(n) = (-1)^n n$, the values of which are $-1, 2, -3, 4, -5, \ldots$. This function oscillates, for it does not tend to a limit, nor to $+\infty$, nor to $-\infty$; and in this case we cannot assign any limit beyond which the numerical value of the terms does not rise. The distinction between these two examples suggests a further definition.

DEFINITION. *If $\phi(n)$ oscillates as n tends to ∞, then $\phi(n)$ will be said to* **oscillate finitely** *or* **infinitely** *according as it is or is not possible to assign a number K such that all the values of $\phi(n)$ are numerically less than K, i.e. $|\phi(n)| < K$ for all values of n.*

These definitions, as well as those of §§ 58 and 60, are further illustrated in the following examples.

Examples XXIV. Consider the behaviour as n tends to ∞ of the following functions:

1. $(-1)^n$, $5+3(-1)^n$, $1000000n^{-1}+(-1)^n$, $1000000(-1)^n+n^{-1}$.

2. $(-1)^n n$, $1000000+(-1)^n n$.

3. $1000000-n$, $(-1)^n(1000000-n)$.

4. $n\{1+(-1)^n\}$. In this case the values of $\phi(n)$ are

$$0,\ 4,\ 0,\ 8,\ 0,\ 12,\ 0,\ 16,\ \ldots.$$

The odd terms are all zero and the even terms tend to $+\infty$: $\phi(n)$ oscillates infinitely.

5. $n^2+(-1)^n 2n$. The second term oscillates infinitely, but the first is very much larger than the second when n is large. In fact $\phi(n) \geqq n^2-2n$ and $n^2-2n = (n-1)^2-1$ is greater than any assigned value Δ if $n > 1+\sqrt{(\Delta+1)}$. Thus $\phi(n) \to +\infty$. It should be observed that in this case $\phi(2k+1)$ is always less than $\phi(2k)$, so that the function progresses to infinity by a continual series of steps forwards and backwards. It does not however 'oscillate' according to our definition of the term.

6. $n^2\{1+(-1)^n\}$, $(-1)^n n^2+n$, $n^3+(-1)^n n^2$.

7. $\sin n\theta\pi$. We have already seen (Exs. XXIII. 9) that $\phi(n)$ oscillates finitely when θ is rational, unless θ is an integer, when $\phi(n) = 0$, $\phi(n) \to 0$.

The case in which θ is irrational is a little more difficult, but we can still prove that $\phi(n)$ oscillates finitely. We may suppose without real loss of generality that $0 < \theta < 1$.

First, since $|\phi(n)| < 1$, $\phi(n)$ either oscillates finitely or tends to a limit. If $\sin n\theta\pi \to l$, then

$$2\cos(n+\tfrac{1}{2})\,\theta\pi\sin\tfrac{1}{2}\theta\pi = \sin(n+1)\,\theta\pi - \sin n\theta\pi \to 0,$$

and therefore $\cos(n+\frac{1}{2})\,\theta\pi \to 0$. Hence

$$(n+\tfrac{1}{2})\,\theta = k_n + \tfrac{1}{2} + \epsilon_n,$$

where k_n is an integer and $\epsilon_n \to 0$; and hence

$$\theta = k_n - k_{n-1} + \epsilon_n - \epsilon_{n-1} = l_n + \eta_n,$$

where l_n is an integer and $\eta_n \to 0$. This is plainly impossible, since θ is constant and lies between 0 and 1.

Prove similarly that $\cos n\theta\pi$ oscillates finitely unless θ is an even integer.

8. It is not possible, unless θ is an integer, that $\sin n\theta\pi$ or $\cos n\theta\pi$ should be nearly equal, for all large n, to one or other of two values a, b. [This may be proved by arguments similar to, but a little more complicated than, those of Ex. 7.]

9. $\sin n\theta\pi + n$, $\sin n\theta\pi + n^{-1}$, $(-1)^n \sin n\theta\pi$.

10. $a\cos n\theta\pi + b\sin n\theta\pi$, $\sin^2 n\theta\pi$, $a\cos^2 n\theta\pi + b\sin^2 n\theta\pi$.

11. $n\sin n\theta\pi$. If n is integral, then $\phi(n) = 0$, $\phi(n) \to 0$. If θ is rational but not integral, or irrational, then $\phi(n)$ oscillates infinitely.

12. $n(a\cos^2 n\theta\pi + b\sin^2 n\theta\pi)$. In this case $\phi(n)$ tends to $+\infty$ if a and b are both positive, but to $-\infty$ if both are negative. Consider the special cases in which $a = 0$, $b > 0$, or $a > 0$, $b = 0$, or $a = 0$, $b = 0$. If a and b have opposite signs, $\phi(n)$ generally oscillates infinitely. Consider any exceptional cases.

13. $\sin n!\,\theta\pi$. If θ has a rational value p/q, then $n!\,\theta$ is certainly integral for all values of n greater than or equal to q. Hence $\phi(n) \to 0$. The case in which θ is irrational cannot be dealt with without the aid of considerations of a much more difficult character.

14. $an - [bn]$, $(-1)^n(an - [bn])$.

15. *The smallest prime factor of n.* When n is a prime, $\phi(n) = n$. When n is even, $\phi(n) = 2$. Thus $\phi(n)$ oscillates infinitely.

16. *The largest prime factor of n.*

17. *The number of days in the year n A.D.*

Examples XXV. 1. If $\phi(n) \to +\infty$ and $\psi(n) \geqq \phi(n)$ for all values of n, then $\psi(n) \to +\infty$.

2. If $\phi(n) \to 0$, and $|\psi(n)| \leqq |\phi(n)|$ for all values of n, then $\psi(n) \to 0$.

3. If $\lim |\phi(n)| = 0$, then $\lim \phi(n) = 0$.

4. If $\phi(n)$ tends to a limit or oscillates finitely, and $|\psi(n)| \leqq |\phi(n)|$ when $n \geqq n_0$, then $\psi(n)$ tends to a limit or oscillates finitely.

5. If $\phi(n)$ tends to $+\infty$ or to $-\infty$ or oscillates infinitely, and

$$|\psi(n)| \geqq |\phi(n)|$$

when $n \geqq n_0$, then $\psi(n)$ tends to $+\infty$ or to $-\infty$ or oscillates infinitely.

6. 'If $\phi(n)$ oscillates and, however great be n_0, we can find values of n greater than n_0 for which $\psi(n) > \phi(n)$, and values of n greater than n_0 for which $\psi(n) < \phi(n)$, then $\psi(n)$ oscillates.' Is this true? If not give an example to the contrary.

7. If $\phi(n) \to l$ as $n \to \infty$, then also $\phi(n+p) \to l$, p being any fixed integer. [This follows at once from the definition. Similarly we see that, if $\phi(n)$ tends to $+\infty$ or to $-\infty$ or oscillates, so also does $\phi(n+p)$.]

8. The same conclusions hold (except in the case of oscillation) if p varies with n but is always numerically less than a fixed positive integer N; or if p varies with n in any way, so long as it is always positive.

9. Determine the least value of n_0 for which it is true that

(a) $n^2+2n > 999999 \quad (n \geqq n_0)$, (b) $n^2+2n > 1000000 \quad (n \geqq n_0)$.

10. Determine the least value of n_0 for which it is true that

(a) $n+(-1)^n > 1000 \quad (n \geqq n_0)$, (b) $n+(-1)^n > 1000000 \quad (n \geqq n_0)$.

11. Determine the least value of n_0 for which it is true that

(a) $n^2+2n > \Delta \quad (n \geqq n_0)$, (b) $n+(-1)^n > \Delta \quad (n \geqq n_0)$,

Δ being any positive number.

[(a) $n_0 = [\sqrt{(\Delta+1)}]$: (b) $n_0 = 1+[\Delta]$ or $2+[\Delta]$, according as $[\Delta]$ is odd or even, i.e. $n_0 = 1+[\Delta]+\frac{1}{2}\{1+(-1)^{[\Delta]}\}$.]

12. Determine the least value of n_0 such that

$$(a)\ \frac{n}{n^2+1} < \cdot 0001, \qquad (b)\ \frac{1}{n}+\frac{(-1)^n}{n^2} < \cdot 000001,$$

when $n \geqq n_0$. [Let us take the latter case. In the first place

$$\frac{1}{n}+\frac{(-1)^n}{n^2} \leqq \frac{n+1}{n^2},$$

and it is easy to see that the least value of n_0, such that $(n+1)/n^2 < \cdot 000001$ when $n \geqq n_0$, is 1000002. But the inequality given is satisfied by $n = 1000001$, and this is the value of n_0 required.]

63. Some general theorems with regard to limits. A. The behaviour of the sum of two functions whose behaviour is known.

THEOREM I. *If $\phi(n)$ and $\psi(n)$ tend to limits a, b, then $\phi(n)+\psi(n)$ tends to the limit $a+b$.*

This is almost obvious*. The argument which the reader will at once form in his mind is roughly this: 'when n is large, $\phi(n)$ is nearly equal to a and $\psi(n)$ to b, and therefore their sum is nearly equal to $a+b$'. It is however desirable to state the argument quite formally.

Let δ be any assigned positive number (e.g. ·001, ·0000001, ...). We require to show that a number n_0 can be found such that

$$|\phi(n)+\psi(n)-a-b|<\delta \quad \text{..............(1)}$$

when $n \geqq n_0$. Now by a proposition proved in Ch. III (more generally indeed than we need here) the modulus of the sum of two numbers is less than or equal to the sum of their moduli. Thus

$$|\phi(n)+\psi(n)-a-b| \leqq |\phi(n)-a|+|\psi(n)-b|.$$

It follows that the desired condition will certainly be satisfied if n_0 can be so chosen that

$$|\phi(n)-a|+|\psi(n)-b|<\delta \quad \text{..............(2)}$$

when $n \geqq n_0$.

Given any positive number δ', we can find n_1 so that $|\phi(n)-a|<\delta'$ for $n \geqq n_1$. We take $\delta' = \frac{1}{2}\delta$, so that $|\phi(n)-a|<\frac{1}{2}\delta$ when $n \geqq n_1$. Similarly we can find n_2 so that $|\psi(n)-b|<\frac{1}{2}\delta$ when $n \geqq n_2$. Now take n_0 to be *the greater of the two numbers* n_1, n_2. Then $|\phi(n)-a|<\frac{1}{2}\delta$ and $|\psi(n)-b|<\frac{1}{2}\delta$ when $n \geqq n_0$, and therefore (2) is satisfied and the theorem is proved.

* There is a certain ambiguity in this phrase which the reader will do well to notice. When one says 'such and such a theorem is almost obvious' one may mean one or other of two things. One may mean 'it is difficult to doubt the truth of the theorem', 'the theorem is such as common sense instinctively accepts', as it accepts, for example, the truth of the propositions '$2+2=4$' or 'the base-angles of an isosceles triangles are equal'. That a theorem is 'obvious' in this sense does not prove that it is true, since the most confident of the intuitive judgments of common sense are often found to be mistaken; and even if the theorem is true, the fact that it is also 'obvious' is no reason for not proving it, if a proof can be found. The object of mathematics is to prove that certain premises imply certain conclusions; and the fact that the conclusions may be as 'obvious' as the premises never detracts from the necessity, and often not even from the interest of the proof.

But sometimes (as for example here) we mean by 'this is almost obvious' something quite different from this. We mean 'a moment's reflection should not only convince the reader of the truth of what is stated, but should also suggest to him the general lines of a rigorous proof'. And often, when a statement is 'obvious' in this sense, one may well omit the proof, not because the proof is unnecessary, but because it is a waste of time to state in detail what the reader can easily supply for himself.

The substance of these remarks was suggested to me many years ago by Prof. Littlewood.

The argument may be stated concisely thus: since $\lim \phi(n) = a$ and $\lim \psi(n) = b$, we can choose n_1, n_2 so that

$$|\phi(n) - a| < \tfrac{1}{2}\delta \quad (n \geqq n_1), \qquad |\psi(n) - b| < \tfrac{1}{2}\delta \quad (n \geqq n_2);$$

and then, if n is not less than either n_1 or n_2,

$$|\phi(n) + \psi(n) - a - b| \leqq |\phi(n) - a| + |\psi(n) - b| < \delta;$$

and therefore $$\lim \{\phi(n) + \psi(n)\} = a + b.$$

64. Results subsidiary to Theorem I. The reader should have no difficulty in verifying the following subsidiary results.

1. *If $\phi(n)$ tends to a limit, but $\psi(n)$ tends to $+\infty$ or to $-\infty$ or oscillates finitely or infinitely, then $\phi(n) + \psi(n)$ behaves like $\psi(n)$.*

2. *If $\phi(n) \to +\infty$, and $\psi(n) \to +\infty$ or oscillates finitely, then $\phi(n) + \psi(n) \to +\infty$.*

In this statement we may obviously change $+\infty$ into $-\infty$ throughout.

3. *If $\phi(n) \to +\infty$ and $\psi(n) \to -\infty$, then $\phi(n) + \psi(n)$ may tend either to a limit or to $+\infty$ or to $-\infty$ or may oscillate either finitely or infinitely.*

These five possibilities are illustrated in order by (i) $\phi(n) = n$, $\psi(n) = -n$, (ii) $\phi(n) = n^2$, $\psi(n) = -n$, (iii) $\phi(n) = n$, $\psi(n) = -n^2$, (iv) $\phi(n) = n + (-1)^n$, $\psi(n) = -n$, (v) $\phi(n) = n^2 + (-1)^n n$, $\psi(n) = -n^2$.

4. *If $\phi(n) \to +\infty$ and $\psi(n)$ oscillates infinitely, then $\phi(n) + \psi(n)$ may tend to $+\infty$ or oscillate infinitely, but cannot tend to a limit, or to $-\infty$, or oscillate finitely.*

For $\psi(n) = \{\phi(n) + \psi(n)\} - \phi(n)$; and, if $\phi(n) + \psi(n)$ behaved in any of the three last ways, it would follow, from the previous results, that $\psi(n) \to -\infty$, which is not the case. As examples of the two cases which are possible, consider (i) $\phi(n) = n^2$, $\psi(n) = (-1)^n n$, (ii) $\phi(n) = n$, $\psi(n) = (-1)^n n^2$. Here again the signs of $+\infty$ and $-\infty$ may be permuted throughout.

5. *If $\phi(n)$ and $\psi(n)$ both oscillate finitely, then $\phi(n) + \psi(n)$ must tend to a limit or oscillate finitely.*

As examples take

(i) $\phi(n) = (-1)^n$, $\psi(n) = (-1)^{n+1}$, (ii) $\phi(n) = \psi(n) = (-1)^n$.

6. *If $\phi(n)$ oscillates finitely, and $\psi(n)$ infinitely, then $\phi(n) + \psi(n)$ oscillates infinitely.*

For $\phi(n)$ is always less in absolute value than a certain constant, say K. On the other hand $\psi(n)$, since it oscillates infinitely, must assume values numerically greater than any assignable number (e.g. $10K$, $100K$, ...). Hence $\phi(n)+\psi(n)$ must assume values numerically greater than any assignable number (e.g. $9K$, $99K$, ...). Hence $\phi(n)+\psi(n)$ must either tend to $+\infty$ or $-\infty$ or oscillate infinitely. But if it tended to $+\infty$, then

$$\psi(n) = \{\phi(n)+\psi(n)\}-\phi(n)$$

would also tend to $+\infty$, in virtue of the preceding results. Thus $\phi(n)+\psi(n)$ cannot tend to $+\infty$, nor, for similar reasons, to $-\infty$: hence it oscillates infinitely.

7. *If both $\phi(n)$ and $\psi(n)$ oscillate infinitely, then $\phi(n)+\psi(n)$ may tend to a limit, or to $+\infty$, or to $-\infty$, or oscillate either finitely or infinitely.*

Suppose, for instance, that $\phi(n) = (-1)^n n$, while $\psi(n)$ is in turn each of the functions $(-1)^{n+1} n$, $\{1+(-1)^{n+1}\} n$, $-\{1+(-1)^n\} n$, $(-1)^{n+1}(n+1)$, $(-1)^n n$. We thus obtain examples of all five possibilities.

The results 1–7 cover all the cases which are really distinct. Before passing on to consider the product of two functions, we may point out that the result of Theorem I may be immediately extended to the sum of three or more functions which tend to limits as $n \to \infty$.

65. B. The behaviour of the product of two functions whose behaviour is known. We can now prove a similar set of theorems concerning the product of two functions. The principal result is the following.

THEOREM II. *If* $\lim \phi(n) = a$ *and* $\lim \psi(n) = b$, *then*

$$\lim \phi(n)\psi(n) = ab.$$

Let $$\phi(n) = a+\phi_1(n), \quad \psi(n) = b+\psi_1(n),$$

so that $\lim \phi_1(n) = 0$ and $\lim \psi_1(n) = 0$. Then

$$\phi(n)\psi(n) = ab + a\psi_1(n) + b\phi_1(n) + \phi_1(n)\psi_1(n).$$

Hence the numerical value of the difference $\phi(n)\psi(n) - ab$ is certainly not greater than the sum of the numerical values of $a\psi_1(n)$, $b\phi_1(n)$, $\phi_1(n)\psi_1(n)$. From this it follows that

$$\lim \{\phi(n)\psi(n) - ab\} = 0,$$

which proves the theorem.

The following is a strictly formal proof. We have

$$|\phi(n)\psi(n)-ab| \leqq |a\psi_1(n)|+|b\phi_1(n)|+|\phi_1(n)||\psi_1(n)|.$$

Assuming that neither a nor b is zero, we may suppose $\delta<3|a||b|$, and choose n_0 so that

$$|\phi_1(n)|<\tfrac{1}{3}\delta/|b|, \quad |\psi_1(n)|<\tfrac{1}{3}\delta/|a|,$$

when $n \geqq n_0$. Then

$$|\phi(n)\psi(n)-ab|<\tfrac{1}{3}\delta+\tfrac{1}{3}\delta+\{\tfrac{1}{9}\delta^2/(|a||b|)\}<\delta.$$

Thus we can choose n_0 so that $|\phi(n)\psi(n)-ab|<\delta$ when $n \geqq n_0$, and the theorem follows. The reader should supply a proof for the case in which at least one of a and b is zero.

We need hardly point out that this theorem, like Theorem I, may be immediately extended to the product of any number of functions of n. There is also a series of subsidiary theorems concerning products analogous to those stated in § 64 for sums. We must distinguish now *six* different ways in which $\phi(n)$ may behave as n tends to ∞. It may (1) tend to a limit *other than zero*, (2) tend to zero, (3a) tend to $+\infty$, (3b) tend to $-\infty$, (4) oscillate finitely, (5) oscillate infinitely. It is not necessary, as a rule, to take account separately of (3a) and (3b), as the results for one case may be deduced from those for the other by a change of sign.

To state these subsidiary theorems at length would occupy more space than we can afford. We select the two which follow as examples, leaving the verification of them to the reader. He will find it an instructive exercise to formulate some of the remaining theorems himself.

(i) *If* $\phi(n)\to+\infty$ and $\psi(n)$ *oscillates finitely, then* $\phi(n)\psi(n)$ *must tend to* $+\infty$ *or to* $-\infty$ *or oscillate infinitely.*

Examples of these three possibilities may be obtained by taking $\phi(n)$ to be n and $\psi(n)$ to be one of the three functions $2+(-1)^n$, $-2-(-1)^n$, $(-1)^n$.

(ii) *If* $\phi(n)$ *and* $\psi(n)$ *oscillate finitely, then* $\phi(n)\psi(n)$ *must tend to a limit (which may be zero) or oscillate finitely.*

For examples, take

(a) $\phi(n)=\psi(n)=(-1)^n$, (b) $\phi(n)=1+(-1)^n$, $\psi(n)=1-(-1)^n$,

and

(c) $\phi(n)=\cos\tfrac{1}{2}n\pi$, $\psi(n)=\sin\tfrac{1}{2}n\pi$.

A particular case of Theorem II which is important is that in which $\psi(n)$ is constant. The theorem then asserts simply that

$\lim k\phi(n) = ka$ if $\lim \phi(n) = a$. To this we may join the subsidiary theorem that if $\phi(n) \to +\infty$ then $k\phi(n) \to +\infty$ or $k\phi(n) \to -\infty$, according as k is positive or negative, unless $k = 0$, when of course $k\phi(n) = 0$ for all values of n and $\lim k\phi(n) = 0$. And if $\phi(n)$ oscillates finitely or infinitely, then so does $k\phi(n)$, unless $k = 0$.

66. C. The behaviour of the difference or quotient of two functions whose behaviour is known. There is, of course, a similar set of theorems for the difference of two given functions, which are obvious corollaries from what precedes. In order to deal with the quotient

$$\frac{\phi(n)}{\psi(n)},$$

we begin with the following theorem.

THEOREM III. *If* $\lim \phi(n) = a$, *and* a *is not zero, then*

$$\lim \frac{1}{\phi(n)} = \frac{1}{a}.$$

Let $$\phi(n) = a + \phi_1(n),$$

so that $\lim \phi_1(n) = 0$. Then

$$\left|\frac{1}{\phi(n)} - \frac{1}{a}\right| = \frac{|\phi_1(n)|}{|a|\,|a+\phi_1(n)|},$$

and it is plain, since $\lim \phi_1(n) = 0$, that we can choose n_0 so that this is smaller than any assigned number δ when $n \geqq n_0$.

From Theorems II and III we can at once deduce the principal theorem for quotients, viz.

THEOREM IV. *If* $\lim \phi(n) = a$ *and* $\lim \psi(n) = b$, *and* b *is not zero, then*

$$\lim \frac{\phi(n)}{\psi(n)} = \frac{a}{b}.$$

The reader will again find it instructive to formulate, prove, and illustrate by examples some of the 'subsidiary theorems' corresponding to Theorems III and IV.

67. THEOREM V. *If $R\{\phi(n), \psi(n), \chi(n), \dots\}$ is any rational function of $\phi(n)$, $\psi(n)$, $\chi(n)$, ..., i.e. any function of the form*

$$P\{\phi(n), \psi(n), \chi(n), \dots\}/Q\{\phi(n), \psi(n), \chi(n), \dots\},$$

where P and Q denote polynomials in $\phi(n)$, $\psi(n)$, $\chi(n)$, ...: and if

$$\lim \phi(n) = a, \quad \lim \psi(n) = b, \quad \lim \chi(n) = c, \quad \dots,$$

and

$$Q(a, b, c, \dots) \neq 0;$$

then

$$\lim R\{\phi(n), \psi(n), \chi(n), \dots\} = R(a, b, c, \dots).$$

For P is a sum of a finite number of terms of the type

$$A\{\phi(n)\}^p \{\psi(n)\}^q \dots,$$

where A is a constant and p, q, ... positive integers. This term, by Theorem II (or rather by its obvious extension to the product of any number of functions), tends to the limit $Aa^p b^q \dots$, and so P tends to the limit $P(a, b, c, \dots)$, by the similar extension of Theorem I. Similarly Q tends to $Q(a, b, c, \dots)$; and the result then follows from Theorem IV.

68. The preceding general theorem may be applied to the following very important particular problem: *what is the behaviour of the most general rational function of n, viz.*

$$S(n) = \frac{a_0 n^p + a_1 n^{p-1} + \dots + a_p}{b_0 n^q + b_1 n^{q-1} + \dots + b_q},$$

as n tends to ∞?*

In order to apply the theorem we transform $S(n)$ by writing it in the form

$$n^{p-q}\left\{\left(a_0 + \frac{a_1}{n} + \dots + \frac{a_p}{n^p}\right) \bigg/ \left(b_0 + \frac{b_1}{n} + \dots + \frac{b_q}{n^q}\right)\right\}.$$

The function in curly brackets is of the form $R\{\phi(n)\}$, where $\phi(n) = 1/n$, and therefore tends, as n tends to ∞, to the limit $R(0) = a_0/b_0$. Now $n^{p-q} \to 0$ if $p < q$; $n^{p-q} = 1$ and $n^{p-q} \to 1$ if $p = q$; and $n^{p-q} \to +\infty$ if $p > q$. Hence, by Theorem II,

$$\lim S(n) = 0 \quad (p < q),$$

$$\lim S(n) = a_0/b_0 \quad (p = q),$$

$$S(n) \to +\infty \quad (p > q,\ a_0/b_0 \textit{ positive}),$$

$$S(n) \to -\infty \quad (p > q,\ a_0/b_0 \textit{ negative}).$$

* We naturally suppose that neither a_0 nor b_0 is zero.

Examples XXVI. 1. What is the behaviour of the functions

$$\left(\frac{n-1}{n+1}\right)^2, \quad (-1)^n\left(\frac{n-1}{n+1}\right)^2, \quad \frac{n^2+1}{n}, \quad (-1)^n\frac{n^2+1}{n},$$

as $n \to \infty$?

2. Which (if any) of the functions

$$1/(\cos^2 \tfrac{1}{2}n\pi + n\sin^2 \tfrac{1}{2}n\pi), \quad 1/\{n(\cos^2 \tfrac{1}{2}n\pi + n\sin^2 \tfrac{1}{2}n\pi)\},$$
$$(n\cos^2 \tfrac{1}{2}n\pi + \sin^2 \tfrac{1}{2}n\pi)/\{n(\cos^2 \tfrac{1}{2}n\pi + n\sin^2 \tfrac{1}{2}n\pi)\}$$

tend to a limit as $n \to \infty$?

3. Denoting by $S(n)$ the general rational function of n considered above, show that in all cases

$$\lim \frac{S(n+1)}{S(n)} = 1, \quad \lim \frac{S\{n+(1/n)\}}{S(n)} = 1.$$

69. Functions of *n* which increase steadily with *n*. A special but particularly important class of functions of n is formed by those whose variation as n tends to ∞ is always in the same direction, that is to say those which always increase (or always decrease) as n increases. Since $-\phi(n)$ always increases if $\phi(n)$ always decreases, it is not necessary to consider the two kinds of functions separately; for theorems proved for one kind can at once be extended to the other.

Definition. *The function $\phi(n)$ will be said to increase steadily with n, or to be an increasing function of n, if $\phi(n+1) \geqq \phi(n)$ for all values of n.*

It is to be observed that we do not exclude the case in which $\phi(n)$ has the *same* value for several values of n; all we exclude is possible *decrease*. Thus the function

$$\phi(n) = 2n + (-1)^n,$$

whose values for $n = 0, 1, 2, 3, 4, \ldots$ are

$$1, 1, 5, 5, 9, 9, \ldots,$$

is said to increase steadily with n. Our definition includes even functions which remain constant from some value of n onwards; thus $\phi(n) = 1$ increases steadily.

If $\phi(n+1) > \phi(n)$ for all n, we say that $\phi(n)$ is a *strictly increasing* function of n.

There is one very important theorem concerning functions of this class.

THEOREM. *If $\phi(n)$ increases steadily with n, then either* (i) $\phi(n)$ *tends to a limit as n tends to* ∞, *or* (ii) $\phi(n) \to +\infty$.

That is to say, while there are in general *five* alternatives as to the behaviour of a function, there are *two* only for this special kind of function.

This theorem is a simple corollary of Dedekind's theorem (§ 17). We divide the real numbers ξ into two classes L and R, putting ξ in L or R according as $\phi(n) \geqq \xi$ for some value of n (and so of course for all greater values), or $\phi(n) < \xi$ for all values of n.

The class L certainly exists; the class R may or may not. If it does not then, given any number Δ, however large, $\phi(n) > \Delta$ for all sufficiently large values of n, and so

$$\phi(n) \to +\infty.$$

If on the other hand R exists, the classes L and R form a section of the real numbers in the sense of § 17. Let a be the number corresponding to the section, and let δ be any positive number. Then $\phi(n) < a + \delta$ for all values of n, and so, since δ is arbitrary, $\phi(n) \leqq a$. On the other hand $\phi(n) > a - \delta$ for some value of n, and so for all sufficiently large values. Thus

$$a - \delta < \phi(n) \leqq a$$

for all sufficiently large values of n; i.e.

$$\phi(n) \to a.$$

It should be observed that in general $\phi(n) < a$ for all values of n; for if $\phi(n)$ is equal to a for any value of n it must be equal to a for all greater values of n. Thus $\phi(n)$ can never be equal to a except in the case in which the values of $\phi(n)$ are ultimately all the same. If this is so, a is the largest member of L; otherwise L has no largest member.

COR. 1. *If $\phi(n)$ increases steadily with n, then it will tend to a limit or to $+\infty$ according as it is or is not possible to find a number K such that $\phi(n) < K$ for all values of n.*

We shall find this corollary very useful later on.

COR. 2. *If $\phi(n)$ increases steadily with n, and $\phi(n) < K$ for all values of n, then $\phi(n)$ tends to a limit and this limit is less than or equal to K.*

It should be noticed that the limit may be equal to K: if e.g. $\phi(n) = 3-(1/n)$, then every value of $\phi(n)$ is less than 3, but the limit is equal to 3.

COR. 3. *If $\phi(n)$ increases steadily with n, and tends to a limit, then*

$$\phi(n) \leqq \lim \phi(n)$$

for all values of n.

The reader should write out for himself the corresponding theorems and corollaries for the case in which $\phi(n)$ *decreases* as n increases.

70. The great importance of these theorems lies in the fact that they give us (what we have so far been without) a means of deciding, in a great many cases, whether a given function of n does or does not tend to a limit as $n \to \infty$, *without requiring us to be able to guess or otherwise infer beforehand what the limit is.* If we know what the limit, if there is one, must be, we can use the test

$$|\phi(n) - l| < \delta \quad (n \geqq n_0):$$

as for example in the case of $\phi(n) = 1/n$, where it is obvious that the limit can only be zero. But suppose we have to determine whether

$$\phi(n) = \left(1+\frac{1}{n}\right)^n$$

tends to a limit. In this case it is not obvious what the limit, if there is one, will be: and it is evident that the test above, which involves l, cannot be used, at any rate directly, to decide whether l exists or not.

Of course the test can sometimes be used indirectly, to prove by means of a *reductio ad absurdum* that l *cannot* exist. If e.g. $\phi(n) = (-1)^n$, it is clear that l would have to be equal to 1 and also equal to -1, which is obviously impossible.

71. Alternative proof of Weierstrass's theorem of § 19. The results of § 69 enable us to give an alternative proof of the important theorem proved in § 19.

If we divide PQ into two equal parts, one at least of them must contain infinitely many points of S. We select the one which does, or, if both do, we select the left-hand half; and we denote the selected half by $P_1 Q_1$ (Fig. 26). If $P_1 Q_1$ is the left-hand half, P_1 is the same point as P.

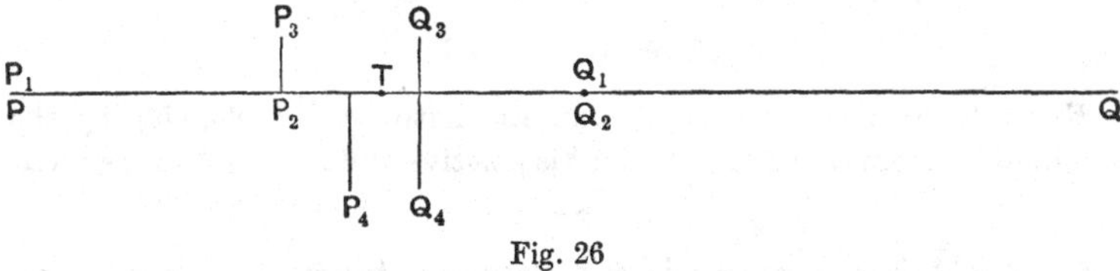

Fig. 26

Similarly, if we divide $P_1 Q_1$ into two halves, one at least of them must contain infinitely many points of S. We select the half $P_2 Q_2$ which does so, or, if both do so, we select the left-hand half. Proceeding in this way we can define a sequence of intervals

$$PQ,\ P_1 Q_1,\ P_2 Q_2,\ P_3 Q_3,\ \ldots,$$

each of which is a half of its predecessor, and each of which contains infinitely many points of S.

The points P, P_1, P_2, ... progress steadily from left to right, and so P_n tends to a limiting position T. Similarly Q_n tends to a limiting position T'. But TT' is plainly less than $P_n Q_n$, whatever the value of n; and $P_n Q_n$, being equal to $PQ/2^n$, tends to zero. Hence T' coincides with T, and P_n and Q_n both tend to T.

Then T is a point of accumulation of S. For suppose that ξ is its coordinate, and consider any interval of the type $(\xi-\delta,\ \xi+\delta)$. If n is sufficiently large, $P_n Q_n$ will lie entirely inside this interval*. Hence $(\xi-\delta,\ \xi+\delta)$ contains infinitely many points of S.

72. The limit of x^n as n tends to ∞. Let us apply the results of §69 to the particularly important case in which $\phi(n) = x^n$. If $x = 1$ then $\phi(n) = 1$, $\lim \phi(n) = 1$, and if $x = 0$ then $\phi(n) = 0$, $\lim \phi(n) = 0$, so that these special cases need not detain us.

First, suppose x positive. Then, since $\phi(n+1) = x\phi(n)$, $\phi(n)$ increases with n if $x > 1$, decreases as n increases if $x < 1$.

If $x > 1$, then x^n must tend either to a limit (which must obviously be greater than 1) or to $+\infty$. Suppose it tends to a limit l.

* This will certainly be the case as soon as $PQ/2^n < \delta$.

Then $\lim \phi(n+1) = \lim \phi(n) = l$, by Exs. xxv. 7; but

$$\lim \phi(n+1) = \lim x\phi(n) = x \lim \phi(n) = xl,$$

and therefore $l = xl$; and this is impossible, since x and l are both greater than 1. Hence

$$x^n \to +\infty \quad (x > 1).$$

Example. The reader may give an alternative proof, showing by the binomial theorem that $x^n > 1 + n\delta$ if δ is positive and $x = 1 + \delta$, and so that

$$x^n \to +\infty.$$

On the other hand x^n is a decreasing function if $x < 1$, and must therefore tend to a limit or to $-\infty$. Since x^n is positive the second alternative may be ignored. Thus $\lim x^n = l$, say, and as above $l = xl$, so that l must be zero. Hence

$$\lim x^n = 0 \quad (0 < x < 1).$$

Example. Prove as in the preceding example that $(1/x)^n$ tends to $+\infty$ if $0 < x < 1$, and deduce that x^n tends to 0.

We have finally to consider the case in which x is negative. If $-1 < x < 0$ and $x = -y$, so that $0 < y < 1$, then it follows from what precedes that $\lim y^n = 0$ and therefore $\lim x^n = 0$. If $x = -1$ it is obvious that x^n oscillates, taking the values -1, 1 alternatively. Finally if $x < -1$, and $x = -y$, so that $y > 1$, then y^n tends to $+\infty$, and therefore x^n takes values, both positive and negative, numerically greater than any assigned number. Hence x^n oscillates infinitely. To sum up:

$$\phi(n) = x^n \to +\infty \quad (x > 1),$$
$$\lim \phi(n) = 1 \quad (x = 1),$$
$$\lim \phi(n) = 0 \quad (-1 < x < 1),$$
$$\phi(n) \textit{ oscillates finitely} \quad (x = -1),$$
$$\phi(n) \textit{ oscillates infinitely} \quad (x < -1).$$

Examples XXVII*. 1. If $\phi(n)$ is positive and $\phi(n+1) \geqq K\phi(n)$, where $K > 1$, for all values of n, then $\phi(n) \to +\infty$.

[For $\qquad \phi(n) \geqq K\phi(n-1) \geqq K^2\phi(n-2) \geqq \dots \geqq K^{n-1}\phi(1),$
from which the conclusion follows at once, since $K^n \to \infty$.]

* These examples are particularly important and several of them will be made use of later in the text. They should therefore be studied very carefully.

2. The same result is true if the conditions are satisfied only when $n \geqq n_0$.

3. If $\phi(n)$ is positive and $\phi(n+1) < K\phi(n)$, where $0 < K < 1$, then $\lim \phi(n) = 0$. This result also is true if the conditions are satisfied only when $n \geqq n_0$.

4. If $|\phi(n+1)| < K|\phi(n)|$ when $n \geqq n_0$, and $0 < K < 1$, then $\lim \phi(n) = 0$.

5. If $\phi(n)$ is positive and $\lim \dfrac{\phi(n+1)}{\phi(n)} = l > 1$, then $\phi(n) \rightarrow +\infty$.

[For we can determine n_0 so that $\{\phi(n+1)\}/\{\phi(n)\} > K > 1$ when $n \geqq n_0$: we may, e.g., take K half-way between 1 and l. Now apply Ex. 1.]

6. If
$$\lim \frac{\phi(n+1)}{\phi(n)} = l, \quad -1 < l < 1,$$
then $\lim \phi(n) = 0$. [This follows from Ex. 4 as Ex. 5 follows from Ex. 1.]

7. Determine the behaviour, as $n \rightarrow \infty$, of $\phi(n) = n^r x^n$, where r is any positive integer.

[If $x = 0$ then $\phi(n) = 0$ for all values of n, and $\phi(n) \rightarrow 0$. In all other cases
$$\frac{\phi(n+1)}{\phi(n)} = \left(\frac{n+1}{n}\right)^r x \rightarrow x.$$
First suppose x positive. Then $\phi(n) \rightarrow +\infty$ if $x > 1$ (Ex. 5) and $\phi(n) \rightarrow 0$ if $x < 1$ (Ex. 6). If $x = 1$, then $\phi(n) = n^r \rightarrow +\infty$. Secondly suppose x negative. Then $|\phi(n)| = n^r |x|^n$ tends to $+\infty$ if $|x| \geqq 1$ and to 0 if $|x| < 1$. Hence $\phi(n)$ oscillates infinitely if $x \leqq -1$ and $\phi(n) \rightarrow 0$ if $-1 < x < 0$.]

8. Discuss $n^{-r}x^n$ in the same way. [The results are the same, except that $\phi(n) \rightarrow 0$ when $x = 1$ or -1.]

9. Draw up a table to show how $n^k x^n$ behaves as $n \rightarrow \infty$, for all real values of x, and all positive and negative integral values of k.

[The reader will observe that the value of k is immaterial except in the special cases when $x = 1$ or -1. Since $\lim \{(n+1)/n\}^k = 1$, whether k be positive or negative, the limit of the ratio $\phi(n+1)/\phi(n)$ depends only on x, and the behaviour of $\phi(n)$ is in general dominated by the factor x^n. The factor n^k only asserts itself when x is numerically equal to 1.]

10. Prove that if x is positive then $\sqrt[n]{x} \rightarrow 1$ as $n \rightarrow \infty$. [Suppose, e.g., $x > 1$. Then x, $\sqrt{x}$, $\sqrt[3]{x}$, ... is a decreasing sequence, and $\sqrt[n]{x} > 1$ for all values of n. Thus $\sqrt[n]{x} \rightarrow l$, where $l \geqq 1$. But if $l > 1$ we can find values of n, as large as we please, for which $\sqrt[n]{x} > l$ or $x > l^n$; and, since $l^n \rightarrow +\infty$ as $n \rightarrow \infty$, this is impossible.]

11. $\sqrt[n]{n} \rightarrow 1$. [For $\sqrt[n+1]{(n+1)} < \sqrt[n]{n}$ if $(n+1)^n < n^{n+1}$ or $(1+n^{-1})^n < n$, which is certainly satisfied if $n \geqq 3$ (see § 73 for a proof). Thus $\sqrt[n]{n}$ decreases

as n increases from 3 onwards, and, since it is always greater than unity, it tends to a limit which is greater than or equal to unity. But if $\sqrt[n]{n} \to l$, where $l > 1$, then $n > l^n$, which is certainly untrue for sufficiently large values of n, since $l^n/n \to +\infty$ with n (Exs. 7, 8).]

12. $x^n/n! \to 0$, for all values of x. [If $u_n = x^n/n!$, then $u_{n+1}/u_n = x/(n+1)$, which tends to zero when $n \to \infty$, so that u_n tends to zero (Ex. 6).]

13. $\sqrt[n]{(n!)} \to \infty$. [For $n! > x^n$, however large be x, for sufficiently large n (Ex. 12).]

14. Show that if $-1 < x < 1$ then

$$u_n = \frac{m(m-1)\dots(m-n+1)}{n!}x^n = \binom{m}{n}x^n$$

tends to zero as $n \to \infty$.

[If m is a positive integer, $u_n = 0$ for $n > m$. Otherwise

$$\frac{u_{n+1}}{u_n} = \frac{m-n}{n+1}x \to -x,$$

unless $x = 0$.]

73. The limit of $\left(1+\dfrac{1}{n}\right)^n$. A more difficult problem which can be solved by the help of § 69 arises when $\phi(n) = (1+n^{-1})^n$.

It follows from the binomial theorem that

$$\left(1+\frac{1}{n}\right)^n = 1 + n.\frac{1}{n} + \frac{n(n-1)}{1.2}\frac{1}{n^2} + \dots + \frac{n(n-1)\dots(n-n+1)}{1.2\dots n}\frac{1}{n^n}$$

$$= 1 + 1 + \frac{1}{1.2}\left(1-\frac{1}{n}\right) + \frac{1}{1.2.3}\left(1-\frac{1}{n}\right)\left(1-\frac{2}{n}\right) + \dots$$

$$+ \frac{1}{1.2\dots n}\left(1-\frac{1}{n}\right)\left(1-\frac{2}{n}\right)\dots\left(1-\frac{n-1}{n}\right).$$

The $(p+1)$th term in this expression, viz.

$$\frac{1}{1.2\dots p}\left(1-\frac{1}{n}\right)\left(1-\frac{2}{n}\right)\dots\left(1-\frac{p-1}{n}\right),$$

is positive and an increasing function of n, and the number of terms also increases with n. Hence $\left(1+\dfrac{1}{n}\right)^n$ increases with n, and so tends to a limit or to $+\infty$, as $n \to \infty$.

But

$$\left(1+\frac{1}{n}\right)^n < 1+1+\frac{1}{1.2}+\frac{1}{1.2.3}+\ldots+\frac{1}{1.2.3\ldots n}$$

$$< 1+1+\frac{1}{2}+\frac{1}{2^2}+\ldots+\frac{1}{2^{n-1}} < 3.$$

Thus $\left(1+\frac{1}{n}\right)^n$ cannot tend to $+\infty$, and so

$$\lim_{n\to\infty}\left(1+\frac{1}{n}\right)^n = e,$$

where e is a number such that $2 < e \leqq 3$.

Example. Find the limit of

$$n^{-n-1}(n+1)^n. \qquad \textit{(Math. Trip. 1934)}$$

74. Some algebraical lemmas. It will be convenient to prove at this stage a number of elementary inequalities which will be useful to us later on.

(i) It is evident that if $\alpha > 1$ and r is a positive integer then

$$r\alpha^r > \alpha^{r-1}+\alpha^{r-2}+\ldots+1.$$

Multiplying both sides of this inequality by $\alpha-1$, we obtain

$$r\alpha^r(\alpha-1) > \alpha^r - 1;$$

and adding $r(\alpha^r-1)$ to each side, and dividing by $r(r+1)$, we obtain

$$\frac{\alpha^{r+1}-1}{r+1} > \frac{\alpha^r-1}{r} \quad (\alpha > 1) \quad \ldots\ldots\ldots\ldots\ldots\ldots(1).$$

Similarly we can prove that

$$\frac{1-\beta^{r+1}}{r+1} < \frac{1-\beta^r}{r} \quad (0 < \beta < 1) \quad \ldots\ldots\ldots\ldots\ldots\ldots(2).$$

It follows that if r and s are positive integers, and $r > s$, then

$$\frac{\alpha^r-1}{r} > \frac{\alpha^s-1}{s}, \quad \frac{1-\beta^r}{r} < \frac{1-\beta^s}{s} \quad \ldots\ldots\ldots\ldots\ldots\ldots(3).$$

Here $0 < \beta < 1 < \alpha$. In particular, when $s = 1$, we have

$$\alpha^r - 1 > r(\alpha-1), \quad 1-\beta^r < r(1-\beta) \ldots\ldots\ldots\ldots\ldots\ldots(4).$$

(ii) The inequalities (3) and (4) have been proved on the supposition that r and s are positive integers. But it is easy to see that they hold under the more general hypothesis that r and s are any positive rational numbers. Let us consider, for example, the first of the inequalities (3). Let $r = a/b$, $s = c/d$, where a, b, c, d are positive integers; so that $ad > bc$. If we put

$\alpha = \gamma^{bd}$, the inequality takes the form

$$(\gamma^{ad} - 1)/ad > (\gamma^{bc} - 1)/bc;$$

and this we have proved already. The same argument applies to the remaining inequalities; and it can evidently be proved in a similar manner that

$$\alpha^s - 1 < s(\alpha - 1), \quad 1 - \beta^s > s(1 - \beta) \quad\text{..................(5)},$$

if s is a positive rational number less than 1.

(iii) In what follows it is to be understood *that all the letters denote positive numbers, that r and s are rational, and that α and r are greater than* 1, *β and s less than* 1. Writing $1/\beta$ for α, and $1/\alpha$ for β, in (4), we obtain

$$\alpha^r - 1 < r\alpha^{r-1}(\alpha - 1), \quad 1 - \beta^r > r\beta^{r-1}(1 - \beta) \quad\text{...........(6)}.$$

Similarly, from (5), we deduce

$$\alpha^s - 1 > s\alpha^{s-1}(\alpha - 1), \quad 1 - \beta^s < s\beta^{s-1}(1 - \beta) \quad\text{...........(7)}.$$

Combining (4) and (6), we see that

$$r\alpha^{r-1}(\alpha - 1) > \alpha^r - 1 > r(\alpha - 1) \quad\text{..................(8)}.$$

Writing x/y for α, we obtain

$$rx^{r-1}(x - y) > x^r - y^r > ry^{r-1}(x - y) \quad\text{..................(9)}$$

if $x > y > 0$. And the same argument, applied to (5) and (7), leads to

$$sx^{s-1}(x - y) < x^s - y^s < sy^{s-1}(x - y) \quad\text{...............(10)}.$$

Examples XXVIII. 1. Verify (9) for $r = 2, 3$, and (10) for $s = \frac{1}{2}, \frac{1}{3}$.

2. Show that (9) and (10) are also true if $y > x > 0$.

3. Show that (9) also holds for $r < 0$.

[A much more complete discussion of the inequalities (9) and (10) will be found in *Inequalities*, Ch. II. See also Appendix I.]

4. If $\phi(n) \to l$, where $l > 0$, as $n \to \infty$, and k is rational, then $\phi^k \to l^k$.

[We may suppose that $k > 0$, in virtue of Theorem III of § 66; and that $\frac{1}{2}l < \phi < 2l$, as is certainly the case from a certain value of n onwards. If $k > 1$,

$$k\phi^{k-1}(\phi - l) > \phi^k - l^k > kl^{k-1}(\phi - l)$$

or

$$kl^{k-1}(l - \phi) > l^k - \phi^k > k\phi^{k-1}(l - \phi),$$

according as $\phi > l$ or $\phi < l$. It follows that the ratio of $|\phi^k - l^k|$ and $|\phi - l|$ lies between $k(\frac{1}{2}l)^{k-1}$ and $k(2l)^{k-1}$. The proof is similar when $0 < k < 1$. The result is still true when $l = 0$, if $k > 0$.]

5. Extend the results of Exs. XXVII. 7, 8, 9 to the case in which r or k are any rational numbers.

75. The limit of $n(\sqrt[n]{x} - 1)$. If in the first inequality (3) of § 74 we put $r = 1/(n-1)$, $s = 1/n$, we see that

$$(n-1)(\sqrt[n-1]{\alpha} - 1) > n(\sqrt[n]{\alpha} - 1)$$

when $\alpha > 1$. Thus if $\phi(n) = n(\sqrt[n]{\alpha} - 1)$ then $\phi(n)$ decreases steadily as n increases. Also $\phi(n)$ is always positive. Hence $\phi(n)$ tends to a limit l as $n \to \infty$, and $l \geqq 0$.

Again if, in the first inequality (7) of § 74, we put $s = 1/n$, we obtain

$$n(\sqrt[n]{\alpha} - 1) > \sqrt[n]{\alpha}\left(1 - \frac{1}{\alpha}\right) > 1 - \frac{1}{\alpha}.$$

Thus $l \geqq 1 - (1/\alpha) > 0$. Hence, if $\alpha > 1$, we have

$$\lim_{n \to \infty} n(\sqrt[n]{\alpha} - 1) = f(\alpha),$$

where $f(\alpha) > 0$.

Next suppose $\beta < 1$, and let $\beta = 1/\alpha$; then $n(\sqrt[n]{\beta} - 1) = -n(\sqrt[n]{\alpha} - 1)/\sqrt[n]{\alpha}$. Now $n(\sqrt[n]{\alpha} - 1) \to f(\alpha)$, and (Exs. XXVII. 10)

$$\sqrt[n]{\alpha} \to 1.$$

Hence, if $\beta = 1/\alpha < 1$, we have

$$n(\sqrt[n]{\beta} - 1) \to -f(\alpha$$

Finally, if $x = 1$, then $n(\sqrt[n]{x} - 1) = 0$ for all values of n.

Thus we arrive at the result: *the limit*

$$\lim n(\sqrt[n]{x} - 1)$$

defines a function of x for all positive values of x. This function $f(x)$ possesses the properties

$$f(1/x) = -f(x), \quad f(1) = 0,$$

and is positive or negative according as $x > 1$ or $x < 1$. Later on we shall be able to identify this function with the *Napierian logarithm* of x.

Example. Prove that $f(xy) = f(x) + f(y)$. [Use the equations

$$f(xy) = \lim n(\sqrt[n]{xy} - 1) = \lim \{n(\sqrt[n]{x} - 1)\sqrt[n]{y} + n(\sqrt[n]{y} - 1)\}.]$$

76. Infinite series. Suppose that $u(n)$ is any function of n defined for all values of n. If we add up the values of $u(\nu)$ for $\nu = 1, 2, \ldots, n$, we obtain another function of n, viz.

$$s(n) = u(1) + u(2) + \ldots + u(n),$$

also defined for all values of n. It is generally most convenient to alter our notation slightly and write this equation in the form

$$s_n = u_1 + u_2 + \ldots + u_n,$$

or, more shortly,

$$s_n = \sum_{\nu=1}^{n} u_\nu.$$

If now we suppose that s_n tends to a limit s when n tends to ∞, we have

$$\lim_{n\to\infty} \sum_{\nu=1}^{n} u_\nu = s.$$

This equation is usually written in one of the forms

$$\sum_{\nu=1}^{\infty} u_\nu = s, \quad u_1+u_2+u_3+\ldots = s,$$

the dots denoting the indefinite continuance of the series of u's.

The meaning of the above equations, expressed roughly, is that by adding more and more of the u's together we get nearer and nearer to the limit s. More precisely, if any small positive number δ is chosen, we can choose $n_0(\delta)$ so that the sum of the first $n_0(\delta)$ terms, or of any greater number of terms, lies between $s-\delta$ and $s+\delta$; or in symbols

$$s-\delta < s_n < s+\delta,$$

if $n \geqq n_0(\delta)$. In these circumstances we shall call the series

$$u_1+u_2+\ldots$$

a *convergent infinite series*, and we shall call s the *sum* of the series, or the *sum of all the terms* of the series.

Thus to say that the series $u_1+u_2+\ldots$ *converges and has the sum* s, or *converges to the sum* s or simply *converges to* s, is merely another way of stating that the sum $s_n = u_1+u_2+\ldots+u_n$ of the first n terms tends to the limit s as $n\to\infty$, and the consideration of such infinite series introduces no new ideas beyond those with which the early part of this chapter should already have made the reader familiar. In fact the sum s_n is merely a function $\phi(n)$, such as we have been considering, expressed in a particular form. Any function $\phi(n)$ may be expressed in this form, by writing

$$\phi(n) = \phi(1)+\{\phi(2)-\phi(1)\}+\ldots+\{\phi(n)-\phi(n-1)\};$$

and it is sometimes convenient to say that $\phi(n)$ *converges* (instead of 'tends') to the limit l, say, as $n\to\infty$.

If $s_n\to+\infty$ or $s_n\to-\infty$, we shall say that the series $u_1+u_2+\ldots$ is *divergent* or *diverges to* $+\infty$, or $-\infty$, as the case may be. These phrases too may be applied to any function $\phi(n)$: thus if

$\phi(n) \to +\infty$ we may say that $\phi(n)$ *diverges to* $+\infty$. If s_n does not tend to a limit or to $+\infty$ or to $-\infty$, then it oscillates finitely or infinitely: in this case we say that the series $u_1 + u_2 + \dots$ oscillates finitely or infinitely*.

77. General theorems concerning infinite series. When we are dealing with infinite series we shall constantly have occasion to use the following general theorems.

(1) If $u_1 + u_2 + \dots$ is convergent and has the sum s, then $a + u_1 + u_2 + \dots$ is convergent and has the sum $a + s$. Similarly $a + b + c + \dots + k + u_1 + u_2 + \dots$ is convergent and has the sum $a + b + c + \dots + k + s$.

(2) If $u_1 + u_2 + \dots$ is convergent and has the sum s, then $u_{m+1} + u_{m+2} + \dots$ is convergent and has the sum

$$s - u_1 - u_2 - \dots - u_m.$$

(3) If any series considered in (1) or (2) diverges or oscillates, then so do the others.

(4) If $u_1 + u_2 + \dots$ is convergent and has the sum s, then $ku_1 + ku_2 + \dots$ is convergent and has the sum ks.

(5) If the first series considered in (4) diverges or oscillates, then so does the second, unless $k = 0$.

(6) If $u_1 + u_2 + \dots$ and $v_1 + v_2 + \dots$ are both convergent, then the series $(u_1 + v_1) + (u_2 + v_2) + \dots$ is convergent and its sum is the sum of the first two series.

All these theorems are almost obvious and may be proved at once from the definitions or by applying the results of §§ 63–66 to the sum $s_n = u_1 + u_2 + \dots + u_n$. Those which follow are of a somewhat different character.

(7) *If* $u_1 + u_2 + \dots$ *is convergent, then* $\lim u_n = 0$.

* The reader should be warned that the words 'divergent' and 'oscillatory' are used differently by different writers. The use of the words here agrees with that of Bromwich's *Infinite series*. In Hobson's *Theory of functions of a real variable* a series is said to oscillate only if it oscillates *finitely*, series which oscillate infinitely being classed as 'divergent'. Many foreign writers use 'divergent' as meaning merely 'not convergent'.

For $u_n = s_n - s_{n-1}$, and s_n and s_{n-1} have the same limit s. Hence $\lim u_n = s - s = 0$.

The reader may be tempted to think that the converse of the theorem s true and that if $\lim u_n = 0$ then the series Σu_n must be convergent. That this is not the case is easily seen from an example. Let the series be

$$1 + \tfrac{1}{2} + \tfrac{1}{3} + \tfrac{1}{4} + \ldots$$

so that $u_n = 1/n$. The sum of the first four terms is

$$1 + \tfrac{1}{2} + \tfrac{1}{3} + \tfrac{1}{4} > 1 + \tfrac{1}{2} + \tfrac{2}{4} = 1 + \tfrac{1}{2} + \tfrac{1}{2}.$$

The sum of the next four terms is $\frac{1}{5} + \frac{1}{6} + \frac{1}{7} + \frac{1}{8} > \frac{4}{8} = \frac{1}{2}$; the sum of the next eight terms is greater than $\frac{8}{16} = \frac{1}{2}$, and so on. The sum of the first

$$4 + 4 + 8 + 16 + \ldots + 2^n = 2^{n+1}$$

terms is greater than

$$2 + \tfrac{1}{2} + \tfrac{1}{2} + \tfrac{1}{2} + \ldots + \tfrac{1}{2} = \tfrac{1}{2}(n + 3),$$

and this increases beyond all limit with n: hence the series diverges to $+\infty$.

(8) *If $u_1 + u_2 + u_3 + \ldots$ is convergent, then so is any series formed by grouping the terms in brackets in any way to form new single terms, and the sums of the two series are the same.*

The reader will be able to supply the proof of this theorem. Here again the converse is not true. Thus $1 - 1 + 1 - 1 + \ldots$ oscillates, while

$$(1 - 1) + (1 - 1) + \ldots$$

or $0 + 0 + 0 + 0 + \ldots$ converges to 0.

(9) *If every term u_n is positive* (*or zero*), *then the series Σu_n must either converge or diverge to $+\infty$. If it converges, its sum must be positive* (unless all the terms are zero, when of course its sum is zero).

For s_n is an increasing function of n, according to the definition of § 69, and we can apply the results of that section to s_n.

(10) *If every term u_n is positive* (*or zero*), *then a necessary and sufficient condition that the series Σu_n should be convergent is that it should be possible to find a number K such that the sum of any number of terms is less than K; and, if K can be so found, then the sum of the series is not greater than K.*

This also follows at once from § 69. It is perhaps hardly necessary to point out that the theorem is not true if the condition that every u_n is positive is not fulfilled. For example

$$1-1+1-1+\dots$$

obviously oscillates, s_n being alternately equal to 1 and to 0.

(11) *If* $u_1+u_2+\dots$, $v_1+v_2+\dots$ *are two series of positive (or zero) terms, and the second series is convergent, and if* $u_n \leqq Kv_n$, *where K is a constant, for all values of n, then the first series is also convergent, and its sum does not exceed K times that of the second.*

For if $v_1+v_2+\dots = t$ then $v_1+v_2+\dots+v_n \leqq t$ for all values of n, and so $u_1+u_2+\dots+u_n \leqq Kt$; which proves the theorem.

Conversely, if Σu_n *is divergent, and* $v_n \geqq Ku_n$, *where* $K>0$, *then* Σv_n *is divergent.*

78. The infinite geometrical series. We shall now consider the 'geometrical' series, whose general term is $u_n = r^{n-1}$. In this case

$$s_n = 1+r+r^2+\dots+r^{n-1} = (1-r^n)/(1-r),$$

except in the special case in which $r=1$, when

$$s_n = 1+1+\dots+1 = n.$$

In the last case $s_n \to +\infty$. In the general case s_n will tend to a limit if and only if r^n does so. Referring to the results of § 72 we see that *the series* $1+r+r^2+\dots$ *is convergent and has the sum* $1/(1-r)$ *if and only if* $-1<r<1$.

If $r \geqq 1$, then $s_n \geqq n$, and so $s_n \to +\infty$; i.e. the series diverges to $+\infty$. If $r=-1$, then $s_n = 1$ or $s_n = 0$ according as n is odd or even: i.e. s_n oscillates finitely. If $r<-1$, then s_n oscillates infinitely. Thus, to sum up, *the series* $1+r+r^2+\dots$ *diverges to* $+\infty$ *if* $r \geqq 1$, *converges to* $1/(1-r)$ *if* $-1<r<1$, *oscillates finitely if* $r=-1$, *and oscillates infinitely if* $r<-1$.

Examples XXIX. 1. Recurring decimals. The commonest example of an infinite geometric series is given by an ordinary recurring decimal. Consider, for example, the decimal $\cdot 217\dot{1}\dot{3}$. This stands, according to the ordinary rules of arithmetic, for

$$\frac{2}{10}+\frac{1}{10^2}+\frac{7}{10^3}+\frac{1}{10^4}+\frac{3}{10^5}+\frac{1}{10^6}+\frac{3}{10^7}+\dots = \frac{217}{1000}+\frac{13}{10^5}\Big/\left(1-\frac{1}{10^2}\right) = \frac{2687}{12375}.$$

The reader should consider where and how any of the general theorems of §77 have been used in this reduction.

2. Show that in general

$$\cdot a_1 a_2 \dots a_m \dot{\alpha}_1 \alpha_2 \dots \dot{\alpha}_n = \frac{a_1 a_2 \dots a_m \alpha_1 \alpha_2 \dots \alpha_n - a_1 a_2 \dots a_m}{99 \dots 900 \dots 0},$$

the denominator containing n 9's and m 0's.

3. Show that a pure recurring decimal is always equal to a proper fraction whose denominator does not contain 2 or 5 as a factor.

4. A decimal with m non-recurring and n recurring decimal figures is equal to a proper fraction whose denominator is divisible by 2^m or 5^m but by no higher power of either.

5. The converses of Exs. 3, 4 are also true. Let $r = p/q$, and suppose first that q is prime to 10. If we divide all powers of 10 by q we can obtain at most q different remainders. It is therefore possible to find two numbers n_1 and n_2, where $n_1 > n_2$, such that 10^{n_1} and 10^{n_2} give the same remainder. Hence $10^{n_1} - 10^{n_2} = 10^{n_2}(10^{n_1-n_2} - 1)$ is divisible by q, and so $10^n - 1$, where $n = n_1 - n_2$, is divisible by q. Hence r may be expressed in the form $P/(10^n - 1)$, or in the form

$$\frac{P}{10^n} + \frac{P}{10^{2n}} + \dots,$$

i.e. as a pure recurring decimal with n figures. If on the other hand $q = 2^\alpha 5^\beta Q$, where Q is prime to 10, and m is the greater of α and β, then $10^m r$ has a denominator prime to 10, and is therefore expressible as the sum of an integer and a pure recurring decimal. But this is not true of $10^\mu r$, for any value of μ less than m; hence the decimal for r has exactly m non-recurring figures.

6. To the results of Exs. 2–5 we must add that of Ex. I. 3. Finally, if we observe that

$$\cdot\dot{9} = \frac{9}{10} + \frac{9}{10^2} + \frac{9}{10^3} + \dots = 1,$$

we see that every terminating decimal can also be expressed as a mixed recurring decimal whose recurring part is composed entirely of 9's. For example, $\cdot 217 = \cdot 216\dot{9}$. Thus every proper fraction can be expressed as a recurring decimal, and conversely.

7. Decimals in general. The expression of irrational numbers as non-recurring decimals. Any decimal, whether recurring or not, corresponds to a definite number between 0 and 1. For the decimal $\cdot a_1 a_2 a_3 a_4 \dots$ stands for the series

$$\frac{a_1}{10} + \frac{a_2}{10^2} + \frac{a_3}{10^3} + \dots.$$

Since all the digits a_r are positive, the sum s_n of the first n terms of this series increases with n, and it is certainly not greater than $\cdot\dot{9}$ or 1. Hence s_n tends to a limit between 0 and 1.

Moreover no two decimals can correspond to the same number (except in the special case noticed in Ex. 6). For suppose that $\cdot a_1 a_2 a_3 \ldots$, $\cdot b_1 b_2 b_3 \ldots$ are two decimals which agree as far as the figures a_{r-1}, b_{r-1}, while $a_r > b_r$. Then $a_r \geqq b_r + 1 > b_r \cdot b_{r+1} b_{r+2} \ldots$ (unless b_{r+1}, b_{r+2}, ... are all 9's), and so

$$\cdot a_1 a_2 \ldots a_r a_{r+1} \ldots > \cdot b_1 b_2 \ldots b_r b_{r+1} \ldots.$$

It follows that the expression of a rational fraction as a recurring decimal (Exs. 2–6) is unique. It also follows that every decimal which does not terminate or recur represents some *irrational* number between 0 and 1. Conversely, any such number can be expressed as such a decimal. For it must lie in one of the intervals

$$0,\ \tfrac{1}{10};\quad \tfrac{1}{10},\ \tfrac{2}{10};\quad \ldots;\quad \tfrac{9}{10},\ 1.$$

If it lies between $\frac{1}{10}r$ and $\frac{1}{10}(r+1)$, then the first figure is r. By sub-dividing this interval into 10 parts we can determine the second figure; and so on. But (Exs. 3, 4) the decimal cannot recur. Thus, for example, the decimal $1{\cdot}414\ldots$, obtained by the ordinary process for the extraction of $\sqrt{2}$, cannot recur.

8. The decimals ·1010010001000010... and ·2020020002000020..., in which the number of zeros between two 1's or 2's increases by one at each stage, represent irrational numbers.

9. The decimal ·11101010001010.... in which the nth figure is 1 if n is prime, and zero otherwise, represents an irrational number. [Since the number of primes is infinite the decimal does not terminate. Nor can it recur: for if it did we could determine m and p so that m, $m+p$, $m+2p$, $m+3p$, ... are all prime numbers; and this is absurd, since the series includes $m+mp$*.]

Examples XXX. 1. The series $r^m + r^{m+1} + \ldots$ is convergent if $-1 < r < 1$, and its sum is $1/(1-r) - 1 - r - \ldots - r^{m-1}$ (§ 77, (2)).

2. The series $r^m + r^{m+1} + \ldots$ is convergent if $-1 < r < 1$, and its sum is $r^m/(1-r)$ (§ 77, (4)). Verify that the results of Exs. 1 and 2 are in agreement.

3. Prove that the series $1 + 2r + 2r^2 + \ldots$ is convergent, and that its sum is $(1+r)/(1-r)$, (α) by writing it in the form $-1 + 2(1 + r + r^2 + \ldots)$, ($\beta$) by writing it in the form $1 + 2(r + r^2 + \ldots)$, (γ) by adding the two series $1 + r + r^2 + \ldots$, $r + r^2 + \ldots$. In each case mention which of the theorems of § 77 are used in your proof.

* All the results of Exs. XXIX may be extended, with suitable modifications, to decimals in any scale of notation. For a fuller discussion see Bromwich, *Infinite series*, Appendix I.

4. Prove that the 'arithmetic' series

$$a+(a+b)+(a+2b)+\dots$$

is always divergent, unless both a and b are zero. Show that, if b is not zero, the series diverges to $+\infty$ or to $-\infty$ according to the sign of b, while if $b=0$ it diverges to $+\infty$ or $-\infty$ according to the sign of a.

5. What is the sum of the series

$$(1-r)+(r-r^2)+(r^2-r^3)+\dots$$

when the series is convergent? [The series converges only if $-1<r\leqq 1$. Its sum is 1, except when $r=1$, when its sum is 0.]

6. Sum the series $r^2+\dfrac{r^2}{1+r^2}+\dfrac{r^2}{(1+r^2)^2}+\dots$. [The series is always convergent. Its sum is $1+r^2$, except when $r=0$, when its sum is 0.]

7. If we assume that $1+r+r^2+\dots$ is convergent, then we can prove that its sum is $1/(1-r)$ by means of § 77, (1) and (4). For if $1+r+r^2+\dots=s$ then

$$s=1+r(1+r^2+\dots)=1+rs.$$

8. Sum the series

$$r+\frac{r}{1+r}+\frac{r}{(1+r)^2}+\dots$$

when it is convergent. [The series is convergent if $-1<1/(1+r)<1$, i.e. if $r<-2$ or if $r>0$, and its sum is $1+r$. It is also convergent when $r=0$, when its sum is 0.]

9. Answer the same question for the series

$$r-\frac{r}{1+r}+\frac{r}{(1+r)^2}-\dots,\qquad r+\frac{r}{1-r}+\frac{r}{(1-r)^2}+\dots,$$

$$1-\frac{r}{1+r}+\left(\frac{r}{1+r}\right)^2-\dots,\qquad 1+\frac{r}{1-r}+\left(\frac{r}{1-r}\right)^2+\dots.$$

10. Consider the convergence of the series

$$(1+r)+(r^2+r^3)+\dots,\qquad (1+r+r^2)+(r^3+r^4+r^5)+\dots,$$

$$1-2r+r^2+r^3-2r^4+r^5+\dots,\qquad (1-2r+r^2)+(r^3-2r^4+r^5)+\dots,$$

and find their sums when they are convergent.

11. If $0\leqq a_n\leqq 1$ then the series $a_0+a_1r+a_2r^2+\dots$ is convergent for $0\leqq r<1$, and its sum is not greater than $1/(1-r)$.

12. If in addition the series $a_0+a_1+a_2+\dots$ is convergent, then the series $a_0+a_1r+a_2r^2+\dots$ is convergent for $0\leqq r\leqq 1$, and its sum is not greater than the lesser of $a_0+a_1+a_2+\dots$ and $1/(1-r)$.

13. The series

$$1+\frac{1}{1}+\frac{1}{1.2}+\frac{1}{1.2.3}+\dots$$

is convergent. [For $1/(1.2\dots n)\leqq 1/2^{n-1}$.]

14. The series

$$1+\frac{1}{1.2}+\frac{1}{1.2.3.4}+\ldots,\quad \frac{1}{1}+\frac{1}{1.2.3}+\frac{1}{1.2.3.4.5}+\ldots$$

are convergent.

15. The general harmonic series

$$\frac{1}{a}+\frac{1}{a+b}+\frac{1}{a+2b}+\ldots,$$

where a and b are positive, diverges to $+\infty$.

[For $u_n = 1/(a+nb) > 1/\{n(a+b)\}$. Now compare with $1+\frac{1}{2}+\frac{1}{3}+\ldots$.]

16. Show that the series

$$(u_0-u_1)+(u_1-u_2)+(u_2-u_3)+\ldots$$

is convergent if and only if u_n tends to a limit as $n\to\infty$.

17. If $u_1+u_2+u_3+\ldots$ is divergent, then so is any series formed by grouping the terms in brackets in any way to form new single terms.

18. Any series, formed by taking a selection of the terms of a convergent series of positive terms, is itself convergent.

79. The representation of functions of a continuous real variable by means of limits. In the preceding sections we have frequently been concerned with limits such as

$$\lim_{n\to\infty} \phi_n(x),$$

and series such as

$$u_1(x)+u_2(x)+\ldots = \lim_{n\to\infty} \{u_1(x)+u_2(x)+\ldots+u_n(x)\},$$

in which the function of n whose limit we are seeking involves, besides n, another variable x. In such cases the limit is of course a function of x. Thus in § 75 we encountered the function

$$f(x) = \lim_{n\to\infty} n(\sqrt[n]{x}-1):$$

and the sum of the geometrical series $1+x+x^2+\ldots$ is a function of x, viz. the function which is equal to $1/(1-x)$ if $-1<x<1$ and is undefined for all other values of x.

Many of the apparently 'arbitrary' or 'unnatural' functions considered in Ch. II are capable of a simple representation of this kind, as will appear from the following examples.

Examples XXXI. 1. $\phi_n(x) = x$. Here n does not appear at all in the expression of $\phi_n(x)$, and $\phi(x) = \lim \phi_n(x) = x$ for all values of x.

2. $\phi_n(x) = x/n$. Here $\phi(x) = \lim \phi_n(x) = 0$ for all values of x.

3. $\phi_n(x) = nx$. If $x > 0$, $\phi_n(x) \to +\infty$; if $x < 0$, $\phi_n(x) \to -\infty$: only when $x = 0$ has $\phi_n(x)$ a limit (viz. 0) as $n \to \infty$. Thus $\phi(x) = 0$ when $x = 0$ and is not defined for any other values of x.

4. $\phi_n(x) = 1/nx$, $nx/(nx+1)$.

5. $\phi_n(x) = x^n$. Here $\phi(x) = 0$, $(-1 < x < 1)$; $\phi(x) = 1$, $(x = 1)$; and $\phi(x)$ is not defined for any other value of x.

6. $\phi_n(x) = x^n(1-x)$. Here $\phi(x)$ differs from the $\phi(x)$ of Ex. 5 in that it has the value 0 when $x = 1$.

7. $\phi_n(x) = x^n/n$. Here $\phi(x)$ differs from the $\phi(x)$ of Ex. 6 in that it has the value 0 when $x = -1$ as well as when $x = 1$.

8. $\phi_n(x) = x^n/(x^n+1)$. [$\phi(x) = 0$, $(-1 < x < 1)$; $\phi(x) = \frac{1}{2}$, $(x = 1)$; $\phi(x) = 1$, $(x < -1$ or $x > 1)$; and $\phi(x)$ is not defined when $x = -1$.]

9. $$\phi_n(x) = \frac{x^n}{x^n - 1}, \quad \frac{1}{x^n+1}, \quad \frac{1}{x^n - 1}, \quad \frac{1}{x^n + x^{-n}}, \quad \frac{1}{x^n - x^{-n}}.$$

10. Prove that, if $x > 0$, the function $(x^n - 1)/(x^n + 1)$ tends to a limit when $n \to \infty$, and that the limit has three different values in the three cases $x < 1$, $x = 1$, and $x > 1$. (*Math. Trip.* 1935)

Discuss also the functions

$$\frac{nx^n - 1}{nx^n + 1}, \quad \frac{x^n - n}{x^n + n}.$$

11. Construct an example in which $\phi(x) = 1$, $(|x| > 1)$; $\phi(x) = -1$, $(|x| < 1)$; and $\phi(x) = 0$, $(x = 1$ and $x = -1)$.

12. $$\phi_n(x) = x\left(\frac{x^{2n} - 1}{x^{2n} + 1}\right)^2, \quad \frac{n}{x^n + x^{-n} + n}.$$

13. $$\phi_n(x) = \frac{x^n f(x) + g(x)}{x^n + 1}.$$

[Here $\phi(x) = f(x)$, $(|x| > 1)$; $\phi(x) = g(x)$, $(|x| < 1)$; $\phi(x) = \frac{1}{2}\{f(x) + g(x)\}$, $(x = 1)$; and $\phi(x)$ is undefined when $x = -1$.]

14. $\phi_n(x) = (2/\pi) \arctan(nx)$. [$\phi(x) = 1$, $(x > 0)$; $\phi(x) = 0$, $(x = 0)$; $\phi(x) = -1$, $(x < 0)$. This function is important in the theory of numbers, and is usually denoted by $\operatorname{sgn} x$.]

15. $\phi_n(x) = \sin nx\pi$. [$\phi(x) = 0$ when x is an integer; and $\phi(x)$ is otherwise undefined (Ex. XXIV. 7).]

16. If $\phi_n(x) = \sin n!x\pi$, then $\phi(x) = 0$ for all rational values of x (Ex. XXIV. 13). [The consideration of irrational values presents greater difficulties.]

17. $\phi_n(x) = (\cos^2 x\pi)^n$. [$\phi(x) = 0$ except when x is integral, when $\phi(x) = 1$.]

18. If $N \geqq 1752$ then the number of days in the year N A.D. is

$$\lim \{365 + (\cos^2 \tfrac{1}{4}N\pi)^n - (\cos^2 \tfrac{1}{100}N\pi)^n + (\cos^2 \tfrac{1}{400}N\pi)^n\}.$$

80. The bounds of a bounded aggregate. Let S be any system or aggregate of real numbers s. If there is a number K such that $s \leqq K$ for every s of S, we say that S is *bounded above.* If there is a number k such that $s \geqq k$ for every s, we say that S is *bounded below.* If S is both bounded above and bounded below, we say simply that S is *bounded.*

Suppose first that S is bounded above (but not necessarily below). There will be an infinity of numbers which possess the property possessed by K; any number greater than K, for example, possesses it. We shall prove that *among these numbers there is a least**, which we shall call M. This number M is not exceeded by any member of S, but every number less than M is exceeded by at least one member of S.

We divide the real numbers ξ into two classes L and R, putting ξ into L or R according as it is or is not exceeded by members of S. Then every ξ belongs to one and one only of the classes L and R. Each class exists; for any number less than any member of S belongs to L, while K belongs to R. Finally, any member of L is less than some member of S, and therefore less than any member of R. Thus the three conditions of Dedekind's theorem (§ 17) are satisfied, and there is a number M dividing the classes.

The number M is the number whose existence we had to prove. In the first place, M cannot be exceeded by any member of S. For if there were such a member s of S, we could write $s = M + \eta$, where η is positive. The number $M + \frac{1}{2}\eta$ would then belong to L, because it is less than s, and to R, because it is greater than M; and this is impossible. On the other hand, any number less than M belongs to L, and is therefore exceeded by at least one member of S. Thus M has all the properties required.

This number M we call the *upper bound* of S, and we may enunciate the following theorem. *Any aggregate S which is bounded above has an upper bound M. No member of S exceeds M; but any number less than M is exceeded by at least one member of S.*

* An infinite aggregate of numbers does not necessarily possess a least member. The set consisting of the numbers

$$1, \frac{1}{2}, \frac{1}{3}, \dots, \frac{1}{n}, \dots,$$

for example, has no least member.

In exactly the same way we can prove the corresponding theorem for an aggregate bounded below (but not necessarily above). *Any aggregate S which is bounded below has a lower bound m. No member of S is less than m; but there is at least one member of S which is less than any number greater than m.*

It will be observed that, when S is bounded above, $M \leqq K$, and when S is bounded below, $m \geqq k$. When S is bounded, $k \leqq m \leqq M \leqq K$.

81. The bounds of a bounded function. Suppose that $\phi(n)$ is a function of the positive integral variable n. The aggregate of all the values $\phi(n)$ defines a set S, to which we may apply all the arguments of § 80. If S is bounded above, or bounded below, or bounded, we say that $\phi(n)$ is bounded above, or bounded below, or bounded. If $\phi(n)$ is bounded above, that is to say if there is a number K such that $\phi(n) \leqq K$ for all values of n, then there is a number M such that

(i) $\phi(n) \leqq M$ *for all values of* n;

(ii) *if* δ *is any positive number then* $\phi(n) > M - \delta$ *for at least one value of* n.

This number M we call the *upper bound* of $\phi(n)$. Similarly, if $\phi(n)$ is bounded below, that is to say if there is a number k such that $\phi(n) \geqq k$ for all values of n, then there is a number m such that

(i) $\phi(n) \geqq m$ *for all values of* n;

(ii) *if* δ *is any positive number then* $\phi(n) < m + \delta$ *for at least one value of* n.

This number m we call the *lower bound* of $\phi(n)$.

If K exists, $M \leqq K$; if k exists, $m \geqq k$; and if both k and K exist, then

$$k \leqq m \leqq M \leqq K.$$

82. The limits of indetermination of a bounded function. Suppose that $\phi(n)$ is a bounded function, and M and m its upper and lower bounds. Let us take any real number ξ, and consider now the relations of inequality which may hold between ξ and the values assumed by $\phi(n)$ for *large* values of n. There are three mutually exclusive possibilities:

(1) $\xi \geqq \phi(n)$ for all sufficiently large values of n;

(2) $\xi \leqq \phi(n)$ for all sufficiently large values of n;

(3) $\xi < \phi(n)$ for an infinity of values of n, and also $\xi > \phi(n)$ for an infinity of values of n.

In case (1) we shall say that ξ is a *superior* number, in case (2) that it is an *inferior* number, and in case (3) that it is an *intermediate* number. It is plain that no superior number can be less than m, and no inferior number greater than M.

Let us consider the aggregate of all superior numbers. It is bounded below, since none of its members is less than m, and has therefore a lower bound, which we shall denote by Λ. Similarly the aggregate of inferior numbers has an upper bound, which we denote by λ.

We call Λ and λ respectively the *upper and lower limits of indetermination of $\phi(n)$ as n tends to infinity*; and write

$$\Lambda = \overline{\lim}\,\phi(n), \quad \lambda = \underline{\lim}\,\phi(n).$$

These numbers have the following properties:

(1) $$m \leqq \lambda \leqq \Lambda \leqq M;$$

(2) Λ and λ are the upper and lower bounds of the aggregate of intermediate numbers, if any such exist;

(3) if δ is any positive number, then $\phi(n) < \Lambda + \delta$ for all sufficiently large values of n, and $\phi(n) > \Lambda - \delta$ for an infinity of values of n;

(4) similarly $\phi(n) > \lambda - \delta$ for all sufficiently large values of n, and $\phi(n) < \lambda + \delta$ for an infinity of values of n;

(5) a necessary and sufficient condition that $\phi(n)$ should tend to a limit is that $\Lambda = \lambda$, and in this case the limit is l, the common value of λ and Λ.

Of these properties, (1) is an immediate consequence of the definitions; and we can prove (2) as follows. If $\Lambda = \lambda = l$, there can be at most one intermediate number, viz. l, and there is nothing to prove. Suppose then that $\Lambda > \lambda$. Any intermediate number ξ is less than any superior and greater than any inferior number, so that $\lambda \leqq \xi \leqq \Lambda$. But if $\lambda < \xi < \Lambda$ then ξ must be intermediate, since it is plainly neither superior nor inferior. Hence there are intermediate numbers as near as we please to either λ or Λ.

To prove (3) we observe that $\Lambda + \delta$ is superior and $\Lambda - \delta$ intermediate or inferior. The result is then an immediate consequence of the definitions; and the proof of (4) is substantially the same.

Finally (5) may be proved as follows. If $\Lambda = \lambda = l$, then

$$l - \delta < \phi(n) < l + \delta$$

for every positive value of δ and all sufficiently large values of n, so that $\phi(n) \to l$. Conversely, if $\phi(n) \to l$, then the inequalities above written hold for all sufficiently large values of n. Hence $l - \delta$ is inferior and $l + \delta$ superior, so that

$$\lambda \geqq l - \delta, \quad \Lambda \leqq l + \delta,$$

and therefore $\Lambda - \lambda \leqq 2\delta$. Since $\Lambda - \lambda \geqq 0$, this can only be true if $\Lambda = \lambda$.

Examples XXXII. 1. Neither Λ nor λ is affected by any alteration in any finite number of values of $\phi(n)$.

2. If $\phi(n) = a$ for all values of n, then $m = \lambda = \Lambda = M = a$.

3. If $\phi(n) = n^{-1}$, then $m = \lambda = \Lambda = 0$ and $M = 1$.

4. If $\phi(n) = (-1)^n$, then $m = \lambda = -1$ and $\Lambda = M = 1$.

5. If $\phi(n) = (-1)^n n^{-1}$, then $m = -1$, $\lambda = \Lambda = 0$, $M = \frac{1}{2}$.

6. If $\phi(n) = (-1)^n (1+n^{-1})$, then $m = -2$, $\lambda = -1$, $\Lambda = 1$, $M = \frac{3}{2}$.

7. Let $\phi(n) = \sin n\theta\pi$, where $\theta > 0$. If θ is an integer, then

$$m = \lambda = \Lambda = M = 0.$$

If θ is rational but not integral a variety of cases arise. Suppose, e.g., that $\theta = p/q$, p and q being positive, odd, and prime to one another, and $q > 1$. Then $\phi(n)$ assumes the cyclical sequence of values

$$\sin\frac{p\pi}{q},\quad \sin\frac{2p\pi}{q},\ldots,\quad \sin\frac{(2q-1)p\pi}{q},\quad \sin\frac{2qp\pi}{q},\ldots.$$

It is easily verified that the numerically greatest and least values of $\phi(n)$ are $\cos(\pi/2q)$ and $-\cos(\pi/2q)$, so that

$$m = \lambda = -\cos\frac{\pi}{2q},\quad \Lambda = M = \cos\frac{\pi}{2q}.$$

The reader may discuss similarly the cases which arise when p and q are not both odd.

The case in which θ is irrational is more difficult: it may be shown that in this case $m = \lambda = -1$ and $\Lambda = M = 1$. It may also be shown that the values of $\phi(n)$ are scattered all over the interval $(-1, 1)$ in such a way that, if ξ is *any* number of the interval, then there is a sequence $n_1, n_2, \ldots$ such that $\phi(n_k) \to \xi$ as $k \to \infty$*.

The results are very similar when $\phi(n)$ is the fractional part of $n\theta$.

83. The general principle of convergence for a bounded function. The results of the preceding sections enable us to formulate a very important necessary and sufficient condition that a bounded function $\phi(n)$ should tend to a limit, a condition usually referred to as *the general principle of convergence* to a limit.

THEOREM 1. *A necessary and sufficient condition that a bounded function $\phi(n)$ should tend to a limit is that, when any positive number δ is given, it should be possible to find a number $n_0(\delta)$ such that*

$$|\phi(n_2) - \phi(n_1)| < \delta$$

for all values of n_1 and n_2 such that $n_2 > n_1 \geqq n_0(\delta)$.

In the first place, the condition is *necessary*. For if $\phi(n) \to l$ then we can find n_0 so that

$$l - \tfrac{1}{2}\delta < \phi(n) < l + \tfrac{1}{2}\delta$$

when $n \geqq n_0$, and so

$$|\phi(n_2) - \phi(n_1)| < \delta \quad \text{.........................(1)}$$

when $n_1 \geqq n_0$ and $n_2 \geqq n_0$.

* A number of simple proofs of this result are given by Hardy and Littlewood, 'Some problems of Diophantine approximation", *Acta mathematica*, vol. XXXVII.

In the second place, the condition is *sufficient*. In order to prove this we have only to show that it involves $\lambda = \Lambda$. But if $\lambda < \Lambda$ then there are, however small δ may be, infinitely many values of n such that $\phi(n) < \lambda + \delta$ and infinitely many such that $\phi(n) > \Lambda - \delta$; and therefore we can find values of n_1 and n_2, each greater than any assigned number n_0, and such that

$$\phi(n_2) - \phi(n_1) > \Lambda - \lambda - 2\delta,$$

which is greater than $\frac{1}{2}(\Lambda - \lambda)$ if δ is small enough. This plainly contradicts the inequality (1). Hence $\lambda = \Lambda$, and so $\phi(n)$ tends to a limit.

84. Unbounded functions. So far we have restricted ourselves to bounded functions; but the 'general principle of convergence' is the same for unbounded as for bounded functions, and the words '*a bounded function*' may be omitted from the enunciation of Theorem 1.

In the first place, if $\phi(n)$ tends to a limit l then it is certainly bounded; for all but a finite number of its values are less than $l + \delta$ and greater than $l - \delta$.

In the second place, if the condition of Theorem 1 is satisfied, we have

$$|\phi(n_2) - \phi(n_1)| < \delta$$

whenever $n_1 \geqq n_0$ and $n_2 \geqq n_0$. Let us choose some particular value n_1 greater than n_0. Then

$$\phi(n_1) - \delta < \phi(n_2) < \phi(n_1) + \delta$$

when $n_2 \geqq n_0$. Hence $\phi(n)$ is bounded; and so the second part of the proof of the preceding section applies also.

The theoretical importance of the 'general principle of convergence' can hardly be overestimated. Like the theorems of § 69, it gives us a means of deciding whether a function $\phi(n)$ tends to a limit or not, without requiring us to be able to tell beforehand what the limit, if it exists, must be; and it has not the limitations inevitable in theorems of such a special character as those of § 69. But in elementary work it is generally possible to dispense with it, and to obtain all we want from these special theorems. And it will be found that, in spite of the importance of the principle, practically no applications are made of it in the chapters which follow*. We will only remark that, if we suppose that

$$\phi(n) = s_n = u_1 + u_2 + \dots + u_n,$$

we obtain at once a necessary and sufficient condition for the convergence of an infinite series, viz.

THEOREM 2. *A necessary and sufficient condition for the convergence of the series $u_1 + u_2 + \dots$ is that, given any positive number δ, it should be possible to find n_0 so that*

$$|u_{n_1+1} + u_{n_1+2} + \dots + u_{n_2}| < \delta$$

for all values of n_1 and n_2 such that $n_2 > n_1 \geqq n_0$.

* A few proofs given in Ch. VIII can be simplified by the use of the principle.

85. Limits of complex functions and series of complex terms. In this chapter we have, up to the present, concerned ourselves only with real functions of n and series all of whose terms are real. There is however no difficulty in extending our ideas and definitions to the case in which the functions or the terms of the series are complex.

Suppose that $\phi(n)$ is complex and equal to

$$\rho(n)+i\sigma(n),$$

where $\rho(n)$, $\sigma(n)$ are real functions of n. Then *if $\rho(n)$ and $\sigma(n)$ converge respectively to limits r and s as $n\to\infty$, we shall say that $\phi(n)$ converges to the limit $l = r+is$*, and write

$$\lim \phi(n) = l.$$

Similarly, when u_n is complex and equal to v_n+iw_n, we shall say that *the series*

$$u_1+u_2+u_3+\dots$$

is convergent and has the sum $l = r+is$, if the series

$$v_1+v_2+v_3+\dots, \quad w_1+w_2+w_3+\dots$$

are convergent and have the sums r, s respectively.

To say that $u_1+u_2+u_3+\dots$ is convergent and has the sum l is of course the same as to say that the sum

$$s_n = u_1+u_2+\dots+u_n = (v_1+v_2+\dots+v_n)+i(w_1+w_2+\dots+w_n)$$

converges to the limit l as $n\to\infty$.

In the case of real functions and series we also gave definitions of *divergence* and *oscillation, finite* or *infinite*. But in the case of complex functions and series, where we have to consider the behaviour both of $\rho(n)$ and of $\sigma(n)$, there are so many possibilities that this is hardly worth while. When it is necessary to make further distinctions of this kind, we shall make them by stating the way in which the real or imaginary parts behave when taken separately.

86. The reader will find no difficulty in proving such theorems as the following, which are obvious extensions of theorems already proved for real functions and series.

(1) If $\lim \phi(n) = l$, then $\lim \phi(n+p) = l$ for any fixed value of p.

(2) If $u_1+u_2+\ldots$ is convergent and has the sum l, then $a+b+c+\ldots+k+u_1+u_2+\ldots$ is convergent and has the sum $a+b+c+\ldots+k+l$, and $u_{p+1}+u_{p+2}+\ldots$ is convergent and has the sum $l-u_1-u_2-\ldots-u_p$.

(3) If $\lim \phi(n) = l$ and $\lim \psi(n) = m$, then

$$\lim \{\phi(n)+\psi(n)\} = l+m.$$

(4) If $\lim \phi(n) = l$, then $\lim k\phi(n) = kl$.

(5) If $\lim \phi(n) = l$ and $\lim \psi(n) = m$, then $\lim \phi(n)\psi(n) = lm$.

(6) If $u_1+u_2+\ldots$ converges to the sum l, and $v_1+v_2+\ldots$ to the sum m, then $(u_1+v_1)+(u_2+v_2)+\ldots$ converges to the sum $l+m$.

(7) If $u_1+u_2+\ldots$ converges to the sum l, then $ku_1+ku_2+\ldots$ converges to the sum kl.

(8) If $u_1+u_2+u_3+\ldots$ is convergent, then $\lim u_n = 0$.

(9) If $u_1+u_2+u_3+\ldots$ is convergent, then so is any series formed by grouping the terms in brackets, and the sums of the two series are the same.

As an example, let us prove theorem (5). Let

$$\phi(n) = \rho(n)+i\sigma(n), \quad \psi(n) = \rho'(n)+i\sigma'(n), \quad l = r+is, \quad m = r'+is'.$$

Then $$\rho(n)\to r, \quad \sigma(n)\to s, \quad \rho'(n)\to r', \quad \sigma'(n)\to s'.$$

But $$\phi(n)\psi(n) = \rho\rho'-\sigma\sigma'+i(\rho\sigma'+\rho'\sigma),$$

and $$\rho\rho'-\sigma\sigma'\to rr'-ss', \quad \rho\sigma'+\rho'\sigma\to rs'+r's;$$

so that $$\phi(n)\psi(n)\to rr'-ss'+i(rs'+r's),$$

i.e. $$\phi(n)\psi(n)\to(r+is)(r'+is') = lm.$$

The following theorems are of a rather different character.

(10) *In order that $\phi(n)=\rho(n)+i\sigma(n)$ should converge to zero as $n\to\infty$, it is necessary and sufficient that*

$$|\phi(n)| = \sqrt{[\{\rho(n)\}^2+\{\sigma(n)\}^2]}$$

should converge to zero.

If $\rho(n)$ and $\sigma(n)$ both converge to zero, then it is plain that $\sqrt{(\rho^2+\sigma^2)}$ does so. The converse follows from the fact that the numerical value of ρ or σ cannot be greater than $\sqrt{(\rho^2+\sigma^2)}$.

(11) *More generally, in order that $\phi(n)$ should converge to a limit l, it is necessary and sufficient that*

$$|\phi(n)-l|$$

should converge to zero.

For $\phi(n)-l$ converges to zero, and we can apply (10).

(12) *Theorems* 1 *and* 2 *of* §§ 83–84 *are still true when* $\phi(n)$ *and* u_n *are complex.*

We have to show that a necessary and sufficient condition that $\phi(n)$ should tend to l is that

$$|\phi(n_2)-\phi(n_1)|<\delta \quad \text{.........................(1)}$$

when $n_2>n_1 \geqq n_0$.

If $\phi(n)\to l$ then $\rho(n)\to r$ and $\sigma(n)\to s$, and so we can find numbers n_0' and n_0'' depending on δ and such that

$$|\rho(n_2)-\rho(n_1)|<\tfrac{1}{2}\delta, \quad |\sigma(n_2)-\sigma(n_1)|<\tfrac{1}{2}\delta,$$

the first inequality holding when $n_2>n_1 \geqq n_0'$, and the second when $n_2>n_1 \geqq n_0''$. Hence

$$|\phi(n_2)-\phi(n_1)| \leqq |\rho(n_2)-\rho(n_1)|+|\sigma(n_2)-\sigma(n_1)|<\delta$$

when $n_2>n_1\geqq n_0$, where n_0 is the greater of n_0' and n_0''. Thus the condition (1) is *necessary*. To prove that it is *sufficient* we have only to observe that

$$|\rho(n_2)-\rho(n_1)| \leqq |\phi(n_2)-\phi(n_1)|<\delta$$

when $n_2>n_1\geqq n_0$. Thus $\rho(n)$ tends to a limit r, and in the same way it may be shown that $\sigma(n)$ tends to a limit s.

87. The limit of z^n as $n\to\infty$, z being any complex number. Let us consider the important case in which $\phi(n)=z^n$. This problem has already been discussed for real values of z in § 72.

If $z^n\to l$ then $z^{n+1}\to l$, by (1) of § 86. But, by (4) of § 86,

$$z^{n+1}=zz^n\to zl,$$

and therefore $l=zl$, which is only possible if (a) $l=0$ or (b) $z=1$. If $z=1$ then $\lim z^n=1$. Apart from this special case the limit, if it exists, can only be zero.

Now if $z=r(\cos\theta+i\sin\theta)$, where r is positive, then

$$z^n=r^n(\cos n\theta+i\sin n\theta),$$

so that $|z^n| = r^n$. Thus $|z^n|$ tends to zero if and only if $r < 1$; and it follows from (10) of § 86 that

$$\lim z^n = 0$$

if and only if $r < 1$. In no other case does z^n converge to a limit, except when $z = 1$ and $z^n \to 1$.

88. The geometric series $1+z+z^2+\dots$ when z is complex. Since

$$s_n = 1+z+z^2+\dots+z^{n-1} = \frac{1-z^n}{1-z},$$

unless $z = 1$, when the value of s_n is n, it follows that *the series $1+z+z^2+\dots$ is convergent if and only if $r = |z| < 1$. And its sum when convergent is* $1/(1-z)$.

Thus if $z = r(\cos\theta + i\sin\theta) = r\operatorname{Cis}\theta$, and $r < 1$, we have

$$1+z+z^2+\dots = \frac{1}{1-r\operatorname{Cis}\theta},$$

or

$$1+r\operatorname{Cis}\theta+r^2\operatorname{Cis}2\theta+\dots = \frac{1}{1-r\operatorname{Cis}\theta} = \frac{1-r\cos\theta+ir\sin\theta}{1-2r\cos\theta+r^2}.$$

Separating the real and imaginary parts, we obtain

$$1+r\cos\theta+r^2\cos 2\theta+\dots = \frac{1-r\cos\theta}{1-2r\cos\theta+r^2},$$

$$r\sin\theta+r^2\sin 2\theta+\dots = \frac{r\sin\theta}{1-2r\cos\theta+r^2},$$

provided $r < 1$. If we change θ into $\theta+\pi$, we see that these results hold also for negative values of r numerically less than 1. Thus they hold when $-1 < r < 1$.

Examples XXXIII. 1. Prove directly that $\phi(n) = r^n\cos n\theta$ converges to 0 when $r < 1$ and to 1 when $r = 1$ and θ is a multiple of 2π. Prove further that if $r = 1$ and θ is not a multiple of 2π, then $\phi(n)$ oscillates finitely; if $r > 1$ and θ is a multiple of 2π, then $\phi(n) \to +\infty$; and if $r > 1$ and θ is not a multiple of 2π, then $\phi(n)$ oscillates infinitely.

2. Establish a similar series of results for $\phi(n) = r^n\sin n\theta$.

3. Prove that $$z^m + z^{m+1} + \dots = \frac{z^m}{1-z},$$

$$z^m + 2z^{m+1} + 2z^{m+2} + \dots = z^m \frac{1+z}{1-z},$$

if and only if $|z| < 1$. Which of the theorems of § 86 do you use?

4. Prove that if $-1 < r < 1$ then

$$1 + 2r\cos\theta + 2r^2\cos 2\theta + \dots = \frac{1-r^2}{1-2r\cos\theta + r^2}.$$

5. The series $$1 + \frac{z}{1+z} + \left(\frac{z}{1+z}\right)^2 + \dots$$

converges to the sum $1\Big/\left(1 - \dfrac{z}{1+z}\right) = 1+z$ if $\left|\dfrac{z}{1+z}\right| < 1$. Show that this condition is equivalent to the condition that z has a real part greater than $-\frac{1}{2}$.

89. The symbols O, o, $\sim$. We conclude this chapter with some definitions which we shall not use till later but whose logical place is here.

Suppose that $f(n)$ and $\phi(n)$ are two functions of n defined at any rate for all sufficiently large values of n, say for $n \geqq n_0$; and that $\phi(n)$ is positive and increases or decreases steadily as n increases, so that $\phi(n)$ tends to zero, or to a positive limit, or to infinity, when $n \to \infty$. In practice $\phi(n)$ will be some simple function such as $1/n$, 1, or n. Then we lay down the following definitions.

(i) *If there is a constant K such that*

$$|f| \leqq K\phi$$

for $n \geqq n_0$, we write $$f = O(\phi).$$

(ii) *If* $$f/\phi \to 0$$

when $n \to \infty$, we write $$f = o(\phi).$$

(iii) *If* $$f/\phi \to l,$$

where $l \neq 0$, we write $$f \sim l\phi.$$

In particular $$f = O(1)$$

means that f is bounded (so that it either tends to a limit or oscillates finitely), and $$f = o(1)$$

means that $f \to 0$.

Thus

$$n = O(n^2), \quad 100n^2+1000n = O(n^2), \quad \sin n\theta\pi = O(1),$$

$$n = o(n^2), \quad 100n^2+1000n = o(n^3), \quad \sin n\theta\pi = o(n),$$

$$n+1 \sim n, \quad 100n^2+1000n \sim 100n^2, \quad n+\sin n\theta\pi \sim n,$$

and
$$\frac{a_0 n^p + a_1 n^{p-1} + \dots + a_p}{b_0 n^q + b_1 n^{q-1} + \dots + b_q} \sim \frac{a_0}{b_0} n^{p-q}$$
if $a_0 \neq 0$, $b_0 \neq 0$.

We add one remark to guard against a possible misapprehension. To say '$f = O(\phi)$' is to assert what is often meant by saying that 'f is not of a higher order of magnitude than ϕ', and does not exclude its being of a lower order (as in the first of the relations above).

So far we have defined (e.g.) '$f(n) = O(1)$', or '$f(n) = o(n)$', but not '$O(1)$' or '$o(n)$' in isolation. We can however make our definitions more elastic. We may agree that $O(\phi)$ or $o(\phi)$ denotes *an unspecified f such that $f = O(\phi)$ or $f = o(\phi)$*; and we can then write, for example,

$$O(1)+O(1) = O(1) = o(n),$$

meaning thereby 'if $f = O(1)$ and $g = O(1)$, then $f+g = O(1)$ and *a fortiori* $f+g = o(n)$'. Or again we may write

$$\sum_{r=1}^{n} O(1) = O(n)$$

meaning that the sum of n terms, each less in absolute value than a constant, is less than a constant multiple of n.

It will be observed that our formulae involving O and o will usually not be reversible. Thus '$o(1) = O(1)$', i.e. 'if $f = o(1)$ then $f = O(1)$', is true, but '$O(1) = o(1)$' is false.

It is easy to formulate a number of general properties of our symbols, such as

(1) $O(\phi)+O(\psi) = O(\phi+\psi)$,

(2) $O(\phi)\,O(\psi) = O(\phi\psi)$,

(3) $O(\phi)\,o(\psi) = o(\phi\psi)$,

(4) if $f \sim \phi$ then $f+o(\phi) \sim \phi$.

Such theorems are immediate corollaries from the definitions.

The usefulness of these definitions, and of the corresponding definitions for functions of a continuous variable, will appear more clearly in later chapters.

MISCELLANEOUS EXAMPLES ON CHAPTER IV

1. The function $\phi(n)$ takes the values 1, 0, 0, 0, 1, 0, 0, 0, 1, ... when $n = 0, 1, 2, \ldots$. Express $\phi(n)$ in terms of n by a formula which does not involve trigonometrical functions. [$\phi(n) = \frac{1}{4}\{1+(-1)^n+i^n+(-i)^n\}$.]

2. If $\phi(n)$ steadily increases, and $\psi(n)$ steadily decreases, as n tends to ∞, and if $\psi(n) > \phi(n)$ for all values of n, then both $\phi(n)$ and $\psi(n)$ tend to limits, and $\lim \phi(n) \leqq \lim \psi(n)$. [This is an immediate corollary from § 69.]

3. Prove that, if

$$\phi(n) = \left(1+\frac{1}{n}\right)^n, \quad \psi(n) = \left(1-\frac{1}{n}\right)^{-n},$$

then $\phi(n+1) > \phi(n)$ and $\psi(n+1) < \psi(n)$. [The first result has already been proved in § 73.]

4. Prove also that $\psi(n) > \phi(n)$ for all values of n: and deduce (by means of the preceding examples) that both $\phi(n)$ and $\psi(n)$ tend to limits as n tends to ∞*.

5. The arithmetic mean of the products of all distinct pairs of positive integers whose sum is n is denoted by S_n. Show that $\lim (S_n/n^2) = 1/6$.
(*Math. Trip.* 1903)

6. If $x_1, x_2, \ldots, x_n$ are positive, $\Sigma x_r = n$, the x_r are not all equal to 1, and m is rational and greater than 1, then $\Sigma x_r^m > n$. (*Math. Trip.* 1934)

[Use the inequality $x^m - 1 > m(x-1)$, true for all positive x but 1 (§ 74).]

7. If $\phi(n)$ is a positive integer for all values of n, and tends to ∞ with n, then $x^{\phi(n)}$ tends to 0 if $0 < x < 1$ and to $+\infty$ if $x > 1$. Discuss the behaviour of $x^{\phi(n)}$, as $n \to \infty$, for other values of x.

8†. If a_n increases or decreases steadily as n increases, then the same is true of $(a_1+a_2+\ldots+a_n)/n$.

9. The function $f(x)$ is increasing and continuous (see Ch. V) for all values of x, and a sequence $x_1, x_2, x_3, \ldots$ is defined by the equation $x_{n+1} = f(x_n)$. Discuss on general graphical grounds the question whether x_n tends to a root of the equation $x = f(x)$. Consider in particular the case in which this equation has only one root, distinguishing the cases in which the curve $y = f(x)$ crosses the line $y = x$ from above to below and from below to above.

* We shall prove in Ch. IX that $\lim \{\psi(n) - \phi(n)\} = 0$, and that therefore each function tends to the limit e.

† Exs. 8–11 and 15 are taken from Bromwich's *Infinite series*.

10. If $x_{n+1} = k/(1+x_n)$, and k and x_1 are positive, then the sequences $x_1, x_3, x_5, \ldots$ and $x_2, x_4, x_6, \ldots$ are one an increasing and the other a decreasing sequence, and each sequence tends to the limit α, the positive root of the equation $x^2+x = k$.

11. If $x_{n+1} = \sqrt{(k+x_n)}$, and k and x_1 are positive, then the sequence $x_1, x_2, x_3, \ldots$ is an increasing or decreasing sequence according as x_1 is less than or greater than α, the positive root of the equation $x^2 = x+k$; and in either case $x_n \to \alpha$ as $n \to \infty$.

12. A sequence of numbers x_n is defined by

$$x_1 = h, \quad x_{n+1} = x_n^2+k,$$

where $0<k<\frac{1}{4}$ and h lies between the roots a and b of the equation

$$x^2-x+k = 0.$$

Prove that $$a<x_{n+1}<x_n<b,$$
and determine the limit of x_n. (*Math. Trip.* 1931)

13. Prove that if $x_1 = \frac{1}{2}\{x+(A/x)\}$, $x_2 = \frac{1}{2}\{x_1+(A/x_1)\}$, and so on, x and A being positive, then $\lim x_n = \sqrt{A}$.

[It can be proved that $\dfrac{x_n-\sqrt{A}}{x_n+\sqrt{A}} = \left(\dfrac{x-\sqrt{A}}{x+\sqrt{A}}\right)^{2^n}$.]

14. A sequence u_n is determined by the relations

$$u_1 = \alpha+\beta, \quad u_n = \alpha+\beta-\frac{\alpha\beta}{u_{n-1}} \quad (n>1),$$

where $\alpha>\beta>0$. Show that

$$u_n = \frac{\alpha^{n+1}-\beta^{n+1}}{\alpha^n-\beta^n},$$

and determine the limit of u_n when $n \to \infty$.

Discuss the case $\alpha = \beta > 0$. (*Math. Trip.* 1933)

15. If x_1, x_2 are positive and $x_{n+1} = \frac{1}{2}(x_n+x_{n-1})$, then the sequences $x_1, x_3, x_5, \ldots$ and $x_2, x_4, x_6, \ldots$ are one a decreasing and the other an increasing sequence, and they have the common limit $\frac{1}{3}(x_1+2x_2)$.

16. If $\lim_{n\to\infty} s_n = l$, then

$$\lim_{n\to\infty} \frac{s_1+s_2+\ldots+s_n}{n} = l.$$

[Let $s_n = l+t_n$. Then we have to prove that $(t_1+t_2+\ldots+t_n)/n$ tends to zero if t_n does so.

We divide the numbers $t_1, t_2, \ldots, t_n$ into two sets $t_1, t_2, \ldots, t_p$ and $t_{p+1}, t_{p+2}, \ldots, t_n$. Here we suppose that p is a function of n which tends to ∞ as $n \to \infty$, but *more slowly than* n, so that $p \to \infty$ and $p/n \to 0$: e.g. we might suppose p to be the integral part of $\sqrt{n}$.

Let δ be any positive number. However small δ may be, we can choose n_0 so that $t_{p+1}, t_{p+2}, \dots, t_n$ are all numerically less than $\frac{1}{2}\delta$ when $n \geqq n_0$, and so

$$|(t_{p+1}+t_{p+2}+\dots+t_n)/n| < \tfrac{1}{2}\delta(n-p)/n < \tfrac{1}{2}\delta.$$

But, if A is the greatest of the moduli of all the numbers $t_1, t_2, \dots$, we have

$$|(t_1+t_2+\dots+t_p)/n| < pA/n,$$

and this also will be less than $\frac{1}{2}\delta$ when $n \geqq n_0$, if n_0 is large enough, since $p/n \to 0$ as $n \to \infty$. Thus

$$|(t_1+t_2+\dots+t_n)/n| \leqq |(t_1+t_2+\dots+t_p)/n| + |(t_{p+1}+t_{p+2}+\dots+t_n)/n| < \delta$$

when $n \geqq n_0$; which proves the theorem.

The reader, if he desires to become expert in dealing with questions about limits, should study the argument above with great care. It is very often necessary, in proving the limit of some given expression to be zero, to split it into two parts which have to be proved to have the limit zero in slightly different ways. When this is the case the proof is never very easy.

The point of the proof is this: we have to prove that $(t_1+t_2+\dots+t_n)/n$ is small when n is large, the t's being small when their suffixes are large. We split up the terms in the brackets into two groups. The terms in the first group are not all small, but their number is small compared with n. The number in the second group is *not* small compared with n, but the terms are all small and their number at any rate less than n, so that their sum is small compared with n. Hence each of the parts into which $(t_1+t_2+\dots+t_n)/n$ has been divided is small when n is large.]

17. If $\phi(n)-\phi(n-1) \to l$ as $n \to \infty$, then $\phi(n)/n \to l$.

[If $\phi(n) = s_1+s_2+\dots+s_n$, then $\phi(n)-\phi(n-1) = s_n$, and the theorem reduces to that proved in the preceding example.]

18. If $s_n = \frac{1}{2}\{1-(-1)^n\}$, so that s_n is equal to 1 or 0 according as n is odd or even, then $(s_1+s_2+\dots+s_n)/n \to \frac{1}{2}$ as $n \to \infty$.

[This example proves that the converse of Ex. 16 is not true: for s_n oscillates as $n \to \infty$.]

19. If c_n, s_n denote the sums of the first n terms of the series

$$\tfrac{1}{2}+\cos\theta+\cos 2\theta+\dots, \quad \sin\theta+\sin 2\theta+\dots,$$

and θ is not a multiple of 2π, then

$$\lim(c_1+c_2+\dots+c_n)/n = 0, \quad \lim(s_1+s_2+\dots+s_n)/n = \tfrac{1}{2}\cot\tfrac{1}{2}\theta.$$

20. A sequence y_n is defined in terms of a sequence x_n by the relations

$$y_0 = x_0, \quad y_n = x_n - \alpha x_{n-1} \qquad (n>0),$$

where $|\alpha| < 1$. Express x_n in terms of the sequence y_n, and show that, if $y_n \to l$, then $x_n \to l/(1-\alpha)$. (*Math. Trip.* 1932)

21. Draw a graph of the function y defined by the equation

$$y = \lim_{n\to\infty} \frac{x^{2n}\sin\frac{1}{2}\pi x + x^2}{x^{2n}+1}. \qquad \textit{(Math. Trip. 1901)}$$

22. The function $$y = \lim_{n\to\infty} \frac{1}{1+n\sin^2 \pi x}$$

is equal to 0 except when x is an integer, and then equal to 1. The function

$$y = \lim_{n\to\infty} \frac{\psi(x)+n\phi(x)\sin^2\pi x}{1+n\sin^2\pi x}$$

is equal to $\phi(x)$ unless x is an integer, and then equal to $\psi(x)$.

23. Show that the graph of the function

$$y = \lim_{n\to\infty} \frac{x^n\phi(x)+x^{-n}\psi(x)}{x^n+x^{-n}}$$

is composed of parts of the graphs of $\phi(x)$ and $\psi(x)$, together with (as a rule) two isolated points. Is y defined when (a) $x = 1$, (b) $x = -1$, (c) $x = 0$?

24. Prove that the function y which is equal to 0 when x is rational, and to 1 when x is irrational, may be represented in the form

$$y = \lim_{m\to\infty} \operatorname{sgn}\{\sin^2(m!\,\pi x)\},$$

where $$\operatorname{sgn} x = \lim_{n\to\infty} \frac{2}{\pi}\arctan(nx),$$

as in Ex. XXXI. 14. [If x is rational then $\sin^2(m!\,\pi x)$, and therefore $\operatorname{sgn}\{\sin^2(m!\,\pi x)\}$, is equal to zero from a certain value of m onwards: if x is irrational then $\sin^2(m!\,\pi x)$ is always positive, and so $\operatorname{sgn}\{\sin^2(m!\,\pi x)\}$ is always equal to 1.]

Prove that y may also be represented in the form

$$1 - \lim_{m\to\infty}\,[\lim_{n\to\infty}\{\cos(m!\,\pi x)\}^{2n}].$$

25. Sum the series

$$\sum_1^\infty \frac{1}{\nu(\nu+1)}, \quad \sum_1^\infty \frac{1}{\nu(\nu+1)\dots(\nu+k)}.$$

[Since

$$\frac{1}{\nu(\nu+1)\dots(\nu+k)} = \frac{1}{k}\left\{\frac{1}{\nu(\nu+1)\dots(\nu+k-1)} - \frac{1}{(\nu+1)(\nu+2)\dots(\nu+k)}\right\},$$

we have

$$\sum_1^n \frac{1}{\nu(\nu+1)\dots(\nu+k)} = \frac{1}{k}\left\{\frac{1}{1.2\dots k} - \frac{1}{(n+1)(n+2)\dots(n+k)}\right\}$$

and so $$\sum_1^\infty \frac{1}{\nu(\nu+1)\dots(\nu+k)} = \frac{1}{k(k!)}.]$$

26. If $|z|<|\alpha|$, then

$$\frac{L}{z-\alpha}=-\frac{L}{\alpha}\left(1+\frac{z}{\alpha}+\frac{z^2}{\alpha^2}+\dots\right);$$

and if $|z|>|\alpha|$, then $\dfrac{L}{z-\alpha}=\dfrac{L}{z}\left(1+\dfrac{\alpha}{z}+\dfrac{\alpha^2}{z^2}+\dots\right)$.

27. **Expansion of $(Az+B)/(az^2+2bz+c)$ in powers of z.** Let α, β be the roots of $az^2+2bz+c=0$, so that $az^2+2bz+c=a(z-\alpha)(z-\beta)$. We shall suppose that A, B, a, b, c are all real, and α and β unequal. It is then easy to verify that

$$\frac{Az+B}{az^2+2bz+c}=\frac{1}{a(\alpha-\beta)}\left(\frac{A\alpha+B}{z-\alpha}-\frac{A\beta+B}{z-\beta}\right).$$

There are two cases, according as $b^2>ac$ or $b^2<ac$.

(1) If $b^2>ac$ then the roots α, β are real and distinct. If $|z|$ is less than either $|\alpha|$ or $|\beta|$, we can expand $1/(z-\alpha)$ and $1/(z-\beta)$ in ascending powers of z (Ex. 26). If $|z|$ is greater than either $|\alpha|$ or $|\beta|$, we must expand in descending powers of z; while if $|z|$ lies between $|\alpha|$ and $|\beta|$ one fraction must be expanded in ascending and one in descending powers of z. The reader should write down the actual results. If $|z|$ is equal to $|\alpha|$ or $|\beta|$ then no such expansion is possible.

(2) If $b^2<ac$ then the roots are conjugate complex numbers (Ch. III, § 43), and we can write

$$\alpha=\rho\operatorname{Cis}\phi,\quad \beta=\rho\operatorname{Cis}(-\phi),$$

where $\rho^2=\alpha\beta=c/a$, $\rho\cos\phi=\frac{1}{2}(\alpha+\beta)=-b/a$, so that $\cos\phi=-\sqrt{(b^2/ac)}$, $\sin\phi=\sqrt{\{1-(b^2/ac)\}}$.

If $|z|<\rho$ then each fraction may be expanded in ascending powers of z. The coefficient of z^n will be found to be

$$\frac{A\rho\sin n\phi+B\sin\{(n+1)\phi\}}{a\rho^{n+1}\sin\phi}.$$

If $|z|>\rho$ we obtain a similar expansion in descending powers, while if $|z|=\rho$ no such expansion is possible.

28. Show that if $|z|<1$ then

$$1+2z+3z^2+\dots+(n+1)z^n+\dots=1/(1-z)^2.$$

[The sum to n terms is $\dfrac{1-z^n}{(1-z)^2}-\dfrac{nz^n}{1-z}$.]

29. Expand $L/(z-\alpha)^2$ in powers of z, ascending or descending according as $|z|<|\alpha|$ or $|z|>|\alpha|$.

30. Show that if $b^2=ac$ and $|az|<|b|$ then

$$\frac{Az+B}{az^2+2bz+c}=\sum_0^\infty p_n z^n,$$

where $p_n = (-a)^n b^{-n-2}\{(n+1)aB - nbA\}$; and find the corresponding expansion, in descending powers of z, which holds when $|az| > |b|$.

31. If $a/(a+bz+cz^2) = 1+p_1z+p_2z^2+\ldots$, then

$$1+p_1^2z+p_2^2z^2+\ldots = \frac{a+cz}{a-cz}\frac{a^2}{a^2-(b^2-2ac)z+c^2z^2}.$$

(*Math. Trip.* 1900)

32. If $\sin 2^n\theta\pi \to l$ when $n\to\infty$, then $l=0$ and θ is a rational number whose denominator is a power of 2.

[Plainly
$$2^n\theta = p_n + c + \eta_n,$$
where p_n is an integer, c a constant, and $\eta_n \to 0$; and hence
$$p_{n+1} - 2p_n - c + \eta_{n+1} - 2\eta_n = 0.$$
Since $p_{n+1}-2p_n$ is an integer, this is only possible if (i) $c=0$, so that $l=0$ and (ii) $p_{n+1} = 2p_n$ and $\eta_{n+1} = 2\eta_n$ from a certain value of n on, say for $n \geqq n_0$. But then
$$2^\nu \eta_{n_0} = \eta_{n_0+\nu} \to 0$$
when $\nu\to\infty$; and this is only possible if $\eta_{n_0} = 0$, so that $2^{n_0}\theta = p_{n_0}$.

It is instructive to consider $\sin a^n\theta\pi$, where a is integral and greater than 2. It is then possible that $l \neq 0$; thus $\sin 9^n\theta\pi \to 1$ when $\theta = \frac{1}{2}$.]

33. If $P(n)$ is a polynomial in n, with integral coefficients and of degree m, and $\sin\{P(n)\theta\pi\} \to 0$, then θ is rational.

[It is best to prove more*, viz. that if
$$P(n)\theta = k_n + a_n + \epsilon_n \quad \ldots\ldots\ldots\ldots\ldots\ldots\ldots\ldots(1),$$
where k_n is an integer, a_n has one or other of any finite system of values, and $\epsilon_n \to 0$, then θ is rational.

In the first place, if we replace n by $n+1$ in (1), subtract, and observe (i) that $P(n+1)-P(n)$ is a polynomial of degree $m-1$ and (ii) that $a_{n+1}-a_n$ has only a finite number of possible values, we obtain an induction from $m-1$ to m. The problem is thus reduced to the case $m=1$, $P(n) = An+B$. In this case (1) gives
$$A\theta = (k_{n+1}-k_n) + (a_{n+1}-a_n) + (\epsilon_{n+1}-\epsilon_n).$$
This is only possible if $\epsilon_{n+1}-\epsilon_n = 0$ for $n \geqq n_0$; and then
$$a_n = a_{n_0} + l_n + (n-n_0)A\theta,$$
where l_n is an integer, for $n\geqq n_0$. Since a_n has only a finite number of values, $l_n + nA\theta$ has only a finite number of values, and therefore θ is rational.]

* This argument is due to Mr Ingham.

CHAPTER V

LIMITS OF FUNCTIONS OF A CONTINUOUS VARIABLE. CONTINUOUS AND DISCONTINUOUS FUNCTIONS

90. Limits as x tends to ∞. We shall now return to functions of a continuous real variable. We shall confine ourselves entirely to *one-valued* functions*, and we shall denote such a function by $\phi(x)$. We suppose x to assume successively all values corresponding to points on our fundamental straight line Λ, starting from some definite point on the line and progressing always to the right. In these circumstances we say that x *tends to infinity*, or to ∞, and write $x \to \infty$. The only difference between the 'tending of n to ∞' discussed in the last chapter, and this 'tending of x to ∞', is that x assumes all values as it tends to ∞, i.e. that the point P which corresponds to x coincides in turn with every point of Λ to the right of its initial position, whereas n tended to ∞ by a series of jumps. We can express this distinction by saying that x tends *continuously* to ∞.

As we explained at the beginning of the last chapter, there is a very close correspondence between functions of x and functions of n. Every function of n may be regarded as a selection from the values of a function of x. In the last chapter we discussed the peculiarities which may characterise the behaviour of a function $\phi(n)$ as n tends to ∞. Now we are concerned with the same problem for a function $\phi(x)$; and the definitions and theorems to which we are led are practically repetitions of those of the last chapter. Thus corresponding to Def. I of § 58 we have:

DEFINITION 1. *The function $\phi(x)$ is said to tend to the limit l as x tends to ∞ if, when any positive number δ, however small, is*

* Thus $\sqrt{x}$ stands in this chapter for the one-valued function $+\sqrt{x}$ and not (as in § 26) for the two-valued function whose values are $+\sqrt{x}$ and $-\sqrt{x}$.

assigned, a number $x_0(\delta)$ *can be chosen such that, for all values of* x *equal to or greater than* $x_0(\delta)$, $\phi(x)$ *differs from* l *by less than* δ, *i.e. if*

$$|\phi(x)-l|<\delta$$

when $x \geqq x_0(\delta)$.

When this is the case we may write

$$\lim_{x\to\infty} \phi(x) = l,$$

or, when there is no risk of ambiguity, simply $\lim \phi(x) = l$, or $\phi(x) \to l$. Similarly we have:

Definition 2. *The function* $\phi(x)$ *is said to tend to* ∞ *with* x *if, when any number* Δ, *however large, is assigned, we can choose a number* $x_0(\Delta)$ *such that*

$$\phi(x) > \Delta$$

when $x \geqq x_0(\Delta)$.

We then write $$\phi(x)\to\infty.$$

Similarly we define $\phi(x)\to-\infty$*. Finally we have:

Definition 3. *If the conditions of neither of the two preceding definitions are satisfied, then* $\phi(x)$ *is said to oscillate as* x *tends to* ∞. *If* $|\phi(x)|$ *is less than some constant* K *when* $x \geqq x_0$†, *then* $\phi(x)$ *is said to oscillate finitely, and otherwise infinitely.*

The reader will remember that in the last chapter we considered very carefully various less formal ways of expressing the facts represented by the formulae $\phi(n)\to l$, $\phi(n)\to\infty$. Similar modes of expression may of course be used in the present case. Thus we may say that $\phi(x)$ is small or nearly equal to l or large when x is large, using the words 'small', 'nearly', 'large' in a sense similar to that in which they were used in Ch. IV.

Examples XXXIV. 1. Consider the behaviour of the following functions as $x\to\infty$: $1/x$, $1+(1/x)$, x^2, x^k, $[x]$, $x-[x]$, $[x]+\sqrt{\{x-[x]\}}$.

The first four functions correspond exactly to functions of n fully discussed in Ch. IV. The graphs of the last three were constructed in

* We shall sometimes find it convenient to write $+\infty$, $x\to+\infty$, $\phi(x)\to+\infty$ instead of ∞, $x\to\infty$, $\phi(x)\to\infty$.

† In the corresponding definition of § 62, we postulated that $|\phi(n)|<K$ for *all* values of n, and not merely when $n \geqq n_0$. But then the two hypotheses would have been equivalent; for if $|\phi(n)|<K$ when $n\geqq n_0$, then $|\phi(n)|\leqq K'$ for all values of n, where K' is the greatest of $|\phi(1)|$, $|\phi(2)|$, ..., $|\phi(n_0-1)|$ and K. Here the matter is not quite so simple, since there are infinitely many values of x less than x_0.

Ch. II (Exs. XVI. 1, 2, 4), and the reader will see at once that $[x] \to \infty$, $x-[x]$ oscillates finitely, and $[x]+\sqrt{\{x-[x]\}} \to \infty$.

One simple remark may be inserted here. The function $\phi(x) = x-[x]$ oscillates between 0 and 1, as is obvious from the form of its graph. It is equal to zero whenever x is an integer, so that the function $\phi(n)$ derived from it is always zero and so tends to the limit zero. The same is true if

$$\phi(x) = \sin x\pi, \quad \phi(n) = \sin n\pi = 0.$$

It is evident that $\phi(x) \to l$ or $\phi(x) \to \infty$ or $\phi(x) \to -\infty$ involves the corresponding property for $\phi(n)$, but that the converse is often untrue.

2. Consider in the same way the functions:

$$\frac{\sin x\pi}{x}, \quad x \sin x\pi, \quad (x \sin x\pi)^2, \quad \tan x\pi, \quad \frac{\tan x\pi}{x}, \quad a\cos^2 x\pi + b\sin^2 x\pi,$$

illustrating your remarks by means of the graphs of the functions.

3. Give a geometrical explanation of Def. 1, analogous to the geometrical explanation of Ch. IV, § 59.

4. If $\phi(x) \to l$, and l is not zero, then $\phi(x) \cos x\pi$ and $\phi(x) \sin x\pi$ oscillate finitely. If $\phi(x) \to \infty$ or $\phi(x) \to -\infty$, then they oscillate infinitely. The graph of either function is a wavy curve oscillating between the curves $y = \phi(x)$ and $y = -\phi(x)$.

5. Discuss the behaviour, as $x \to \infty$, of the function

$$y = f(x) \cos^2 x\pi + F(x) \sin^2 x\pi,$$

where $f(x)$ and $F(x)$ are some pair of simple functions (e.g. x and x^2). [The graph of y is a curve oscillating between the curves $y = f(x)$, $y = F(x)$.]

91. Limits as x tends to $-\infty$. The reader will have no difficulty in framing for himself definitions of the meaning of the assertions 'x tends to $-\infty$', or '$x \to -\infty$' and

$$\lim_{x \to -\infty} \phi(x) = l, \quad \phi(x) \to \infty, \quad \phi(x) \to -\infty.$$

In fact, if $x = -y$ and $\phi(x) = \phi(-y) = \psi(y)$, then y tends to ∞ as x tends to $-\infty$, and the question of the behaviour of $\phi(x)$ as x tends to $-\infty$ is the same as that of the behaviour of $\psi(y)$ as y tends to ∞.

92. Theorems corresponding to those of Ch. IV, §§ 63–69. The theorems concerning the sums, products, and quotients of functions proved in Ch. IV are all true (with obvious verbal alterations which the reader will have no difficulty in supplying) for functions of the continuous variable x. Not only the enunciations but the proofs remain substantially the same.

The definition which corresponds to that of § 69 is as follows: *the function $\phi(x)$ will be said to increase steadily with x if $\phi(x_2) \geqq \phi(x_1)$ whenever $x_2 > x_1$.* In many cases, of course, this condition is only satisfied from a definite value of x onwards, i.e. when $x_2 > x_1 \geqq x_0$. The theorem which follows in that section requires no alteration but that of n into x: and the proof is the same, except for obvious verbal changes.

If $\phi(x_2) > \phi(x_1)$, the possibility of equality being excluded, whenever $x_2 > x_1$, then $\phi(x)$ will be said to be *steadily and strictly increasing*, or simply *strictly increasing*. We shall find that the distinction is often important (cf. §§ 109, 110).

The reader should consider whether or no the following functions increase steadily with x (or at any rate increase steadily from a certain value of x onwards): $x^2 - x$, $x + \sin x$, $x + 2\sin x$, $x^2 + 2\sin x$, $[x]$, $[x] + \sin x$, $[x] + \sqrt{\{x - [x]\}}$. All these functions tend to ∞ when $x \to \infty$.

93. Limits as x tends to 0. Let $\phi(x)$ be a function of x such that $\lim\limits_{x\to\infty} \phi(x) = l$, and let $y = 1/x$. Then

$$\phi(x) = \phi(1/y) = \psi(y),$$

say. When x tends to ∞, y tends to the limit 0, and $\psi(y)$ tends to the limit l.

Let us now dismiss x and consider $\psi(y)$ simply as a function of y. We are for the moment concerned only with those values of y which correspond to large positive values of x, that is to say with small positive values of y. And $\psi(y)$ has the property that by making y sufficiently small we can make $\psi(y)$ differ by as little as we please from l. To put the matter more precisely, the statement expressed by $\lim \phi(x) = l$ means that, when any positive number δ, however small, is assigned, we can choose x_0 so that $|\phi(x) - l| < \delta$ for all values of x greater than or equal to x_0. But this is the same thing as saying that we can choose $y_0 = 1/x_0$ so that $|\psi(y) - l| < \delta$ for all positive values of y less than or equal to y_0.

We are thus led to the following definitions:

A. *If, when any positive number δ, however small, is assigned, we can choose $y_0(\delta)$ so that*

$$|\phi(y) - l| < \delta$$

when $0 < y \leqq y_0(\delta)$, *then we say that* $\phi(y)$ *tends to the limit* l *as* y *tends to* 0 *by positive values, and we write*

$$\lim_{y \to +0} \phi(y) = l.$$

B. *If, when any number* Δ, *however large, is assigned, we can choose* $y_0(\Delta)$ *so that*

$$\phi(y) > \Delta$$

when $0 < y \leqq y_0(\Delta)$, *then we say that* $\phi(y)$ *tends to* ∞ *as* y *tends to* 0 *by positive values, and we write*

$$\phi(y) \to \infty.$$

We define in a similar way the meaning of '$\phi(y)$ tends to the limit l as y tends to 0 by negative values', or '$\lim \phi(y) = l$ when $y \to -0$'. We have in fact only to alter $0 < y \leqq y_0(\delta)$ to $-y_0(\delta) \leqq y < 0$ in definition A. There is of course a corresponding analogue of definition B, and similar definitions in which

$$\phi(y) \to -\infty$$

as $y \to +0$ or $y \to -0$.

If $\lim_{y \to +0} \phi(y) = l$ and $\lim_{y \to -0} \phi(y) = l$, we write simply

$$\lim_{y \to 0} \phi(y) = l.$$

This case is so important that it is worth while to give a formal definition.

If, when any positive number δ, *however small, is assigned, we can choose* $y_0(\delta)$ *so that, for all values of* y *different from zero but numerically less than or equal to* $y_0(\delta)$, $\phi(y)$ *differs from* l *by less than* δ, *then we say that* $\phi(y)$ *tends to the limit* l *as* y *tends to* 0, *and write*

$$\lim_{y \to 0} \phi(y) = l.$$

So also, if $\phi(y) \to \infty$ as $y \to +0$ and also as $y \to -0$, we say that $\phi(y) \to \infty$ as $y \to 0$. We define similarly the statement that $\phi(y) \to -\infty$ as $y \to 0$.

Finally, if $\phi(y)$ does not tend to a limit, or to ∞, or to $-\infty$, as $y \to +0$, we say that $\phi(y)$ oscillates as $y \to +0$, finitely or infinitely as the case may be; and we define oscillation as $y \to -0$ similarly.

The preceding definitions have been stated in terms of a variable denoted by y: what letter is used is of course immaterial, and we may suppose x written instead of y throughout them.

94. Limits as x tends to a. Suppose next that $\phi(y) \to l$ as $y \to 0$, and write

$$y = x - a, \quad \phi(y) = \phi(x-a) = \psi(x).$$

If $y \to 0$ then $x \to a$ and $\psi(x) \to l$, and we are naturally led to write

$$\lim_{x \to a} \psi(x) = l,$$

or simply $\lim \psi(x) = l$ or $\psi(x) \to l$, and to say that $\psi(x)$ *tends to the limit l as x tends to a.* The meaning of this equation may be formally and directly defined as follows: *if, given δ, we can always determine $\epsilon(\delta)$ so that*

$$|\phi(x) - l| < \delta$$

when $0 < |x-a| \leqq \epsilon(\delta)$, *then*

$$\lim_{x \to a} \phi(x) = l.$$

By restricting ourselves to values of x greater than a, i.e. by replacing $0 < |x-a| \leqq \epsilon(\delta)$ by $a < x \leqq a + \epsilon(\delta)$, we define '$\phi(x)$ tends to l when x approaches a from the right', which we may write as

$$\lim_{x \to a+0} \phi(x) = l.$$

In the same way we can define the meaning of

$$\lim_{x \to a-0} \phi(x) = l.$$

Thus $\lim\limits_{x \to a} \phi(x) = l$ is equivalent to the two assertions

$$\lim_{x \to a+0} \phi(x) = l, \quad \lim_{x \to a-0} \phi(x) = l.$$

We can give similar definitions referring to the cases in which $\phi(x) \to \infty$ or $\phi(x) \to -\infty$ as $x \to a$ through values greater or less than a; but it is probably unnecessary to dwell further on these definitions, since they are quite similar to those stated above in the special case when $a = 0$, and we can always discuss the behaviour of $\phi(x)$ as $x \to a$ by putting $x - a = y$ and supposing that $y \to 0$.

95. Steadily increasing or decreasing functions. If there is a number ϵ such that $\phi(x') \leqq \phi(x'')$ whenever $a-\epsilon < x' < x'' < a+\epsilon$, then $\phi(x)$ will be said to *increase steadily in the neighbourhood of* $x = a$.

Suppose first that $x < a$, and put $y = 1/(a-x)$. Then $y \to \infty$ as $x \to a-0$, and $\phi(x) = \psi(y)$ is a steadily increasing function of y, never greater than $\phi(a)$. It follows from § 92 that $\phi(x)$ tends to a limit not greater than $\phi(a)$. We shall write

$$\lim_{x \to a-0} \phi(x) = \phi(a-0)^*.$$

We can define $\phi(a-0)$ in a similar manner; and it is clear that

$$\phi(a-0) \leqq \phi(a) \leqq \phi(a+0).$$

It is obvious that similar considerations may be applied to *decreasing* functions.

If $\phi(x') < \phi(x'')$, the possibility of equality being excluded, whenever $a-\epsilon < x' < x'' < a+\epsilon$, then $\phi(x)$ will be said to be *steadily increasing in the stricter sense*.

96. Limits of indetermination and the principle of convergence. All of the argument of §§ 80–84 may be applied to functions of a continuous variable x which tends to a limit a. In particular, if $\phi(x)$ is *bounded* in an interval including a (i.e. if we can find ϵ, H, and K so that $H < \phi(x) < K$ when $a-\epsilon \leqq x \leqq a+\epsilon$)†, then we can define λ and Λ, the lower and upper limits of indetermination of $\phi(x)$ as $x \to a$, and prove that $\lambda = \Lambda = l$ is a necessary and sufficient condition for $\phi(x)$ to tend to l. We can also establish the analogue of the principle of convergence, i.e. prove that *a necessary and sufficient condition that $\phi(x)$ should tend to a limit is that, when δ is given, we can choose $\epsilon(\delta)$ so that $|\phi(x_2)-\phi(x_1)| < \delta$ if $0 < |x_2-a| < |x_1-a| \leqq \epsilon(\delta)$*. Similarly, *a necessary and sufficient condition that $\phi(x)$ should tend to a limit when $x \to \infty$ is that $|\phi(x_2)-\phi(x_1)| < \delta$ if $x_2 > x_1 \geqq X(\delta)$*.

Examples XXXV. 1. If $\phi(x) \to l$, $\psi(x) \to l'$, as $x \to a$, then

$$\phi(x)+\psi(x) \to l+l', \quad \phi(x)\psi(x) \to ll', \quad \phi(x)/\psi(x) \to l/l',$$

unless in the last case $l' = 0$.

[We saw in § 92 that the theorems of Ch. IV, §§ 63 *et seq.* hold also for functions of x when $x \to \infty$ or $x \to -\infty$. By putting $x = 1/y$ we may extend them to functions of y, when $y \to 0$, and by putting $y = z-a$ to functions of z, when $z \to a$.

* It will of course be understood that $\phi(a-0)$ has no meaning other than that of a conventional abbreviation for the limit on the left-hand side.

We can use the symbols $\phi(a+0)$ and $\phi(a-0)$ whenever the limits by which they are defined exist; but they will not usually satisfy inequalities like those in the text.

† See § 103.

The reader should also prove them directly from the formal definition given above. Thus, in order to obtain a direct proof of the first result he need only take the proof of Theorem I of § 63 and write throughout x for n, a for ∞ and $0<|x-a|\leqq\epsilon$ for $n\geqq n_0$.]

2. If m is a positive integer, then $x^m\to 0$ as $x\to 0$.

3. If m is a negative integer, then $x^m\to+\infty$ as $x\to+0$, while $x^m\to-\infty$ or $x^m\to+\infty$ as $x\to-0$, according as m is odd or even. If $m=0$, then $x^m=1$ and $x^m\to 1$.

4. $$\lim_{x\to 0}(a+bx+cx^2+\dots+kx^m)=a.$$

5. $$\lim_{x\to 0}\frac{a+bx+\dots+kx^m}{\alpha+\beta x+\dots+\kappa x^\mu}=\frac{a}{\alpha},$$

unless $\alpha=0$. If $\alpha=0$ and $a\neq 0$, $\beta\neq 0$, then the function tends to $+\infty$ or $-\infty$, as $x\to+0$, according as a and β have like or unlike signs; the case is reversed if $x\to-0$. The case in which both a and α vanish is considered in Ex. xxxvi. 5. Discuss the cases which arise when $a\neq 0$ and more than one of the first coefficients in the denominator vanish.

6. $\lim\limits_{x\to a} x^m=a^m$, if m is any positive or negative integer, except when $a=0$ and m is negative [If $m>0$, put $x=y+a$ and apply Ex. 4. When $m<0$, the result follows from Ex. 1 above. It follows at once that $\lim P(x)=P(a)$, if $P(x)$ is any polynomial.]

7. $\lim\limits_{x\to a} R(x)=R(a)$, if R denotes any rational function and a is not one of the roots of its denominator.

8. Show that $\lim\limits_{x\to a} x^m=a^m$ for all rational values of m, except when $a=0$ and m is negative. [This follows at once, when a is positive, from the inequalities (9) or (10) of § 74. For $|x^m-a^m|<H|x-a|$, where H is the greater of the absolute values of mx^{m-1} and ma^{m-1} (cf. Ex. xxviii. 4). If a is negative we write $x=-y$ and $a=-b$. Then

$$\lim x^m=\lim(-1)^m y^m=(-1)^m b^m=a^m.]$$

97. The reader will probably fail to see at first that any proof of such results as those of Exs. 4, 5, 6, 7, 8 above is necessary. He may ask 'why not simply put $x=0$, or $x=a$? Of course we then get a, a/α, a^m, $P(a)$, $R(a)$'. It is very important that he should see exactly where he is wrong. We shall therefore consider this point carefully before passing on to any further examples.

The statement $$\lim_{x\to 0}\phi(x)=l$$

is a statement about the values of $\phi(x)$ when x has any value *distinct from but differing by little from zero** It is *not* a statement about the *value of* $\phi(x)$ *when* $x = 0$. When we make the statement we assert that, when x is *nearly* equal to zero, $\phi(x)$ is nearly equal to l. We assert nothing whatever about what happens when x is *actually* equal to 0. So far as we know, $\phi(x)$ may not be defined at all for $x = 0$; or it may have some value other than l. For example, consider the function defined for all values of x by the equation $\phi(x) = 0$. It is obvious that

$$\lim \phi(x) = 0 \quad \text{.......................(1)}.$$

Now consider the function $\psi(x)$ which differs from $\phi(x)$ only in that $\psi(x) = 1$ when $x = 0$. Then

$$\lim \psi(x) = 0 \quad \text{.......................(2)},$$

for, when x is nearly equal to zero, $\psi(x)$ is not only nearly but exactly equal to zero. But $\psi(0) = 1$. The graph of this function consists of the axis of x, with the point $x = 0$ left out, and one isolated point, viz. the point $(0, 1)$. The equation (2) expresses the fact that if we move along the graph towards the axis of y, from either side, then the ordinate of the curve, being always equal to zero, tends to the limit zero. This fact is in no way affected by the position of the isolated point $(0, 1)$.

The reader may object to this example on the score of artificiality: but it is easy to write down simple formulae representing functions which behave precisely like this near $x = 0$. One is

$$\psi(x) = [1-x^2],$$

where $[1-x^2]$ denotes as usual the greatest integer not greater than $1-x^2$. For if $x = 0$ then $\psi(x) = [1] = 1$; while if $0 < x < 1$, or $-1 < x < 0$, then $0 < 1-x^2 < 1$ and so $\psi(x) = [1-x^2] = 0$.

Or again, let us consider the function

$$y = x/x$$

already discussed in Ch. II, § 24, (2). This function is equal to 1 for all values of x save $x = 0$. It is *not* equal to 1 when $x = 0$: it is

* Thus in Def. A of § 93 we make a statement about values of y such that $0 < y \leq y_0$, the first of these inequalities being inserted expressly in order to exclude the value $y = 0$.

in fact not defined at all for $x = 0$. For when we say that $\phi(x)$ is defined for $x = 0$ we mean (as we explained in Ch. II, *l.c.*) that we can calculate its value for $x = 0$ by putting $x = 0$ in the formula which defines $\phi(x)$. In this case we cannot. When we put $x = 0$ in $\phi(x)$ we obtain $0/0$, which is meaningless. The reader may object 'divide numerator and denominator by x', but this is impossible when $x = 0$. Thus $y = x/x$ is a function which differs from $y = 1$ solely in that it is not defined for $x = 0$. None the less

$$\lim (x/x) = 1,$$

for x/x is equal to 1 so long as x differs from zero, however small the difference may be.

Similarly $\phi(x) = \{(x+1)^2 - 1\}/x = x + 2$ so long as x is not equal to zero, but is undefined when $x = 0$. None the less $\lim \phi(x) = 2$.

On the other hand there is of course nothing to prevent the limit of $\phi(x)$ as x tends to zero from being equal to $\phi(0)$, the value of $\phi(x)$ for $x = 0$. Thus if $\phi(x) = x$, then $\phi(0) = 0$ and $\lim \phi(x) = 0$.

Examples XXXVI. 1. $\lim_{x \to a} \dfrac{x^2 - a^2}{x - a} = 2a$.

2. $\lim_{x \to a} \dfrac{x^m - a^m}{x - a} = ma^{m-1}$, if m is any integer (zero included).

3. Show that the result of Ex. 2 remains true for all rational values of m, provided a is positive. [This follows at once from the inequalities (9) and (10) of § 74.]

4. $\lim_{x \to 1} \dfrac{x^7 - 2x^5 + 1}{x^3 - 3x^2 + 2} = 1$. [Observe that $x - 1$ is a factor of both numerator and denominator.]

5. Discuss the behaviour of

$$\phi(x) = \frac{a_0 x^m + a_1 x^{m+1} + \dots + a_k x^{m+k}}{b_0 x^n + b_1 x^{n+1} + \dots + b_l x^{n+l}},$$

where $a_0 \neq 0$, $b_0 \neq 0$, as x tends to 0 by positive or negative values.

[If $m > n$, $\lim \phi(x) = 0$. If $m = n$, $\lim \phi(x) = a_0/b_0$. If $m < n$ and $n - m$ is even, $\phi(x) \to +\infty$ or $\phi(x) \to -\infty$ according as $a_0/b_0 > 0$ or $a_0/b_0 < 0$. If $m < n$ and $n - m$ is odd, then $\phi(x) \to +\infty$ as $x \to +0$ and $\phi(x) \to -\infty$ as $x \to -0$, or $\phi(x) \to -\infty$ as $x \to +0$ and $\phi(x) \to +\infty$ as $x \to -0$, according as $a_0/b_0 > 0$ or $a_0/b_0 < 0$.]

6. If a and b are positive, then

$$\lim_{x\to+0}\frac{x}{a}\left[\frac{b}{x}\right]=\frac{b}{a},\quad \lim_{x\to+0}\frac{b}{x}\left[\frac{x}{a}\right]=0.$$

How do the functions behave when $x\to 0$ through negative values?

7*. $\lim\sqrt{(1+x)}=\lim\sqrt{(1-x)}=1$. [Put $1+x=y$ or $1-x=y$, and use Ex. xxxv. 8.]

8. $\lim\{\sqrt{(1+x)}-\sqrt{(1-x)}\}/x=1$. [Multiply numerator and denominator by $\sqrt{(1+x)}+\sqrt{(1-x)}$.]

9. Consider the behaviour of $\{\sqrt{(1+x^m)}-\sqrt{(1-x^m)}\}/x^n$ as $x\to 0$, m and n being positive integers.

10. $\lim\frac{1}{x}\{\sqrt{(1+x+x^2)}-1\}=\frac{1}{2}$.

11. $\lim\frac{\sqrt{(1+x)}-\sqrt{(1+x^2)}}{\sqrt{(1-x^2)}-\sqrt{(1-x)}}=1$.

12. Draw a graph of the function

$$y=\left\{\frac{1}{x-1}+\frac{1}{x-\frac{1}{2}}+\frac{1}{x-\frac{1}{3}}+\frac{1}{x-\frac{1}{4}}\right\}\Big/\left\{\frac{1}{x-1}+\frac{1}{x-\frac{1}{2}}+\frac{1}{x-\frac{1}{3}}+\frac{1}{x-\frac{1}{4}}\right\}.$$

Has it a limit as $x\to 0$? [Here $y=1$ except for $x=1,\frac{1}{2},\frac{1}{3},\frac{1}{4}$, when y is not defined, and $y\to 1$ as $x\to 0$.]

13. $\lim\frac{\sin x}{x}=1$.

[It may be deduced from the definitions of the trigonometrical ratios† that if x is positive and less than $\frac{1}{2}\pi$ then

$$\sin x<x<\tan x$$

or

$$\cos x<\frac{\sin x}{x}<1$$

or

$$0<1-\frac{\sin x}{x}<1-\cos x=2\sin^2\tfrac{1}{2}x.$$

But $2\sin^2\frac{1}{2}x<2(\frac{1}{2}x)^2<\frac{1}{2}x^2$. Hence

$$\lim_{x\to+0}\left(1-\frac{\sin x}{x}\right)=0,\quad \lim_{x\to+0}\frac{\sin x}{x}=1.$$

Since $\frac{\sin x}{x}$ is an even function, the result follows.]

* In the examples which follow it is to be assumed that limits as $x\to 0$ are required, unless (as in Exs. 19, 22) the contrary is explicitly stated.

† The proofs of the inequalities which are used here depend on certain properties of the 'area' of a sector of a circle which are usually taken as geometrically intuitive; for example, that the area of the sector is greater than that of the triangle inscribed in the sector. The justification of these assumptions must be postponed to Ch. VII.

14. $\lim \dfrac{1-\cos x}{x^2} = \frac{1}{2}$. 15. $\lim \dfrac{\sin \alpha x}{x} = \alpha$. Is this true if $\alpha = 0$?

16. $\lim \dfrac{\arcsin x}{x} = 1$. 17. $\lim \dfrac{\tan \alpha x}{x} = \alpha$, $\lim \dfrac{\arctan \alpha x}{x} = \alpha$.

18. $\lim \dfrac{\operatorname{cosec} x - \cot x}{x} = \frac{1}{2}$. 19. $\lim\limits_{x\to 1} \dfrac{1+\cos \pi x}{\tan^2 \pi x} = \frac{1}{2}$.

20. How do the functions $\sin(1/x)$, $(1/x)\sin(1/x)$, $x\sin(1/x)$ behave as $x \to 0$? [The first oscillates finitely, the second infinitely, the third tends to the limit 0. None is defined when $x = 0$. See Exs. xv. 6, 7, 8.]

21. Does the function

$$y = \left(\sin\frac{1}{x}\right)\Big/\left(\sin\frac{1}{x}\right)$$

tend to a limit as x tends to 0? [*No*. The function is equal to 1 except when $\sin(1/x) = 0$; i.e. when $x = 1/\pi, 1/2\pi, \dots, -1/\pi, -1/2\pi, \dots$. For these values the formula for y assumes the meaningless form 0/0, and y is therefore not defined for an infinity of values of x near $x = 0$.]

22. Prove that if m is any integer then $[x] \to m$ and $x-[x] \to 0$ as $x \to m+0$, and $[x] \to m-1$, $x-[x] \to 1$ as $x \to m-0$.

98. The symbols O, o, $\sim$: orders of smallness and greatness. The definitions of § 89 may be extended, with obvious changes, to functions of a continuous variable which tends to infinity or to a limit. Thus $f = O(\phi)$, when $x \to \infty$, means that $|f| < K\phi$ for $x \geqq x_0$; $f = o(\phi)$ means that $f/\phi \to 0$; and $f \sim l\phi$, where $l \neq 0$, means that $f/\phi \to l$. Similarly $f = O(\phi)$, when $x \to a$, means that $|f| < K\phi$ for all x differing from but sufficiently near to a.

Thus

$$x+x^2 = O(x^2),\quad x = o(x^2),\quad x+x^2 \sim x^2,\quad \sin x = O(1),\quad x^{-\frac{1}{2}} = o(1)$$

when $x \to \infty$, and

$$x+x^2 = O(x),\ x^2 = o(x),\ x+x^2 \sim x,\ \sin(1/x) = O(1),\ x^{\frac{1}{2}} = o(1)$$

when $x \to 0$.

Suppose, to fix our ideas, that $x \to 0$. The functions

$$x,\ x^2,\ x^3,\ \dots$$

form a scale each member of which tends to zero more rapidly than the preceding member, since

$$x^m = o(x^{m-1}), \quad x^{m+1} = o(x^m)$$

for every positive integer m; and it is natural to use them to measure the 'order of smallness' of any function which tends to zero with x. If

$$\phi(x) \sim lx^m,$$

where $l \neq 0$, when $x \to 0$, we say that $\phi(x)$ is of the mth order of smallness when x is small*.

This scale is of course in no way complete. Thus $\phi(x) = x^{\frac{7}{5}}$ tends to zero more rapidly than x but more slowly than x^2. We might attempt to make it more complete by including fractional orders of smallness; we might say, for example, that $x^{\frac{7}{5}}$ is of the $\frac{7}{5}$th order. We shall however see in Ch. IX that even then our scale would be quite incomplete.

We define orders of greatness similarly. Thus $\phi(x)$ is of the mth order of greatness if $\phi(x)/x^{-m} = x^m\phi(x)$ tends to a limit l, not zero, when $x \to 0$.

These definitions refer to the case in which $x \to 0$. There are naturally corresponding definitions when $x \to \infty$ or $x \to a$. Thus, if $x^m\phi(x)$ tends to a limit, not 0, when $x \to \infty$, we say that $\phi(x)$ is of the mth order of smallness for large x; and if $(x-a)^m\phi(x)$ tends to a limit, not 0, when $x \to a$, we say that $\phi(x)$ is of the mth order of greatness for x near a.

Many of the results of the last set of examples may be restated in the language of this section. Thus

$$\sin\alpha x \sim \alpha x, \quad 1 - \cos x \sim \tfrac{1}{2}x^2, \quad \operatorname{cosec} x - \cot x \sim \tfrac{1}{2}x;$$

the second function is of the second order of smallness and the others of the first.

* We might say, more generally, that $\phi(x)$ is of the mth order of smallness if there are positive constants A, B such that

$$A\,|x|^m \leqq |\phi(x)| \leqq B\,|x|^m.$$

The definition in the text is sufficiently general for our purpose.

99. Continuous functions of a real variable. The reader has no doubt some idea as to what is meant by a *continuous curve*. Thus he would call the curve C in Fig. 27 continuous, the curve C' generally continuous but discontinuous for $x = \xi'$ and $x = \xi''$.

Either of these curves may be regarded as the graph of a function $\phi(x)$. It is natural to call a function *continuous* if its graph is a continuous curve, and otherwise discontinuous. Let us take this as a provisional definition and try to distinguish more precisely some of the properties which are involved in it.

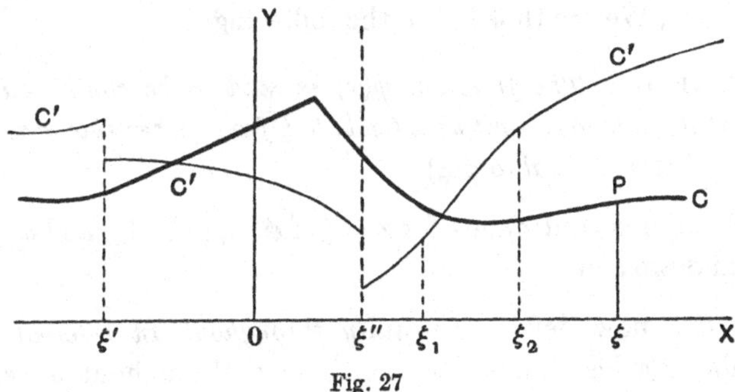

Fig. 27

In the first place it is evident that the property of the function $y = \phi(x)$ of which C is the graph may be analysed into some property possessed by the curve at each of its points. To be able to define continuity *for all values of* x we must first define continuity *for any particular value of* x. Let us therefore fix on some particular value of x, say the value $x = \xi$ corresponding to the point of P of the graph. What are the characteristic properties of $\phi(x)$ associated with this value of x?

In the first place $\phi(x)$ *is defined for* $x = \xi$. This is obviously essential. If $\phi(\xi)$ were not defined there would be a point missing from the curve.

Secondly $\phi(x)$ *is defined for all values of* x *near* $x = \xi$; i.e. we can find an interval, including $x = \xi$ in its interior, for all points of which $\phi(x)$ is defined.

Thirdly *if x approaches the value ξ from either side, then $\phi(x)$ approaches the limit $\phi(\xi)$.*

The properties thus defined are far from exhausting those of the picture of a curve by the eye of common sense, a picture which is a generalisation from particular curves such as straight lines and circles. But they are the simplest and most fundamental properties: and the graph of any function which has these properties would, so far as drawing it is practically possible, satisfy our geometrical feeling of what a continuous curve should be. We therefore select them as embodying the mathematical notion of continuity. We are thus led to the following

DEFINITION. *The function $\phi(x)$ is said to be continuous for $x = \xi$ if it tends to a limit as x tends to ξ from either side, and each of these limits is equal to $\phi(\xi)$.*

Thus $\phi(x)$ is continuous for $x = \xi$ if $\phi(\xi)$, $\phi(\xi - 0)$, and $\phi(\xi + 0)$ exist and are equal.

We can now define *continuity throughout an interval*. The function $\phi(x)$ is said to be continuous throughout a certain interval of values of x if it is continuous for all values of x in that interval. It is said to be *continuous everywhere* if it is continuous for every value of x. Thus $[x]$ is continuous in the interval $(\epsilon, 1 - \epsilon)$, where ϵ is any positive number less than $\frac{1}{2}$, but is not continuous for $x = 0$ or $x = 1$, or in any interval including either of these points; and 1 and x are continuous everywhere.

If we recur to the definitions of a limit we see that our definition is equivalent to '*$\phi(x)$ is continuous for $x = \xi$ if, given δ, we can choose $\epsilon(\delta)$ so that $|\phi(x) - \phi(\xi)| < \delta$ if $0 \leqq |x - \xi| \leqq \epsilon(\delta)$*'.

We have often to consider functions defined only in an interval (a, b). In this case it is convenient to make a slight and natural change in our definition of continuity at the particular points a and b. We shall say that $\phi(x)$ is continuous for $x = a$ if $\phi(a + 0)$ exists and is equal to $\phi(a)$, and for $x = b$ if $\phi(b - 0)$ exists and is equal to $\phi(b)$.

100. The definition of continuity given in the last section may be illustrated geometrically as follows. Draw the two horizontal lines $y = \phi(\xi) - \delta$ and $y = \phi(\xi) + \delta$. Then $|\phi(x) - \phi(\xi)| < \delta$ expresses the fact that the point on the curve corresponding to x lies between these two lines. Similarly $|x - \xi| \leqq \epsilon$ expresses the fact that x lies in the interval $(\xi - \epsilon, \xi + \epsilon)$. Thus our definition asserts

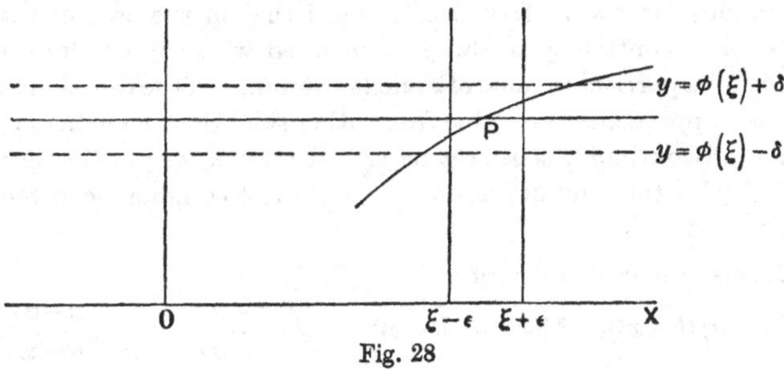

Fig. 28

that if we draw two such horizontal lines, no matter how close together, we can always cut off a vertical strip of the plane by two vertical lines in such a way that all that part of the curve which is contained in the strip lies between the two horizontal lines. This is evidently true of the curve C (Fig. 27), whatever value ξ may have.

We shall now discuss the continuity of some special types of functions. Some of the results which follow were (as we pointed out at the time) taken for granted in Ch. II.

Examples XXXVII. 1. The sum or product of two functions continuous at a point is continuous at that point. The quotient is also continuous unless the denominator vanishes at the point. [This follows at once from Ex. XXXV. 1.]

2. Any polynomial is continuous for all values of x. Any rational fraction is continuous except for values of x for which the denominator vanishes. [This follows from Exs. XXXV. 6, 7.]

3. $\sqrt{x}$ is continuous for all positive values of x (Ex. XXXV. 8). It is not defined when $x < 0$, but is continuous for $x = 0$ in virtue of the remark made at the end of § 99. The same is true of $x^{m/n}$, where m and n are any positive integers of which n is even.

4. The function $x^{m/n}$ is continuous for all values of x if n is odd.

5. $1/x$ is not continuous for $x = 0$. It has no value for $x = 0$, nor does it tend to a limit as $x \to 0$. In fact $1/x \to +\infty$ or $1/x \to -\infty$ according as $x \to 0$ by positive or negative values.

6. Discuss the continuity of $x^{-m/n}$, where m and n are positive integers, for $x = 0$.

7. The standard rational function $R(x) = P(x)/Q(x)$ is discontinuous for $x = a$, where a is any root of $Q(x) = 0$. Thus $(x^2+1)/(x^2-3x+2)$ is discontinuous for $x = 1$. It will be noticed that in the case of rational functions a discontinuity is always associated with (a) a failure of the definition for a particular value of x and (b) a tending of the function to $+\infty$ or $-\infty$ as x approaches this value from either side. Such a particular kind of point of discontinuity is usually described as an *infinity* of the function. An 'infinity' is the kind of discontinuity of most common occurrence in ordinary work.

8. Discuss the continuity of

$$\sqrt{\{(x-a)(b-x)\}},\quad \sqrt[3]{\{(x-a)(b-x)\}},\quad \sqrt{\left(\frac{x-a}{b-x}\right)},\quad \sqrt[3]{\left(\frac{x-a}{b-x}\right)}.$$

9. $\sin x$ and $\cos x$ are continuous for all values of x.

[Thus $$\sin(x+h) - \sin x = 2\sin\tfrac{1}{2}h\cos(x+\tfrac{1}{2}h),$$
which is numerically less than the numerical value of h.]

10. For what values of x are $\tan x$, $\cot x$, $\sec x$, and $\operatorname{cosec} x$ continuous or discontinuous?

11. If $f(y)$ is continuous for $y = \eta$, and $\phi(x)$ is a continuous function of x which is equal to η when $x = \xi$, then $f\{\phi(x)\}$ is continuous for $x = \xi$.

12. If $\phi(x)$ is continuous for any particular value of x, then any polynomial in $\phi(x)$, such as $a\{\phi(x)\}^m + \dots$, is also continuous.

13. Discuss the continuity of

$$(a\cos^2 x + b\sin^2 x)^{-1},\quad \sqrt{(2+\cos x)},\quad \sqrt{(1+\sin x)},\quad (1+\sin x)^{-\frac{1}{2}}.$$

14. $\sin(1/x)$, $x\sin(1/x)$, and $x^2\sin(1/x)$ are continuous except for $x = 0$.

15. The function which is equal to $x\sin(1/x)$ except when $x = 0$, and to zero when $x = 0$, is continuous for all values of x.

16. $[x]$ and $x - [x]$ are discontinuous for all integral values of x.

17. For what (if any) values of x are the following functions discontinuous: $[x^2]$, $[\sqrt{x}]$, $\sqrt{(x-[x])}$, $[x]+\sqrt{(x-[x])}$, $[2x]$, $[x]+[-x]$?

18. **Classification of discontinuities.** Some of the preceding examples suggest a classification of different types of discontinuity.

(1) Suppose that $\phi(x)$ tends to a limit when $x \to a$ either by values less than or by values greater than a. Denote these limits, as in § 95, by $\phi(a-0)$

and $\phi(a+0)$ respectively. Then, for continuity, it is necessary and sufficient that $\phi(x)$ should be defined for $x = a$, and that

$$\phi(a-0) = \phi(a) = \phi(a+0).$$

Discontinuity may arise in a variety of ways.

(α) $\phi(a-0)$ may be equal to $\phi(a+0)$, but $\phi(a)$ may not be defined, or may differ from $\phi(a-0)$ and $\phi(a+0)$. Thus if $\phi(x) = x \sin(1/x)$ and $a = 0$, $\phi(0-0) = \phi(0+0) = 0$, but $\phi(r)$ is not defined for $x = 0$. Or if $\phi(x) = [1-x^2]$ and $a = 0$, $\phi(0-0) = \phi(0+0) = 0$, but $\phi(0) = 1$.

(β) $\phi(a-0)$ and $\phi(a+0)$ may be unequal. In this case $\phi(a)$ may be equal to one or to neither, or be undefined. The first case is illustrated by $\phi(x) = [x]$, for which $\phi(0-0) = -1$, $\phi(0+0) = \phi(0) = 0$; the second by $\phi(x) = [x]-[-x]$, for which $\phi(0-0) = -1$, $\phi(0+0) = 1$, $\phi(0) = 0$; and the third by $\phi(x) = [x]+x\sin(1/x)$, for which $\phi(0-0) = -1$, $\phi(0+0) = 0$, and $\phi(0)$ is undefined.

In any of these cases we say that $\phi(x)$ has a *simple discontinuity* at $x = a$. And to these cases we may add those in which $\phi(x)$ is defined onlv on one side of $x = a$, and $\phi(a-0)$ or $\phi(a+0)$, as the case may be, exists, but $\phi(x)$ is either not defined when $x = a$ or has a value different from $\phi(a-0)$ or $\phi(a+0)$.

It is plain from § 95 that *a function which increases or decreases steadily in the neighbourhood of $x = a$ can have at most a simple discontinuity for $c = a$.*

(2) It may happen that $\phi(x)$ tends to a limit, or to $+\infty$, or to $-\infty$, when x tends to a from either side, and that it tends to $+\infty$ or to $-\infty$ when x tends to a from one side at least. This is so, for example, if $\phi(x)$ is $1/x$ or $1/x^2$, or if it is $1/x$ for positive and zero for negative x. In such cases we say that $x = a$ is an *infinity* of $\phi(x)$. We also include the cases in which $\phi(x)$ tends to $+\infty$ or $-\infty$ on one side of a and is not defined on the other.

(3) Any point of discontinuity which is not a point of simple discontinuity or an infinity is called a point of *oscillatory discontinuity*. Thus $x = 0$ is an oscillatory discontinuity of $\sin(1/x)$.

19. What is the nature of the discontinuities at $x = 0$ of the functions

$$\frac{\sin x}{x},\quad [x]+[-x],\quad \operatorname{cosec} x,\quad \sqrt{\left(\frac{1}{x}\right)},\quad \sqrt[3]{\left(\frac{1}{x}\right)},\quad \operatorname{cosec}\frac{1}{x},\quad \frac{\sin(1/x)}{\sin(1/x)}?$$

20. The function which is equal to 1 when x is rational and to 0 when x is irrational (Ch. II, Ex. XVI. 10) is discontinuous for all values of x. So too is any function which is defined only for rational or for irrational values of x.

21. The function which is equal to x when x is irrational and to $\sqrt{\{(1+p^2)/(1+q^2)\}}$ when x is a rational fraction p/q (Ch. II, Ex. XVI. 11) is discontinuous for all negative and for positive rational values of x, but continuous for positive irrational values.

22. For what points are the functions considered in Ch. IV, Exs. XXXI discontinuous, and what is the nature of their discontinuities? [Consider, e.g., the function $y = \lim x^n$ (Ex. 5). Here y is defined only when $-1 < x \leqq 1$: it is equal to 0 when $-1 < x < 1$ and to 1 when $x = 1$. The points $x = 1$ and $x = -1$ are points of simple discontinuity.]

101. The fundamental property of a continuous function. The 'continuous curve' pictured by common sense has another characteristic property. Let A and B be two points on the graph of $\phi(x)$ whose coordinates are x_0, $\phi(x_0)$ and x_1, $\phi(x_1)$, and let λ be a straight line passing between A and B. Then it would seem that, if the graph is continuous, it must cut λ.

It is clear that, if we consider this property as an intrinsic geometrical property of continuous curves, there is no real loss of generality in supposing λ to be parallel to the axis of x. In this case the ordinates of A and B cannot be equal: let us suppose, for definiteness, that $\phi(x_1) > \phi(x_0)$. And let λ be the line $y = \eta$, where $\phi(x_0) < \eta < \phi(x_1)$. Then to say that the graph of $\phi(x)$ must cut λ is the same thing as to say that there is a value of x between x_0 and x_1 for which $\phi(x) = \eta$.

We conclude then that a continuous function $\phi(x)$ should possess the following property: *if*

$$\phi(x_0) = y_0, \quad \phi(x_1) = y_1,$$

and $y_0 < \eta < y_1$, *then there is a value of* x *between* x_0 *and* x_1 *for which* $\phi(x) = \eta$. In other words, *as* x *varies from* x_0 *to* x_1, y *must assume at least once every value between* y_0 *and* y_1.

We shall now prove that if $\phi(x)$ is a continuous function of x in the sense defined in § 99 then it does in fact possess this property. There is a certain range of values of x, to the right of x_0, for which $\phi(x) < \eta$. For $\phi(x_0) < \eta$, and so $\phi(x)$ is certainly less than η if $\phi(x) - \phi(x_0)$ is numerically less than $\eta - \phi(x_0)$. But since $\phi(x)$ is continuous for $x = x_0$, this condition is certainly satisfied if x is near enough to x_0. Similarly there is a certain range of values, to the left of x_1, for which $\phi(x) > \eta$.

Let us divide the values of x between x_0 and x_1 into two classes L, R as follows:

(1) in the class L we put all values ξ of x such that $\phi(x) < \eta$ when $x = \xi$ and for all values of x between x_0 and ξ;

(2) in the class R we put all the other values of x, i.e. all numbers ξ such that either $\phi(\xi) \geqq \eta$ or there is a value of x between x_0 and ξ for which $\phi(x) \geqq \eta$.

Then it is evident that these two classes satisfy all the conditions imposed upon the classes L, R of § 17, and so constitute a section of the real numbers. Let ξ_0 be the number corresponding to the section.

First suppose $\phi(\xi_0) > \eta$, so that ξ_0 belongs to the upper class: and let $\phi(\xi_0) = \eta + k$, say. Then $\phi(\xi') < \eta$, and so

$$\phi(\xi_0) - \phi(\xi') > k$$

for all values of ξ' less than ξ_0, which contradicts the condition of continuity for $x = \xi_0$.

Next suppose $\phi(\xi_0) = \eta - k < \eta$. Then, if ξ' is any number greater than ξ_0, either $\phi(\xi') \geqq \eta$ or we can find a number ξ'' between ξ_0 and ξ' such that $\phi(\xi'') \geqq \eta$. In either case we can find a number as near to ξ_0 as we please and such that the corresponding values of $\phi(x)$ differ by more than k. And this again contradicts the hypothesis that $\phi(x)$ is continuous for $x = \xi_0$.

Hence $\phi(\xi_0) = \eta$, and the theorem is established. It should be observed that we have proved more than is asserted explicitly in the theorem; we have proved in fact that ξ_0 is the *least* value of x for which $\phi(x) = \eta$. It is not obvious, or indeed generally true, that there is a least among the values of x for which a function assumes a given value, though this is true for continuous functions.

It is easy to see that the converse of the theorem just proved is not true. Thus such a function as the function $\phi(x)$ whose graph is represented by Fig. 29 obviously assumes at least once every value between $\phi(x_0)$ and

$\phi(x_1)$: yet $\phi(x)$ is discontinuous. Indeed it is not even true that $\phi(x)$ must be continuous when it assumes each value *once and once only*. Thus let $\phi(x)$ be defined as follows from $x = 0$ to $x = 1$. If $x = 0$ let $\phi(x) = 0$; if $0 < x < 1$ let $\phi(x) = 1 - x$; and if $x = 1$ let $\phi(x) = 1$. The graph of the function is shown in Fig. 30; it includes the points O, C but not the points A, B. It is clear that, as x varies from 0 to 1, $\phi(x)$ assumes once and once only every value between $\phi(0) = 0$ and $\phi(1) = 1$; but $\phi(x)$ is discontinuous for $x = 0$ and $x = 1$.

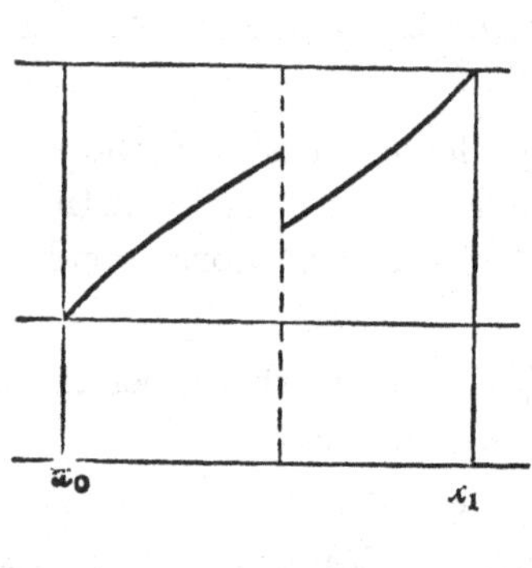

Fig. 29

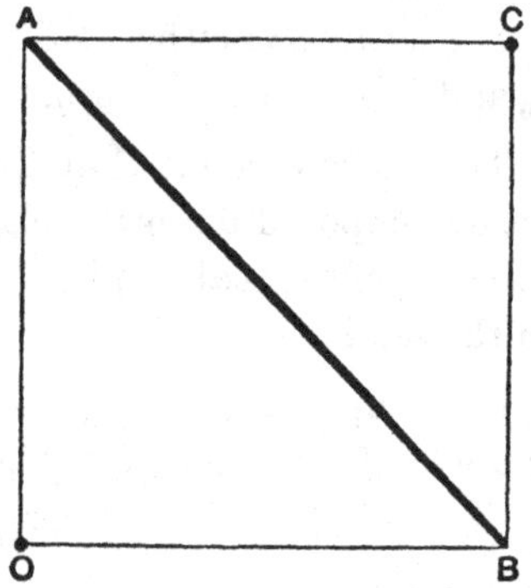

Fig. 30

The curves which occur in elementary mathematics are usually composed of *a finite number of pieces along which y always varies in the same direction*. It is easy to show that if $y = \phi(x)$ always varies in the same direction, i.e. steadily increases or decreases, as x varies from x_0 to x_1, then the two notions of continuity are really equivalent, i.e. that if $\phi(x)$ takes every value between $\phi(x_0)$ and $\phi(x_1)$ then it must be a continuous function in the sense of § 99. For let ξ be any value of x between x_0 and x_1. When $x \to \xi$ through values less than ξ, $\phi(x)$ tends to the limit $\phi(\xi - 0)$ (§ 95). Similarly, when $x \to \xi$ through values greater than ξ, $\phi(x)$ tends to the limit $\phi(\xi + 0)$. The function will be continuous for $x = \xi$ if and only if

$$\phi(\xi - 0) = \phi(\xi) = \phi(\xi + 0).$$

But if either of these equations is untrue, say the first, then it is evident that $\phi(x)$ never assumes any value which lies between $\phi(\xi - 0)$ and $\phi(\xi)$, which is contrary to our assumption. Thus $\phi(x)$ must be continuous.

102. Further properties of continuous functions. In this and the following sections we prove a series of important general theorems.

THEOREM 1. *Suppose that $\phi(x)$ is continuous for $x = \xi$, and that $\phi(\xi)$ is positive. Then we can determine a positive number ϵ such that $\phi(x)$ is positive throughout the interval $(\xi-\epsilon, \xi+\epsilon)$.*

For, taking $\delta = \frac{1}{2}\phi(\xi)$ in the fundamental inequality of p. 186, we can choose ϵ so that

$$|\phi(x)-\phi(\xi)| < \tfrac{1}{2}\phi(\xi)$$

throughout $(\xi-\epsilon, \xi+\epsilon)$, and then

$$\phi(x) \geqq \phi(\xi) - |\phi(x)-\phi(\xi)| > \tfrac{1}{2}\phi(\xi) > 0,$$

so that $\phi(x)$ is positive. There is plainly a corresponding theorem referring to negative values of $\phi(x)$.

THEOREM 2. *If $\phi(x)$ is continuous for $x = \xi$, and $\phi(x)$ vanishes for values of x as near as we please, or assumes, for values of x as near to ξ as we please, both positive and negative values, then $\phi(\xi) = 0$.*

This is an obvious corollary of Theorem 1. If $\phi(\xi)$ is not zero, it must be positive or negative; and if it were, for example, positive, $\phi(x)$ would be positive for all values of x sufficiently near to ξ, which contradicts the hypotheses of the theorem.

103. The range of values of a continuous function. Let us consider a function $\phi(x)$ about which we shall assume at present only that it is defined for every value of x in an interval (a, b).

The values assumed by $\phi(x)$ for values of x in (a, b) form an aggregate S to which we can apply the arguments of § 80, as we applied them in § 81 to the aggregate of values of a function of n. If there is a number K such that $\phi(x) \leqq K$, for all values of x in question, we say that $\phi(x)$ is *bounded above*. In this case $\phi(x)$ possesses an *upper bound* M: no value of $\phi(x)$ exceeds M, but any number less than M is exceeded by at least one value of $\phi(x)$. Similarly we define '*bounded below*', '*lower bound*', '*bounded*', as applied to functions of a continuous variable x.

THEOREM 1. *If $\phi(x)$ is continuous throughout (a, b), then it is bounded in (a, b).*

We can certainly determine an interval (a, ξ), extending to the right from a, in which $\phi(x)$ is bounded. For since $\phi(x)$ is continuous for $x = a$, we can, given any positive number δ, determine an interval (a, ξ) throughout which $\phi(x)$ lies between $\phi(a) - \delta$ and $\phi(a) + \delta$; and obviously $\phi(x)$ is bounded in this interval.

Now divide the points ξ of the interval (a, b) into two classes L, R, putting ξ in L if $\phi(x)$ is bounded in (a, ξ), and in R if this is not the case. It follows from what precedes that L certainly exists: what we propose to prove is that R does not. Suppose that R does exist, and let β be the number corresponding to the section whose lower and upper classes are L and R. Since $\phi(x)$ is continuous for $x = \beta$, we can, however small δ may be, determine an interval $(\beta - \eta, \beta + \eta)$* throughout which

$$\phi(\beta) - \delta < \phi(x) < \phi(\beta) + \delta.$$

Thus $\phi(x)$ is bounded in $(\beta - \eta, \beta + \eta)$. Now $\beta - \eta$ belongs to L. Therefore $\phi(x)$ is bounded in $(a, \beta - \eta)$; and therefore it is bounded in the whole interval $(a, \beta + \eta)$. But $\beta + \eta$ belongs to R and so $\phi(x)$ is *not* bounded in $(a, \beta + \eta)$. This contradiction shows that R does not exist, and so that $\phi(x)$ is bounded in the whole interval (a, b).

THEOREM 2. *If $\phi(x)$ is continuous throughout (a, b), and M and m are its upper and lower bounds, then $\phi(x)$ assumes the values M and m at least once each in the interval.*

For, given any positive number δ, we can find a value of x for which $M - \phi(x) < \delta$ or $1/\{M - \phi(x)\} > 1/\delta$. Hence $1/\{M - \phi(x)\}$ is not bounded, and therefore, by Theorem 1, is not continuous. But $M - \phi(x)$ is a continuous function, and so $1/\{M - \phi(x)\}$ is continuous at any point at which its denominator does not vanish (Ex. XXXVII. 1). There must therefore be a point at which the denominator vanishes, and at this point $\phi(x) = M$. Similarly it may be shown that there is a point at which $\phi(x) = m$.

The proof just given is indirect, and it may be well, in view of the great importance of the theorem, to indicate alternative lines of proof. It will however be convenient to postpone these for a moment†.

* If $\beta = b$ we must replace this interval by $(\beta - \eta, \beta)$, and $\beta + \eta$ by β, throughout the argument which follows.

† See § 105.

Examples XXXVIII. 1. If $\phi(x) = 1/x$ except when $x = 0$, and $\phi(x) = 0$ when $x = 0$, then $\phi(x)$ has neither an upper nor a lower bound in any interval which includes $x = 0$ in its interior, such as the interval $(-1, +1)$.

2. If $\phi(x) = 1/x^2$ except when $x = 0$, and $\phi(x) = 0$ when $x = 0$, then $\phi(x)$ has the lower bound 0, but no upper bound, in the interval $(-1, +1)$.

3. Let $\phi(x) = \sin(1/x)$ except when $x = 0$, and $\phi(x) = 0$ when $x = 0$. Then $\phi(x)$ is discontinuous for $x = 0$. In any interval $(-\delta, +\delta)$ the lower bound is -1 and the upper bound $+1$, and each of these values is assumed by $\phi(x)$ an infinity of times.

4. Let $\phi(x) = x - [x]$. This function is discontinuous for all integral values of x. In the interval $(0, 1)$ its lower bound is 0 and its upper bound 1. It is equal to 0 when $x = 0$ or $x = 1$, but it is never equal to 1. Thus $\phi(x)$ never assumes a value equal to its upper bound.

5. Let $\phi(x) = 0$ when x is irrational, and $\phi(x) = q$ when x is a rational fraction p/q. Then $\phi(x)$ has the lower bound 0, but no upper bound, in any interval (a, b). But if $\phi(x) = (-1)^p q$ when $x = p/q$ then $\phi(x)$ has neither an upper nor a lower bound in any interval.

104. The oscillation of a function in an interval. Let $\phi(x)$ be any function bounded throughout (a, b), and M and m its upper and lower bounds. We shall now use the notation $M(a, b)$, $m(a, b)$ for M, m, in order to exhibit explicitly the dependence of M and m on a and b, and we shall write

$$O(a, b) = M(a, b) - m(a, b).$$

This number $O(a, b)$, the difference between the upper and lower bounds of $\phi(x)$ in (a, b), we shall call the *oscillation of* $\phi(x)$ *in* (a, b). The simplest of the properties of the functions $M(a, b)$, $m(a, b)$, $O(a, b)$ are as follows.

(1) *If* $a \leqq c \leqq b$, *then* $M(a, b)$ *is equal to the greater of* $M(a, c)$ *and* $M(c, b)$, *and* $m(a, b)$ *to the lesser of* $m(a, c)$ *and* $m(c, b)$.

(2) $M(a, b)$ *is an increasing*, $m(a, b)$ *a decreasing, and* $O(a, b)$ *an increasing function of* b.

(3) $O(a, b) \leqq O(a, c) + O(c, b)$.

The first two theorems are almost immediate consequences of our definitions. Let μ be the greater of $M(a, c)$ and $M(c, b)$, and let δ be any positive number. Then $\phi(x) \leqq \mu$ throughout (a, c) and (c, b), and therefore throughout (a, b); and $\phi(x) > \mu - \delta$ somewhere in (a, c) or in (c, b), and therefore somewhere in (a, b). Hence $M(a, b) = \mu$. The proposition concerning m may be proved similarly. Thus (1) is proved, and (2) is an obvious corollary.

Suppose now that M_1 is the greater and M_2 the less of $M(a, c)$ and $M(c, b)$, and that m_1 is the less and m_2 the greater of $m(a, c)$ and $m(c, b)$. Then, since c belongs to both intervals, $\phi(c)$ is not greater than M_2 nor less than m_2. Hence $M_2 \geqq m_2$, whether these numbers correspond to the same one of the intervals (a, c) and (c, b) or not, and

$$O(a, b) = M_1 - m_1 \leqq M_1 + M_2 - m_1 - m_2.$$

But $$O(a, c) + O(c, b) = M_1 + M_2 - m_1 - m_2;$$

and (3) follows.

105. Alternative proofs of Theorem 2 of § 103. The most straightforward proof of Theorem 2 of § 103 is as follows. Let ξ be any number of the interval (a, b). The function $M(a, \xi)$ increases steadily with ξ and never exceeds M. We can therefore construct a section of the numbers ξ by putting ξ in L or in R according as $M(a, \xi) < M$ or $M(a, \xi) = M$. Let β be the number corresponding to the section. If $a < \beta < b$, we have

$$M(a, \beta - \eta) < M, \quad M(a, \beta + \eta) = M$$

for all positive values of η, and so

$$M(\beta - \eta, \beta + \eta) = M,$$

by (1) of § 104. Hence $\phi(x)$ assumes, for values of x as near as we please to β, values as near as we please to M, and so, since $\phi(x)$ is continuous, $\phi(\beta)$ must be equal to M.

If $\beta = a$ then $M(a, a + \eta) = M$. And if $\beta = b$ then $M(a, b - \eta) < M$, and so $M(b - \eta, b) = M$. In either case the argument may be completed as before.

The theorem may also be proved by the method of repeated bisection used in § 71. If M is the upper bound of $\phi(x)$ in an interval PQ, and PQ is divided into two equal parts, then it is possible to find a half P_1Q_1 in which the upper bound of $\phi(x)$ is also M. Proceeding as in § 71, we construct a sequence of intervals PQ, P_1Q_1, P_2Q_2, ... in each of which the upper bound of $\phi(x)$ is M. These intervals, as in § 71, converge to a point T, and it is easily proved that the value of $\phi(x)$ at this point is M.

106. Sets of intervals on a line. The Heine-Borel theorem. We shall now proceed to prove some theorems concerning the oscillations of a function which are of particular importance, as we shall see later, in the theory of integration. These theorems depend upon a general theorem concerning intervals on a line.

Suppose that we are given a *set of intervals* in a straight line, that is to say an aggregate each of whose members is an interval (α, β). We make no restriction as to the nature of these intervals; they may be finite or infinite in number; they may or may not overlap*; and any number of them may be included in others.

It is worth while in passing to give a few examples of sets of intervals to which we shall have occasion to return later.

(i) If the interval $(0, 1)$ is divided into n equal parts, then the n intervals thus formed define a finite set of non-overlapping intervals which just cover up the line.

(ii) We take every point ξ of the interval $(0, 1)$, and associate with ξ the interval $(\xi - \epsilon, \xi + \epsilon)$, where ϵ is a positive number less than 1, except that with 0 we associate $(0, \epsilon)$ and with 1 we associate $(1 - \epsilon, 1)$, and in general we reject any part of any interval which projects outside the interval $(0, 1)$. We thus define an infinite set of intervals, and it is obvious that many of them overlap with one another.

(iii) We take the rational points p/q of the interval $(0, 1)$, and associate with p/q the interval

$$\left(\frac{p}{q} - \frac{\epsilon}{q^3},\ \frac{p}{q} + \frac{\epsilon}{q^3}\right),$$

where ϵ is positive and less than 1. We regard 0 as $0/1$ and 1 as $1/1$: in these two cases we reject the part of the interval which lies outside $(0, 1)$. We obtain thus an infinite set of intervals, which plainly overlap with one another, since there are an infinity of rational points, other than p/q, in the interval associated with p/q.

The Heine-Borel theorem. *Suppose that we are given an interval (a, b), and a set of intervals I each of whose members is included in (a, b). Suppose further that I possesses the following properties:*

(i) *every point of (a, b), other than a or b, lies inside*† *at least one interval of I;*

* The word *overlap* is used in its obvious sense: two intervals overlap if they have points in common which are not end points of either. Thus $(0, \frac{2}{3})$ and $(\frac{1}{3}, 1)$ overlap. A pair of intervals such as $(0, \frac{1}{2})$ and $(\frac{1}{2}, 1)$ may be said to *abut*.

† That is to say 'in and not at an end of'.

(ii) *a is the left-hand end point, and b the right-hand end point, of at least one interval of I.*

Then it is possible to choose A FINITE NUMBER *of intervals from the set I which form a set of intervals possessing the properties* (i) *and* (ii).

It is sufficient to prove that *there exists a positive number l such that any sub-interval of (a, b) of length l is a sub-interval of at least one interval of I.* For then we have only to choose a finite number of (overlapping) sub-intervals of (a, b) of length l and possessing the properties (i) and (ii). Each of these sub-intervals is contained in at least one interval of I, and the result follows.

We give two proofs of the existence of the number l, the first of which is indirect.*

(a) Suppose that no such number l exists. Then there is a sequence of sub-intervals of (a, b),

$$(c_n, c_n + 2^{-n}),$$

which are not sub-intervals of any interval of I. By Weierstrass's theorem (§ 19), the set of points c_n has at least one point of accumulation, c, say, in (a, b), and (by (ii)) c does not coincide with a or b. c is an interior point of at least one interval (x', x'') of I, and for some sufficiently large n the interval $(c_n, c_n + 2^{-n})$ is a sub-interval of (x', x''). Thus we obtain a contradiction, and the number l must exist.

(b) Let a' be the mid-point of an interval of I which has a as left-hand end point, b' the mid-point of an interval of I which has b as right-hand end point. We may suppose that these two intervals do not overlap (else there is nothing to prove). For any given point x of (a', b'), let $\rho(x)$ be the upper bound of numbers λ such that the interval $(x-\lambda, x+\lambda)$ is a sub-interval of at least one member of I. $\rho(x)$ is clearly positive in (a', b')†; we shall show that $\rho(x)$ is continuous in the (closed) interval

* Both proofs were suggested to me by Mr A. S. Besicovitch.

† Including the end points a', b'. This is no longer true in the whole interval (a, b).

$(a'\,b')$. For suppose $(x-\lambda, x+\lambda)$ is a sub-interval of a member of I, and take x' near x. Then plainly the interval $(x'-\lambda', x'+\lambda')$, where $\lambda' = \lambda - |x-x'|$, is contained in $(x-\lambda, x+\lambda)$, and so in a member of I. Hence

$$\rho(x') \geqslant \lambda' = \lambda - |x-x'|.$$

Since λ can be as near to $\rho(x)$ as we please, we must therefore have also

$$\rho(x') \geqslant \rho(x) - |x-x'|,$$

and similarly

$$\rho(x) \geqslant \rho(x') - |x-x'|.$$

Thus

$$|\rho(x) - \rho(x')| \leqslant |x-x'|,$$

and $\rho(x)$ is continuous.

Hence $\rho(x)$ attains its lower bound in (a', b'), m say, $m>0$, and we may take l as the minimum of (i) $2(a'-a)$, (ii) $2(b-b')$, and (iii) some (any) number less than $2m$*.

It is instructive to consider the examples of p. 197 in the light of this theorem.

(i) Here the conditions of the theorem are not satisfied; the points $1/n$, $2/n$, $3/n$, ... do not lie inside any interval of I.

(ii) Here the conditions of the theorem are satisfied. The set of intervals

$$(0, 2\epsilon),\ (\epsilon, 3\epsilon),\ (2\epsilon, 4\epsilon),\ \ldots,\ (1-2\epsilon, 1),$$

associated with the points ϵ, 2ϵ, 3ϵ, ..., $1-\epsilon$, possesses the properties required.

(iii) In this case we can prove, by using the theorem, that there are, if ϵ is small enough, points of $(0, 1)$ which do not lie in any interval of I.

If every point of $(0, 1)$ lay inside an interval of I (with the obvious reservation as to the end points), then we could find a finite number of intervals of I possessing the same property and having therefore a total length greater than 1. Now there are two intervals, of total length 2ϵ, for which $q=1$, and $q-1$ intervals, of total length $2\epsilon(q-1)/q^3$, associated with any other value of q. The sum of any finite number of intervals of I can therefore not be greater than 2ϵ times that of the series

$$1 + \frac{1}{2^3} + \frac{2}{3^3} + \frac{3}{4^3} + \ldots,$$

which will be shown to be convergent in Ch. VIII. Hence it follows that, if ϵ is small enough, the supposition that every point of $(0, 1)$ lies inside an interval of I leads to a contradiction.

* Not $2m$. For any given x, $\rho(x)$ need not be an attained bound of the numbers λ.

The reader may be tempted to think that this proof is needlessly elaborate, and that the existence of points of the interval, not in any interval of I, follows at once from the fact that the sum of all these intervals is less than 1. But the theorem to which he would be appealing is (when the set of intervals is infinite) far from obvious, and can only be proved rigorously by some such use of the Heine-Borel theorem as is made in the text.

107. The oscillation of a continuous function. We shall now apply the Heine-Borel theorem to the proof of two important theorems concerning the oscillation of a continuous function.

THEOREM I. *If $\phi(x)$ is continuous throughout the interval (a, b), then we can divide (a, b) into a finite number of sub-intervals (a, x_1), (x_1, x_2), ..., (x_n, b), in each of which the oscillation of $\phi(x)$ is less than an assigned positive number δ.*

Let ξ be any number between a and b. Since $\phi(x)$ is continuous for $x = \xi$, we can determine an interval $(\xi - \epsilon, \xi + \epsilon)$ such that the oscillation of $\phi(x)$ in this interval is less than δ. It is indeed obvious that there are an infinity of such intervals corresponding to every ξ and every δ, for if the condition is satisfied for any particular value of ϵ, then it is satisfied *a fortiori* for any smaller value. What values of ϵ are admissible will naturally depend upon ξ; we have at present no reason for supposing that a value of ϵ admissible for one value of ξ will be admissible for another. We shall call the intervals thus associated with ξ *the δ-intervals of ξ.*

If $\xi = a$ then we can determine an interval $(a, a + \epsilon)$, and so an infinity of such intervals, having the same property. These we call the δ-intervals of a, and we can define in a similar manner the δ-intervals of b.

Consider now the set I of intervals formed by taking all the δ-intervals of all points of (a, b). It is plain that this set satisfies the conditions of the Heine-Borel theorem; every point interior to the interval is interior to at least one interval of I, and a and b are end points of at least one such interval. We can therefore determine a set I' which is formed by a finite number of intervals of I, and which possesses the same property as I itself.

The intervals which compose the set I' will in general overlap, as in Fig. 32. But their end points obviously divide up (a, b) into a finite set of intervals I'' each of which is included in an interval of I', and in each of which the oscillation of $\phi(x)$ is less than δ. Thus Theorem I is proved.

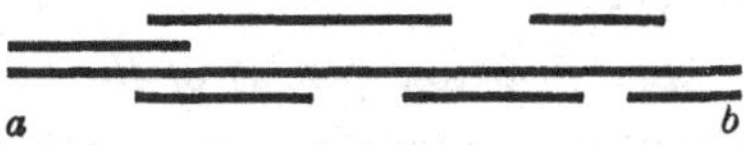

a b

Fig. 32

THEOREM II. *Given any positive number δ, we can find a number η such that, if the interval (a, b) is divided in any manner into sub-intervals of length less than η, then the oscillation of $\phi(x)$ in each of them will be less than δ.*

Take $\delta_1 < \frac{1}{2}\delta$, and construct, as in Theorem I, a finite set of sub-intervals j in each of which the oscillation of $\phi(x)$ is less than δ_1. Let η be the length of the least of these sub-intervals j. If now we divide (a, b) into parts each of length less than η, then any such part must lie entirely within at most two successive sub-intervals j. Hence, in virtue of (3) of § 104, the oscillation of $\phi(x)$, in one of the parts of length less than η, cannot exceed twice the greatest oscillation of $\phi(x)$ in a sub-interval j, and is therefore less than $2\delta_1$, and therefore than δ.

This theorem is of fundamental importance in the theory of definite integrals (Ch. VII). It is impossible, without the use of this or some similar theorem, to prove that a function continuous throughout an interval necessarily possesses an integral over that interval.

108. Continuous functions of several variables. The notions of continuity and discontinuity may be extended to functions of several independent variables (Ch. II, §§ 31 *et seq.*). Their application to such functions, however, raises questions much more complex and difficult than those which we have considered in this chapter. It would be impossible for us to discuss these questions in any detail here; but we shall, in the sequel, require to know what is meant by a continuous function of two variables, and we accordingly give the following definition. It is a straightforward generalisation of the last form of the definition of § 99.

The function $\phi(x,y)$ *of the two variables* x *and* y *is said to be* **continuous** *for* $x = \xi$, $y = \eta$ *if, given any positive number* δ, *however small, we can choose* $\epsilon(\delta)$ *so that*

$$|\phi(x,y) - \phi(\xi,\eta)| < \delta$$

when $0 \leqq |x-\xi| \leqq \epsilon(\delta)$ *and* $0 \leqq |y-\eta| \leqq \epsilon(\delta)$; *that is to say if we can draw a square, whose sides are parallel to the axes of coordinates and of length* $2\epsilon(\delta)$, *whose centre is the point* (ξ,η), *and which is such that the value of* $\phi(x,y)$ *at any point inside it or on its boundary differs from* $\phi(\xi,\eta)$ *by less than* δ*.

This definition of course presupposes that $\phi(x,y)$ is defined at all points of the square in question, and in particular at the point (ξ,η). Another method of stating the definition is this: $\phi(x,y)$ is *continuous for* $x = \xi$, $y = \eta$ *if* $\phi(x,y) \to \phi(\xi,\eta)$ *when* $x \to \xi$ *and* $y \to \eta$ *in any manner*. This statement is apparently simpler; but it contains phrases the precise meaning of which has not yet been explained and can only be explained by the help of inequalities like those which occur in our original statement.

It is easy to prove that the sums, the products, and in general the quotients of continuous functions of two variables are themselves continuous. A polynomial in two variables is continuous for all values of the variables; and the ordinary functions of x and y which occur in every-day analysis are *generally* continuous, i.e. are continuous except for pairs of values of x and y connected by special relations.

The reader should observe carefully that to assert the continuity of $\phi(x,y)$ with respect to the two variables x and y is to assert much more than its continuity with respect to each variable considered separately. It is plain that if $\phi(x,y)$ is continuous with respect to x and y, then it is continuous with respect to x (or y) when any fixed value is assigned to y (or x). But the converse is by no means true. Suppose, for example, that

$$\phi(x,y) = \frac{2xy}{x^2+y^2}$$

* The reader should draw a figure to illustrate the definition.

when neither x nor y is zero, and $\phi(x, y) = 0$ when either x or y is zero. Then if y has any fixed value, zero or not, $\phi(x, y)$ is a continuous function of x, and in particular continuous for $x = 0$; for its value when $x = 0$ is zero, and it tends to the limit zero as $x \to 0$. In the same way it may be shown that $\phi(x, y)$ is a continuous function of y. But $\phi(x, y)$ is *not* a continuous function of x *and* y for $x = 0$, $y = 0$. Its value when $x = 0$, $y = 0$ is zero; but if x and y tend to zero along the straight line $y = \alpha x$, then

$$\phi(x, y) = \frac{2\alpha}{1+\alpha^2}, \quad \lim \phi(x, y) = \frac{2\alpha}{1+\alpha^2},$$

which may have any value between -1 and 1.

109. Implicit functions. We have already, in Ch. II, met with the idea of an *implicit function*. Thus, if x and y are connected by the relation

$$y^5 - xy - y - x = 0 \quad \text{.........................(1)},$$

then y is an 'implicit function' of x.

But it is far from obvious that such an equation as this does really define a function y of x, or several such functions. In Ch. II we were content to take this for granted. We are now in a position to consider whether the assumption we made then was justified.

We shall find the following terminology useful. Suppose that it is possible to surround a point (a, b), as in § 108, with a square throughout which a certain condition is satisfied. We shall call such a square a *neighbourhood* of (a, b), and say that the condition in question is satisfied *in the neighbourhood of* (a, b), or *near* (a, b), meaning by this simply that it is possible to find *some* square throughout which the condition is satisfied. It is obvious that similar language may be used when we are dealing with a single variable, the square being replaced by an interval on a line.

THEOREM. *If* (i) $f(x, y)$ *is a continuous function of* x *and* y *in the neighbourhood of* (a, b),

(ii) $f(a, b) = 0$,

(iii) $f(x, y)$ *is, for all values of* x *in the neighbourhood of* a, *a steadily increasing function of* y, *in the stricter sense of* § 95,

then (1) *there is a unique function* $y = \phi(x)$ *which, when substituted in the equation* $f(x, y) = 0$, *satisfies it identically for all values of* x *in the neighbourhood of* a,

(2) $\phi(x)$ *is continuous for all values of* x *in the neighbourhood of* a.

In the figure the square represents a 'neighbourhood' of (a, b) throughout which the conditions (i) and (iii) are satisfied, and P the point (a, b). If we take Q and R as in the figure, it follows from (iii) that $f(x, y)$ is positive at Q and negative at R. This being so, and $f(x,y)$ being continuous at Q and at R, we can draw lines QQ' and RR' parallel to OX, so that $R'Q'$ is parallel to OY and $f(x,y)$ is positive at all points of QQ' and negative at all points of RR'. In particular $f(x,y)$ is positive at Q' and negative at R', and therefore, in virtue of (iii) and § 101 vanishes once and only once at a point P' on $R'Q'$. The same construction gives us a unique point at which $f(x,y) = 0$ on each ordinate between RQ and $R'Q'$. It is obvious, moreover, that the same construction can be carried out to the left of RQ. The aggregate of points such as P' gives us the graph of the required function $y = \phi(x)$.

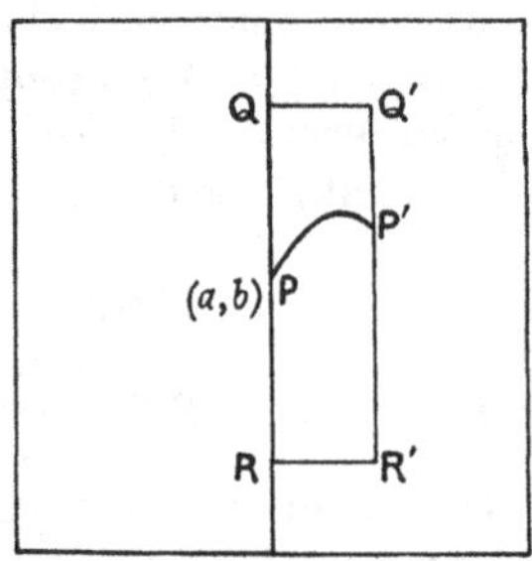

Fig. 33

It remains to prove that $\phi(x)$ is continuous. This is most simply effected by using the idea of the 'limits of indetermination' of $\phi(x)$ as $x \to \alpha$ (§ 96). Suppose that $x \to a$, and let λ and Λ be the limits of indetermination of $\phi(x)$ as $x \to a$. It is evident that the points (a, λ) and (a, Λ) lie on QB. Moreover, we can find a sequence of values of x such that $\phi(x) \to \lambda$ when $x \to a$ through the values of the sequence; and since $f\{x, \phi(x)\} = 0$, and $f(x,y)$ is a continuous function of x and y, we have

$$f(a, \lambda) = 0.$$

Hence $\lambda = b$; and similarly $\Lambda = b$. Thus $\phi(x)$ tends to the limit b as $x \to a$, and so $\phi(x)$ is continuous for $x = a$. It is evident that we can show in exactly the same way that $\phi(x)$ is continuous for any value of x in the neighbourhood of a.

It is clear that the truth of the theorem would not be affected if we were to change 'increasing' to 'decreasing' in condition (iii).

As an example, let us consider the equation (1), taking $a = 0$, $b = 0$. It is evident that the conditions (i) and (ii) are satisfied. Moreover

$$f(x, y) - f(x, y') = (y - y')\,(y^4 + y^3y' + y^2y'^2 + yy'^3 + y'^4 - x - 1)$$

has, when x, y, and y' are sufficiently small, the sign opposite to that of $y - y'$. Hence condition (iii) (with 'decreasing' for 'increasing') is satisfied. It follows that there is one and only one continuous function y which satisfies the equation (1) identically and vanishes with x.

The same conclusion would follow if the equation were

$$y^2 - xy - y - x = 0.$$

The function in question is in this case

$$y = \tfrac{1}{2}\{1 + x - \sqrt{(1 + 6x + x^2)}\},$$

where the square root is positive. The second root, in which the sign of the square root is changed, does not satisfy the condition of vanishing with x.

There is one point in the proof which the reader should be careful to observe. We supposed that the hypotheses of the theorem were satisfied 'in the neighbourhood of (a, b)', that is to say throughout a certain square $\xi-\epsilon \leqq x \leqq \xi+\epsilon$, $\eta-\epsilon \leqq y \leqq \eta+\epsilon$. The conclusion holds 'in the neighbourhood of $x = a$', that is to say throughout a certain interval $\xi-\epsilon_1 \leqq x \leqq \xi+\epsilon_1$. There is nothing to show that the ϵ_1 of the conclusion is the ϵ of the hypotheses, and indeed this is generally untrue.

110. Inverse functions. Suppose in particular that $f(x, y)$ is of the form $F(y)-x$. We then obtain the following theorem.

If $F(y)$ is a function of y, continuous and steadily increasing (or decreasing) in the stricter sense of § 95, *in the neighbourhood of $y = b$, and $F(b) = a$, then there is a unique continuous function $y = \phi(x)$ which is equal to b when $x = a$ and satisfies the equation $F(y) = x$ identically in the neighbourhood of $x = a$.*

The function thus defined is called the *inverse function* of $F(y)$.

Suppose for example that $y^3 = x$, $a = 0$, $b = 0$. Then all the conditions of the theorem are satisfied. The inverse function is $x = \sqrt[3]{y}$.

If we had supposed that $y^2 = x$, then the conditions of the theorem would not have been satisfied, for y^2 is not a steadily increasing function of y in any interval which includes $y = 0$: it decreases when y is negative and increases when y is positive. And in this case the conclusion of the theorem does not hold, for $y^2 = x$ defines *two* functions of x, viz. $y = \sqrt{x}$ and $y = -\sqrt{x}$, both of which vanish when $x = 0$, and each of which is defined only for positive values of x, so that the equation has sometimes two solutions and sometimes none. The reader should consider the more general equations

$$y^{2n} = x, \quad y^{2n+1} = x$$

in the same way. Another interesting example is given by the equation

$$y^5 - y - x = 0,$$

already considered in Ex. XIV. 7.

Similarly the equation $\sin y = x$

has just one solution which vanishes with x, viz. the value of $\arcsin x$ which vanishes with x. There are of course an infinity of solutions, given by the other values of $\arcsin x$ (cf. Ex. XV. 10), which do not satisfy this condition.

So far we have considered only what happens in the neighbourhood of a particular value of x. Let us suppose now that $F(y)$ is positive and steadily increasing (or decreasing) throughout an interval (a, b). Given any point ξ of (a, b), we can determine an interval i including ξ, and a unique and continuous inverse function $\phi_i(x)$ defined throughout i.

From the set I of intervals i we can, in virtue of the Heine-Borel theorem, pick out a finite sub-set covering up the whole interval (a, b); and it is plain that the finite set of functions $\phi_i(x)$, corresponding to the sub-set of intervals i thus selected, define together a unique inverse function $\phi(x)$ continuous throughout (a, b).

We thus obtain the theorem: *if $x = F(y)$, where $F(y)$ is continuous and increases steadily and strictly from A to B as y increases from a to b, then there is a unique inverse function $y = \phi(x)$ which is continuous and increases steadily and strictly from a to b as x increases from A to B.*

It is worth while to show how this theorem can be obtained directly without the help of the more difficult theorem of § 109. Suppose that $A < \xi < B$, and consider the class of values of y such that (i) $a < y < b$ and (ii) $F(y) \leqq \xi$. This class has an upper bound η, and plainly $F(\eta) \leqq \xi$. If $F(\eta)$ were less than ξ, we could find a value of y such that $y > \eta$ and $F(y) < \xi$, and η would not be the upper bound of the class considered. Hence $F(\eta) = \xi$. The equation $F(y) = \xi$ has therefore a unique solution $y = \eta = \phi(\xi)$, say; and plainly η increases steadily and continuously with ξ, which proves the theorem.

MISCELLANEOUS EXAMPLES ON CHAPTER V

1. Show that in general

$$\frac{ax^n + bx^{n-1} + \dots + k}{Ax^n + Bx^{n-1} + \dots + K} = \alpha + \frac{\beta}{x}(1 + \eta),$$

where $\alpha = a/A$, $\beta = (bA - aB)/A^2$, and η is of the first order of smallness when x is large. Indicate any exceptional cases.

2. Determine α, β, and γ so that

$$\frac{ax^2 + bx + c}{Ax^2 + Bx + C} = \alpha + \frac{\beta}{x} + \frac{\gamma}{x^2}(1 + \eta),$$

where η is of the first order of smallness when x is large. Indicate any exceptional cases.

3. Show that, if $P(x)$ is a polynomial $ax^n + bx^{n-1} + \dots + k$ whose first coefficient a is positive, then $P(x+h) - P(x)$ and

$$P(x + 2h) - 2P(x + h) + P(x)$$

increase steadily from a certain value of x onwards.

4. Prove that

$$P(x+h) - P(x) \sim nhax^{n-1}, \quad P(x + 2h) - 2P(x + h) + P(x) \sim n(n-1)h^2ax^{n-2},$$

when $x \to \infty$.

5. Show that $\lim_{x\to\infty} \sqrt{x}\{\sqrt{(x+a)}-\sqrt{x}\} = \frac{1}{2}a.$

[Use the formula $\sqrt{(x+a)}-\sqrt{x} = a/\{\sqrt{(x+a)}+\sqrt{x}\}$.]

6. Show that $\sqrt{(x+a)} = \sqrt{x}+\frac{1}{2}ax^{-\frac{1}{2}}(1+\eta)$, where η is of the first order of smallness when x is large.

7. Find values of α and β such that $\sqrt{(ax^2+2bx+c)}-\alpha x-\beta$ has the limit zero as $x\to\infty$; and prove that

$$\lim x\{\sqrt{(ax^2+2bx+c)}-\alpha x-\beta\}=(ac-b^2)/2a\sqrt{a}.$$

8. Evaluate $\lim_{x\to\infty} x^3\{\sqrt{[x^2+\sqrt{(x^4+1)}]}-x\sqrt{2}\}.$

9. Prove that $\sec x-\tan x\to 0$ as $x\to\frac{1}{2}\pi$.

10. Prove that $\phi(x) = 1-\cos(1-\cos x)$ is of the fourth order of smallness when x is small; and find the limit of $\phi(x)/x^4$ as $x\to 0$.

11. Prove that $\phi(x) = x\sin(\sin x)-\sin^2 x$ is of the sixth order of smallness when x is small; and find the limit of $\phi(x)/x^6$ as $x\to 0$.

12. From a point P on a radius OA of a circle, produced beyond the circle, a tangent PT is drawn to the circle, touching it in T, and TN is drawn perpendicular to OA. Show that $NA/AP\to 1$ as P moves up to A.

13. Tangents are drawn to a circular arc at its middle point and its extremities; Δ is the area of the triangle formed by the chord of the arc and the two tangents at the extremities, and Δ' the area of that formed by the three tangents. Show that $\Delta/\Delta'\to 4$ as the length of the arc tends to zero.

14. For what values of a does $\{a+\sin(1/x)\}/x$ tend to (1) ∞, (2) $-\infty$, as $x\to 0$? [To ∞ if $a>1$, to $-\infty$ if $a<-1$: otherwise the function oscillates.]

15. If $\phi(x) = 1/q$ when $x = p/q$, and $\phi(x) = 0$ when x is irrational, then $\phi(x)$ is continuous for all irrational and discontinuous for all rational values of x.

16. Show that the function whose graph is drawn in Fig. 30 may be represented by either of the formulae

$$1-x+[x]-[1-x],\quad 1-x-\lim_{n\to\infty}(\cos^{2n+1}\pi x).$$

17. Show that the function $\phi(x)$ which is equal to 0 when $x = 0$, to $\frac{1}{2}-x$ when $0<x<\frac{1}{2}$, to $\frac{1}{2}$ when $x = \frac{1}{2}$, to $\frac{3}{2}-x$ when $\frac{1}{2}<x<1$, and to 1 when $x = 1$, assumes every value between 0 and 1 once and once only as x increases from 0 to 1, but is discontinuous for $x = 0$, $x = \frac{1}{2}$, and $x = 1$. Show also that the function may be represented by the formula

$$\tfrac{1}{2}-x+\tfrac{1}{2}[2x]-\tfrac{1}{2}[1-2x].$$

18. Let $\phi(x) = x$ when x is rational and $\phi(x) = 1-x$ when x is irrational. Show that $\phi(x)$ assumes every value between 0 and 1 once and once only as x increases from 0 to 1, but is discontinuous for every value of x except $x = \frac{1}{2}$.

19. Prove that a function which is increasing at every point of (a, b) is an increasing function in (a, b).

Show that a function which is 'increasing on the right' at every point of (a, b) is not necessarily an increasing function in (a, b), but is so if it is continuous. (*Math. Trip.* 1926)

[We say that '$\phi(x)$ is increasing at x' when (i) $\phi(x') \geqq \phi(x)$ for all x' of some interval to the right of x and (ii) $\phi(x') \leqq \phi(x)$ for all x' of some interval to the left of x. When (i) alone is given, we say that $\phi(x)$ is 'increasing on the right'.

We have to prove that $\phi(x_2) \geqq \phi(x_1)$ if $a \leqq x_1 < x_2 \leqq b$. We divide the points ξ of (x_1, b) into two classes L and R, L if $\phi(x') \geqq \phi(x_1)$ for all x' of (x_1, ξ), and R in the contrary case, and denote by β the number corresponding to the section. The conclusion will follow if $\beta = b$ (i.e. if R does not exist).

If $\beta < b$ and $\phi(\beta) \geqq \phi(x_1)$, then we can, by (i), find an interval to the right of β in which $\phi(x) \geqq \phi(\beta) \geqq \phi(x_1)$, and this contradicts the definition of β. Hence $\phi(\beta) < \phi(x_1)$ if $\beta < b$. So far we have used (i) only.

If (ii) also is true, then there are points to the left of β at which $\phi(x) \leqq \phi(\beta) < \phi(x_1)$, and this again contradicts the definition of β. Hence $\beta = b$, as required. The same conclusion follows if (i) only is given but $\phi(x)$ is continuous; for then $\phi(x) < \phi(x_1)$ for values of x to the left of but sufficiently near to β.

The example $a = 0$, $b = 2$, $f(x) = x$ for $0 \leqq x < 1$, $f(x) = x-1$ for $1 \leqq x \leqq 2$, shows that the conclusion does not follow from (i) alone.]

20. As x increases from $-\frac{1}{2}\pi$ to $\frac{1}{2}\pi$, $y = \sin x$ is continuous and steadily increases, in the stricter sense, from -1 to 1. Deduce the existence of a function $x = \arcsin y$ which is a continuous and steadily increasing function of y from $y = -1$ to $y = 1$.

21. Show that the numerically least value of $\arctan y$ is continuous for all values of y and increases steadily from $-\frac{1}{2}\pi$ to $\frac{1}{2}\pi$ as y varies through all real values.

22. Examine whether the equation

$$x+y+P(x, y) = 0,$$

where $P(x, y)$ is a polynomial containing no term of degree less than 2, defines a unique function vanishing at $x = 0$ and continuous in the neighbourhood of $x = 0$. (*Math. Trip.* 1936)

23. Discuss, on the lines of §§ 109–110, the solution of the equations

$$y^2-y-x=0,\quad y^4-y^2-x^2=0,\quad y^4-y^2+x^2=0$$

in the neighbourhood of $x=0$, $y=0$.

24. If $ax^2+2bxy+cy^2+2dx+2ey=0$ and $\Delta=2bde-ae^2-cd^2$, then one value of y is given by $y=\alpha x+\beta x^2+\gamma x^3+O(x^4)$, where

$$\alpha=-d/e,\quad \beta=\Delta/2e^3,\quad \gamma=(cd-be)\Delta/2e^5.$$

[If $y-\alpha x=\eta$, then

$$-2e\eta=ax^2+2bx(\eta+\alpha x)+c(\eta+\alpha x)^2=Ax^2+2Bx\eta+C\eta^2,$$

say. It is evident that η is of the second order of smallness, $x\eta$ of the third, and η^2 of the fourth; and $-2e\eta=Ax^2-(AB/e)x^3$, the error being of the fourth order.]

25. If $x=ay+by^2+cy^3$, then one value of y is given by

$$y=\alpha x+\beta x^2+\gamma x^3+O(x^4),$$

where $\alpha=1/a$, $\beta=-b/a^3$, $\gamma=(2b^2-ac)/a^5$.

26. If $x=ay+by^n$, where n is an integer greater than unity, then one value of y is given by $y=\alpha x+\beta x^n+\gamma x^{2n-1}+O(x^{3n-2})$, where $\alpha=1/a$, $\beta=-b/a^{n+1}$, $\gamma=nb^2/a^{2n+1}$.

27. Show that the least positive root of the equation $xy=\sin x$ is a continuous function of y throughout the interval $(0, 1)$, and decreases steadily from π to 0 as y increases from 0 to 1. [The function is the inverse of $(\sin x)/x$: apply § 110.]

28. The least positive root of $xy=\tan x$ is a continuous function of y throughout the interval $(1,\infty)$, and increases steadily from 0 to $\frac{1}{2}\pi$ as y increases from 1 towards ∞.

29. A function $\phi(x)$ is said to be *upper semi-continuous* at x if

$$\phi(x')<\phi(x)+\delta$$

for every positive δ and all x' of an interval (depending on x and δ) round x. Prove that a function upper semi-continuous at all points of (a,b) has an upper bound, which it attains, in (a,b). (*Math. Trip.* 1924)

[To prove the existence of an upper bound M, replace 'bounded' by 'bounded above' in the proof of Theorem 1 of § 103. To prove that $\phi(x)$ attains the value M, make corresponding changes in the argument of § 105. We find that $\phi(x)$ assumes, near β, values as near as we please to M, and this contradicts the inequality $\phi(x)<\phi(\beta)+\delta$ if $\phi(\beta)<M$ and δ is sufficiently small.

We can define lower semi-continuity similarly by an inequality

$$\phi(x')>\phi(x)-\delta.$$

A lower semi-continuous function has an attained lower bound. A function both upper and lower semi-continuous is continuous.]

CHAPTER VI

DERIVATIVES AND INTEGRALS

111. Derivatives or differential coefficients. We return to the consideration of the properties which we naturally associate with the notion of a curve. The first and most obvious property is, as we saw in the last chapter, that which gives a curve its appearance of connectedness, and which we embodied in our definition of a continuous function.

The ordinary curves which occur in elementary geometry, such as straight lines, circles and conic sections, have much more 'regularity' than is implied by mere continuity. In particular they have a definite *direction* at every point; there is a *tangent* at every point of the curve. The tangent to a curve at P is defined, in elementary geometry, as 'the limiting position of the chord PQ, when Q moves up towards coincidence with P'. Let us consider what is implied in the assumption of the existence of such a limiting position.

In the figure (Fig. 34) P is a fixed point on the curve $y = \phi(x)$, and Q a variable point; PM, QN are parallel to OY and PR to

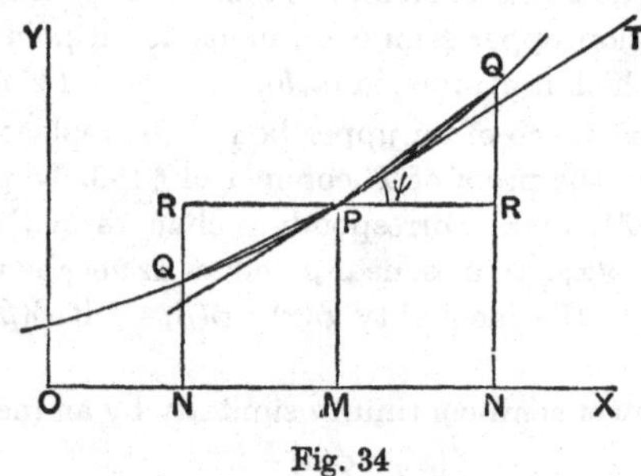

Fig. 34

OX. We denote the coordinates of P by x, y and those of Q by

$x+h$, $y+k$: h will be positive or negative according as N lies to the right or left of M.

We have assumed that there is a tangent to the curve at P, or that there is a definite 'limiting position' of the chord PQ. Suppose that PT, the tangent at P, makes an angle ψ with OX. Then to say that PT is the limiting position of PQ is equivalent to saying that the limit of the angle QPR, when Q approaches P along the curve from either side, is ψ. We have now to distinguish two cases, a general case and an exceptional one.

The general case is that in which ψ is not equal to $\frac{1}{2}\pi$, so that PT is not parallel to OY. In this case RPQ tends to the limit ψ, and

$$RQ/PR = \tan RPQ$$

tends to the limit $\tan\psi$. Now

$$\frac{RQ}{PR} = \frac{NQ-MP}{MN} = \frac{\phi(x+h)-\phi(x)}{h};$$

and so
$$\lim_{h\to 0} \frac{\phi(x+h)-\phi(x)}{h} = \tan\psi \quad \text{..............(1).}$$

The reader should be careful to note that in all these equations all lengths are regarded as affected with the proper sign, so that (e.g.) RQ is negative in the figure when Q lies to the left of P; and that the convergence to the limit is unaffected by the sign of h.

Thus the assumption that the curve which is the graph of $\phi(x)$ has a tangent at P, which is not perpendicular to the axis of x, implies that $\{\phi(x+h)-\phi(x)\}/h$ *tends to a limit when h tends to zero.*

This of course implies that both of

$$\{\phi(x+h)-\phi(x)\}/h, \quad \{\phi(x-h)-\phi(x)\}/(-h)$$

tend to limits when $h\to 0$ by positive values only, and that the two limits are equal. If these limits exist but are not equal, then the curve has an angle at the particular point considered, as in Fig. 35.

Now let us suppose that the curve has (like the circle or ellipse) a tangent at every point of its length, or at any rate every portion of its length which corresponds to a certain range of variation of x. Further let us suppose this tangent never perpendicular to the axis of x: this would restrict us, when the curve is a circle, to an arc less than a semi-circle. Then (1) is true for all values of x which fall inside this range. To each such value of x corresponds a value of $\tan\psi$; $\tan\psi$ is a function of x which is defined for all values of x in the range of values under consideration. We shall call this function the *derivative* of $\phi(x)$, and we shall denote it by

$$\phi'(x).$$

Another name for the derivative of $\phi(x)$ is the *differential coefficient* of $\phi(x)$; and the operation of calculating $\phi'(x)$ from $\phi(x)$ is generally known as *differentiation*. This terminology is firmly established for historical reasons: see § 116.

Before we proceed to consider the special case mentioned above, in which $\psi = \frac{1}{2}\pi$, we shall illustrate our definition by some general remarks and particular illustrations.

112. Some general remarks. (1) The existence of a derived function $\phi'(x)$ for all values of x in the interval $a \leqq x \leqq b$ implies that $\phi(x)$ is continuous at every point of this interval. For it is evident that $\{\phi(x+h)-\phi(x)\}/h$ cannot tend to a limit unless $\lim \phi(x+h) = \phi(x)$, and it is this which is the property denoted by continuity.

(2) It is natural to ask whether the converse is true, i.e. whether every continuous curve has a definite tangent at every point, and every function a differential coefficient for every value of x for which it is continuous*. The answer is obviously *No*: it is sufficient to consider the curve formed by two straight lines meeting to form an angle (Fig. 35). The reader will see at once that in this case $\{\phi(x+h)-\phi(x)\}/h$ has the limit $\tan\beta$ when $h\to 0$ by positive values and the limit $\tan\alpha$ when $h\to 0$ by negative values.

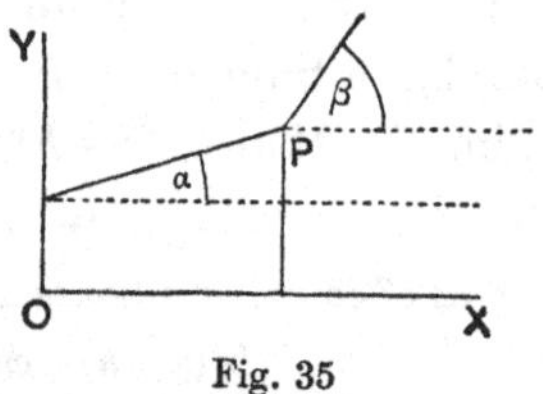

Fig. 35

* We leave out of account the exceptional case (which we have still to examine) in which the curve has a tangent perpendicular to OX: apart from this possibility the two forms of the question are equivalent.

This is of course a case in which a curve might reasonably be said to have *two* directions at a point. But the following example, although a little more difficult, shows that there are cases in which a continuous curve cannot be said to have either one direction or several directions at one of its points. Draw the graph (Fig. 14, p. 56) of the function $x\sin(1/x)$. The function is not defined for $x = 0$, and so is discontinuous for $x = 0$. On the other hand the function defined by the equations

$$\phi(x) = x\sin(1/x) \quad (x \neq 0), \qquad \phi(x) = 0 \quad (x = 0)$$

is continuous for $x = 0$ (Exs. XXXVII. 14, 15), and the graph of this function is a continuous curve.

But $\phi(x)$ has no derivative for $x = 0$. For $\phi'(0)$ would be, by definition, $\lim\{\phi(h)-\phi(0)\}/h$ or $\lim\sin(1/h)$; and no such limit exists.

It has been shown that a continuous function of x may have no derivative for *any* value of x, but the proof of this is much more difficult. The reader who is interested in the question may be referred to Bromwich's *Infinite series* (1st edition), pp. 490–1, or Hobson's *Theory of functions of a real variable* (2nd edition), vol. II. pp. 411–12.

(3) The notion of a derivative or differential coefficient was suggested to us by geometrical considerations. But there is nothing geometrical in the notion itself. The derivative $\phi'(x)$ of a function $\phi(x)$ may be defined, without any reference to any kind of geometrical representation of $\phi(x)$, by the equation

$$\phi'(x) = \lim_{h \to 0} \frac{\phi(x+h)-\phi(x)}{h};$$

and $\phi(x)$ has or has not a derivative, for any particular value of x, according as this limit does or does not exist. The geometry of curves is merely one of many departments of mathematics in which the idea of a derivative finds an application.

Another important application is in dynamics. Suppose that a particle is moving in a straight line in such a way that at time t its distance from a fixed point on the line is $\phi(t)$. Then the 'velocity of the particle at time t' is by definition the limit of

$$\frac{\phi(t+h)-\phi(t)}{h}$$

as $h \to 0$. The notion of 'velocity' is merely a special case of that of the derivative of a function.

Examples XXXIX. 1. If $\phi(x)$ is a constant, then $\phi'(x) = 0$. Interpret this result geometrically.

2. If $\phi(x) = ax+b$, then $\phi'(x) = a$. Prove this (i) from the formal definition and (ii) by geometrical considerations.

3. If $\phi(x) = x^m$, where m is a positive integer, then $\phi'(x) = mx^{m-1}$.

[For
$$\phi'(x) = \lim \frac{(x+h)^m - x^m}{h}$$
$$= \lim \left\{ mx^{m-1} + \frac{m(m-1)}{1.2} x^{m-2}h + \dots + h^{m-1} \right\}.$$

The reader should observe that this method cannot be applied to $x^{p/q}$, where p/q is a rational fraction, because $(x+h)^{p/q}$ cannot be expressed as a finite series of powers of h. We shall show later on (§ 119) that the result of this example holds for all rational values of m. Meanwhile the reader will find it instructive to determine $\phi'(x)$ when m has some special fractional value (e.g. $\frac{1}{2}$), by means of some special device.]

4. If $\phi(x) = \sin x$, then $\phi'(x) = \cos x$; and if $\phi(x) = \cos x$, then $\phi'(x) = -\sin x$.

[For example, if $\phi(x) = \sin x$, we have
$$\frac{\phi(x+h) - \phi(x)}{h} = \frac{2 \sin \frac{1}{2}h}{h} \cos (x + \tfrac{1}{2}h),$$
the limit of which, when $h \to 0$, is $\cos x$, since $\lim \cos (x + \frac{1}{2}h) = \cos x$ (the cosine being a continuous function) and $\lim \{(\sin \frac{1}{2}h)/\frac{1}{2}h\} = 1$ (Ex. XXXVI. 13).]

5. **Equations of the tangent and normal to a curve** $y = \phi(x)$. The tangent to the curve at the point (x_0, y_0) is the line through (x_0, y_0) which makes with OX an angle ψ, where $\tan \psi = \phi(x_0)$. Its equation is therefore
$$y - y_0 = (x - x_0)\,\phi'(x_0);$$
and the equation of the normal (the perpendicular to the tangent at the point of contact) is
$$(y - y_0)\,\phi'(x_0) + x - x_0 = 0.$$
We have assumed that the tangent is not parallel to the axis of y. In this special case it is obvious that the tangent and normal are $x = x_0$ and $y = y_0$ respectively.

6. Write down the equations of the tangent and normal at any point of the parabola $x^2 = 4ay$. Show that if $x_0 = 2a/m$, $y_0 = a/m^2$, then the tangent at (x_0, y_0) is $x = my + (a/m)$.

113. We have seen that if $\phi(x)$ is not continuous for a value of x then it cannot have a derivative for that value of x. Thus such functions as $1/x$ or $\sin(1/x)$, which are not defined for

$x = 0$, and so necessarily discontinuous for $x = 0$, cannot have derivatives for $x = 0$. Or again the function $[x]$, which is discontinuous for every integral value of x, has no derivative for any such value of x.

Example. Since $[x]$ is constant between every two integral values of x, its derivative, whenever it exists, has the value zero. Thus the derivative of $[x]$, which we may represent by $[x]'$, is a function equal to zero for all values of x save integral values and undefined for integral values. It is interesting to note that the function $1 - \dfrac{\sin \pi x}{\sin \pi x}$ has exactly the same properties.

We saw also in Ex. XXXVII. 7 that the types of discontinuity which occur most commonly, when we are dealing with the very simplest functions, such as polynomials or rational or trigonometrical functions, are associated with a relation of the type

$$\phi(x) \to +\infty$$

or $\phi(x) \to -\infty$. In all these cases, as in such cases as those considered above, there is no derivative for certain special values of x.

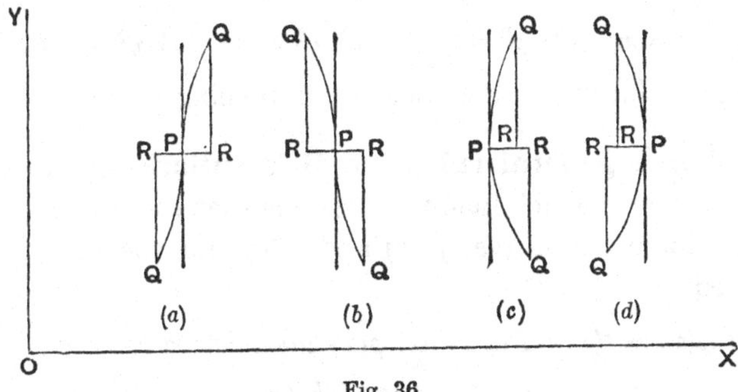

Fig. 36

Thus *all discontinuities of a function $\phi(x)$ are also discontinuities of its derivative $\phi'(x)$.* But the converse is not true, as we may easily see if we return to the geometrical point of view of § 111 and consider the special case, hitherto left aside, in which the graph of $\phi(x)$ has a tangent parallel to OY. This case may be sub-divided into a number of cases, of which the most typical are shown in Fig. 36. In cases (c) and (d) the function is two valued on one side of P and not defined on the other. In such cases we may consider the two sets of values of $\phi(x)$, which occur on one side of P or the other, as defining distinct functions $\phi_1(x)$ and $\phi_2(x)$, the upper part of the curve corresponding to $\phi_1(x)$.

The reader will easily convince himself that in (a)

$$\{\phi(x+h)-\phi(x)\}/h \to +\infty,$$

as $h \to 0$, and in (b)

$$\{\phi(x+h)-\phi(x)\}/h \to -\infty;$$

while in (c)

$$\{\phi_1(x+h)-\phi_1(x)\}/h \to +\infty, \quad \{\phi_2(x+h)-\phi_2(x)\}/h \to -\infty,$$

and in (d)

$$\{\phi_1(x+h)-\phi_1(x)\}/h \to -\infty, \quad \{\phi_2(x+h)-\phi_2(x)\}/h \to +\infty,$$

though of course in (c) only positive and in (d) only negative values of h can be considered, a fact which by itself would preclude the existence of a derivative.

We can obtain examples of these four cases by considering the functions defined by the equations

$$(a)\ y^3 = x, \quad (b)\ y^3 = -x, \quad (c)\ y^2 = x, \quad (d)\ y^2 = -x,$$

the special value of x under consideration being $x = 0$.

114. Some general rules for differentiation. Throughout the theorems which follow we assume that the functions $f(x)$ and $F(x)$ have derivatives $f'(x)$ and $F'(x)$ for the values of x considered.

(1) *If* $\phi(x) = f(x) + F(x)$, *then* $\phi(x)$ *has a derivative*

$$\phi'(x) = f'(x) + F'(x).$$

(2) *If* $\phi(x) = kf(x)$, *where* k *is a constant, then* $\phi(x)$ *has a derivative*

$$\phi'(x) = kf'(x).$$

We leave it as an exercise to the reader to deduce these results from the general theorems stated in Ex. XXXV. 1.

(3) *If* $\phi(x) = f(x)\,F(x)$, *then* $\phi(x)$ *has a derivative*

$$\phi'(x) = f(x)\,F'(x) + f'(x)\,F(x).$$

For
$$\phi'(x) = \lim \frac{f(x+h)F(x+h) - f(x)F(x)}{h}$$
$$= \lim \left\{ f(x+h)\frac{F(x+h)-F(x)}{h} + F(x)\frac{f(x+h)-f(x)}{h} \right\}$$
$$= f(x)F'(x) + F(x)f'(x).$$

(4) *If* $\phi(x) = \dfrac{1}{f(x)}$ *and* $f(x) \neq 0$, *then* $\phi(x)$ *has a derivative*
$$\phi'(x) = -\frac{f'(x)}{\{f(x)\}^2}.$$

For
$$\phi'(x) = \lim \frac{1}{h}\left\{\frac{f(x)-f(x+h)}{f(x+h)f(x)}\right\} = -\frac{f'(x)}{\{f(x)\}^2}.$$

(5) *If* $\phi(x) = \dfrac{f(x)}{F(x)}$ *and* $F(x) \neq 0$, *then* $\phi(x)$ *has a derivative*
$$\phi'(x) = \frac{f'(x)F(x) - f(x)F'(x)}{\{F(x)\}^2}.$$

This follows at once from (3) and (4).

(6) *If* $\phi(x) = F\{f(x)\}$, *then* $\phi(x)$ *has a derivative*
$$\phi'(x) = F'\{f(x)\}f'(x).$$

The proof of this theorem requires a little care*.

We write $f(x) = y$, $f(x+h) = y+k$, so that $k \to 0$ when $h \to 0$ and
$$k/h \to f'(x) \qquad (1).$$

We must now distinguish two cases.

(*a*) Suppose that $f'(x) \neq 0$, and that h is small, but not zero. Then $k \neq 0$, because of (1), and
$$\frac{\phi(x+h)-\phi(x)}{h} = \frac{F(y+k)-F(y)}{k}\frac{k}{h} \to F'(y)f'(x).$$

(*b*) Suppose that $f'(x) = 0$, and that h is small, but not zero. There are now two possibilities. If $k = 0$,† then
$$\frac{\phi(x+h)-\phi(x)}{h} = 0.$$

* The proofs in many text-books (and in the first three editions of this book) are inaccurate. See a note by Prof. H. S. Carslaw in vol. XXIX of the *Bulletin of the American Mathematical Society*.

† The fallacy in the inaccurate proofs lies in overlooking this possibility.

If $k \neq 0$, then

$$\frac{\phi(x+h)-\phi(x)}{h} = \frac{F(y+k)-F(y)}{k}\frac{k}{h}.$$

The first factor is nearly $F'(y)$, and the second is small, because $k/h \to 0$. Hence $\{\phi(x+h)-\phi(x)\}/h$ is small in any case, and

$$\frac{\phi(x+h)-\phi(x)}{h} \to 0 = F'(y)f'(x).$$

Our last theorem requires a few words of preliminary explanation. Suppose that $x = \psi(y)$, where $\psi(y)$ is continuous and steadily increasing (or decreasing), in the stricter sense of § 95, in a certain interval of values of y. Then we may write $y = \phi(x)$, where ϕ is the 'inverse' function (§ 110) of ψ.

(7) *If $y = \phi(x)$, where ϕ is the inverse function of ψ, so that $x = \psi(y)$, and $\psi(y)$ has a derivative $\psi'(y)$ which is not equal to zero, then $\phi(x)$ has a derivative*

$$\phi'(x) = \frac{1}{\psi'(y)}.$$

For if $\phi(x+h) = y+k$, then $k \to 0$ as $h \to 0$, and

$$\phi'(x) = \lim_{h\to 0}\frac{\phi(x+h)-\phi(x)}{(x+h)-x} = \lim_{k\to 0}\frac{(y+k)-y}{\psi(y+k)-\psi(y)} = \frac{1}{\psi'(y)}.$$

115. Derivatives of complex functions. So far we have supposed that $y = \phi(x)$ is a *real* function of x. If y is a complex function $\phi(x)+i\psi(x)$, then we define the derivative of y as being $\phi'(x)+i\psi'(x)$. The reader will have no difficulty in seeing that Theorems (1)–(5) above retain their validity when $\phi(x)$ is complex. Theorems (6) and (7) have also analogues for complex functions, but these depend upon the general notion of a 'function of a complex variable', a notion which we have encountered at present only in a few particular cases.

116. The notation of the differential calculus. We have already explained that what we call a *derivative* is often called a *differential coefficient*. Not only a different name but a different notation is often used; the derivative of the function $y = \phi(x)$ is often denoted by one or other of the expressions

$$D_x y, \quad \frac{dy}{dx}.$$

Of these the last is the most usual and convenient: the reader must however be careful to remember that dy/dx does not mean 'a certain number dy divided by another number dx': it means 'the result of a certain operation D_x or d/dx applied to $y = \phi(x)$', the operation being that of forming the quotient $\{\phi(x+h) - \phi(x)\}/h$ and making $h \to 0$.

Of course a notation at first sight so peculiar would not have been adopted without some reason, and the reason was as follows. The denominator h of the fraction $\{\phi(x+h) - \phi(x)\}/h$ is the difference of the values $x+h$, x of the independent variable x; similarly the numerator is the difference of the corresponding values $\phi(x+h)$, $\phi(x)$ of the dependent variable y. These differences may be called the *increments* of x and y respectively, and denoted by δx and δy. Then the fraction is $\delta y/\delta x$, and it is for many purposes convenient to denote the limit of the fraction, which is the same thing as $\phi'(x)$, by dy/dx. But this notation must for the present be regarded as purely symbolical. The dy and dx which occur in it cannot be separated, and standing by themselves they would mean nothing: in particular dy and dx do not mean $\lim \delta y$ and $\lim \delta x$, these limits being both zero. The reader will have to become familiar with this notation, but so long as it puzzles him he can avoid it by writing the differential coefficient in the form $D_x y$, or using the notation $\phi(x)$, $\phi'(x)$, as we have done in the preceding sections of this chapter.

In Ch. VII, however, we shall show how it is possible to define the symbols dx and dy in such a way that they have an independent meaning and that the derivative dy/dx is actually their quotient.

The theorems of § 114 may of course at once be translated into this notation. They may be stated as follows:

(1) *if* $y = y_1 + y_2$, *then* $\dfrac{dy}{dx} = \dfrac{dy_1}{dx} + \dfrac{dy_2}{dx}$;

(2) *if* $y = ky_1$, *then* $\dfrac{dy}{dx} = k\dfrac{dy_1}{dx}$;

(3) *if* $y = y_1 y_2$, *then* $\dfrac{dy}{dx} = y_1\dfrac{dy_2}{dx} + y_2\dfrac{dy_1}{dx}$;

(4) *if* $y = \dfrac{1}{y_1}$, *then* $\dfrac{dy}{dx} = -\dfrac{1}{y_1^2}\dfrac{dy_1}{dx}$;

(5) *if* $y = \dfrac{y_1}{y_2}$, *then* $\dfrac{dy}{dx} = \left(y_2\dfrac{dy_1}{dx} - y_1\dfrac{dy_2}{dx}\right)\Big/ y_2^2$;

(6) *if y is a function of x, and z a function of y, then*

$$\frac{dz}{dx} = \frac{dz}{dy}\frac{dy}{dx};$$

(7) $$\frac{dy}{dx} = 1\Big/\left(\frac{dx}{dy}\right).$$

Examples XL. 1. If $y = y_1 y_2 y_3$, then

$$\frac{dy}{dx} = y_2 y_3 \frac{dy_1}{dx} + y_3 y_1 \frac{dy_2}{dx} + y_1 y_2 \frac{dy_3}{dx};$$

and if $y = y_1 y_2 \dots y_n$, then

$$\frac{dy}{dx} = \sum_{r=1}^{n} y_1 y_2 \dots y_{r-1} y_{r+1} \dots y_n \frac{dy_r}{dx}.$$

In particular, if $y = z^n$, then $dy/dx = nz^{n-1}(dz/dx)$; and if $y = x^n$, then $dy/dx = nx^{n-1}$, as was proved otherwise in Ex. XXXIX. 3.

2. If $y = y_1 y_2 \dots y_n$, then

$$\frac{1}{y}\frac{dy}{dx} = \frac{1}{y_1}\frac{dy_1}{dx} + \frac{1}{y_2}\frac{dy_2}{dx} + \dots + \frac{1}{y_n}\frac{dy_n}{dx}.$$

In particular, if $y = z^n$, then $\dfrac{1}{y}\dfrac{dy}{dx} = \dfrac{n}{z}\dfrac{dz}{dx}$.

117. Standard forms. We shall now investigate more systematically the forms of the derivatives of a few of the simplest types of functions.

A. Polynomials. If $\phi(x) = a_0 x^n + a_1 x^{n-1} + \dots + a_n$, then

$$\phi'(x) = na_0 x^{n-1} + (n-1)a_1 x^{n-2} + \dots + a_{n-1}.$$

It is sometimes more convenient to use for the standard form of a polynomial of degree n in x what is known as the *binomial form*, viz.

$$a_0 x^n + \binom{n}{1} a_1 x^{n-1} + \binom{n}{2} a_2 x^{n-2} + \dots + a_n.$$

In this case

$$\phi'(x) = n\left\{a_0 x^{n-1} + \binom{n-1}{1} a_1 x^{n-2} + \binom{n-1}{2} a_2 x^{n-3} + \dots + a_{n-1}\right\}.$$

The binomial form of $\phi(x)$ is often written symbolically as

$$(a_0, a_1, \dots, a_n \between x, 1)^n;$$

and then $$\phi'(x) = n(a_0, a_1, \dots, a_{n-1} \between x, 1)^{n-1}.$$

We shall see later that $\phi(x)$ can always be expressed as the product of n factors in the form

$$\phi(x) = a_0(x-\alpha_1)(x-\alpha_2)\dots(x-\alpha_n),$$

where the α's are real or complex numbers. Then

$$\phi'(x) = a_0 \Sigma (x-\alpha_2)(x-\alpha_3)\dots(x-\alpha_n),$$

the notation implying that we form all possible products of $n-1$ factors, and add them all together. This form of the result holds even if several of the numbers α are equal; but of course then some of the terms on the right-hand side are repeated. The reader will easily verify that if

$$\phi(x) = a_0(x-\alpha_1)^{m_1}(x-\alpha_2)^{m_2}\dots(x-\alpha_\nu)^{m_\nu},$$

then $$\phi'(x) = a_0 \Sigma m_1 (x-\alpha_1)^{m_1-1}(x-\alpha_2)^{m_2}\dots(x-\alpha_\nu)^{m_\nu}.$$

Examples XLI. 1. Show that if $\phi(x)$ is a polynomial then $\phi'(x)$ is the coefficient of h in the expansion of $\phi(x+h)$ in powers of h.

2. If $\phi(x)$ is divisible by $(x-\alpha)^2$, then $\phi'(x)$ is divisible by $x-\alpha$: and generally, if $\phi(x)$ is divisible by $(x-\alpha)^m$, then $\phi'(x)$ is divisible by $(x-\alpha)^{m-1}$.

3. Conversely, if $\phi(x)$ and $\phi'(x)$ are *both* divisible by $x-\alpha$, then $\phi(x)$ is divisible by $(x-\alpha)^2$; and if $\phi(x)$ is divisible by $x-\alpha$ and $\phi'(x)$ by $(x-\alpha)^{m-1}$, then $\phi(x)$ is divisible by $(x-\alpha)^m$.

4. Show how to determine as completely as possible the multiple roots of $P(x) = 0$, where $P(x)$ is a polynomial, with their degrees of multiplicity, by means of the elementary algebraical operations.

[If H_1 is the highest common factor of P and P', H_2 the highest common factor of H_1 and P'', H_3 that of H_2 and P''', and so on, then the roots of $H_1H_3/H_2^2 = 0$ are the *double* roots of $P = 0$, the roots of $H_2H_4/H_3^2 = 0$ the *treble* roots, and so on. But it may not be possible to complete the solution of $H_1H_3/H_2^2 = 0$, $H_2H_4/H_3^2 = 0$, ... Thus if $P(x) = (x-1)^3(x^5-x-7)^2$, then $H_1H_3/H_2^2 = x^5-x-7$ and $H_2H_4/H_3^2 = x-1$; and we cannot solve the first equation.]

5. Find all the roots, with their degrees of multiplicity, of

$$x^4+3x^3-3x^2-11x-6=0,\quad x^6+2x^5-8x^4-14x^3+11x^2+28x+12=0.$$

6. If $ax^2+2bx+c$ has a double root, i.e. is of the form $a(x-\alpha)^2$, then $2(ax+b)$ must be divisible by $x-\alpha$, so that $\alpha=-b/a$. This value of x must satisfy $ax^2+2bx+c=0$. Verify that the condition thus arrived at is $ac-b^2=0$.

7. The equation $1/(x-a)+1/(x-b)+1/(x-c)=0$ can have a pair of equal roots only if $a=b=c$. (*Math. Trip.* 1905)

8. Show that
$$ax^3+3bx^2+3cx+d=0$$
has a double root if $G^2+4H^3=0$, where $H=ac-b^2$, $G=a^2d-3abc+2b^3$.

[Put $ax+b=y$, when the equation reduces to $y^3+3Hy+G=0$. This must have a root in common with $y^2+H=0$.]

9. The reader may verify that if α, β, γ, δ are the roots of
$$ax^4+4bx^3+6cx^2+4dx+e=0,$$
then the equation whose roots are
$$\tfrac{1}{12}a\{(\alpha-\beta)(\gamma-\delta)-(\gamma-\alpha)(\beta-\delta)\},$$
and two similar expressions formed by permuting α, β, γ cyclically, is
$$4y^3-g_2y-g_3=0,$$
where $g_2=ae-4bd+3c^2,\quad g_3=ace+2bcd-ad^2-eb^2-c^3.$

It is clear that if two of α, β, γ, δ are equal then two of the roots of this cubic will be equal. Using the result of Ex. 8 we deduce that $g_2^3-27g_3^2=0$.

10. **Rolle's theorem for polynomials.** *If $\phi(x)$ is any polynomial, then between any pair of roots of $\phi(x)=0$ lies a root of $\phi'(x)=0$.*

A proof of this theorem applying to more general functions will be given later. The following is an algebraical proof valid for polynomials only. We suppose that α, β are two successive roots, repeated respectively m and n times, so that
$$\phi(x)=(x-\alpha)^m(x-\beta)^n\theta(x),$$
where $\theta(x)$ is a polynomial which has the same sign, say the positive sign, for $\alpha\leqq x\leqq\beta$. Then
$$\begin{aligned}\phi'(x)&=(x-\alpha)^m(x-\beta)^n\theta'(x)+\{m(x-\alpha)^{m-1}(x-\beta)^n+n(x-\alpha)^m(x-\beta)^{n-1}\}\theta(x)\\&=(x-\alpha)^{m-1}(x-\beta)^{n-1}[(x-\alpha)(x-\beta)\theta'(x)+\{m(x-\beta)+n(x-\alpha)\}\theta(x)]\\&=(x-\alpha)^{m-1}(x-\beta)^{n-1}F(x),\end{aligned}$$
say. Now $F(\alpha)=m(\alpha-\beta)\theta(\alpha)$ and $F(\beta)=n(\beta-\alpha)\theta(\beta)$, which have opposite signs. Hence $F(x)$, and so $\phi'(x)$, vanishes for some value of x between α and β.

118. B. Rational functions. If

$$R(x) = \frac{P(x)}{Q(x)},$$

where P and Q are polynomials, it follows at once from § 114 (5) that

$$R'(x) = \frac{P'(x)\,Q(x) - P(x)\,Q'(x)}{\{Q(x)\}^2},$$

and this formula enables us to write down the derivative of any rational function. The form in which we obtain it, however, may or may not be the simplest possible. It will be the simplest possible if $Q(x)$ and $Q'(x)$ have no common factor, i.e. if $Q(x)$ has no repeated factor. But if $Q(x)$ has a repeated factor, then the expression which we obtain for $R'(x)$ will be capable of further reduction.

It is very often convenient, in differentiating a rational function, to employ the method of partial fractions. We shall suppose that $Q(x)$, as in § 117, is expressed in the form

$$a_0(x-\alpha_1)^{m_1}(x-\alpha_2)^{m_2}\dots(x-\alpha_\nu)^{m_\nu}.$$

Then it is proved in treatises on algebra* that $R(x)$ can be expressed in the form

$$\Pi(x) + \frac{A_{1,1}}{x-\alpha_1} + \frac{A_{1,2}}{(x-\alpha_1)^2} + \dots + \frac{A_{1,m_1}}{(x-\alpha_1)^{m_1}}$$
$$+ \frac{A_{2,1}}{x-\alpha_2} + \frac{A_{2,2}}{(x-\alpha_2)^2} + \dots + \frac{A_{2,m_2}}{(x-\alpha_2)^{m_2}} + \dots,$$

where $\Pi(x)$ is a polynomial; i.e. as the sum of a polynomial and the sum of a number of terms of the type

$$\frac{A}{(x-\alpha)^p},$$

where α is a root of $Q(x) = 0$. We know already how to find the derivative of the polynomial: and it follows at once from Theorem (4) of § 114, or, if α is complex, from its extension indicated in § 115, that the derivative of the rational function last written is

$$-\frac{pA(x-\alpha)^{p-1}}{(x-\alpha)^{2p}} = -\frac{pA}{(x-\alpha)^{p+1}}.$$

* See, e.g., Chrystal's *Algebra*, 2nd edition, vol. I. pp. 151 *et seq.*

We are now able to write down the derivative of the general rational function $R(x)$, in the form

$$\Pi'(x)-\frac{A_{1,1}}{(x-\alpha_1)^2}-\frac{2A_{1,2}}{(x-\alpha_1)^3}-\dots-\frac{A_{2,1}}{(x-\alpha_2)^2}-\frac{2A_{2,2}}{(x-\alpha_2)^3}-\dots.$$

Incidentally we have proved that *the derivative of* x^m *is* mx^{m-1}, *for all integral values of* m *positive or negative*, except when m is negative and $x = 0$.

The method explained in this section is particularly useful when we have to differentiate a rational function several times (see Exs. XLV).

Examples XLII. 1. Prove that

$$\frac{d}{dx}\left(\frac{x}{1+x^2}\right)=\frac{1-x^2}{(1+x^2)^2},\quad \frac{d}{dx}\left(\frac{1-x^2}{1+x^2}\right)=-\frac{4x}{(1+x^2)^2}.$$

2. Prove that

$$\frac{d}{dx}\left(\frac{ax^2+2bx+c}{Ax^2+2Bx+C}\right)=2\frac{(ax+b)(Bx+C)-(bx+c)(Ax+B)}{(Ax^2+2Bx+C)^2}.$$

3. If Q has a factor $(x-\alpha)^m$, then the denominator of R' (when R' is reduced to its lowest terms) is divisible by $(x-\alpha)^{m+1}$ but by no higher power of $x-\alpha$.

4. In no case can the denominator of R' have a *simple* factor $x-\alpha$. Hence a rational function whose denominator contains a simple factor cannot be the derivative of a rational function. For example, $1/x$ is not the derivative of a rational function.

119. C. Algebraical functions. The results of the preceding sections, together with Theorem (6) of § 114, enable us to obtain the derivative of any explicit algebraical function.

The most important such function is x^m, where m is a rational number. We have seen already (§ 118) that the derivative of this function is mx^{m-1} when m is an integer positive or negative; and we shall now prove that this result is true (provided that $x \neq 0$) for all rational values of m. Suppose that $y = x^m = x^{p/q}$, where p and q are integers and q positive; and let $z = x^{1/q}$, so that $x = z^q$ and $y = z^p$. Then

$$\frac{dy}{dx}=\left(\frac{dy}{dz}\right)\bigg/\left(\frac{dx}{dz}\right)=\frac{p}{q}z^{p-q}=mx^{m-1}.$$

This result may also be deduced as a corollary from Ex. XXXVI. 3. For, if $\phi(x) = x^m$, we have

$$\phi'(x) = \lim_{h\to 0} \frac{(x+h)^m - x^m}{h} = \lim_{\xi\to x} \frac{\xi^m - x^m}{\xi - x} = mx^{m-1}.$$

It is clear that the more general formula

$$\frac{d}{dx}(ax+b)^m = ma(ax+b)^{m-1}$$

holds also for all rational values of m.

The differentiation of *implicit* algebraical functions involves certain theoretical difficulties to which we shall return in Ch. VII. But there is no practical difficulty in the actual calculation of the derivative of such a function: the method to be adopted will be illustrated sufficiently by an example. Suppose that y is given by the equation

$$x^3 + y^3 - 3axy = 0.$$

Differentiating with respect to x, we find

$$x^2 + y^2\frac{dy}{dx} - a\left(y + x\frac{dy}{dx}\right) = 0$$

and so

$$\frac{dy}{dx} = -\frac{x^2 - ay}{y^2 - ax}.$$

Examples XLIII. 1. Find the derivatives of

$$\sqrt{\left(\frac{1+x}{1-x}\right)},\quad \sqrt{\left(\frac{ax+b}{cx+d}\right)},\quad \sqrt{\left(\frac{ax^2+2bx+c}{Ax^2+2Bx+C}\right)},\quad (ax+b)^m(cx+d)^n.$$

2. Prove that

$$\frac{d}{dx}\left\{\frac{x}{\sqrt{(a^2+x^2)}}\right\} = \frac{a^2}{(a^2+x^2)^{\frac{3}{2}}},\quad \frac{d}{dx}\left\{\frac{x}{\sqrt{(a^2-x^2)}}\right\} = \frac{a^2}{(a^2-x^2)^{\frac{3}{2}}}.$$

3. Find the differential coefficient of y when

(i) $ax^2 + 2hxy + by^2 + 2gx + 2fy + c = 0$, (ii) $x^5 + y^5 - 5ax^2y^2 = 0$.

120. D. Transcendental functions. We have already proved (Ex. XXXIX. 4) that

$$D_x \sin x = \cos x,\quad D_x \cos x = -\sin x.$$

By means of Theorems (4) and (5) of § 114, the reader will easily verify that

$$D_x \tan x = \sec^2 x,\qquad D_x \cot x = -\operatorname{cosec}^2 x,$$
$$D_x \sec x = \tan x \sec x,\qquad D_x \operatorname{cosec} x = -\cot x \operatorname{cosec} x.$$

And by means of Theorem (7) we can determine the derivatives of the inverse trigonometrical functions. The reader should verify the following formulae:

$$D_x \operatorname{arc} \sin x = \pm \frac{1}{\sqrt{(1-x^2)}}, \qquad D_x \operatorname{arc} \cos x = \mp \frac{1}{\sqrt{(1-x^2)}},$$

$$D_x \operatorname{arc} \tan x = \frac{1}{1+x^2}, \qquad D_x \operatorname{arc} \cot x = -\frac{1}{1+x^2},$$

$$D_x \operatorname{arc} \sec x = \pm \frac{1}{x\sqrt{(x^2-1)}}, \quad D_x \operatorname{arc} \operatorname{cosec} x = \mp \frac{1}{x\sqrt{(x^2-1)}}.$$

In the case of the inverse sine and cosecant the ambiguous sign is the same as that of $\cos(\operatorname{arc} \sin x)$, in the case of the inverse cosine and secant the same as that of $\sin(\operatorname{arc} \cos x)$.

The more general formulae

$$D_x \operatorname{arc} \sin(x/a) = \pm \frac{1}{\sqrt{(a^2-x^2)}}, \quad D_x \operatorname{arc} \tan(x/a) = \frac{a}{x^2+a^2},$$

which are easily deduced from Theorems (6) and (7) of § 114, are also of considerable importance. In the first of them the ambiguous sign is the same as that of $a \cos\{\operatorname{arc} \sin(x/a)\}$, since

$$a\sqrt{\{1-(x^2/a^2)\}} = \pm\sqrt{(a^2-x^2)}$$

according as a is positive or negative.

Finally, by means of Theorem (6) of § 114, we can differentiate composite functions involving symbols both of algebraical and trigonometrical functionality, and so write down the derivative of any such function as occurs in the following examples.

Examples XLIV*. 1. Find the derivatives of

$$\cos^m x, \quad \sin^m x, \quad \cos x^m, \quad \sin x^m, \quad \cos(\sin x), \quad \sin(\cos x),$$

$$\sqrt{(a^2\cos^2 x + b^2 \sin^2 x)}, \quad \frac{\cos x \sin x}{\sqrt{(a^2\cos^2 x + b^2\sin^2 x)}},$$

$$x \operatorname{arc} \sin x + \sqrt{(1-x^2)}, \quad (1+x) \operatorname{arc} \tan \sqrt{x} - \sqrt{x}.$$

2. Differentiate

$$\operatorname{arc} \sin(1-x^2)^{\frac{1}{2}}, \quad \tan \operatorname{arc} \sin x, \quad \operatorname{arc} \tan \frac{\cos x}{1+\sin x}, \quad \operatorname{arc} \tan \frac{a+b\cos x}{b+a\cos x}.$$

(*Math. Trip.* 1926, 1929, 1930)

* In these examples m is a rational number and $a, b, \ldots, \alpha, \beta, \ldots$ have such values that the functions which involve them are real. An ambiguous sign is sometimes omitted.

3. Differentiate

$$\arcsin x + \arccos x, \quad \arctan x + \operatorname{arc\,cot} x, \quad \arctan\left(\frac{a+x}{1-ax}\right)$$

and explain the simplicity of the results.

4. Differentiate

$$\frac{1}{\sqrt{(ac-b^2)}} \arctan \frac{ax+b}{\sqrt{(ac-b^2)}}, \quad -\frac{1}{\sqrt{(-a)}} \arcsin \frac{ax+b}{\sqrt{(b^2-ac)}}.$$

5. Show that each of the functions

$$2\arcsin\sqrt{\left(\frac{x-\beta}{\alpha-\beta}\right)}, \quad 2\arctan\sqrt{\left(\frac{x-\beta}{\alpha-x}\right)}, \quad \arcsin\frac{2\sqrt{\{(\alpha-x)(x-\beta)\}}}{\alpha-\beta}$$

has the derivative $\dfrac{1}{\sqrt{\{(\alpha-x)(x-\beta)\}}}$.

6. Prove that

$$\frac{d}{d\theta}\left\{\arccos\sqrt{\left(\frac{\cos 3\theta}{\cos^3\theta}\right)}\right\} = \sqrt{\left(\frac{3}{\cos\theta\cos 3\theta}\right)}.$$

(*Math. Trip.* 1904)

7. Show that

$$\frac{1}{\sqrt{\{C(Ac-aC)\}}}\frac{d}{dx}\left[\arccos\sqrt{\left\{\frac{C(ax^2+c)}{c(Ax^2+C)}\right\}}\right] = \frac{1}{(Ax^2+C)\sqrt{(ax^2+c)}}.$$

8. Each of the functions

$$\frac{1}{\sqrt{(a^2-b^2)}}\arccos\left(\frac{a\cos x+b}{a+b\cos x}\right), \quad \frac{2}{\sqrt{(a^2-b^2)}}\arctan\left\{\sqrt{\left(\frac{a-b}{a+b}\right)}\tan\tfrac{1}{2}x\right\}$$

has the derivative $1/(a+b\cos x)$.

9. If $X = a+b\cos x+c\sin x$, and

$$y = \frac{1}{\sqrt{(a^2-b^2-c^2)}}\arccos\frac{aX-a^2+b^2+c^2}{X\sqrt{(b^2+c^2)}},$$

then $dy/dx = 1/X$.

10. Prove that the derivative of $F[f\{\phi(x)\}]$ is

$$F'[f\{\phi(x)\}]f'\{\phi(x)\}\phi'(x),$$

and extend the result to still more complicated cases.

11. If u and v are functions of x, then

$$D_x \arctan\frac{u}{v} = \frac{vD_x u - uD_x v}{u^2+v^2}.$$

12. The derivative of $y = (\tan x + \sec x)^m$ is $my\sec x$.

13. The derivative of $y = \cos x + i\sin x$ is iy.

14. Differentiate $x\cos x$, $(\sin x)/x$. Show that the values of x for which the tangents to the curves $y = x\cos x$, $y = (\sin x)/x$ are parallel to the axis of x are roots of $\cot x = x$, $\tan x = x$ respectively.

15. It is easy to see (cf. Ex. XVII. 5) that the equation $\sin x = ax$, where a is positive, has no real roots except $x = 0$ if $a \geqq 1$, and a finite number of roots, which increases as a diminishes, if $a < 1$. Prove that the values of a for which the number of roots changes are the values of $\cos\xi$, where ξ is a positive root of the equation $\tan\xi = \xi$. [The values required are the values of a for which $y = ax$ touches $y = \sin x$.]

16. If $\phi(x) = x^2\sin(1/x)$ when $x \neq 0$, and $\phi(0) = 0$, then

$$\phi'(x) = 2x\sin(1/x) - \cos(1/x)$$

when $x \neq 0$, and $\phi'(0) = 0$. And $\phi'(x)$ is discontinuous for $x = 0$ (cf. § 112, (2)).

17. Find the equations of the tangent and normal at the point (x_0, y_0) of the circle $x^2+y^2 = a^2$, and reduce them to the forms $xx_0+yy_0 = a^2$ and $xy_0-yx_0 = 0$.

18. Find the equations of the tangent and normal at any point of the ellipse $(x/a)^2+(y/b)^2 = 1$ and the hyperbola $(x/a)^2-(y/b)^2 = 1$.

19. The equations of the tangent and normal to the curve $x = \phi(t)$, $y = \psi(t)$, at the point whose parameter is t, are

$$\frac{x-\phi(t)}{\phi'(t)} = \frac{y-\psi(t)}{\psi'(t)}, \quad \{x-\phi(t)\}\phi'(t)+\{y-\psi(t)\}\psi'(t) = 0.$$

121. Repeated differentiation. We may form a new function $\phi''(x)$ from $\phi'(x)$ just as we formed $\phi'(x)$ from $\phi(x)$. This function is called the *second derivative* or *second differential coefficient* of $\phi(x)$. The second derivative of $y = \phi(x)$ may also be written in any of the forms

$$D_x^2 y, \quad \left(\frac{d}{dx}\right)^2 y, \quad \frac{d^2y}{dx^2}.$$

In the same way we may define the *nth derivative* or *nth differential coefficient* of $y = \phi(x)$, which may be written in any of the forms

$$\phi^{(n)}(x), \quad D_x^n y, \quad \left(\frac{d}{dx}\right)^n y, \quad \frac{d^ny}{dx^n}.$$

But it is only in a few cases that it is easy to write down a general formula for the nth differential coefficient of a given function. Some of these cases will be found in the examples which follow.

Examples XLV. 1. If $\phi(x) = x^m$, then

$$\phi^{(n)}(x) = m(m-1)\dots(m-n+1)x^{m-n}.$$

This result enables us to write down the nth derivative of any polynomial.

2. If $\phi(x) = (ax+b)^m$, then

$$\phi^{(n)}(x) = m(m-1)\dots(m-n+1)a^n(ax+b)^{m-n}.$$

In these two examples m may have any rational value. If m is a positive integer, and $n > m$, then $\phi^{(n)}(x) = 0$.

3. The formula

$$\left(\frac{d}{dx}\right)^n \frac{A}{(x-\alpha)^p} = (-1)^n \frac{p(p+1)\dots(p+n-1)A}{(x-\alpha)^{p+n}}$$

enables us to write down the nth derivative of any rational function expressed in the standard form as a sum of partial fractions.

4. Prove that the nth derivative of $1/(1-x^2)$ is

$$\tfrac{1}{2}(n!)\{(1-x)^{-n-1} + (-1)^n(1+x)^{-n-1}\}.$$

5. Find the nth derivatives of

$$\frac{x+1}{x^2-4}, \quad \frac{x^4}{(x-1)(x-2)}, \quad \frac{4x}{(x-1)^2(x+2)}.$$

(*Math. Trip.* 1930, 1933, 1934)

6. Show that the value of $\left(\dfrac{d}{dx}\right)^n \dfrac{x^3}{x^2-1}$

for $x = 0$ is 0 if n is even and $-n!$ if n is odd and greater than 1.

(*Math. Trip.* 1935)

7. **Leibniz's theorem.** If y is a product uv, and we can form the first n derivatives of u and v, then we can form the nth derivative of y by means of *Leibniz's theorem*, which gives the rule

$$(uv)_n = u_n v + \binom{n}{1} u_{n-1} v_1 + \binom{n}{2} u_{n-2} v_2 + \dots + \binom{n}{r} u_{n-r} v_r + \dots + u v_n,$$

where suffixes indicate differentiations, so that u_n, for example, denotes the nth derivative of u. To prove the theorem we observe that

$$(uv)_1 = u_1 v + u v_1,$$

$$(uv)_2 = u_2 v + 2u_1 v_1 + u v_2,$$

and so on. It is obvious that by repeating this process we arrive at a formula of the type

$$(uv)_n = u_n v + a_{n,1} u_{n-1} v_1 + a_{n.2} u_{n-2} v_2 + \dots + a_{n,r} u_{n-r} v_r + \dots + u v_n.$$

Let us assume that $a_{n,r} = \binom{n}{r}$ for $r = 1, 2, \dots, n-1$, and show that if this is so then $a_{n+1,r} = \binom{n+1}{r}$ for $r=1, 2, \ldots, n$. It will then follow by the principle of mathematical induction that $a_{n,r} = \binom{n}{r}$ for all values of n and r in question.

When we form $(uv)_{n+1}$ by differentiating $(uv)_n$, it is clear that the coefficient of $u_{n+1-r}v_r$ is

$$a_{n,r}+a_{n,r-1} = \binom{n}{r}+\binom{n}{r-1} = \binom{n+1}{r}.$$

This establishes the theorem.

8. The nth derivative of $x^m f(x)$ is

$$\frac{m!}{(m-n)!}x^{m-n}f(x)+n\frac{m!}{(m-n+1)!}x^{m-n+1}f'(x)$$
$$+\frac{n(n-1)}{1.2}\frac{m!}{(m-n+2)!}x^{m-n+2}f''(x)+\ldots,$$

the series being continued for $n+1$ terms or until it terminates.

9. Prove that $D_x^n \cos x = \cos(x+\tfrac{1}{2}n\pi)$, $D_x^n \sin x = \sin(x+\tfrac{1}{2}n\pi)$.

10. Find the nth derivatives of

$$\cos^2 x \sin x, \quad \cos x \cos 2x \cos 3x, \quad x^3 \cos x.$$

(*Math. Trip.* 1925, 1930, 1934)

11. If $y = A\cos mx + B\sin mx$, then $D_x^2 y + m^2 y = 0$. And if

$$y = A\cos mx + B\sin mx + P_n(x),$$

where $P_n(x)$ is a polynomial of degree n, then $D_x^{n+3}y + m^2 D_x^{n+1}y = 0$.

12. If $x^2 D_x^2 y + x D_x y + y = 0$, then

$$x^2 D_x^{n+2}y + (2n+1)xD_x^{n+1}y + (n^2+1)D_x^n y = 0.$$

[Differentiate n times by Leibniz's theorem.]

13. If U_n denotes the nth derivative of $(Lx+M)/(x^2-2Bx+C)$, then

$$\frac{x^2-2Bx+C}{(n+1)(n+2)}U_{n+2}+\frac{2(x-B)}{n+1}U_{n+1}+U_n = 0.$$

(*Math. Trip.* 1900)

[First obtain the equation when $n=0$; then differentiate n times by Leibniz's theorem.]

14. Show that if $u = \arctan x$ then

$$(1+x^2)\frac{d^2u}{dx^2}+2x\frac{du}{dx} = 0,$$

and hence determine the values of all the derivatives of u for $x = 0$.

(*Math. Trip.* 1931)

15. **The nth derivatives of $a/(a^2+x^2)$ and $x/(a^2+x^2)$.** Since

$$\frac{a}{a^2+x^2} = \frac{1}{2i}\left(\frac{1}{x-ai} - \frac{1}{x+ai}\right), \quad \frac{x}{a^2+x^2} = \frac{1}{2}\left(\frac{1}{x-ai} + \frac{1}{x+ai}\right),$$

we have

$$D_x^n\left(\frac{a}{a^2+x^2}\right) = \frac{(-1)^n n!}{2i}\left\{\frac{1}{(x-ai)^{n+1}} - \frac{1}{(x+ai)^{n+1}}\right\},$$

and a similar formula for $D_x^n\{x/(a^2+x^2)\}$. If $\rho = \sqrt{(x^2+a^2)}$, and θ is the numerically smallest angle whose cosine and sine are x/ρ and a/ρ, then $x+ai = \rho \operatorname{Cis}\theta$ and $x-ai = \rho \operatorname{Cis}(-\theta)$, and so

$$\begin{aligned} D_x^n \frac{a}{a^2+x^2} &= \tfrac{1}{2}(-1)^{n-1} n!\, i\rho^{-n-1}[\operatorname{Cis}\{(n+1)\theta\} - \operatorname{Cis}\{-(n+1)\theta\}] \\ &= (-1)^n n!(x^2+a^2)^{-\frac{1}{2}(n+1)} \sin\{(n+1)\arctan(a/x)\}. \end{aligned}$$

Similarly

$$D_x^n \frac{x}{a^2+x^2} = (-1)^n n!(x^2+a^2)^{-\frac{1}{2}(n+1)} \cos\{(n+1)\arctan(a/x)\}.$$

16. Prove that

$$D_x^n \frac{\cos x}{x} = \{P_n \cos(x+\tfrac{1}{2}n\pi) + Q_n \sin(x+\tfrac{1}{2}n\pi)\}\, x^{-n-1},$$

$$D_x^n \frac{\sin x}{x} = \{P_n \sin(x+\tfrac{1}{2}n\pi) - Q_n \cos(x+\tfrac{1}{2}n\pi)\}\, x^{-n-1},$$

where P_n and Q_n are polynomials in x of degree n and $n-1$ respectively.

17. Establish the formulae

$$\frac{dx}{dy} = 1\Big/\left(\frac{dy}{dx}\right), \quad \frac{d^2x}{dy^2} = -\frac{d^2y}{dx^2}\Big/\left(\frac{dy}{dx}\right)^3, \quad \frac{d^3x}{dy^3} = -\left\{\frac{d^3y}{dx^3}\frac{dy}{dx} - 3\left(\frac{d^2y}{dx^2}\right)^2\right\}\Big/\left(\frac{dy}{dx}\right)^5.$$

122. Some general theorems concerning derived functions. In what follows the distinction between a 'closed' and an 'open' interval will often be important. A *closed* interval (a, b) is the set of x for which $a \leqq x \leqq b$. An *open* interval is the set for which $a < x < b$ (i.e. the closed interval minus its end points)*.

We shall be concerned with functions continuous in the closed interval (a, b) and differentiable in the open interval (a, b). In other words we shall assume that our function $\phi(x)$ satisfies the following conditions:

(1) $\phi(x)$ is continuous for $a \leqq x \leqq b$, continuity at the end points

* We might define a *half-closed* interval by inequalities $a < x \leqq b$, or $a \leqq x < b$, but we shall not use the phrase.

of the interval being understood in the sense explained at the end of § 99;

(2) $\phi'(x)$ exists for every x for which $a < x < b$.

It may seem odd to use the closed interval in one condition and the open one in the other, but we shall find the distinction important. It is plain that, if we know nothing of $\phi(x)$ outside (a, b), we cannot extend condition (2) to cover the end points without new definitions.

We begin by a theorem concerning a particular value of x.

THEOREM A. *If $\phi'(x_0) > 0$ then $\phi(x) < \phi(x_0)$ for all values of x less than x_0 but sufficiently near to x_0, and $\phi(x) > \phi(x_0)$ for all values of x greater than x_0 but sufficiently near to x_0.*

For $\{\phi(x_0+h) - \phi(x_0)\}/h$ converges to a positive limit $\phi'(x_0)$ as $h \to 0$. This can only be the case if $\phi(x_0+h) - \phi(x_0)$ and h have the same sign for sufficiently small values of h, and this is what the theorem states. Of course from a geometrical point of view the result is intuitive, the inequality $\phi'(x) > 0$ expressing the fact that the tangent to the curve $y = \phi(x)$ makes a positive acute angle with the axis of x. The reader can formulate for himself the corresponding theorem for the case in which $\phi'(x) < 0$.

We shall express the conclusion of Theorem A by saying that *$\phi(x)$ is strictly increasing at $x = x_0$.**

The theorem which follows is generally known as Rolle's theorem, and is particularly important.

THEOREM B. *If $\phi(x)$ is continuous in the closed, and differentiable in the open, interval, and its values at the end points a, b are equal, then there is a point of the open interval at which $\phi'(x) = 0$.*

We may suppose that

$$\phi(a) = 0, \quad \phi(b) = 0;$$

if $\phi(a) = \phi(b) = k$, and $k \neq 0$, we consider $\phi(x) - k$ instead of $\phi(x)$.

There are two possibilities. If $\phi(x) = 0$ throughout (a, b), then $\phi'(x) = 0$ for $a < x < b$, and there is nothing to prove.

If on the other hand $\phi(x)$ is not always zero, there are values of x for which it is positive or negative. Suppose, for example, that

* Compare Ex. 19 on p. 208.

it is sometimes positive. Then $\phi(x)$ has an upper bound M in (a, b), and $\phi(x) = M$ for some ξ of (a, b), by Theorem 2 of § 103; and it is plain that ξ is not a or b. If $\phi'(\xi)$ were positive or negative, there would, by Theorem A, be values of x near ξ (on one side or the other) at which $\phi(x) > M$, in contradiction to the definition of M. Hence $\phi'(\xi) = 0$.

COR. 1. *If $\phi(x)$ is continuous in the closed, and differentiable in the open, interval, and $\phi'(x) > 0$ for every x of the open interval, then $\phi(x)$ is an increasing function of x throughout the interval, in the stricter sense of* § 95.

We have to prove that $\phi(x_1) < \phi(x_2)$ for $a \leqq x_1 < x_2 \leqq b$. We suppose first that $a < x_1 < x_2 < b$.

If $\phi(x_1) = \phi(x_2)$ then, by Theorem B, there is an x between x_1 and x_2 for which $\phi'(x) = 0$, in contradiction to our assumptions.

If $\phi(x_1) > \phi(x_2)$ then, by Theorem A, there is an x_3 near to and greater than x_1 for which $\phi(x_3) > \phi(x_1) > \phi(x_2)$; and therefore, after § 101, an x_4 between x_3 and x_2 for which $\phi(x_4) = \phi(x_1)$; and therefore, by Theorem B, an x between x_1 and x_4 for which $\phi'(x) = 0$, again in contradiction to our assumptions.

It follows that $\phi(x_1) < \phi(x_2)$.

It remains to extend the inequality to the cases in which $x_1 = a$ or $x_2 = b$. It follows from what we have proved that

$$\phi(x) < \phi(x')$$

if $a < x < x' < b$, so that $\phi(x)$ decreases strictly as x approaches a from the right. Hence

$$\phi(a) = \lim_{x \to a+0} \phi(x) < \phi(x').$$

Similarly $\phi(x') < \phi(b)$.

COR. 2. *If $\phi'(x) > 0$ throughout the interval (a, b), and $\phi(a) \geqq 0$, then $\phi(x)$ is positive throughout the interval (a, b).*

The reader should compare the first of these corollaries very carefully with Theorem A. If, as in Theorem A, we assume only that $\phi'(x)$ is positive *at a single point* $x = x_0$, then we can prove that $\phi(x_1) < \phi(x_2)$ when x_1 and x_2 are sufficiently near to x_0 and $x_1 < x_0 < x_2$. For $\phi(x_1) < \phi(x_0)$ and $\phi(x_2) > \phi(x_0)$, by Theorem A. But this does not prove that there is any interval including x_0 throughout which $\phi(x)$ is a steadily increasing function, for the assumption that x_1 and x_2 lie on opposite sides of x_0 is essential to our conclusion. We shall return to this point, and illustrate it by an actual example, in a moment (§ 125).

123. Maxima and minima. We shall say that the value $\phi(\xi)$ assumed by $\phi(x)$ when $x = \xi$ is a *maximum* if $\phi(\xi)$ is greater than any other value assumed by $\phi(x)$ in the immediate neighbourhood of $x = \xi$, i.e. if we can find an interval $(\xi-\epsilon, \xi+\epsilon)$ of values of x such that $\phi(\xi) > \phi(x)$ when $\xi - \epsilon < x < \xi$ and when

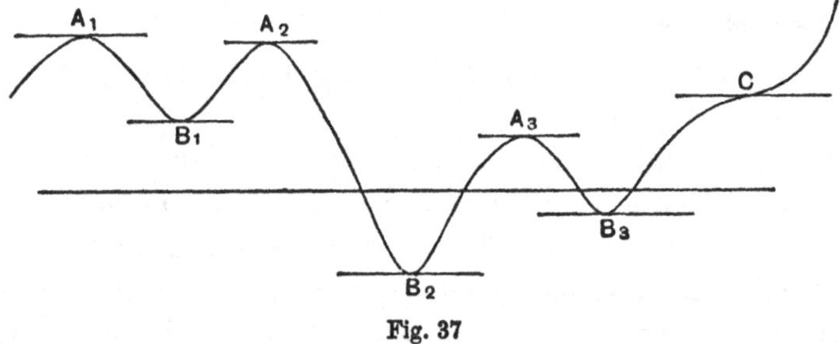

Fig. 37

$\xi < x < \xi + \epsilon$; and we define a *minimum* in a similar manner. Thus in the figure the points A correspond to maxima, the points B to minima of the function whose graph is there shown. It is to be observed that the fact that A_3 corresponds to a maximum and B_1 to a minimum is in no way inconsistent with the fact that the value of the function is greater at B_1 than at A_3.

THEOREM C. *A* **necessary** *condition for a maximum or minimum value of a differentiable function $\phi(x)$ at $x = \xi$ is that $\phi'(\xi) = 0$.*

This follows at once from Theorem A. That the condition is not *sufficient* is evident from a glance at the point C in the figure. Thus if $y = x^3$ then $\phi'(x) = 3x^2$, which vanishes when $x = 0$. But $x = 0$ does not give either a maximum or a minimum of x^3, as is obvious from the form of the graph of x^3 (Fig. 10, p. 47).

But *there will certainly be a maximum at $x = \xi$ if $\phi'(\xi) = 0$, $\phi'(x) > 0$ for all values of x less than but near to ξ, and $\phi'(x) < 0$ for all values of x greater than but near to ξ*; and if the signs of these two inequalities are reversed there will certainly be a minimum. For then we can (by Cor. 1 of § 122) determine an interval $(\xi-\epsilon, \xi)$ throughout which $\phi(x)$ increases with x, and an interval $(\xi, \xi+\epsilon)$ throughout which it decreases as x increases.

This result may also be stated thus. If the sign of $\phi'(x)$ changes at $x = \xi$ from positive to negative, then $x = \xi$ gives a maximum of $\phi(x)$; and if the sign of $\phi'(x)$ changes in the opposite sense, then $x = \xi$ gives a minimum.

A maximum, as we have defined it, is a *strict* maximum; $\phi(\xi) > \phi(x)$ for all x near ξ. We might relax our definition, and require only that $\phi(\xi) \geqq \phi(x)$ for all x near ξ. With this definition, for example, a constant would have a maximum (and a minimum) for every value of the variable. Theorem C would still be true.

A maximum or minimum is often called an 'extremal' or 'turning' value.

124. There is another way of stating conditions for a maximum or minimum which is often useful. Let us assume that $\phi(x)$ has a second derivative $\phi''(x)$: this of course does not follow from the existence of $\phi'(x)$, any more than the existence of $\phi'(x)$ follows from that of $\phi(x)$; but in such cases as we are likely to meet with at present the condition is generally satisfied.

THEOREM D. *If* $\phi'(\xi) = 0$ *and* $\phi''(\xi) \neq 0$, *then* $\phi(x)$ *has a maximum or minimum at* $x = \xi$, *a maximum if* $\phi''(\xi) < 0$, *a minimum if* $\phi''(\xi) > 0$.

Suppose, e.g., that $\phi''(\xi) < 0$. Then, by Theorem A, $\phi'(x)$ is positive when x is less than ξ but sufficiently near to ξ, and negative when x is greater than ξ but sufficiently near to ξ. Thus $x = \xi$ gives a maximum.

125. In what has preceded we have assumed that $\phi(x)$ has a derivative for all values of x in the interval under consideration. If this condition is not fulfilled, the theorems cease to be true. Thus Theorem B fails for the function

$$y = 1 - \sqrt{(x^2)},$$

where the square root is to be taken positive. The graph of this function is shown in Fig. 38. Here $\phi(-1) = \phi(1) = 0$: but $\phi'(x)$, as is evident from the figure, is equal to 1 if x is negative and to -1 if x is positive, and never vanishes. There is no derivative for $x = 0$, and no tangent to the graph at P. And in this case $x = 0$ obviously gives a maximum of $\phi(x)$, but the test for a maximum fails.

The bare existence of the derivative $\phi'(x)$, however, is all that we have assumed. And there is one assumption in particular that we have not made, and that is that $\phi'(x)$ *itself is a continuous function*. This raises an interesting point. *Can* a function $\phi(x)$ have a derivative for all values of x which is not itself continuous? In other words, can a curve have a tangent at every point, and yet the direction of the tangent not vary continuously? Common sense may incline at first to the answer *No*; but it is not difficult to show that this answer is wrong.

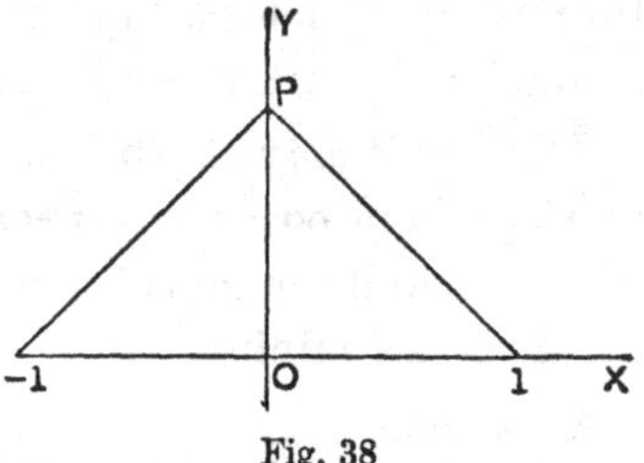

Fig. 38

Consider the function $\phi(x)$ defined, when $x \neq 0$, by the equation

$$\phi(x) = x^2 \sin(1/x);$$

and suppose that $\phi(0) = 0$. Then $\phi(x)$ is continuous for all values of x. If $x \neq 0$ then

$$\phi'(x) = 2x \sin(1/x) - \cos(1/x);$$

while

$$\phi'(0) = \lim_{h \to 0} \frac{h^2 \sin(1/h)}{h} = 0.$$

Thus $\phi'(x)$ exists for all values of x. But $\phi'(x)$ is discontinuous for $x = 0$; for $2x \sin(1/x)$ tends to 0 as $x \to 0$, and $\cos(1/x)$ oscillates between the limits of indetermination -1 and 1, so that $\phi'(x)$ oscillates between the same limits.

What is practically the same example enables us also to illustrate the point referred to at the end of § 122. Let

$$\phi(x) = x^2 \sin(1/x) + \alpha x,$$

where $0 < \alpha < 1$, when $x \neq 0$, and $\phi(0) = 0$. Then $\phi'(0) = \alpha > 0$. Thus the conditions of Theorem A of § 122 are satisfied. But if $x \neq 0$ then

$$\phi'(x) = 2x \sin(1/x) - \cos(1/x) + \alpha,$$

which oscillates between the limits of indetermination $\alpha - 1$ and $\alpha + 1$ as $x \to 0$. Since $\alpha - 1 < 0$, we can find values of x, as near to 0 as we like, for which $\phi'(x) < 0$; and it is therefore impossible to find any interval, including $x = 0$, throughout which $\phi(x)$ is a steadily increasing function of x.

It is, however, impossible that $\phi'(x)$ should have what was called in Ch. V (Ex. XXXVII. 18) a 'simple' discontinuity. If $\phi'(x) \to a$ when $x \to +0$, $\phi'(x) \to b$ when $x \to -0$, and $\phi'(0) = c$, then $a = b = c$, and $\phi'(x)$ is continuous for $x = 0$. For a proof see § 126, Ex. XLVII. 5.

Examples XLVI. 1. Verify Theorem B when $\phi(x) = (x-a)^m(x-b)^n$ or $\phi(x) = (x-a)^m(x-b)^n(x-c)^p$, where m, n, p are positive integers and $a<b<c$.

[The first function vanishes for $x = a$ and $x = b$. And

$$\phi'(x) = (x-a)^{m-1}(x-b)^{n-1}\{(m+n)x-mb-na\}$$

vanishes for $x = (mb+na)/(m+n)$, which lies between a and b. In the second case we have to verify that the quadratic equation

$$(m+n+p)x^2-\{m(b+c)+n(c+a)+p(a+b)\}x+mbc+nca+pab = 0$$

has roots between a and b and between b and c.]

2. Show that $x-\sin x$ is an increasing function throughout any interval of values of x, and that $\tan x - x$ increases as x increases from $-\frac{1}{2}\pi$ to $\frac{1}{2}\pi$. For what values of a is $ax-\sin x$ a steadily increasing or decreasing function of x?

3. Show that $x/(\sin x)$ increases steadily from $x = 0$ to $x = \frac{1}{2}\pi$.

(*Math. Trip.* 1927)

4. Show that $\tan x - x$ increases from $x = \frac{1}{2}\pi$ to $x = \frac{3}{2}\pi$, from $x = \frac{3}{2}\pi$ to $x = \frac{5}{2}\pi$, and so on, and deduce that there is one and only one root of the equation $\tan x = x$ in each of these intervals (cf. Ex. XVII. 4).

5. Deduce from Ex. 2 that $\sin x - x<0$ if $x>0$, from this that $\cos x - 1 + \frac{1}{2}x^2 > 0$, and from this that $\sin x - x + \frac{1}{6}x^3 > 0$. And, generally, prove that if

$$C_{2m} = \cos x - 1 + \frac{x^2}{2!} - \dots - (-1)^m \frac{x^{2m}}{2m!},$$

$$S_{2m+1} = \sin x - x + \frac{x^3}{3!} - \dots - (-1)^m \frac{x^{2m+1}}{(2m+1)!},$$

and $x>0$, then C_{2m} and S_{2m+1} are positive or negative according as m is odd or even.

6. If $f(x)$ and $f''(x)$ are continuous and have the same sign at every point of an interval (a, b), then this interval can include at most one root of either of the equations $f(x) = 0$, $f'(x) = 0$.

7. The functions u, v and their derivatives u', v' are continuous throughout a certain interval of values of x, and $uv'-u'v$ never vanishes at any point of the interval. Show that between any two roots of $u = 0$ lies one of $v = 0$, and conversely. Verify the theorem when $u = \cos x$, $v = \sin x$.

[If v does not vanish between two roots of $u = 0$, say α and β, then the function u/v is continuous throughout the interval (α, β) and vanishes at its extremities. Hence $(u/v)' = (u'v-uv')/v^2$ must vanish between α and β, which contradicts our hypotheses.]

8. Find the largest and smallest values of x^3-18x^2+96x in the interval $(0, 9)$. (*Math. Trip.* 1931)

9. Discuss the maxima and minima of the function $(x-a)^m(x-b)^n$, where m and n are any positive integers, considering the different cases which occur according as m and n are odd or even. Sketch the graph of the function.

10. Show that the function $(x+5)^2(x^3-10)$ has a minimum when $x=1$, and investigate its other turning values. (*Math. Trip.* 1936)

11. Show that
$$\left(\alpha-\frac{1}{\alpha}-x\right)(4-3x^2)$$
has just one maximum and just one minimum, and that the difference between them is
$$\frac{4}{9}\left(\alpha+\frac{1}{\alpha}\right)^3.$$
What is the least value of this difference for different values of α? (*Math. Trip.* 1933)

12. Show that $(ax+b)/(cx+d)$ has no maxima or minima, whatever values a, b, c, d may have. Draw a graph of the function.

13. Discuss the maxima and minima of the function
$$y=(ax^2+2bx+c)/(Ax^2+2Bx+C),$$
when the denominator has complex roots.

[We may suppose a and A positive. The derivative vanishes if
$$(ax+b)(Bx+C)-(Ax+B)(bx+c)=0 \quad \ldots\ldots\ldots\ldots(1).$$
This equation must have real roots. For if not the derivative would always have the same sign, and this is impossible, since y is continuous for all values of x, and $y\to a/A$ as $x\to+\infty$ or $x\to-\infty$. It is easy to verify that the curve cuts the line $y=a/A$ in one and only one point, and that it lies above this line for large positive values of x, and below it for large negative values, or *vice versa*, according as $b/a>B/A$ or $b/a<B/A$. Thus the algebraically greater root of (1) gives a maximum if $b/a>B/A$, a minimum in the contrary case.]

14. The maximum and minimum values themselves are the values of λ for which $ax^2+2bx+c-\lambda(Ax^2+2Bx+C)$ is a perfect square. [This is the condition that $y=\lambda$ should touch the curve.]

15. If $Ax^2+2Bx+C=0$ has real roots, then it is convenient to proceed as follows. We have
$$y-\frac{a}{A}=\frac{2\lambda x+\mu}{A(Ax^2+2Bx+C)},$$

where $\lambda = bA - aB$, $\mu = cA - aC$. Writing further ξ for $2\lambda x + \mu$ and η for $(A/4\lambda^2)(Ay - a)$, we obtain an equation of the form

$$\eta = \xi/\{(\xi - p)(\xi - q)\}.$$

A minimum of y, considered as a function of x, corresponds to a minimum of η considered as a function of ξ, and *vice versa*, and similarly for a maximum.

The derivative of η with respect to ξ vanishes if

$$(\xi - p)(\xi - q) - \xi(\xi - p) - \xi(\xi - q) = 0,$$

or if $\xi^2 = pq$. Thus there are two roots of the derivative if p and q have the same sign, none if they have opposite signs. In the latter case the form of the graph of η is as shown in Fig. 39*a*.

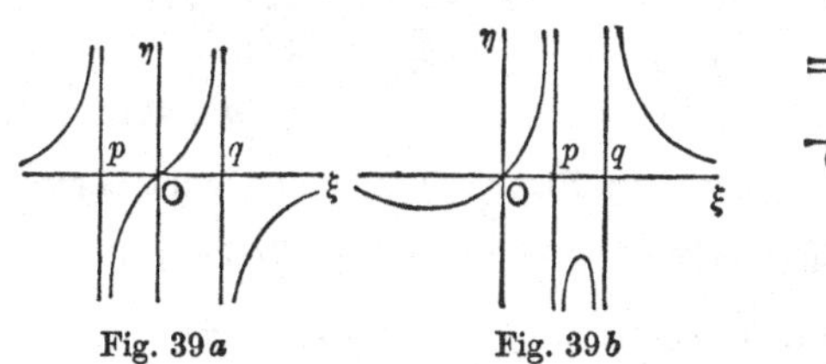

Fig. 39*a* Fig. 39*b*

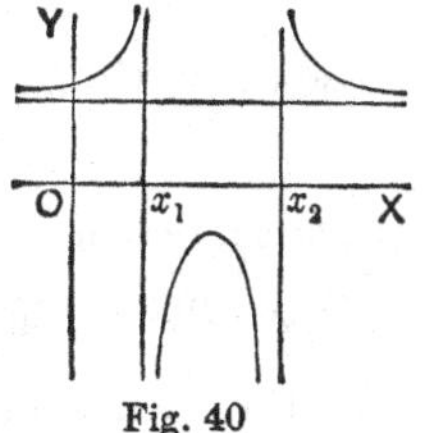

Fig. 40

When p and q are positive the general form of the graph is as shown in Fig. 39*b*, and it is easy to see that $\xi = \sqrt{(pq)}$ gives a maximum and $\xi = -\sqrt{(pq)}$ a minimum.

The preceding discussion fails if $\lambda = 0$, i.e. if $a/A = b/B$. But in this case we have

$$y - (a/A) = \mu/\{A(Ax^2 + 2Bx + C)\} = \mu/\{A^2(x - x_1)(x - x_2)\},$$

say, and $dy/dx = 0$ gives the single value $x = \frac{1}{2}(x_1 + x_2)$. On drawing a graph it becomes clear that this value gives a maximum or minimum according as μ is positive or negative. The graph shown in Fig. 40 corresponds to the former case.

16. Show that $(x - \alpha)(x - \beta)/(x - \gamma)$ assumes all real values as x varies, if γ lies between α and β, and otherwise assumes all values except those included in an interval of length $4\sqrt{(|\alpha - \gamma||\beta - \gamma|)}$.

17. Show that
$$y = \frac{x^2 + 2x + c}{x^2 + 4x + 3c}$$
can assume any real value if $0 < c < 1$, and draw a graph of the function in this case. (*Math. Trip.* 1910)

18. The graph of
$$y = \frac{ax + b}{(x - 1)(x - 4)}$$
has a turning value at the point $(2, -1)$. Find a and b, and show that the turning value is a maximum. Sketch the curve. (*Math. Trip.* 1930)

19. Determine the function of the form $(ax^2+2bx+c)/(Ax^2+2Bx+C)$ which has turning values 2 and 3 when $x=1$ and $x=-1$ respectively, and has the value 2·5 when $x=0$. (*Math. Trip.* 1908)

20. The maximum and minimum of $(x+a)(x+b)/(x-a)(x-b)$, where a and b are positive, are

$$-\left(\frac{\sqrt{a}+\sqrt{b}}{\sqrt{a}-\sqrt{b}}\right)^2, \quad -\left(\frac{\sqrt{a}-\sqrt{b}}{\sqrt{a}+\sqrt{b}}\right)^2.$$

21. The maximum value of $(x-1)^2/(x+1)^3$ is $\frac{2}{27}$.

22. Discuss the maxima and minima of

$$\frac{x(x-1)}{x^2+3x+3}, \quad \frac{x^4}{(x-1)(x-3)^3}, \quad \frac{(x-1)^2(3x^2-2x-37)}{(x+5)^2(3x^2-14x-1)}.$$

(*Math. Trip.* 1898)

[If the last function be denoted by $P(x)/Q(x)$, it will be found that

$$P'Q-PQ' = 72(x-7)(x-3)(x-1)(x+1)(x+2)(x+5).]$$

23. Find the maxima and minima of $a\cos x+b\sin x$. Verify the result by expressing the function in the form $A\cos(x-\alpha)$.

24. Show that $\sin(x+a)/\sin(x+b)$ has no maxima or minima. Draw a graph of the function.

25. Show that the function

$$\frac{\sin^2 x}{\sin(x+a)\sin(x+b)} \quad (0<a<b<\pi)$$

has an infinity of minima equal to 0 and of maxima equal to

$$-4\frac{\sin a\sin b}{\sin^2(a-b)}.$$

(*Math. Trip.* 1909)

26. The least value of $a^2\sec^2 x+b^2\operatorname{cosec}^2 x$ is $(a+b)^2$ when $ab\geqq 0$.

27. Show that $\tan 3x\cot 2x$ cannot lie between $\frac{1}{9}$ and $\frac{3}{2}$.

28. Show that the maxima and minima of $\sin mx\operatorname{cosec} x$, where m is an integer, are given by $\tan mx = m\tan x$; and deduce that

$$\sin^2 mx \leqq m^2\sin^2 x.$$

(*Math. Trip.* 1926)

[Observe that

$$\frac{\sin^2 mx}{\sin^2 x} = m^2\frac{\cos^2 mx}{\cos^2 x} = m^2\frac{1+\tan^2 x}{1+\tan^2 mx} = m^2\frac{1+\tan^2 x}{1+m^2\tan^2 x}$$

at a maximum or minimum.]

29. Find the maxima and minima of the function y defined by

$$\frac{ay+b}{cy+d} = \sin^2 x + 2\cos x + 1,$$

where $ad \neq bc$. (*Math. Trip.* 1928)

30. Show that, if the sum of the lengths of the hypotenuse and another side of a right-angled triangle is given, then the area of the triangle is a maximum when the angle between those sides is $60°$.

(*Math. Trip.* 1909)

31. A line is drawn through a fixed point (a, b) to meet the axes OX, OY in P and Q. Show that the minimum values of PQ, $OP+OQ$, and $OP.OQ$ are respectively $(a^{\frac{2}{3}}+b^{\frac{2}{3}})^{\frac{3}{2}}$, $(\sqrt{a}+\sqrt{b})^2$, and $4ab$.

32. A tangent to an ellipse meets the axes in P and Q. Show that the least value of PQ is equal to the sum of the semi-axes of the ellipse.

33. A lane runs at a right angle out of a road 18 feet wide. How many feet wide is the lane if it is just possible to carry a pole 45 feet long from the road into the lane, keeping it horizontal? (*Math. Trip.* 1934)

34. Two points A and B lie on a line on opposite sides of and equidistant from a fixed point O of the line; and P is a fixed point not on the line. Show that $AP+BP$ increases with AB. (*Math. Trip.* 1934)

35. Find the lengths and directions of the axes of the conic

$$ax^2+2hxy+by^2 = 1.$$

[The length r of the semi-diameter which makes an angle θ with the axis of x is given by

$$1/r^2 = a\cos^2\theta + 2h\cos\theta\sin\theta + b\sin^2\theta.$$

The condition for a maximum or minimum value of r is $\tan 2\theta = 2h/(a-b)$. Eliminating θ between these two equations, we find

$$\{a-(1/r^2)\}\{b-(1/r^2)\} = h^2.]$$

36. The greatest value of $ax+by$, where x and y are positive and $x^2+xy+y^2 = 3\kappa^2$, is

$$2\kappa\sqrt{(a^2-ab+b^2)}.$$

[If $ax+by$ is a maximum, then $a+b(dy/dx) = 0$. The relation between x and y gives $(2x+y)+(x+2y)(dy/dx) = 0$. Equate the two values of dy/dx.]

37. The greatest value of $x^m y^n$, where x and y are positive and $x+y = k$, is

$$m^m n^n k^{m+n}/(m+n)^{m+n}.$$

38. If θ and ϕ are acute angles connected by the relation

$$a\sec\theta + b\sec\phi = c,$$

where a, b, c are positive, then $a\cos\theta + b\cos\phi$ is a minimum when $\theta = \phi$.

126. The mean value theorem. We can proceed now to the proof of another general theorem of great importance, a theorem commonly known as '*the mean value theorem*' or '*the theorem of the mean*'.

THEOREM. *If $\phi(x)$ is continuous in the closed interval (a, b), and differentiable in the open interval, then there is a value ξ of x between a and b, such that*

$$\phi(b) - \phi(a) = (b-a)\,\phi'(\xi).$$

Before we give a strict proof of this theorem, which is one of the most important theorems in the differential calculus, it will be well to point out its obvious geometrical meaning. This is simply that, if the curve APB (see Fig. 41) has a tangent at all points of its length, then there must be a point, such as P, where the tangent is parallel to AB. For $\phi'(\xi)$ is the tangent of the angle which the tangent at P makes with OX, and $\{\phi(b) - \phi(a)\}/(b-a)$ the tangent of the angle which AB makes with OX.

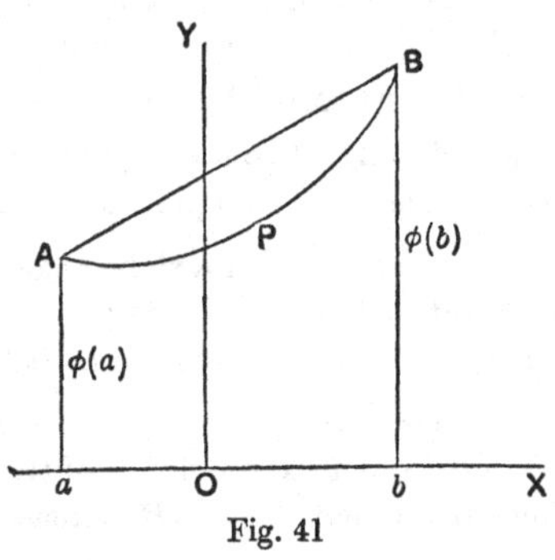

Fig. 41

It is easy to give a strict proof. Consider the function

$$\phi(b) - \phi(x) - \frac{b-x}{b-a}\{\phi(b) - \phi(a)\},$$

which vanishes when $x = a$ and $x = b$. It follows from Theorem B of § 122 that there is a value ξ for which its derivative vanishes. But this derivative is

$$\frac{\phi(b) - \phi(a)}{b-a} - \phi'(x);$$

which proves the theorem. It should be observed again that it has not been assumed that $\phi'(x)$ is continuous.

It is often convenient to express the mean value theorem in the form

$$\phi(b) = \phi(a) + (b-a)\,\phi'\{a + \theta(b-a)\},$$

where θ is a number lying between 0 and 1. Of course $a + \theta(b-a)$

is merely another way of writing 'some number ξ between a and b'. If we put $b = a+h$, we obtain

$$\phi(a+h) = \phi(a)+h\phi'(a+\theta h),$$

which is the form in which the theorem is most often quoted.

Examples XLVII. 1. Show that

$$\phi(b)-\phi(x)-\frac{b-x}{b-a}\{\phi(b)-\phi(a)\}$$

is the difference between the ordinates of a point on the curve and the corresponding point on the chord.

2. Verify the theorem when $\phi(x) = x^2$ and when $\phi(x) = x^3$.

[In the latter case we have to prove that $(b^3-a^3)/(b-a) = 3\xi^2$, where $a<\xi<b$; i.e. that if $\frac{1}{3}(b^2+ab+a^2) = \xi^2$, then ξ lies between a and b.]

3. Find the ξ of the mean value theorem when

$$f(x) = x(x-1)(x-2), \quad a = 0, \quad b = \tfrac{1}{2}.$$

(*Math. Trip.* 1935)

4. Prove Cor. 1 of § 122 by means of the mean value theorem. Prove also that if $\phi'(x) \geqq 0$ then $\phi(x)$ is an increasing function in the weaker sense.

5. Establish the theorem stated at the end of § 125 by means of the mean value theorem.

[Since $\phi'(0) = c$, we can find a small positive value of x such that $\{\phi(x)-\phi(0)\}/x$ is nearly equal to c; and therefore, by the theorem, a small positive value of ξ such that $\phi'(\xi)$ is nearly equal to c, which is inconsistent with $\lim_{x\to+0}\phi'(x) = a$, unless $a = c$. Similarly $b = c$.]

6. Use the mean value theorem to prove Theorem (6) of § 114, assuming that the derivatives which occur are continuous.

[We have

$$\begin{aligned} F\{f(x+h)\}-F\{f(x)\} &= F\{f(x)+hf'(\xi)\}-F\{f(x)\} \\ &= hf'(\xi)\,F'(\eta), \end{aligned}$$

where ξ lies between x and $x+h$ and η between $f(x)$ and $f(x)+hf'(\xi)$.]

7. Prove that if

$$\frac{a_0}{n+1}+\frac{a_1}{n}+\dots+\frac{a_{n-1}}{2}+a_n = 0,$$

then the equation $a_0x^n+a_1x^{n-1}+\dots+a_{n-1}x+a_n = 0$ has at least one root between 0 and 1. (*Math. Trip.* 1929)

127. The mean value theorem gives a proof of a theorem fundamental in the theory of integration: *if* $\phi'(x) = 0$ *for all* x *of an interval, then* $\phi(x)$ *is constant in that interval.*

For, if a and b are any two values of x in the interval, then

$$\phi(b)-\phi(a) = (b-a)\,\phi'\{a+\theta(b-a)\} = 0.$$

An immediate corollary is that if $\phi'(x) = \psi'(x)$ in an interval, then the functions $\phi(x)$ and $\psi(x)$ differ in that interval by a constant.

128. Cauchy's mean value theorem. There is a generalisation of the mean value theorem, due to Cauchy, which is of considerable importance in applications*.

If (i) $\phi(x)$ *and* $\psi(x)$ *are continuous in the closed interval* (a, b), *and differentiable in the open interval*; (ii) $\psi(b) \neq \psi(a)$; *and* (iii) $\phi'(x)$ *and* $\psi'(x)$ *never vanish for the same value of* x, *then there is a* ξ *between* a *and* b *for which*

$$\frac{\phi(b)-\phi(a)}{\psi(b)-\psi(a)} = \frac{\phi'(\xi)}{\psi'(\xi)}.$$

This reduces to the mean value theorem when $\psi(x) = x$, in which case the subsidiary conditions are satisfied automatically.

The proof is a straightforward generalisation of that of § 126. The function

$$\phi(b)-\phi(x)-\frac{\phi(b)-\phi(a)}{\psi(b)-\psi(a)}\{\psi(b)-\psi(x)\}$$

vanishes for $x = a$ and for $x = b$. Hence its derivative vanishes for a ξ between a and b; i.e.

$$\phi'(\xi) = \frac{\phi(b)-\phi(a)}{\psi(b)-\psi(a)}\psi'(\xi)$$

for some such ξ. If $\psi'(\xi)$ were zero, $\phi'(\xi)$ would be so also, in contradiction to our assumptions. Hence $\psi'(\xi) \neq 0$, and the theorem follows when we divide by $\psi'(\xi)$.

The assumption that ϕ' and ψ' never vanish for the same x is essential. Suppose for example that

$$a = -1, \quad b = 1, \quad \phi = x^2, \quad \psi = x^3.$$

* See § 154.

Then $\phi(b)-\phi(a) = 0$, $\psi(b)-\psi(a) = 2$, and the result can only be true if $\phi'(\xi) = 0$, i.e. if $\xi = 0$, in which case $\psi'(\xi)$ also vanishes and the formula becomes meaningless.

129. A theorem of Darboux. We proved in § 101 that if $\phi(x)$ is continuous in (a, b) then it assumes, somewhere in (a, b), every value between $\phi(a)$ and $\phi(b)$. There are other classes of functions which possess this property, and in particular the class of derivatives. If $\phi'(x)$ is the derivative of a function $\phi(x)$, then (whether continuous or not) it has the property stated.

If $\phi(x)$ is differentiable for $a \leqq x \leqq b$, $\phi'(a) = \alpha$, $\phi'(b) = \beta$, and γ lies between α and β, then there is a ξ between a and b for which $\phi'(\xi) = \gamma$.

Suppose, for example, that $\alpha < \gamma < \beta$, and let

$$\psi(x) = \phi(x) - \gamma(x-a).$$

Then $\psi(x)$ is continuous, and therefore attains its lower bound in (a, b) at some point ξ of (a, b). This point ξ cannot be a or b, because

$$\psi'(a) = \alpha - \gamma < 0, \quad \psi'(b) = \beta - \gamma > 0.$$

Hence $\psi(x)$ has a minimum* at a ξ between a and b, and $\psi'(\xi) = 0$, i.e. $\phi'(\xi) = \gamma$.

130. Integration. We have seen how we can find the derivative of a given function $\phi(x)$ in a variety of cases, including those of the commonest occurrence. It is natural to consider the converse question, that of *determining a function whose derivative is a given function.*

Suppose that $\psi(x)$ is the given function. Then we wish to determine a function such that $\phi'(x) = \psi(x)$. A little reflection shows us that this question may really be analysed into three parts.

(1) In the first place we want to know whether such a function as $\phi(x)$ *exists*. This question must be distinguished carefully from the question whether (supposing that there is such a function) we can find any simple formula to express it.

(2) We want to know whether it is possible that more than one such function should exist, i.e. whether the solution of our problem is *unique*; and whether, if it is not, there is any simple relation between the different solutions which will enable us to express all of them in terms of any particular one.

* Not necessarily a strict minimum; but see the penultimate paragraph of § 123.

(3) If there is a solution, we want to know *how to find an actual expression for it.*

It will throw light on the nature of these three questions if we compare them with the three corresponding questions which arise with regard to the differentiation of functions.

(1) A function $\phi(x)$ may have a derivative for all values of x, like x^m, where m is a positive integer, or $\sin x$. It may have one except for certain special values of x, like $\tan x$ or $\sec x$. Or it may never have one, like the function of Ex. XXXVII. 20, which is not even continuous for any x.

The last function is discontinuous for every x, and $\tan x$ and $\sec x$ have derivatives except where they are discontinuous. The example of $\sqrt[3]{x}$ shows that a continuous function may have no derivative for special values of x, here $x = 0$. Whether there are continuous functions which *never* have derivatives, or continuous curves which never have tangents, is a further question which is at present beyond us. Common sense says *No*: but, as we have already stated in § 112, this is one of the cases in which higher mathematics has proved common sense to be mistaken.

But at any rate it is clear that the question 'has $\phi(x)$ a derivative $\phi'(x)$?' is one which has to be answered differently in different circumstances. And we may expect that the converse question 'is there a function $\phi(x)$ of which $\psi(x)$ is the derivative?' will have different answers too. We have already seen that there are cases in which the answer is *No*: thus if $\psi(x)$ is the function which is equal to a, b, or c according as x is less than, equal to, or greater than 0, then the answer is *No* (Ex. XLVII. 5), unless $a = b = c$.

This is a case in which the given function is discontinuous. In what follows, however, we shall generally suppose $\psi(x)$ continuous. And then the answer is *Yes*: *if $\psi(x)$ is continuous, then there is always a function $\phi(x)$ such that $\phi'(x) = \psi(x)$.* The proof of this will be given in Ch. VII.

(2) The second question presents no difficulties. In the case of differentiation we have a direct definition of the derivative which

makes it clear from the beginning that there cannot possibly be more than one. In the converse problem the answer is almost equally simple. It is that if $\phi(x)$ is one solution of the problem then $\phi(x)+C$ is another, for any value of the constant C, and that all possible solutions are comprised in the form $\phi(x)+C$. This follows at once from § 127.

(3) The practical problem of finding $\phi'(x)$ is a fairly simple one when $\phi(x)$ is any function defined by some finite combination of the ordinary functional symbols. The converse problem is much more difficult. The nature of the difficulties will appear more clearly later on.

DEFINITIONS. *If $\psi(x)$ is the derivative of $\phi(x)$, then we call $\phi(x)$ an* **integral** *or* **integral function** *of $\psi(x)$. The operation of forming $\phi(x)$ from $\psi(x)$ we call* **integration.**

We shall use the notation

$$\phi(x) = \int \psi(x)\, dx.$$

It is hardly necessary to point out that $\int \ldots dx$ like $\dfrac{d}{dx}$ must, at present at any rate, be regarded purely as a symbol of operation: the $\int$ and the dx mean no more when taken by themselves than do the d and dx of the other operative symbol.

131. The practical problem of integration. The results of the earlier part of this chapter enable us to write down at once the integrals of some of the commonest functions. Thus

$$\int x^m\, dx = \frac{x^{m+1}}{m+1}, \quad \int \cos x\, dx = \sin x, \quad \int \sin x\, dx = -\cos x \quad \ldots(1).$$

These formulae must be understood as meaning that the function on the right-hand side is *one* integral of that under the sign of integration. The *most general* integral is of course obtained by adding to the former a constant C, known as the *arbitrary constant* of integration.

There is however one case of exception to the first formula, that in which $m = -1$. In this case the formula becomes meaningless, as is only to be expected, since we have seen already (Ex. XLII. 4) that $1/x$ cannot be the derivative of any polynomial or rational fraction.

That there really is a function $F(x)$ such that $D_x F(x) = 1/x$ will be proved in the next chapter. For the present we shall be content to assume its existence. This function $F(x)$ is certainly not a polynomial or rational function; and it can be proved that it is not an algebraical function. It can indeed be proved that $F(x)$ is an essentially new function, independent of any of the classes of functions which we have considered yet, that is to say incapable of expression by means of any finite combination of the functional symbols corresponding to them. The proof of this is too detailed to be inserted in this book; but some further discussion of the subject will be found in Ch. IX, where the properties of $F(x)$ are investigated systematically.

Suppose first that x is positive. Then we shall write

$$\int \frac{dx}{x} = \log x \qquad \text{.......................(2)},$$

and we shall call the function on the right-hand side of this equation the *logarithmic function*: it is defined so far only for positive values of x.

Next suppose x negative. Then $-x$ is positive, and so $\log(-x)$ is defined by what precedes. Also

$$\frac{d}{dx}\log(-x) = \frac{-1}{-x} = \frac{1}{x},$$

so that, when x is negative,

$$\int \frac{dx}{x} = \log(-x) \text{.......................(3)}.$$

The formulae (2) and (3) may be united in the formula

$$\int \frac{dx}{x} = \log(\pm x) = \log|x| \qquad \text{..............(4)},$$

where the ambiguous sign is to be chosen so that $\pm x$ is positive: these formulae hold for all real values of x other than $x = 0$.

The most fundamental of the properties of $\log x$ which will be proved in Ch. IX are expressed by the equations

$$\log 1 = 0, \quad \log(1/x) = -\log x, \quad \log xy = \log x + \log y,$$

of which the second is an obvious deduction from the first and third. It is not really necessary, for the purposes of this chapter, to assume the truth of any of these formulae; but they sometimes enable us to write our formulae in a more compact form than would otherwise be possible.

It follows from the last of the formulae that $\log x^2$ is equal to $2\log x$ if $x>0$ and to $2\log(-x)$ if $x<0$, and in either case to $2\log|x|$. Thus (4) is equivalent to

$$\int \frac{dx}{x} = \tfrac{1}{2}\log x^2 \quad \text{.............................(5).}$$

The five formulae (1)–(3) are the five most fundamental *standard forms* of the integral calculus. To them should be added two more, viz.

$$\int \frac{dx}{1+x^2} = \arctan x, \quad \int \frac{dx}{\sqrt{(1-x^2)}} = \pm \arcsin x^* \quad \text{...(6).}$$

132. Polynomials. All the general theorems of § 114 may also be stated as theorems in integration. Thus we have, to begin with, the formulae

$$\int \{f(x)+F(x)\}\,dx = \int f(x)\,dx + \int F(x)\,dx \quad \text{.........(1),}$$

$$\int kf(x)\,dx = k\int f(x)\,dx \quad \text{..................(2).}$$

Here it is assumed, of course, that the arbitrary constants are adjusted properly. Thus the formula (1) asserts that the sum of *any* integral of $f(x)$ and *any* integral of $F(x)$ is *an* integral of $f(x)+F(x)$.

These theorems enable us to write down at once the integral of any function of the form $\Sigma A_\nu f_\nu(x)$, the sum of a finite number of constant multiples of functions whose integrals are known. In particular we can write down the integral of any polynomial: thus

$$\int (a_0x^n + a_1x^{n-1} + \dots + a_n)\,dx = \frac{a_0x^{n+1}}{n+1} + \frac{a_1x^n}{n} + \dots + a_nx.$$

* See § 120 for the rule for determining the ambiguous sign.

133. Rational functions. It is natural to turn our attention next to rational functions. Let us suppose $R(x)$ to be any rational function expressed in the standard form of § 118, viz. as the sum of a polynomial $\Pi(x)$ and a number of terms of the form $A/(x-\alpha)^p$.

We can at once write down the integrals of the polynomial and of all the other terms except those for which $p = 1$, since

$$\int \frac{A}{(x-\alpha)^p}\,dx = -\frac{A}{p-1}\frac{1}{(x-\alpha)^{p-1}},$$

whether α be real or complex (§ 118).

The terms for which $p = 1$ present rather more difficulty. It follows immediately from Theorem (6) of § 114 that

$$\int F'\{f(x)\}f'(x)\,dx = F\{f(x)\} \quad \text{..............(3).}$$

In particular, if we take $f(x) = ax+b$, where a and b are real, and write $\phi(x)$ for $F(x)$ and $\psi(x)$ for $F'(x)$, so that $\phi(x)$ is an integral of $\psi(x)$, we obtain

$$\int \psi(ax+b)\,dx = \frac{1}{a}\phi(ax+b) \quad \text{..............(4).}$$

Thus, for example,

$$\int \frac{dx}{ax+b} = \frac{1}{a}\log|ax+b|,$$

and in particular, if α is real,

$$\int \frac{dx}{x-\alpha} = \log|x-\alpha|.$$

We can therefore write down the integrals of all the terms in $R(x)$ for which $p = 1$ and α is real. There remain the terms for which $p = 1$ and α is complex.

In order to deal with these we shall introduce a restrictive hypothesis, viz. that all the coefficients in $R(x)$ are real. Then if $\alpha = \gamma+\delta i$ is a root of $Q(x) = 0$, of multiplicity m, so is its conjugate $\bar{\alpha} = \gamma-\delta i$; and if a partial fraction $A_p/(x-\alpha)^p$ occurs in the expression of $R(x)$, so does $\bar{A}_p/(x-\bar{\alpha})^p$, where $\bar{A}_p$ is conjugate to A_p. This follows from the nature of the algebraical processes by means of which the partial fractions can be found, and which are explained at length in treatises on algebra*.

Thus, if a term $(\lambda+\mu i)/(x-\gamma-\delta i)$ occurs in the expression of $R(x)$ in partial fractions, so will a term $(\lambda-\mu i)/(x-\gamma+\delta i)$; and the sum of these two terms is

$$\frac{2\{\lambda(x-\gamma)-\mu\delta\}}{(x-\gamma)^2+\delta^2}.$$

This fraction is in reality the most general fraction of the form

$$\frac{Ax+B}{ax^2+2bx+c},$$

where $b^2 < ac$. The reader will easily verify the equivalence of the two forms, the formulae which express λ, μ, γ, δ in terms of A, B, a, b, c being

$$\lambda = \frac{A}{2a}, \quad \mu = -\frac{D}{2a\sqrt{\Delta}}, \quad \gamma = -\frac{b}{a}, \quad \delta = \frac{\sqrt{\Delta}}{a},$$

where $\Delta = ac-b^2$, and $D = aB-bA$.

If in (3) we suppose $F\{f(x)\}$ to be $\log|f(x)|$, we obtain

$$\int\frac{f'(x)}{f(x)}dx = \log|f(x)| \quad \ldots\ldots\ldots\ldots\ldots\ldots(5);$$

and if we further suppose that $f(x) = (x-\lambda)^2+\mu^2$, we obtain

$$\int\frac{2(x-\lambda)}{(x-\lambda)^2+\mu^2}dx = \log\{(x-\lambda)^2+\mu^2\}.$$

And, in virtue of the equations (6) of § 131 and (4) above, we have

$$\int\frac{-2\delta\mu}{(x-\lambda)^2+\mu^2}dx = -2\delta\arctan\left(\frac{x-\lambda}{\mu}\right).$$

These two formulae enable us to integrate the sum of the two terms which we have been considering in the expression of $R(x)$; and we are thus enabled to write down the integral of any real rational function, if all the factors of its denominator can be determined. The integral of any such function is composed of *the*

* See, for example, Chrystal's *Algebra*, 2nd edition, vol. I, pp. 151–9.

sum of a polynomial, a number of rational functions of the type

$$-\frac{A}{p-1}\frac{1}{(x-\alpha)^{p-1}},$$

a number of logarithmic functions, and a number of inverse tangents.

It only remains to add that if α is complex then the rational function just written always occurs in conjunction with another in which A and α are replaced by the complex numbers conjugate to them, and that the sum of the two functions is a real rational function.

Examples XLVIII. 1. Prove that

$$\int\frac{Ax+B}{ax^2+2bx+c}\,dx = \frac{A}{2a}\log|X| + \frac{D}{2a\sqrt{(-\Delta)}}\log\left|\frac{ax+b-\sqrt{(-\Delta)}}{ax+b+\sqrt{(-\Delta)}}\right|$$

(where $X = ax^2+2bx+c$) if $\Delta < 0$, and

$$\int\frac{Ax+B}{ax^2+2bx+c}\,dx = \frac{A}{2a}\log|X| + \frac{D}{a\sqrt{\Delta}}\arctan\left(\frac{ax+b}{\sqrt{\Delta}}\right)$$

if $\Delta > 0$, Δ and D having the same meanings as on p. 251.

2. In the particular case in which $ac = b^2$ the integral is

$$-\frac{D}{a(ax+b)} + \frac{A}{a}\log|ax+b|.$$

3. Show that if the roots of $Q(x) = 0$ are all real and distinct, and $P(x)$ is of lower degree than $Q(x)$, then

$$\int R(x)\,dx = \Sigma\frac{P(\alpha)}{Q'(\alpha)}\log|x-\alpha|,$$

the summation applying to all the roots α of $Q(x) = 0$.

[The form of the partial fraction corresponding to α may be deduced from the facts that

$$\frac{Q(x)}{x-\alpha} \to Q'(\alpha), \quad (x-\alpha)\,R(x) \to \frac{P(\alpha)}{Q'(\alpha)}.]$$

4. If all the roots of $Q(x)$ are real and α is a double root, the other roots being simple roots, and $P(x)$ is of lower degree than $Q(x)$, then the integral is $A/(x-\alpha) + A'\log|x-\alpha| + \Sigma B\log|x-\beta|$, where

$$A = -\frac{2P(\alpha)}{Q''(\alpha)}, \quad A' = \frac{2\{3P'(\alpha)\,Q''(\alpha) - P(\alpha)\,Q'''(\alpha)\}}{3\{Q''(\alpha)\}^2}, \quad B = \frac{P(\beta)}{Q'(\beta)},$$

and the summation applies to all roots β of $Q(x) = 0$ other than α.

5. Calculate $$\int \frac{dx}{\{(x-1)(x^2+1)\}^2}.$$

[The expression in partial fractions is

$$\frac{1}{4(x-1)^2}-\frac{1}{2(x-1)}-\frac{i}{8(x-i)^2}+\frac{2-i}{8(x-i)}+\frac{i}{8(x+i)^2}+\frac{2+i}{8(x+i)},$$

and the integral is

$$-\frac{1}{4(x-1)}-\frac{1}{4(x^2+1)}-\tfrac{1}{2}\log|x-1|+\tfrac{1}{4}\log(x^2+1)+\tfrac{1}{4}\arc\tan x.]$$

6. Integrate

$$\frac{x}{(x-a)(x-b)(x-c)},\quad \frac{x}{(x-a)^2(x-b)},\quad \frac{x}{(x-a)^2(x-b)^2},\quad \frac{x}{(x-a)^3},$$

$$\frac{x}{(x^2+a^2)(x^2+b^2)},\quad \frac{x^2}{(x^2+a^2)(x^2+b^2)},\quad \frac{x^2-a^2}{x^2(x^2+a^2)},\quad \frac{x^2-a^2}{x(x^2+a^2)^2}.$$

7. Integrate

$$\frac{x}{(x-1)(x^2+1)},\quad \frac{x}{1+x^3},\quad \frac{x^3}{(x-1)^2(x^3+1)}.$$

(*Math. Trip.* 1924, 1926, 1934)

8. Prove the formulae

$$\int\frac{dx}{1+x^4}=\frac{1}{4\sqrt{2}}\left\{\log\left(\frac{1+x\sqrt{2}+x^2}{1-x\sqrt{2}+x^2}\right)+2\arc\tan\left(\frac{x\sqrt{2}}{1-x^2}\right)\right\},$$

$$\int\frac{x^2\,dx}{1+x^4}=\frac{1}{4\sqrt{2}}\left\{-\log\left(\frac{1+x\sqrt{2}+x^2}{1-x\sqrt{2}+x^2}\right)+2\arc\tan\left(\frac{x\sqrt{2}}{1-x^2}\right)\right\},$$

$$\int\frac{dx}{1+x^2+x^4}=\frac{1}{4\sqrt{3}}\left\{\sqrt{3}\log\left(\frac{1+x+x^2}{1-x+x^2}\right)+2\arc\tan\left(\frac{x\sqrt{3}}{1-x^2}\right)\right\}.$$

134. Note on the practical integration of rational functions. The analysis of § 133 gives us a general method by which we can find the integral of any real rational function $R(x)$, *provided we can solve the equation* $Q(x)=0$. In simple cases (as in Ex. 5 above) the application of the method is fairly easy. In more complicated cases the labour involved is sometimes prohibitive, and other devices have to be used. It is not part of the purpose of this book to go into practical problems of integration in detail. The reader who desires fuller information may be referred to Goursat's *Cours d'analyse*, 3rd edition, vol. I, pp. 246 *et seq.*, Bertrand's *Calcul intégral*, and Bromwich's tract *Elementary integrals* (Bowes and Bowes, 1911).

If the equation $Q(x) = 0$ cannot be solved algebraically, then the method of partial fractions naturally fails and recourse must be had to other methods*.

135. Algebraical functions. We naturally pass on next to the question of the integration of *algebraical* functions. We have to consider the problem of integrating y, where y is an algebraical function of x. It is however convenient to consider an apparently more general integral, viz.

$$\int R(x, y)\,dx,$$

where $R(x, y)$ is any rational function of x and y. The greater generality of this form is only apparent, since the function $R(x, y)$ is itself an algebraical function of x. The choice of this form is dictated simply by motives of convenience: such a function as

$$\frac{px+q+\sqrt{(ax^2+2bx+c)}}{px+q-\sqrt{(ax^2+2bx+c)}}$$

is more conveniently regarded as a rational function of x and the simple algebraical function $\sqrt{(ax^2+2bx+c)}$, than directly as itself an algebraical function of x.

136. Integration by substitution and rationalisation. It follows from equation (3) of § 133 that if $\int \psi(x)\,dx = \phi(x)$ then

$$\int \psi\{f(t)\}f'(t)\,dt = \phi\{f(t)\} \quad \text{.................(1)}.$$

This equation supplies us with a method for determining the integral of $\psi(x)$ in a large number of cases in which the form of the integral is not directly obvious. It may be stated as a rule as follows: *put $x = f(t)$, where $f(t)$ is any function of a new variable t which it may be convenient to choose; multiply by $f'(t)$, and determine (if possible) the integral of $\psi\{f(t)\}f'(t)$; express the result in terms of x.* It will often be found that the function of t to which we are led by

* See the author's tract "The integration of functions of a single variable" (*Cambridge Tracts in Mathematics*, No. 2, 2nd edition, 1916). This does not often happen in practice.

the application of this rule is one whose integral can easily be calculated. This is always so, for example, if it is a rational function, and it is very often possible to choose the relation between x and t so that this shall happen. Thus the integral of $R(\sqrt{x})$, where R denotes a rational function, is reduced by the substitution $x = t^2$ to the integral of $2tR(t^2)$, i.e. to the integral of a rational function of t. This method of integration is called *integration by rationalisation*.

Its application to the problem immediately under consideration is obvious. *If we can find a variable t such that x and y are both rational functions of t, say $x = R_1(t)$, $y = R_2(t)$, then*

$$\int R(x,y)\,dx = \int R\{R_1(t), R_2(t)\}\,R_1'(t)\,dt,$$

and the latter integral, being that of a rational function of t, can be calculated by the methods of § 133.

It is important to know when we can find an auxiliary variable t connected with x and y in this manner, but we cannot discuss the general problem here*. We must confine ourselves to a few simple special cases.

137. Integrals connected with conics. Let us suppose that x and y are connected by an equation of the form

$$ax^2 + 2hxy + by^2 + 2gx + 2fy + c = 0;$$

in other words that the graph of y, considered as a function of x, is a conic. Suppose that (ξ, η) is any point on the conic, and let $x - \xi = X$, $y - \eta = Y$. If the relation between x and y is expressed in terms of X and Y, it assumes the form

$$aX^2 + 2hXY + bY^2 + 2GX + 2FY = 0,$$

where $F = h\xi + b\eta + f$, $G = a\xi + h\eta + g$. In this equation put $Y = tX$. It will then be found that X and Y, and therefore x and y, are rational functions of t. The actual formulae are

$$x - \xi = -\frac{2(G + Ft)}{a + 2ht + bt^2}, \quad y - \eta = -\frac{2t(G + Ft)}{a + 2ht + bt^2}.$$

* See the tract referred to on p. 234.

Hence the process of rationalisation described in the preceding section can be carried out.

The reader should verify that

$$hx+by+f=-\tfrac{1}{2}(a+2ht+bt^2)\frac{dx}{dt},$$

so that

$$\int\frac{dx}{hx+by+f}=-2\int\frac{dt}{a+2ht+bt^2}.$$

When $h^2>ab$ it is in some ways advantageous to proceed as follows. The conic is a hyperbola whose asymptotes are parallel to the lines

$$ax^2+2hxy+by^2=0,$$

or

$$b(y-\mu x)(y-\mu' x)=0,$$

say. If we put $y-\mu x=t$, we obtain

$$y-\mu x=t,\quad y-\mu' x=-\frac{2gx+2fy+c}{bt},$$

and it is clear that x and y can be calculated from these equations as rational functions of t. We shall illustrate this process by an application to an important special case.

138. The integral $\int\frac{dx}{\sqrt{(ax^2+2bx+c)}}$. Suppose in particular that $y^2=ax^2+2bx+c$, where $a>0$. It will be found that, if we put $y+x\sqrt{a}=t$, we obtain

$$2\frac{dx}{dt}=\frac{(t^2+c)\sqrt{a}+2bt}{(t\sqrt{a}+b)^2},\quad 2y=\frac{(t^2+c)\sqrt{a}+2bt}{t\sqrt{a}+b},$$

and so

$$\int\frac{dx}{y}=\int\frac{dt}{t\sqrt{a}+b}=\frac{1}{\sqrt{a}}\log\left|x\sqrt{a}+y+\frac{b}{\sqrt{a}}\right| \quad\text{............(1).}$$

If in particular $a=1$, $b=0$, $c=\alpha^2$, or $a=1$, $b=0$, $c=-\alpha^2$, we obtain

$$\int\frac{dx}{\sqrt{(x^2+\alpha^2)}}=\log\{x+\sqrt{(x^2+\alpha^2)}\},\quad \int\frac{dx}{\sqrt{(x^2-\alpha^2)}}=\log|x+\sqrt{(x^2-\alpha^2)}| \quad\text{......(2),}$$

equations whose truth may be verified immediately by differentiation. With these formulae should be associated the third formula

$$\int\frac{dx}{\sqrt{(\alpha^2-x^2)}}=\arcsin\frac{x}{\alpha} \quad\text{........................(3),}$$

which corresponds to a case of the general integral of this section in which $a<0$. In (3) it is supposed that $\alpha>0$, if $\alpha<0$ then the integral is $\arcsin(x/|\alpha|)$ (cf. § 120). In practice we should evaluate the general integral by reducing it (as in the next section) to one or other of these standard forms.

The formula (3) appears very different from the formulae (2): the reader will hardly be in a position to appreciate the connection between them until he has read Ch. X.

139. The integral $\int \frac{\lambda x+\mu}{\sqrt{(ax^2+2bx+c)}}dx$. This integral can be integrated in all cases by means of the results of the preceding sections. It is most convenient to proceed as follows. Since

$$\lambda x+\mu = \frac{\lambda}{a}(ax+b)+\mu-\frac{\lambda b}{a},$$

$$\int \frac{ax+b}{\sqrt{(ax^2+2bx+c)}}dx = \sqrt{(ax^2+2bx+c)},$$

we have

$$\int \frac{(\lambda x+\mu)\,dx}{\sqrt{(ax^2+2bx+c)}} = \frac{\lambda}{a}\sqrt{(ax^2+2bx+c)}+\left(\mu-\frac{\lambda b}{a}\right)\int \frac{dx}{\sqrt{(ax^2+2bx+c)}}.$$

In the last integral a may be positive or negative. If a is positive we put $xa^{\frac{1}{2}}+ba^{-\frac{1}{2}} = t$, when we obtain

$$\frac{1}{\sqrt{a}}\int \frac{dt}{\sqrt{(t^2+\kappa)}},$$

where $\kappa = (ac-b^2)/a$. If a is negative we write A for $-a$ and put $xA^{\frac{1}{2}}-bA^{-\frac{1}{2}} = t$, when we obtain

$$\frac{1}{\sqrt{(-a)}}\int \frac{dt}{\sqrt{(-\kappa-t^2)}}.$$

It thus appears that in any case the calculation of the integral may be made to depend on that of the integral considered in § 138, and that this integral may be reduced to one or other of the three forms

$$\int \frac{dt}{\sqrt{(t^2+\alpha^2)}},\quad \int \frac{dt}{\sqrt{(t^2-\alpha^2)}},\quad \int \frac{dt}{\sqrt{(\alpha^2-t^2)}}.$$

140. The integral $\int(\lambda x+\mu)\sqrt{(ax^2+2bx+c)}\,dx$. In the same way we find

$$\int(\lambda x+\mu)\sqrt{(ax^2+2bx+c)}\,dx$$

$$=\frac{\lambda}{3a}(ax^2+2bx+c)^{\frac{3}{2}}+\left(\mu-\frac{\lambda b}{a}\right)\int\sqrt{(ax^2+2bx+c)}\,dx;$$

and the last integral may be reduced to one or other of the three forms

$$\int\sqrt{(t^2+\alpha^2)}\,dt,\quad \int\sqrt{(t^2-\alpha^2)}\,dt,\quad \int\sqrt{(\alpha^2-t^2)}\,dt.$$

In order to obtain these integrals it is convenient to introduce at this point another general theorem in integration.

141. Integration by parts. The theorem of *integration by parts* is merely another way of stating the rule for the differentiation of a product proved in § 114. It follows at once from Theorem (3) of § 114 that

$$\int f'(x)\,F(x)\,dx=f(x)\,F(x)-\int f(x)\,F'(x)\,dx.$$

It may happen that the function which we wish to integrate is expressible in the form $f'(x)\,F(x)$, and that $f(x)\,F'(x)$ can be integrated. Suppose, for example, that $\phi(x)=x\psi(x)$, where $\psi(x)$ is the second derivative of a known function $\chi(x)$. Then

$$\int\phi(x)\,dx=\int x\chi''(x)\,dx=x\chi'(x)-\int\chi'(x)\,dx=x\chi'(x)-\chi(x).$$

We can illustrate the working of this method of integration by applying it to the integrals of the last section. Taking

$$f(x)=ax+b,\quad F(x)=\sqrt{(ax^2+2bx+c)}=y,$$

we obtain

$$a\int y\,dx=(ax+b)\,y-\int\frac{(ax+b)^2}{y}\,dx=(ax+b)\,y-a\int y\,dx+(ac-b^2)\int\frac{dx}{y},$$

so that

$$\int y\,dx=\frac{(ax+b)\,y}{2a}+\frac{ac-b^2}{2a}\int\frac{dx}{y};$$

and we have seen already (§ 138) how to determine the last integral.

Examples XLIX. 1. Prove that if $a > 0$ then

$$\int \sqrt{(x^2+a^2)}\,dx = \tfrac{1}{2}x\sqrt{(x^2+a^2)} + \tfrac{1}{2}a^2 \log\{x+\sqrt{(x^2+a^2)}\},$$

$$\int \sqrt{(x^2-a^2)}\,dx = \tfrac{1}{2}x\sqrt{(x^2-a^2)} - \tfrac{1}{2}a^2 \log|x+\sqrt{(x^2-a^2)}|,$$

$$\int \sqrt{(a^2-x^2)}\,dx = \tfrac{1}{2}x\sqrt{(a^2-x^2)} + \tfrac{1}{2}a^2 \arcsin\frac{x}{a}.$$

2. Calculate the integrals $\int \frac{dx}{\sqrt{(a^2-x^2)}}$, $\int \sqrt{(a^2-x^2)}\,dx$ by means of the substitution $x = a\sin\theta$, and verify that the results agree with those obtained in § 138 and Ex. 1.

3. Prove, by means of the substitutions $ax+b = 1/t$ and $x = 1/u$, that (in the notation of §§ 133 and 141)

$$\int \frac{dx}{y^3} = \frac{ax+b}{\Delta y}, \quad \int \frac{x\,dx}{y^3} = -\frac{bx+c}{\Delta y}.$$

4. Calculate $\int \frac{dx}{\sqrt{\{(x-a)(b-x)\}}}$, where $b > a$, in three ways, viz. (i) by the methods of the preceding sections, (ii) by the substitution

$$(b-x)/(x-a) = t^2,$$

and (iii) by the substitution $x = a\cos^2\theta + b\sin^2\theta$; and verify that the results agree.

5. Integrate $\frac{x^3}{(x^2+1)^3}$ by the substitutions

$$(a)\ x = \tan\theta, \quad (b)\ u = x^2+1,$$

and verify the agreement of the results. (*Math. Trip.* 1933)

6. Integrate

$$\frac{1}{x(1+x^5)}, \quad \frac{1}{(a+x)\sqrt{(c+x)}}, \quad \frac{x^2+1}{x\sqrt{(4x^2+1)}}, \quad \frac{x}{\sqrt{(x^2+x+1)}}, \quad \frac{1}{x^6\sqrt{(x^2+a^2)}}.$$

(*Math. Trip.* 1923, 1925, 1927, 1929)

7. Show, by means of the substitution $2x+a+b = \tfrac{1}{2}(a-b)(t^2+t^{-2})$, or by multiplying numerator and denominator by $\sqrt{(x+a)}-\sqrt{(x+b)}$, that if $a > b$ then

$$\int \frac{dx}{\sqrt{(x+a)}+\sqrt{(x+b)}} = \tfrac{1}{2}\sqrt{(a-b)}\left(t+\frac{1}{3t^3}\right).$$

8. Find a substitution which will reduce $\int \frac{dx}{(x+a)^{\frac{3}{2}}+(x-a)^{\frac{3}{2}}}$ to the integral of a rational function. (*Math. Trip.* 1899)

9. Show that $\int R\{x, \sqrt[n]{(ax+b)}\}\,dx$ is reduced, by the substitution $ax+b = y^n$, to the integral of a rational function.

10. Prove that

$$\int f''(x)\,F(x)\,dx = f'(x)\,F(x) - f(x)\,F'(x) + \int f(x)\,F''(x)\,dx$$

and generally

$$\int f^{(n)}(x)\,F(x)\,dx = f^{(n-1)}(x)\,F(x) - f^{(n-2)}(x)\,F'(x) + \dots + (-1)^n \int f(x)\,F^{(n)}(x)\,dx.$$

11. The integral $\int (1+x)^p x^q\,dx$, where p and q are rational, can be found in three cases, viz. (i) if p is an integer, (ii) if q is an integer, and (iii) if $p+q$ is an integer. [In case (i) put $x = u^s$, where s is the denominator of q; in case (ii) put $1+x = t^s$, where s is the denominator of p; and in case (iii) put $1+x = xt^s$, where s is the denominator of p.]

12. The integral $\int x^m(ax^n+b)^q\,dx$ can be reduced to the preceding integral by the substitution $ax^n = bt$. [In practice it is often most convenient to calculate a particular integral of this kind by a 'formula of reduction' (cf. Misc. Ex. 55, p. 282).]

13. The integral $\int R\{x, \sqrt{(ax+b)}, \sqrt{(cx+d)}\}\,dx$ can be reduced to that of a rational function by the substitution

$$4x = -\frac{b}{a}\left(t+\frac{1}{t}\right)^2 - \frac{d}{c}\left(t-\frac{1}{t}\right)^2.$$

14. Reduce $\int R(x,y)\,dx$, where $y^2(x-y) = x^2$, to the integral of a rational function. [Putting $y = tx$, we obtain $x = 1/\{t^2(1-t)\}$, $y = 1/\{t(1-t)\}$.]

15. Reduce the integral in the same way when (*a*) $y(x-y)^2 = x$, (*b*) $(x^2+y^2)^2 = a^2(x^2-y^2)$. [In case (*a*) put $x-y=t$: in case (*b*) put $x^2+y^2 = t(x-y)$, when we obtain

$$x = a^2t(t^2+a^2)/(t^4+a^4), \quad y = a^2t(t^2-a^2)/(t^4+a^4).]$$

16. If $y(x-y)^2 = x$, then $\displaystyle\int \frac{dx}{x-3y} = \tfrac{1}{2}\log\{(x-y)^2-1\}$.

17. If $(x^2+y^2)^2 = 2c^2(x^2-y^2)$, then $\displaystyle\int \frac{dx}{y(x^2+y^2+c^2)} = -\frac{1}{c^2}\log\left(\frac{x^2+y^2}{x-y}\right)$.

142. The general integral $\int R(x, y)\,dx$, where $y^2 = ax^2 + 2bx + c$. The most general integral associated, in the manner of § 137, with the special conic $y^2 = ax^2 + 2bx + c$ is

$$\int R(x, \sqrt{X})\,dx \quad \text{......................(1)},$$

where $X = y^2 = ax^2 + 2bx + c$. We suppose that R is a *real* function.

The subject of integration is of the form P/Q, where P and Q are polynomials in x and $\sqrt{X}$. It may therefore be reduced to the form

$$\frac{A + B\sqrt{X}}{C + D\sqrt{X}} = \frac{(A + B\sqrt{X})(C - D\sqrt{X})}{C^2 - D^2 X} = E + F\sqrt{X},$$

where A, B, ... are rational functions of x. The only new problem which arises is that of the integration of a function of the form $F\sqrt{X}$, or, what is the same thing, $G/\sqrt{X}$, where G is a rational function of x. And the integral

$$\int \frac{G}{\sqrt{X}}\,dx \quad \text{......................(2)}$$

can always be evaluated by splitting up G into partial fractions. When we do this, integrals of three different types may arise.

(i) In the first place there may be integrals of the type

$$\int \frac{x^m}{\sqrt{X}}\,dx \quad \text{......................(3)},$$

where m is a positive integer The cases in which $m = 0$ or $m = 1$ have been disposed of in § 139. In order to calculate the integrals corresponding to larger values of m we observe that

$$\frac{d}{dx}(x^{m-1}\sqrt{X}) = (m-1)x^{m-2}\sqrt{X} + \frac{(ax+b)x^{m-1}}{\sqrt{X}} = \frac{\alpha x^m + \beta x^{m-1} + \gamma x^{m-2}}{\sqrt{X}},$$

where α, β, γ are constants whose values may be easily calculated. It is clear that, when we integrate this equation, we obtain a relation between three successive integrals of the type (3). Since we know the values of the integral for $m = 0$ and $m = 1$, we can calculate in turn its values for all other values of m.

(ii) In the second place there may be integrals of the type

$$\int \frac{dx}{(x-p)^m\sqrt{X}} \quad \text{......................(4)},$$

where p is real. If we make the substitution $x - p = 1/t$, then this integral is reduced to an integral in t of the type (3).

(iii) Finally, there may be integrals corresponding to complex roots of the denominator of G. We shall confine ourselves to the simplest case, that in which all such roots are simple roots. In this case (cf. § 133) a pair of conjugate complex roots of G gives rise to an integral of the type

$$\int \frac{Lx+M}{(Ax^2+2Bx+C)\sqrt{(ax^2+2bx+c)}}\,dx \quad \text{...............(5)}.$$

In order to evaluate this integral we put

$$x = \frac{\mu t+\nu}{t+1},$$

where μ and ν are so chosen that

$$a\mu\nu + b(\mu+\nu) + c = 0, \quad A\mu\nu + B(\mu+\nu) + C = 0;$$

so that μ and ν are the roots of the equation

$$(aB-bA)\,\xi^2 - (cA-aC)\,\xi + (bC-cB) = 0.$$

This equation has certainly real roots, for it is the same equation as equation (1) of Ex. XLVI. 13; and it is therefore certainly possible to find real values of μ and ν fulfilling our requirements.

It will be found, on carrying out the substitution, that the integral (5) assumes the form

$$H\int \frac{t\,dt}{(\alpha t^2+\beta)\sqrt{(\gamma t^2+\delta)}} + K\int \frac{dt}{(\alpha t^2+\beta)\sqrt{(\gamma t^2+\delta)}} \quad \text{.........(6)}.$$

The second of these integrals is rationalised by the substitution

$$\frac{t}{\sqrt{(\gamma t^2+\delta)}} = u,$$

which gives $$\int \frac{dt}{(\alpha t^2+\beta)\sqrt{(\gamma t^2+\delta)}} = \int \frac{du}{\beta + (\alpha\delta-\beta\gamma)\,u^2}.$$

Finally, if we put $t = 1/u$ in the first of the integrals (6), it is transformed into an integral of the second type, and may therefore be calculated in the manner just explained, viz. by putting $u/\sqrt{(\gamma+\delta u^2)} = v$, i.e. $1/\sqrt{(\gamma t^2+\delta)} = v$*.

Examples L. 1. Evaluate

$$\int \frac{dx}{x\sqrt{(x^2+2x+3)}}, \quad \int \frac{dx}{(x-1)\sqrt{(x^2+1)}}, \quad \int \frac{dx}{(x+1)\sqrt{(1+2x-x^2)}}.$$

2. Prove that

$$\int \frac{dx}{(x-p)\sqrt{\{(x-p)(x-q)\}}} = \frac{2}{q-p}\sqrt{\left(\frac{x-q}{x-p}\right)}.$$

* The method of integration explained here fails if $a/A = b/B$; but then the integral may be reduced by the substitution $ax+b=t$. For further information concerning the integration of algebraical functions see Stolz, *Grundzüge der Differential-und-intergralrechnung*, vol. I, pp. 331 *et seq.*, or Bromwich's tract quoted on p. 253. An alternative method of reduction has been given by Greenhill: see his *A chapter in the integral calculus*, pp. 12 *et seq.*, and the author's tract quoted on p. 254.

3. If $ag^2+ch^2=-\nu<0$, then

$$\int\frac{dx}{(hx+g)\sqrt{(ax^2+c)}}=-\frac{1}{\sqrt{\nu}}\arctan\left[\frac{\sqrt{\{\nu(ax^2+c)\}}}{ch-agx}\right].$$

4. Show that $\int\frac{dx}{(x-x_0)\,y}$, where $y^2=ax^2+2bx+c$, may be expressed in one or other of the forms

$$-\frac{1}{y_0}\log\left|\frac{axx_0+b(x+x_0)+c+yy_0}{x-x_0}\right|,\quad \frac{1}{z_0}\arctan\left\{\frac{axx_0+b(x+x_0)+c}{yz_0}\right\},$$

according as $ax_0^2+2bx_0+c$ is positive and equal to y_0^2 or negative and equal to $-z_0^2$.

5. Show by means of the substitution $y=\sqrt{(ax^2+2bx+c)}/(x-p)$ that

$$\int\frac{dx}{(x-p)\sqrt{(ax^2+2bx+c)}}=\int\frac{dy}{\sqrt{(\lambda y^2-\mu)}},$$

where $\lambda=ap^2+2bp+c$, $\mu=ac-b^2$. [This method of reduction is elegant but less straightforward than that explained in § 142.]

6. Show that the integral

$$\int\frac{dx}{x\sqrt{(3x^2+2x+1)}}$$

is rationalised by the substitution $x=(1+y^2)/(3-y^2)$.

(*Math. Trip.* 1911)

7. Calculate $$\int\frac{(x+1)\,dx}{(x^2+4)\sqrt{(x^2+9)}}.$$

8. Calculate $$\int\frac{dx}{(5x^2+12x+8)\sqrt{(5x^2+2x-7)}}.$$

[Apply the method of § 142. The equation satisfied by μ and ν is $\xi^2+3\xi+2=0$, so that $\mu=-2$, $\nu=-1$, and the appropriate substitution is $x=-(2t+1)/(t+1)$. This reduces the integral to

$$-\int\frac{dt}{(4t^2+1)\sqrt{(9t^2-4)}}-\int\frac{t\,dt}{(4t^2+1)\sqrt{(9t^2-4)}}.$$

The first of these integrals may be rationalised by putting $t/\sqrt{(9t^2-4)}=u$ and the second by putting $1/\sqrt{(9t^2-4)}=v$.]

9. Calculate

$$\int\frac{(x+1)\,dx}{(2x^2-2x+1)\sqrt{(3x^2-2x+1)}},\quad \int\frac{(x-1)\,dx}{(2x^2-6x+5)\sqrt{(7x^2-22x+19)}}.$$

(*Math. Trip.* 1911)

10. Show that the integral $\int R(x, y)\,dx$, where $y^2 = ax^2 + 2bx + c$, is rationalised by the substitution $t = (x-p)/(y+q)$, where (p, q) is any point on the conic $y^2 = ax^2 + 2bx + c$. [The integral is of course also rationalised by the substitution $t = (x-p)/(y-q)$: cf. § 137.]

143. Transcendental functions. Owing to the great variety of the different classes of transcendental functions, the theory of their integration is a good deal less systematic than that of the integration of rational or algebraical functions. We shall consider in order a few classes of transcendental functions whose integrals can always be found.

144. Polynomials in cosines and sines of multiples of x. We can always integrate any function which is the sum of a finite number of terms such as

$$A \cos^m ax \sin^{m'} ax \cos^n bx \sin^{n'} bx \ldots,$$

where m, m', n, n', ... are positive integers and a, b, ... any real numbers. For such a term can be expressed as the sum of a finite number of terms of the types

$$\alpha \cos\{(pa + qb + \ldots)x\}, \quad \beta \sin\{(pa + qb + \ldots)x\};$$

and the integrals of these terms can be written down at once.

Examples LI. 1. Integrate $\sin^3 x \cos^2 2x$. In this case we use the formulae

$$\sin^3 x = \tfrac{1}{4}(3 \sin x - \sin 3x), \quad \cos^2 2x = \tfrac{1}{2}(1 + \cos 4x).$$

Multiplying these two expressions and replacing $\sin x \cos 4x$, for example, by $\frac{1}{2}(\sin 5x - \sin 3x)$, we obtain

$$\tfrac{1}{16}\int (7 \sin x - 5 \sin 3x + 3 \sin 5x - \sin 7x)\,dx$$
$$= -\tfrac{7}{16}\cos x + \tfrac{5}{48}\cos 3x - \tfrac{3}{80}\cos 5x + \tfrac{1}{112}\cos 7x.$$

The integral may of course be obtained in different forms by different methods. For example

$$\int \sin^3 x \cos^2 2x\,dx = \int (4 \cos^4 x - 4 \cos^2 x + 1)(1 - \cos^2 x) \sin x\,dx,$$

which reduces, on making the substitution $\cos x = t$, to

$$\int (4t^6 - 8t^4 + 5t^2 - 1)\,dt = \tfrac{4}{7}\cos^7 x - \tfrac{8}{5}\cos^5 x + \tfrac{5}{3}\cos^3 x - \cos x.$$

It may be verified that this expression and that obtained above differ only by a constant.

2. Integrate by any method $\cos ax \cos bx$, $\sin ax \sin bx$, $\cos ax \sin bx$, $\cos^2 x$, $\sin^3 x$, $\cos^4 x$, $\cos x \cos 2x \cos 3x$, $\cos^3 2x \sin^2 3x$, $\cos^5 x \sin^7 x$. [In cases of this kind it is sometimes convenient to use a formula of reduction (Misc. Ex. 55, p. 282).]

145. The integrals $\int x^n \cos x\,dx$, $\int x^n \sin x\,dx$ and associated integrals. The method of integration by parts enables us to generalise the preceding results. For

$$\int x^n \cos x\,dx = \quad x^n \sin x - n\int x^{n-1} \sin x\,dx,$$

$$\int x^n \sin x\,dx = -x^n \cos x + n\int x^{n-1} \cos x\,dx,$$

and the integrals can be calculated completely by a repetition of this process whenever n is a positive integer. It follows that we can always calculate $\int x^n \cos ax\,dx$ and $\int x^n \sin ax\,dx$ if n is a positive integer; and so, by a process similar to that of the preceding paragraph, we can calculate

$$\int P(x, \cos ax, \sin ax, \cos bx, \sin bx, \ldots)\,dx,$$

where P is any polynomial.

Examples LII. 1. Integrate $x \sin x$, $x^2 \cos x$, $x^2 \cos^2 x$, $x^2 \sin^2 x \sin^2 2x$, $x \sin^2 x \cos^4 x$, $x^3 \sin^3 \frac{1}{3}x$.

2. Find polynomials P and Q such that

$$\int \{(3x-1)\cos x + (1-2x)\sin x\}\,dx = P\cos x + Q\sin x.$$

3. Prove that $\int x^n \cos x\,dx = P_n \cos x + Q_n \sin x$, where

$$P_n = nx^{n-1} - n(n-1)(n-2)x^{n-3} + \ldots, \quad Q_n = x^n - n(n-1)x^{n-2} + \ldots.$$

146. Rational functions of $\cos x$ and $\sin x$. The integral of any rational function of $\cos x$ and $\sin x$ may be calculated by the substitution $\tan\frac{1}{2}x = t$. For

$$\cos x = \frac{1-t^2}{1+t^2}, \quad \sin x = \frac{2t}{1+t^2}, \quad \frac{dx}{dt} = \frac{2}{1+t^2},$$

so that the integral is reduced to that of a rational function of t. But other substitutions are sometimes more convenient.

Examples LIII. 1. Prove that

$$\int \sec x\,dx = \log|\sec x + \tan x|, \quad \int \operatorname{cosec} x\,dx = \log|\tan\tfrac{1}{2}x|.$$

[Another form of the first integral is $\log|\tan(\frac{1}{4}\pi + \frac{1}{2}x)|$; a third form is $\frac{1}{2}\log|(1+\sin x)/(1-\sin x)|$.]

2. $\int \tan x\,dx = -\log|\cos x|$, $\int \cot x\,dx = \log|\sin x|$, $\int \sec^2 x\,dx = \tan x$,

$\int \operatorname{cosec}^2 x\,dx = -\cot x$, $\int \tan x \sec x\,dx = \sec x$, $\int \cot x \operatorname{cosec} x\,dx = -\operatorname{cosec} x$.

[These integrals are included in the general form, but there is no need to use a substitution, as the results follow at once from § 120 and equation (5) of § 133.]

3. Show that the integral of $1/(a+b\cos x)$, where $a+b$ is positive, may be expressed in one or other of the forms

$$\frac{2}{\sqrt{(a^2-b^2)}}\arctan\left\{t\sqrt{\left(\frac{a-b}{a+b}\right)}\right\}, \quad \frac{1}{\sqrt{(b^2-a^2)}}\log\left|\frac{\sqrt{(b+a)}+t\sqrt{(b-a)}}{\sqrt{(b+a)}-t\sqrt{(b-a)}}\right|,$$

where $t = \tan\frac{1}{2}x$, according as $a^2 > b^2$ or $a^2 < b^2$. If $a^2 = b^2$, then the integral reduces to a constant multiple of that of $\sec^2\frac{1}{2}x$ or $\operatorname{cosec}^2\frac{1}{2}x$, and its value may be written down at once. Deduce the forms of the integral when $a+b$ is negative.

4. Show that if y is defined in terms of x by means of the equation

$$(a+b\cos x)(a-b\cos y) = a^2-b^2,$$

where a is positive and $a^2 > b^2$, then as x varies from 0 to π one value of y also varies from 0 to π. Show also that

$$\sin x = \frac{\sqrt{(a^2-b^2)}\sin y}{a-b\cos y}, \quad \frac{\sin x}{a+b\cos x}\frac{dx}{dy} = \frac{\sin y}{a-b\cos y};$$

and deduce that if $0 < x < \pi$ then

$$\int\frac{dx}{a+b\cos x} = \frac{1}{\sqrt{(a^2-b^2)}}\arccos\left(\frac{a\cos x+b}{a+b\cos x}\right).$$

Show that this result agrees with that of Ex. 3.

5. Show how to integrate $1/(a+b\cos x+c\sin x)$. [Express $b\cos x+c\sin x$ in the form $\sqrt{(b^2+c^2)}\cos(x-\alpha)$.]

6. Integrate $(a+b\cos x+c\sin x)/(\alpha+\beta\cos x+\gamma\sin x)$. [Determine λ, μ, ν so that

$$a+b\cos x+c\sin x = \lambda+\mu(\alpha+\beta\cos x+\gamma\sin x)+\nu(-\beta\sin x+\gamma\cos x).$$

Then the integral is

$$\mu x+\nu\log|\alpha+\beta\cos x+\gamma\sin x|+\lambda\int\frac{dx}{\alpha+\beta\cos x+\gamma\sin x}.]$$

7. Integrate $1/(a\cos^2 x+2b\cos x\sin x+c\sin^2 x)$. [The subject of integration may be expressed in the form $1/(A+B\cos 2x+C\sin 2x)$, where $A=\frac{1}{2}(a+c)$, $B=\frac{1}{2}(a-c)$, $C=b$: but the integral may be calculated more simply by putting $\tan x=t$, when we obtain

$$\int\frac{\sec^2 x\,dx}{a+2b\tan x+c\tan^2 x}=\int\frac{dt}{a+2bt+ct^2}.]$$

147. Integrals involving arc sin x, arc tan x, **and** log x. The integrals of the inverse sine and tangent and of the logarithm can easily be calculated by integration by parts. Thus

$$\int\arcsin x\,dx = x\arcsin x-\int\frac{x\,dx}{\sqrt{(1-x^2)}} = x\arcsin x+\sqrt{(1-x^2)},$$

$$\int\arctan x\,dx = x\arctan x-\int\frac{x\,dx}{1+x^2} = x\arctan x-\tfrac{1}{2}\log(1+x^2),$$

$$\int\log x\,dx = x\log x-\int dx = x(\log x-1).$$

Generally, we can integrate $\phi(x)$, the inverse function of $f(x)$, if we can integrate $f(x)$; for the substitution $y=f(x)$ gives

$$\int\phi(y)\,dy=\int xf'(x)\,dx = xf(x)-\int f(x)\,dx.$$

Integrals of the form

$$\int P(x,\arcsin x)\,dx,\quad \int P(x,\log x)\,dx,$$

where P is a polynomial, can always be calculated. In the first case, for example, we have to calculate a number of integrals of the type $\int x^m(\arcsin x)^n\,dx$. Making the substitution $x = \sin y$, we obtain $\int y^n \sin^m y \cos y dy$, which can be found by the method of § 145. In the second case we have to calculate a number of integrals of the type $\int x^m(\log x)^n\,dx$. Integrating by parts we obtain

$$\int x^m(\log x)^n\,dx = \frac{x^{m+1}(\log x)^n}{m+1} - \frac{n}{m+1}\int x^m(\log x)^{n-1}\,dx,$$

and we can complete the calculation by repeating the argument.

Example. Integrate $x^n \log x$, $x^n \log(1+x)$, $x^8 \arctan x^3$, and $x^{-n} \log x$. (*Math. Trip.* 1924, 1929, 1934)

148. Areas of plane curves. One of the most important applications of the processes of integration which have been explained in the preceding sections is to the calculation of *areas* of plane curves. Suppose that P_0PP' (Fig. 42) is the graph of a continuous curve $y = \phi(x)$ which lies wholly above the axis of x, P being the point (x, y) and P' the point $(x+h, y+k)$, and h being either positive or negative (positive in the figure). The problem is that of calculating the *area* $ONPP_0$.

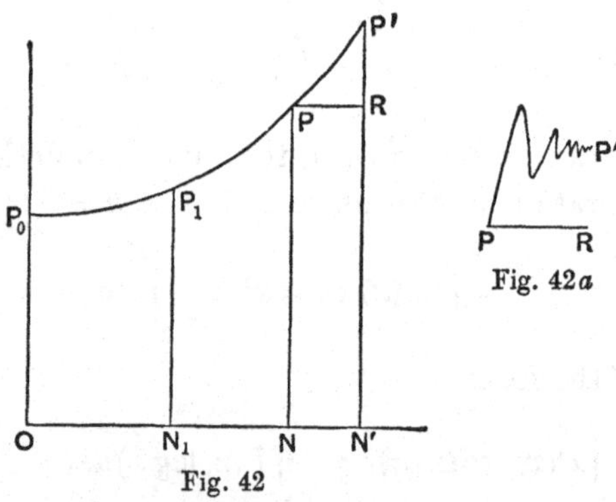

Fig. 42

Fig. 42a

The notion of an 'area' is one which requires very careful mathematical analysis, and we shall return to it in Ch. VII. For the present we shall take it for granted. We shall assume that any such region as $ONPP_0$ has associated with it a positive number $(ONPP_0)$, which we call its area, and that these areas possess the obvious properties indicated by common sense, e.g. that

$$(PRP') + (NN'RP) = (NN'P'P), \quad (N_1NPP_1) < (ONPP_0),$$

and so on.

It is plain, if we take all this for granted, that the area $ONPP_0$ is a function of x; we denote it by $\Phi(x)$. Also $\Phi(x)$ is a *continuous* function. For

$$\begin{aligned}\Phi(x+h) - \Phi(x) &= (NN'P'P) \\ &= (NN'RP) + (PRP') = h\phi(x) + (PRP').\end{aligned}$$

As the figure is drawn, the area PRP' is less than hk. This is not necessarily true in general, because it is not necessarily the case (see for example Fig. 42a) that the arc PP' should rise or fall steadily from P to P'. But the area PRP' is always less than $|h|\lambda(h)$, where $\lambda(h)$ is the greatest distance of any point of the arc PP' from PR. Moreover, since $\phi(x)$ is a continuous function, $\lambda(h) \to 0$ as $h \to 0$. Thus

$$\Phi(x+h) - \Phi(x) = h\{\phi(x) + \mu(h)\},$$

where $|\mu(h)| \leqq \lambda(h)$ and $\lambda(h) \to 0$ as $h \to 0$. From this it follows that $\Phi(x)$ is continuous. Moreover

$$\Phi'(x) = \lim_{h\to 0} \frac{\Phi(x+h) - \Phi(x)}{h} = \lim_{h\to 0} \{\phi(x) + \mu(h)\} = \phi(x).$$

Thus *the ordinate of the curve is the derivative of the area, and the area is the integral of the ordinate.*

We are thus able to formulate a rule for determining the area $ONPP_0$. *Calculate $\Phi(x)$, the integral of $\phi(x)$. This involves an arbitrary constant, which we suppose so chosen that $\Phi(0) = 0$. Then the area required is $\Phi(x)$.*

If it were the area N_1NPP_1 which was wanted, we should determine the constant so that $\Phi(x_1) = 0$, where x_1 is the abscissa of P_1. If the curve lay below the axis of x, $\Phi(x)$ would be negative, and the area would be the absolute value of $\Phi(x)$.

149. Lengths of plane curves. The notion of *length* also requires very careful analysis, and is rather more difficult than that of area. In fact the assumption that P_0P (Fig. 42) has a definite length, which we may denote by $S(x)$, does not suffice for our purposes, as did the corresponding assumption about areas. We cannot even prove that $S(x)$ is continuous, i.e. that $\lim\{S(P')-S(P)\}=0$. This looks obvious enough in the larger figure, but less so in such a case as is shown in the smaller figure. Indeed it is not possible to proceed further, with any degree of rigour, without a careful analysis of just what is meant by the length of a curve.

It is however easy to see what the *formula* must be. Let us suppose that the curve has a tangent whose direction varies continuously, so that $\phi'(x)$ is continuous. Then the assumption that the curve has a length leads to the equation

$$\frac{S(x+h)-S(x)}{h}=\frac{\{PP'\}}{h}=\frac{PP'}{h}\cdot\frac{\{PP'\}}{PP'},$$

where $\{PP'\}$ is the arc whose chord is PP'. Now

$$PP'=\sqrt{(PR^2+RP'^2)}=h\sqrt{\left(1+\frac{k^2}{h^2}\right)},$$

and
$$k=\phi(x+h)-\phi(x)=h\phi'(\xi),$$

where ξ lies between x and $x+h$. Hence

$$\lim(PP'/h)=\lim\sqrt{\{1+[\phi'(\xi)]^2\}}=\sqrt{\{1+[\phi'(x)]^2\}}.$$

If also we assume that

$$\lim\{PP'\}/PP'=1,$$

we obtain the result

$$S'(x)=\lim\frac{S(x+h)-S(x)}{h}=\sqrt{\{1+[\phi'(x)]^2\}},$$

and so
$$S(x)=\int\sqrt{\{1+[\phi'(x)]^2\}}\,dx.$$

Examples LIV. 1. Calculate the area of the segment cut off from the parabola $y=x^2/4a$ by the ordinate $x=\xi$, and the length of the arc which bounds it.

2. Show that the area of the ellipse $(x^2/a^2)+(y^2/b^2) = 1$ is πab.

3. The area included between the curve $x = y^2(1-x)$ and the line $x = 1$ is π. (*Math. Trip.* 1926)

4. Sketch the curve $(1+x^2)\,y^2 = x^2(1-x^2)$, and prove that the area of a loop is $\frac{1}{2}(\pi-2)$. (*Math. Trip.* 1934)

5. Trace the curve $a^4y^2 = x^5(2a-x)$, and show that its area is $\frac{5}{4}\pi a^2$. (*Math. Trip.* 1923)

6. Prove that the area between the curve

$$\left(\frac{x}{a}\right)^{\frac{2}{3}} + \frac{y}{b} = 1$$

and the segment $(-a, a)$ of the axis of x is $\frac{4}{5}ab$. (*Math. Trip.* 1930)

7. Find the area bounded by the curve $y = \sin x$ and the segment of the axis of x from $x = 0$ to $x = 2\pi$. [Here $\Phi(x) = -\cos x$, and the difference between the values of $-\cos x$ for $x = 0$ and $x = 2\pi$ is zero. The explanation of this is of course that between $x = \pi$ and $x = 2\pi$ the curve lies below the axis of x, and so the corresponding part of the area is counted negative in applying the method. The area from $x = 0$ to $x = \pi$ is $-\cos\pi+\cos 0 = 2$; and the whole area required, when every part is counted positive, is twice this, i.e. is 4.]

8. Suppose that the coordinates of any point on a curve are expressed as functions of a parameter t by equations of the type $x = \phi(t)$, $y = \psi(t)$, ϕ and ψ being functions of t with continuous derivatives. Prove that if x steadily increases as t varies from t_0 to t_1, then the area of the region bounded by the corresponding portion of the curve, the axis of x, and the two ordinates corresponding to t_0 and t_1, is, apart from sign, $A(t_1)-A(t_0)$, where

$$A(t) = \int \psi(t)\,\phi'(t)\,dt = \int y\frac{dx}{dt}\,dt.$$

9. Suppose that C is a closed curve formed of a single loop and not met by any parallel to either axis in more than two points. And suppose that the coordinates of any point P on the curve can be expressed as in Ex. 8 in terms of t, and that, as t varies from t_0 to t_1, P moves in the same direction round the curve and returns after a single circuit to its original position. Show that the area of the loop is equal to the difference of the initial and final values of any one of the integrals

$$-\int y\frac{dx}{dt}\,dt, \quad \int x\frac{dy}{dt}\,dt, \quad \tfrac{1}{2}\int\left(x\frac{dy}{dt}-y\frac{dx}{dt}\right)dt,$$

this difference being taken positively.

10. Apply the result of Ex. 9 to determine the areas of the curves given by

(i) $\dfrac{x}{a} = \dfrac{1-t^2}{1+t^2}, \quad \dfrac{y}{a} = \dfrac{2t}{1+t^2},$ (ii) $x = a\cos^3 t, \quad y = b\sin^3 t.$

11. Find the area of the loop of the curve $x^3+y^3 = 3axy$. [Putting $y = tx$, we obtain $x = 3at/(1+t^3)$, $y = 3at^2/(1+t^3)$. As t varies from 0 towards ∞ the loop is described once. Also

$$\tfrac{1}{2}\int\left(y\frac{dx}{dt}-x\frac{dy}{dt}\right)dt = -\tfrac{1}{2}\int x^2\frac{d}{dt}\left(\frac{y}{x}\right)dt = -\tfrac{1}{2}\int\frac{9a^2t^2}{(1+t^3)^2}\,dt = \frac{3a^2}{2(1+t^3)},$$

which tends to 0 as $t\to\infty$. Thus the area of the loop is $\frac{3}{2}a^2$.]

12. Find the area of the loop of the curve $x^5+y^5 = 5ax^2y^2$.

13. The area of the curve

$$x = a\cos t+b\sin t+c, \quad y = a'\cos t+b'\sin t+c',$$

where $ab'-a'b>0$, is $\pi(ab'-a'b)$. (*Math. Trip.* 1927)

14. Prove that the area of a loop of the curve $x = a\sin 2t$, $y = a\sin t$ is $\frac{4}{3}a^2$. (*Math. Trip.* 1908)

15. Trace the curve $x = \cos 2t, \quad y = \sin 3t$

and find the area of the loop. Obtain the Cartesian equation of the curve, and explain why the graph traced from this equation differs from the other graph. (*Math. Trip.* 1928)

[It is assumed, in the usual theory of curves defined by parametric equations, that $x'(t)$ and $y'(t)$ do not vanish simultaneously; and a value of t for which this happens corresponds to some sort of singularity of the curve. In this case $x'(t)$ and $y'(t)$ both vanish for $t = \pm\frac{1}{2}\pi$, when $x = -1$, $y = \mp 1$. If, for example, t increases from 0 to $\frac{1}{2}\pi$, (x, y) moves along the first graph from $(1, 0)$ to $(-1, -1)$, but then turns back and retraces its course.

The Cartesian equation is obtained by eliminating $\tau = \sin t$ from the equations $x = 1-2\tau^2$, $y = 3\tau-4\tau^3$; and only the part of the second graph for which $|\tau|\leq 1$ belongs to the first.]

16. The arc of the ellipse given by $x = a\sin t$, $y = b\cos t$, between the points $t = t_1$ and $t = t_2$, is $F(t_2)-F(t_1)$, where

$$F(t) = a\int\sqrt{(1-e^2\sin^2 t)}\,dt,$$

e being the eccentricity. [This integral cannot be evaluated in terms of such functions as are at present at our disposal.]

17. The coordinates of a point on a cycloid are given by

$$x = a(t+\sin t), \quad y = a(1+\cos t);$$

and the points for which $t = -\frac{1}{2}\pi$ and $t = \frac{1}{2}\pi$ are P and Q. Calculate the area between the arc PQ of the curve and the lines OP, OQ.

(*Math. Trip.* 1934)

18. **Polar coordinates.** Show that the area bounded by the curve $r = f(\theta)$, where $f(\theta)$ is a one-valued function of θ, and the radii $\theta = \theta_1$, $\theta = \theta_2$, is $F(\theta_2) - F(\theta_1)$, where $F(\theta) = \frac{1}{2}\int r^2\,d\theta$. And the length of the corresponding arc of the curve is $\Phi(\theta_2) - \Phi(\theta_1)$, where

$$\Phi(\theta) = \int \sqrt{\left\{r^2 + \left(\frac{dr}{d\theta}\right)^2\right\}}\,d\theta.$$

Hence determine (i) the area and the perimeter of the circle $r = 2a\sin\theta$; (ii) the area between the parabola $r = \frac{1}{2}l\sec^2\frac{1}{2}\theta$ and its latus rectum, and the length of the corresponding arc of the parabola; (iii) the area of the limaçon $r = a + b\cos\theta$, distinguishing the cases in which $a > b$, $a = b$, and $a < b$; and (iv) the areas of the ellipses $1/r^2 = a\cos^2\theta + 2h\cos\theta\sin\theta + b\sin^2\theta$ and $\dfrac{l}{r} = 1 + e\cos\theta$. [In the last case we are led to the integral $\displaystyle\int\frac{d\theta}{(1+e\cos\theta)^2}$, which may be calculated (cf. Ex. LIII. 4) by the help of the substitution

$$(1+e\cos\theta)(1-e\cos\phi) = 1-e^2.]$$

19. Trace the curve $2\theta = (a/r) + (r/a)$, and show that the area bounded by the radius vector $\theta = \beta$, and the two branches which touch at the point $r = a$, $\theta = 1$, is $\frac{2}{3}a^2(\beta^2-1)^{\frac{3}{2}}$. (*Math. Trip.* 1900)

20. Trace the curve whose equation is

$$r^2(a^2 + b^2\tan^2\tfrac{1}{2}\theta) = a^4,$$

where $a > b > 0$, and prove that its area is $\pi a^3/(a+b)$. (*Math. Trip.* 1932)

21. A curve is given by an equation $p = f(r)$, r being the radius vector and p the perpendicular from the origin on to the tangent. Show that the calculation of the area of the region bounded by an arc of the curve and two radii vectores depends upon that of the integral $\frac{1}{2}\displaystyle\int\frac{pr\,dr}{\sqrt{(r^2-p^2)}}$.

MISCELLANEOUS EXAMPLES ON CHAPTER VI

1. A function $f(x)$ is defined as being equal to $1+x$ when $x \leqq 0$, to x when $0 < x < 1$, to $2-x$ when $1 \leqq x \leqq 2$, and to $3x - x^2$ when $x > 2$. Discuss the continuity of $f(x)$ and the existence and continuity of $f'(x)$ for $x = 0$, $x = 1$, and $x = 2$. (*Math. Trip.* 1908)

2. Denoting a, $ax+b$, $ax^2+2bx+c$, ... by u_0, u_1, u_2, ..., show that $u_0^2u_3-3u_0u_1u_2+2u_1^3$ and $u_0u_4-4u_1u_3+3u_2^2$ are independent of x.

3. If $a_0, a_1, \dots, a_{2n}$ are constants and $U_r = (a_0, a_1, \dots, a_r \between x, 1)^r$, then

$$U_0U_{2n}-2nU_1U_{2n-1}+\frac{2n(2n-1)}{1.2}U_2U_{2n-2}-\dots+U_{2n}U_0$$

is independent of x. (*Math. Trip.* 1896)

[Differentiate and use the relation $U_r' = rU_{r-1}$.]

4. The first three derivatives of the function $\arcsin(\mu\sin x)-x$, where $\mu>1$, are positive when $0 \leqq x \leqq \frac{1}{2}\pi$.

5. The constituents of a determinant are functions of x. Show that its differential coefficient is the sum of the determinants formed by differentiating the constituents of one row only, leaving the rest unaltered.

6. If f_1, f_2, f_3, f_4 are polynomials of degree not greater than 4, then

$$\begin{vmatrix} f_1 & f_2 & f_3 & f_4 \\ f_1' & f_2' & f_3' & f_4' \\ f_1'' & f_2'' & f_3'' & f_4'' \\ f_1''' & f_2''' & f_3''' & f_4''' \end{vmatrix}$$

is also a polynomial of degree not greater than 4. [Differentiate five times, using the result of Ex. 5, and rejecting vanishing determinants.]

7. If $yz=1$ and $y_r = (1/r!)D_x^r y$, $z_s = (1/s!)D_x^s z$, then

$$\frac{1}{z^3}\begin{vmatrix} z & z_1 & z_2 \\ z_1 & z_2 & z_3 \\ z_2 & z_3 & z_4 \end{vmatrix} = \frac{1}{y^2}\begin{vmatrix} y_2 & y_3 \\ y_3 & y_4 \end{vmatrix}.$$

(*Math. Trip.* 1905)

8. If $W(y,z,u) = \begin{vmatrix} y & z & u \\ y' & z' & u' \\ y'' & z'' & u'' \end{vmatrix}$, dashes denoting differentiations with respect to x, then

$$W(y,z,u) = y^3 W\left(1, \frac{z}{y}, \frac{u}{y}\right).$$

9. If $$ax^2+2hxy+by^2+2gx+2fy+c=0,$$

then $$\frac{dy}{dx} = -\frac{ax+hy+g}{hx+by+f}, \quad \frac{d^2y}{dx^2} = \frac{abc+2fgh-af^2-bg^2-ch^2}{(hx+by+f)^3}.$$

10. If $y^3+3yx+2x^3=0$, then $x^2(1+x^3)y''-\frac{3}{2}xy'+y=0$.

(*Math. Trip.* 1903)

11. Verify that the differential equation $y = \phi\{\psi(y_1)\} + \phi\{x - \psi(y_1)\}$, where y_1 is the derivative of y, and ψ is the function inverse to ϕ', is satisfied by $y = \phi(c) + \phi(x-c)$ or by $y = 2\phi(\frac{1}{2}x)$.

12. Verify that the differential equation $y = \{x/\psi(y_1)\}\,\phi\{\psi(y_1)\}$, where the notation is the same as that of Ex. 11, is satisfied by $y = c\phi(x/c)$ or by $y = \beta x$, where $\beta = \phi(\alpha)/\alpha$ and α is any root of the equation

$$\phi(\alpha) - \alpha\phi'(\alpha) = 0.$$

13. If $ax+by+c = 0$, then $y_2 = 0$ (suffixes denoting differentiations with respect to x). We may express this by saying that *the general differential equation of all straight lines is* $y_2 = 0$. Find the general differential equations of (i) all circles with their centres on the axis of x, (ii) all parabolas with their axes along the axis of x, (iii) all parabolas with their axes parallel to the axis of y, (iv) all circles, (v) all parabolas, (vi) all conics.

[The equations are (i) $1+y_1^2+yy_2 = 0$, (ii) $y_1^2+yy_2 = 0$, (iii) $y_3 = 0$, (iv) $(1+y_1^2)\,y_3 = 3y_1y_2^2$, (v) $5y_3^2 = 3y_2y_4$, (vi) $9y_2^2y_5 - 45y_2y_3y_4 + 40y_3^3 = 0$. In each case we have only to write down the general equation of the curves in question, and differentiate until we have enough equations to eliminate all the arbitrary constants.]

14. Show that the general differential equations of all parabolas and of all conics are respectively

$$D_x^2(y_2^{-\frac{2}{3}}) = 0, \quad D_x^3(y_2^{-\frac{2}{3}}) = 0.$$

[The equation of a conic may be put in the form

$$y = ax+b \pm \surd(px^2+2qx+r).$$

From this we deduce

$$y_2 = \pm(pr-q^2)\,(px^2+2qx+r)^{-\frac{3}{2}}.$$

If the conic is a parabola, then $p = 0$.]

15. Denoting $\dfrac{dy}{dx}, \dfrac{1}{2!}\dfrac{d^2y}{dx^2}, \dfrac{1}{3!}\dfrac{d^3y}{dx^3}, \dfrac{1}{4!}\dfrac{d^4y}{dx^4}, \ldots$ by $t, a, b, c, \ldots$ and $\dfrac{dx}{dy}, \dfrac{1}{2!}\dfrac{d^2x}{dy^2}, \dfrac{1}{3!}\dfrac{d^3x}{dy^3}, \dfrac{1}{4!}\dfrac{d^4x}{dy^4}, \ldots$ by $\tau, \alpha, \beta, \gamma, \ldots$, show that

$$4ac - 5b^2 = (4\alpha\gamma - 5\beta^2)/\tau^8, \quad bt - a^2 = -(\beta\tau - \alpha^2)/\tau^6.$$

Establish similar formulae for the functions $a^2d - 3abc - 2b^3$, $(1+t^2)\,b - 2a^2t$, $2ct - 5ab$.

16. If $y = \cos(m \arcsin x)$, and y_n is the nth derivative of y, then

$$(1-x^2)\,y_{n+2} - (2n+1)\,xy_{n+1} + (m^2-n^2)\,y_n = 0.$$

(*Math. Trip.* 1930)

[Prove first when $n = 0$, and differentiate n times by Leibniz's theorem.]

17. Prove the formula

$$vD_x^n u = D_x^n(uv) - nD_x^{n-1}(uD_x v) + \frac{n(n-1)}{1.2} D_x^{n-2}(uD_x^2 v) - \dots,$$

where n is any positive integer. [Use the method of induction.]

18. Show that

$$\left(\frac{d}{dx}\right)^{2n} \frac{\sin x}{x} = \frac{2n!}{x^{2n+1}} \{S_{2n-1}(x)\cos x - C_{2n}(x)\sin x\},$$

where $C_{2n}(x)$ and $S_{2n-1}(x)$ are defined as in Ex. XLVI. 5.

(*Math. Trip.* 1936)

19. Prove that

$$\left(\frac{d}{dx}\right)^{2n} \cos^{2\nu} x = \frac{(-1)^n}{2^{2\nu-1}} \sum_{r=0}^{\nu-1} \binom{2\nu}{r} (2\nu-2r)^{2n} \cos 2(\nu-r)x.$$

(*Math. Trip.* 1928)

20. If $y = (1-x^2)^{-\frac{1}{2}} \arcsin x$, where $-1 < x < 1$ and $-\frac{1}{2}\pi < \arcsin x < \frac{1}{2}\pi$, then

$$(1-x^2)\, y_{n+1} - (2n+1)\, xy_n - n^2 y_{n-1} = 0,$$

suffixes denoting differentiations with respect to x. (*Math. Trip.* 1933)

21. If $y = (\arcsin x)^2$, then

$$(1-x^2)\, y_{n+1} - (2n-1)\, xy_n - (n-1)^2 y_{n-1} = 0.$$

Hence find the values of all the derivatives of y for $x = 0$.

(*Math. Trip.* 1930)

22. A curve is given by

$$x = a(2\cos t + \cos 2t), \quad y = a(2\sin t - \sin 2t).$$

Prove (i) that the equations of the tangent and normal, at the point P whose parameter is t, are

$$x\sin\tfrac{1}{2}t + y\cos\tfrac{1}{2}t = a\sin\tfrac{3}{2}t, \quad x\cos\tfrac{1}{2}t - y\sin\tfrac{1}{2}t = 3a\cos\tfrac{3}{2}t;$$

(ii) that the tangent at P meets the curve in the points Q, R whose parameters are $-\frac{1}{2}t$ and $\pi - \frac{1}{2}t$; (iii) that $QR = 4a$; (iv) that the tangents at Q and R are at right angles and intersect on the circle $x^2+y^2 = a^2$; (v) that the normals at P, Q, and R are concurrent and intersect on the circle $x^2+y^2 = 9a^2$; (vi) that the equation of the curve is

$$(x^2+y^2+12ax+9a^2)^2 = 4a(2x+3a)^3.$$

Sketch the form of the curve.

23. Show that the equations which define the curve of Ex. 22 may be replaced by $\xi/a = 2u+u^{-2}$, $\eta/a = 2u^{-1}+u^2$, where $\xi = x+yi$, $\eta = x-yi$, $u = \operatorname{Cis} t$. Show that the tangent and normal, at the point defined by u, are

$$u^2\xi - u\eta = a(u^3-1), \quad u^2\xi + u\eta = 3a(u^3+1),$$

and deduce the properties (ii)–(v) of Ex. 22.

24. Show that the condition that $x^4+4px^3-4qx-1 = 0$ should have equal roots may be expressed in the form $(p+q)^{\frac{2}{3}}-(p-q)^{\frac{2}{3}} = 1$.

(*Math. Trip.* 1898)

25. The roots of a cubic $f(x) = 0$ are α, β, γ in ascending order of magnitude. Show that if (α, β) and (β, γ) are each divided into six equal sub-intervals, then a root of $f'(x) = 0$ will fall in the fourth interval from β on each side. What will be the nature of the cubic in the two cases when a root of $f'(x) = 0$ falls at a point of division? (*Math. Trip.* 1907)

26. If $\phi(x)$ is a polynomial, and λ is real, then there is a root of

$$\phi'(x)+\lambda\phi(x) = 0$$

between any pair of roots of $\phi(x) = 0$.

[Argue as in Ex. XLI. 10.]

27. If α and β are successive roots of $\phi = 0$, then the number of roots of $\phi'+\lambda\phi = 0$ between α and β (each counted according to its multiplicity) is odd.

If the roots of $\phi = 0$ are all real, then those of $\phi'+\lambda\phi = 0$ are all real; and if the former are also all simple, so also are the latter.

(*Math. Trip.* 1933)

28. Deduce from Ex. 27 that

$$\left(\frac{d}{dx}\right)^n (x^2-1)^n$$

has n real simple zeros, all lying between -1 and 1. (*Math. Trip.* 1933)

29. Investigate the maxima and minima of $f(x)$, and the real roots of $f(x) = 0$, $f(x)$ being either of the functions

$$x-\sin x-\tan\alpha(1-\cos x), \quad x-\sin x-(\alpha-\sin\alpha)-\tan\tfrac{1}{2}\alpha(\cos\alpha-\cos x),$$

and α an angle between 0 and π. Show that in the first case the condition for a double root is that $\tan\alpha-\alpha$ should be a multiple of π.

30. Show that by choice of the ratio $\lambda:\mu$ we can make the roots of $\lambda(ax^2+bx+c)+\mu(a'x^2+b'x+c') = 0$ real and having a difference of any magnitude, unless the roots of the two quadratics are all real and interlace; and that in the excepted case the roots are always real, but there is a lower limit for the magnitude of their difference. (*Math. Trip.* 1895)

[Consider the form of the graph of the function

$$(ax^2+bx+c)/(a'x^2+b'x+c'):$$

cf. Exs. XLVI. 13 *et seq.*]

31. Prove that

$$\pi < \frac{\sin\pi x}{x(1-x)} \leqq 4$$

when $0<x<1$, and draw the graph of the function.

32. Draw the graph of the function

$$\pi \cot \pi x - \frac{1}{x} - \frac{1}{x-1}.$$

33. Sketch the general form of the graph of y, given that

$$\frac{dy}{dx} = \frac{(6x^2+x-1)(x-1)^2(x+1)^3}{x^2}. \quad (\textit{Math. Trip.}\ 1908)$$

34. A sheet of paper is folded over so that one corner just reaches the opposite side. Show how the paper must be folded to make the length of the crease a maximum.

35. The greatest acute angle at which the ellipse $(x^2/a^2)+(y^2/b^2) = 1$ can be cut by a concentric circle is $\arctan\{(a^2-b^2)/2ab\}$.

(*Math. Trip.* 1900)

36. In a triangle the area Δ and the semi-perimeter s are fixed. Show that any maximum or minimum of one of the sides is a root of the equation $s(x-s)x^2+4\Delta^2 = 0$. Discuss the reality of the roots of this equation, and whether they correspond to maxima or minima.

[The equations $a+b+c = 2s$, $s(s-a)(s-b)(s-c) = \Delta^2$ determine a and b as functions of c. Differentiate with respect to c, and suppose that $da/dc = 0$. It will be found that $b = c$, $s-b = s \quad c = \frac{1}{2}a$, from which we deduce that $s(a-s)a^2+4\Delta^2 = 0$.

This equation has three real roots if $s^4 > 27\Delta^2$, and one if $s^4 < 27\Delta^2$. In an equilateral triangle (the triangle of minimum perimeter for a given area) $s^4 = 27\Delta^2$; thus it is impossible that $s^4 < 27\Delta^2$. Hence the equation in a has three real roots, and, since their sum is positive and their product negative, two roots are positive and the third negative. Of the two positive roots one corresponds to a maximum and one to a minimum.]

37. The area of the greatest equilateral triangle which can be drawn with its sides passing through three given points A, B, C is

$$2\Delta + \frac{a^2+b^2+c^2}{2\sqrt{3}},$$

a, b, c being the sides and Δ the area of ABC. (*Math. Trip.* 1899)

38. If Δ, Δ' are the areas of the two maximum isosceles triangles which can be described with their vertices at the origin and their base angles on the cardioid $r = a(1+\cos\theta)$, then $256\Delta\Delta' = 25a^4\sqrt{5}$. (*Math. Trip.* 1907)

39. Find the limiting values which $(x^2-4y+8)/(y^2-6x+3)$ approaches as the point (x,y) on the curve $x^2y-4x^2-4xy+y^2+16x-2y-7 = 0$ approaches the position $(2, 3)$. (*Math. Trip.* 1903)

[If we take (2, 3) as a new origin, the equation of the curve becomes $\xi^2\eta-\xi^2+\eta^3=0$, and the function given becomes

$$(\xi^2+4\xi-4\eta)/(\eta^2+6\eta-6\xi).$$

If we put $\eta=t\xi$, we obtain $\xi=(1-t^2)/t$, $\eta=1-t^2$. The curve has a loop branching at the origin, which corresponds to the two values $t=-1$ and $t=1$. Expressing the given function in terms of t, and making t tend to -1 or 1, we obtain the limiting values $-\frac{3}{2}$, $-\frac{5}{8}$.]

40. If
$$f(x)=\frac{1}{\sin x-\sin a}-\frac{1}{(x-a)\cos a},$$
then
$$\frac{d}{da}\{\lim_{x\to a} f(x)\}-\lim_{x\to a} f'(x)=\tfrac{3}{4}\sec^3 a-\tfrac{5}{12}\sec a.$$

(*Math. Trip.* 1896)

41. Show that if $\phi(x)=1/(1+x^2)$ then $\phi^{(n)}(x)=Q_n(x)/(1+x^2)^{n+1}$, where $Q_n(x)$ is a polynomial of degree n. Show also that

(i) $Q_{n+1}=(1+x^2)Q_n'-2(n+1)xQ_n$,

(ii) $Q_{n+2}+2(n+2)xQ_{n+1}+(n+2)(n+1)(1+x^2)Q_n=0$,

(iii) $(1+x^2)Q_n''-2nxQ_n'+n(n+1)Q_n=0$,

(iv) $Q_n=(-1)^n n!\left\{(n+1)x^n-\dfrac{(n+1)n(n-1)}{3!}x^{n-2}+\dots\right\}$,

(v) all the roots of $Q_n=0$ are real and separated by those of $Q_{n-1}=0$.

42. If $f(x)$, $\phi(x)$, and $\psi(x)$ satisfy the conditions of §§ 126–8 concerning continuity and differentiability, then there is a value of ξ, lying between a and b, such that

$$\begin{vmatrix} f(a) & \phi(a) & \psi(a) \\ f(b) & \phi(b) & \psi(b) \\ f'(\xi) & \phi'(\xi) & \psi'(\xi) \end{vmatrix}=0.$$

[Consider the function formed by replacing the constituents of the third row by $f(x)$, $\phi(x)$, $\psi(x)$. This theorem reduces to the mean value theorem (§ 126) when $\phi(x)=x$ and $\psi(x)=1$.]

43. Deduce the theorem of § 128 from Ex. 42. [Take $\psi(x)=x$.]

44. If $\phi(x)$ and $\psi(x)$ satisfy the conditions of § 128, and $\phi'(x)$ never vanishes, then

$$\frac{\phi(\xi)-\phi(a)}{\psi(b)-\psi(\xi)}=\frac{\phi'(\xi)}{\psi'(\xi)}$$

for some ξ of (a, b).

(*Math. Trip.* 1928)

[Apply Rolle's theorem to $\{\phi(x)-\phi(a)\}\{\psi(b)-\psi(x)\}$.]

45. If $\phi(x)$ is continuous for $a \leqq x < b$, $\phi''(x)$ exists, and $\phi''(x) > 0$ for $a < x < b$, then

$$\frac{\phi(x)-\phi(a)}{x-a}$$

increases steadily and strictly for $a < x < b$. (*Math. Trip.* 1933)

46. The functions $f(x)$ and $g(x)$ are continuous for $0 \leqq x \leqq a$ and differentiable for $0 < x < a$; $f(0) = 0$, $g(0) = 0$; and $f'(x)$ and $g'(x)$ are positive. Prove that

(i) if $f'(x)$ increases with x, then $f(x)/x$ increases with x,

(ii) if $f'(x)/g'(x)$ increases with x, then $f(x)/g(x)$ increases with x.

Prove that the functions

$$\frac{x}{\sin x}, \quad \frac{\frac{1}{2}x^2}{1-\cos x}, \quad \frac{\frac{1}{6}x^3}{x-\sin x}, \ldots$$

increase steadily in the interval $0 < x < \frac{1}{2}\pi$. (*Math. Trip.* 1934)

[See Hardy, Littlewood, and Pólya, *Inequalities*, p. 106.]

47. A function $f(x)$ possesses a differential coefficient $f'(\xi)$ at $x = \xi$. Prove that

$$\phi(h,k) = \frac{f(\xi+h)-f(\xi-k)}{h+k} - f'(\xi)$$

tends to zero if h and k tend to zero simultaneously in any way through positive values.

Prove also that, if $f'(x)$ is continuous in an interval including ξ, then we can omit the word 'positive', and suppose only that $h+k \neq 0$.

Finally prove, by considering the function

$$f(0) = 0, \quad f(x) = 1\Big/\left[\frac{1}{x^2}\right] \qquad (x \neq 0),$$

that we cannot remove the restriction in the general case.

(*Math. Trip.* 1923)

[For the first part, use the identity

$$\phi(h,k) = \frac{h}{h+k}\left\{\frac{f(\xi+h)-f(\xi)}{h} - f'(\xi)\right\} + \frac{k}{h+k}\left\{\frac{f(\xi-k)-f(\xi)}{-k} - f'(\xi)\right\}$$

and the inequalities $h < h+k$, $k < h+k$. For the second, use the mean value theorem. For the third, take

$$\xi = 0, \quad h = \left(n - \frac{1}{n}\right)^{-\frac{1}{2}}, \quad k = -n^{-\frac{1}{2}},$$

where n is a positive integer.]

48. If $\phi'(x) \to a$ as $x \to \infty$, and $a \neq 0$, then $\phi(x) \sim ax$. If $a = 0$, then $\phi(x) = o(x)$. If $\phi'(x) \to \infty$, then $\phi(x) \to \infty$. [Use the mean value theorem.]

49. If $\phi(x) \to a$ as $x \to \infty$, then $\phi'(x)$ cannot tend to any limit other than zero.

50. If $\phi(x)+\phi'(x)\to a$ as $x\to\infty$, then $\phi(x)\to a$ and $\phi'(x)\to 0$.

[Let $\phi(x) = a+\psi(x)$, so that $\psi(x)+\psi'(x)\to 0$. If $\psi'(x)$ is of constant sign, say positive, for all sufficiently large values of x, then $\psi(x)$ steadily increases and must tend to a limit l or to ∞. If $\psi(x)\to\infty$ then $\psi'(x)\to-\infty$, which contradicts our hypothesis. If $\psi(x)\to l$ then $\psi'(x)\to -l$, and this is impossible (Ex. 49) unless $l=0$. Similarly we may dispose of the case in which $\psi'(x)$ is ultimately negative. If $\psi'(x)$ changes sign for values of x which surpass all limit, then these are the maxima and minima of $\psi(x)$. If x has a large value corresponding to a maximum or minimum of $\psi(x)$, then $\psi(x)+\psi'(x)$ is small and $\psi'(x)=0$, so that $\psi(x)$ is small. *A fortiori* the other values of $\psi(x)$ are small when x is large.]

51. Show how to reduce $\int R\left\{x, \sqrt{\left(\dfrac{ax+b}{mx+n}\right)}, \sqrt{\left(\dfrac{cx+d}{mx+n}\right)}\right\}dx$ to the integral of a rational function. [Put $mx+n = 1/t$ and use Ex. XLIX. 13.]

52. Calculate the integrals:

$$\int\sqrt{\left(\frac{x-1}{x+1}\right)}\frac{dx}{x},\quad \int\frac{x\,dx}{\sqrt{(1+x)}-\sqrt[3]{(1+x)}},\quad \int\sqrt{\left\{a^2+\sqrt{\left(b^2+\frac{c}{x}\right)}\right\}}\,dx,$$

$$\int\frac{5\cos x+6}{2\cos x+\sin x+3}\,dx,\quad \int\frac{dx}{(2-\sin^2 x)(2+\sin x-\sin^2 x)},\quad \int\operatorname{cosec} x\sqrt{(\sec 2x)}\,dx,$$

$$\int\frac{dx}{\sqrt{\{(1+\sin x)(2+\sin x)\}}},\quad \int\frac{x+\sin x}{1+\cos x}\,dx,\quad \int\operatorname{arc\,sec} x\,dx,\quad \int(\arcsin x)^2\,dx,$$

$$\int x\arcsin x\,dx,\quad \int\frac{x\arcsin x}{\sqrt{(1-x^2)}}\,dx,\quad \int\frac{\arctan x}{(1+x^2)^{\frac{3}{2}}}\,dx,\quad \int\frac{\log(\alpha^2+\beta^2x^2)}{x^2}\,dx.$$

53. Calculate
$$\int\frac{x-1}{x+1}\frac{dx}{\sqrt{\{x(x^2+x+1)\}}}$$
by means of the substitution $u^2 = x+1+x^{-1}$. (*Math. Trip.* 1931)

54. Prove that

$$\int\frac{dx}{x^{2n+1}\sqrt{(1-x^2)}} = \frac{1.3\ldots(2n-1)}{2.4\ldots 2n}\left[\log\frac{1-\sqrt{(1-x^2)}}{x}\right.$$
$$\left.-\left\{\frac{1}{x^2}+\frac{2}{3}\frac{1}{x^4}+\ldots+\frac{2.4\ldots(2n-2)}{3.5\ldots(2n-1)}\frac{1}{x^{2n}}\right\}\sqrt{(1-x^2)}\right],$$

$$\int\frac{dx}{x^{2n+2}\sqrt{(1-x^2)}} = -\frac{2.4\ldots 2n}{3.5\ldots(2n+1)}$$
$$\times\left\{\frac{1}{x}+\frac{1}{2}\frac{1}{x^3}+\ldots+\frac{1.3\ldots(2n-1)}{2.4\ldots 2n}\frac{1}{x^{2n+1}}\right\}\sqrt{(1-x^2)},$$

n being a positive integer. (*Math. Trip.* 1931)

55. Formulae of reduction. (i) Show that

$$2(n-1)(q-\tfrac{1}{4}p^2)\int\frac{dx}{(x^2+px+q)^n}$$
$$=\frac{x+\tfrac{1}{2}p}{(x^2+px+q)^{n-1}}+(2n-3)\int\frac{dx}{(x^2+px+q)^{n-1}}.$$

[Put $x+\tfrac{1}{2}p=t$, $q-\tfrac{1}{4}p^2=\lambda$: then we obtain

$$\int\frac{dt}{(t^2+\lambda)^n}=\frac{1}{\lambda}\int\frac{dt}{(t^2+\lambda)^{n-1}}-\frac{1}{\lambda}\int\frac{t^2\,dt}{(t^2+\lambda)^n}$$
$$=\frac{1}{\lambda}\int\frac{dt}{(t^2+\lambda)^{n-1}}+\frac{1}{2\lambda(n-1)}\int t\frac{d}{dt}\left\{\frac{1}{(t^2+\lambda)^{n-1}}\right\}dt,$$

and the result follows on integrating by parts.

A formula such as this is called a *formula of reduction*. It is most useful when n is a positive integer. We can then express $\int\frac{dx}{(x^2+px+q)^n}$ in terms of $\int\frac{dx}{(x^2+px+q)^{n-1}}$, and so evaluate the integral for every value of n in turn.]

(ii) Show that if $I_{p,q}=\int x^p(1+x)^q\,dx$ then

$$(p+1)I_{p,q}=x^{p+1}(1+x)^q-qI_{p+1,q-1},$$

and obtain a similar formula connecting $I_{p,q}$ with $I_{p-1,q+1}$. Show also, by means of the substitution $x=-y/(1+y)$, that

$$I_{p,q}=(-1)^{p+1}\int y^p(1+y)^{-p-q-2}\,dy.$$

(iii) If $u_n=\int\frac{dx}{(x^2+1)^n}$, then

$$(2n-2)u_n-(2n-3)u_{n-1}=x(x^2+1)^{-(n-1)}.$$

(*Math. Trip.* 1935)

(iv) If $I_{m,n}=\int\frac{x^m\,dx}{(x^2+1)^n}$, then

$$2(n-1)I_{m,n}=-x^{m-1}(x^2+1)^{-(n-1)}+(m-1)I_{m-2,n-1}.$$

(v) If $I_n=\int x^n\cos\beta x\,dx$ and $J_n=\int x^n\sin\beta x\,dx$, then

$$\beta I_n=x^n\sin\beta x-nJ_{n-1},\quad \beta J_n=-x^n\cos\beta x+nI_{n-1}.$$

(vi) If $I_n=\int\cos^n x\,dx$ and $J_n=\int\sin^n x\,dx$, then

$$nI_n=\sin x\cos^{n-1}x+(n-1)I_{n-2},\quad nJ_n=-\cos x\sin^{n-1}x+(n-1)J_{n-2}.$$

(vii) If $I_n=\int\tan^n x\,dx$, then $(n-1)(I_n+I_{n-2})=\tan^{n-1}x$.

(viii) If $I_{m,n}=\int \cos^m x \sin^n x\,dx$, then

$$(m+n)\,I_{m,n}=-\cos^{m+1}x\sin^{n-1}x+(n-1)\,I_{m,n-2}$$
$$=\cos^{m-1}x\sin^{n+1}x+(m-1)\,I_{m-2,n}.$$

[We have

$$(m+1)\,I_{m,n}=-\int \sin^{n-1}x\frac{d}{dx}(\cos^{m+1}x)\,dx$$
$$=-\cos^{m+1}x\sin^{n-1}x+(n-1)\int \cos^{m+2}x\sin^{n-2}x\,dx$$
$$=-\cos^{m+1}x\sin^{n-1}x+(n-1)(I_{m,n-2}-I_{m,n}),$$

which leads to the first reduction formula.]

(ix) Connect $I_{m,n}=\int \sin^m x \sin nx\,dx$ with $I_{m-2,n}$. (*Math. Trip.* 1897)

(x) If $I_{m,n}=\int x^m \operatorname{cosec}^n x\,dx$, then

$$(n-1)(n-2)\,I_{m,n}=(n-2)^2\,I_{m,n-2}+m(m-1)\,I_{m-2,n-2}$$
$$-x^{m-1}\operatorname{cosec}^{n-1}x\{m\sin x+(n-2)\,x\cos x\}.$$

(*Math. Trip.* 1896)

(xi) If $I_n=\int (a+b\cos x)^{-n}\,dx$, then

$$(n-1)(a^2-b^2)\,I_n=-b\sin x(a+b\cos x)^{-(n-1)}+(2n-3)\,aI_{n-1}-(n-2)I_{n-2}.$$

(xii) If $I_n=\int (a\cos^2 x+2h\cos x\sin x+b\sin^2 x)^{-n}\,dx$, then

$$4n(n+1)(ab-h^2)\,I_{n+2}-2n(2n+1)(a+b)\,I_{n+1}+4n^2 I_n=-\frac{d^2 I_n}{dx^2}.$$

(*Math. Trip.* 1898)

(xiii) If $I_{m,n}=\int x^m(\log x)^n\,dx$, then $(m+1)I_{m,n}=x^{m+1}(\log x)^n-nI_{m,n-1}$.

56. If n is a positive integer, then the value of $\int x^m(\log x)^n\,dx$ is

$$x^{m+1}\left\{\frac{(\log x)^n}{m+1}-\frac{n(\log x)^{n-1}}{(m+1)^2}+\frac{n(n-1)(\log x)^{n-2}}{(m+1)^3}-\dots+\frac{(-1)^n n!}{(m+1)^{n+1}}\right\}.$$

57. The area of the curve given by

$$x=\cos\phi+\frac{\sin\alpha\sin\phi}{1-\cos^2\alpha\sin^2\phi},\quad y=\sin\phi-\frac{\sin\alpha\cos\phi}{1-\cos^2\alpha\sin^2\phi},$$

where α is a positive acute angle, is $\frac{1}{2}\pi(1+\sin\alpha)^2/\sin\alpha$.

(*Math. Trip.* 1904)

58. The projection of a chord of a circle of radius a on a diameter is of constant length $2a\cos\beta$; show that the locus of the middle point of the chord consists of two loops, and that the area of either is $a^2(\beta - \cos\beta\sin\beta)$.

(*Math. Trip.* 1903)

59. Show that the length of a quadrant of the curve $(x/a)^{\frac{2}{3}} + (y/b)^{\frac{2}{3}} = 1$ is $(a^2+ab+b^2)/(a+b)$. (*Math. Trip.* 1911)

60. A point A is inside a circle of radius a, at a distance b from the centre. Show that the locus of the foot of the perpendicular drawn from A to a tangent to the circle encloses an area $\pi(a^2+\frac{1}{2}b^2)$.

(*Math. Trip.* 1909)

61. Prove that, if $(a, b, c, f, g, h)(x, y, 1)^2 = 0$ is the equation of a conic, then

$$\int \frac{dx}{(lx+my+n)(hx+by+f)} = \alpha \log \frac{PT}{PT'} + \beta,$$

where PT, PT' are the perpendiculars from the point P of the conic whose coordinates are x and y on to the tangents at the ends of the chord $lx+my+n = 0$, and α, β are constants. (*Math. Trip.* 1902)

62. Show that
$$\int \frac{ax^2+2bx+c}{(Ax^2+2Bx+C)^2}\,dx$$
will be a rational function of x if and only if one or other of $AC - B^2$ and $aC + cA - 2bB$ is zero*.

63. Show that a necessary and sufficient condition that

$$\int \frac{f(x)}{\{F(x)\}^2}\,dx,$$

where f and F are polynomials of which the latter has no repeated factor, should be a rational function of x, is that $f'F' - fF''$ should be divisible by F. (*Math. Trip.* 1910)

64. Show that
$$\int \frac{\alpha\cos x + \beta\sin x + \gamma}{(1-e\cos x)^2}\,dx$$
is a rational function of $\cos x$ and $\sin x$ if and only if $\alpha e + \gamma = 0$; and determine the integral when this condition is satisfied.

(*Math. Trip.* 1910)

* See the author's tract quoted on p. 254.

CHAPTER VII

ADDITIONAL THEOREMS IN THE DIFFERENTIAL AND INTEGRAL CALCULUS

150. Higher mean value theorems. We proved in § 126 that if $f(x)$ is continuous for $a \leqq x \leqq b$, and has a derivative $f'(x)$ for $a < x < b$, then

$$f(b) - f(a) = (b-a)f'(\xi),$$

where $a < \xi < b$; or that

$$f(a+h) - f(a) = hf'(a+\theta_1 h) \quad \text{..............(1)},$$

where $0 < \theta_1 < 1$.

We now impose further restrictions on $f(x)$. We suppose that $f'(x)$ is continuous for $a \leqq x \leqq b$, and that $f''(x)$ exists for $a < x < b$; and we consider the function

$$f(b) - f(x) - (b-x)f'(x) - \left(\frac{b-x}{b-a}\right)^2 \{f(b) - f(a) - (b-a)f'(a)\}.$$

This function vanishes when $x = a$ and when $x = b$, and its derivative is

$$\frac{2(b-x)}{(b-a)^2}\{f(b) - f(a) - (b-a)f'(a) - \tfrac{1}{2}(b-a)^2 f''(x)\};$$

and this derivative must vanish for some value of x between a and b. Hence there is a value ξ of x, between a and b, and therefore representable as $a + \theta_2(b-a)$, where $0 < \theta_2 < 1$, for which

$$f(b) = f(a) + (b-a)f'(a) + \tfrac{1}{2}(b-a)^2 f''(\xi).$$

If we put $b = a + h$, we obtain the equation

$$f(a+h) = f(a) + hf'(a) + \tfrac{1}{2}h^2 f''(a+\theta_2 h) \quad \text{.........(2)},$$

which is the standard form of what may be called *the mean value theorem of the second order*.

We have assumed about $f'(x)$ what we assumed about $\phi(x)$ in § 126, that is to say continuity in the closed and differentiability in the open interval (a, b). Incidentally we have assumed the existence of $f'(a)$ and $f'(b)$, and this assumption is, as it stands, one involving values of $f(x)$ for values of x outside (a, b), to the left of a and the right of b. It may happen in applications that $f(x)$ is not defined outside (a, b). We must then understand that $f'(a)$, for example, is defined with reference only to values of x in (a, b), i.e. that

$$f'(a) = \lim_{h \to +0} \frac{f(a+h) - f(a)}{h}.$$

This convention is parallel to that which we laid down concerning continuity at the end of § 99.

The same point arises, for higher derivatives, in the next theorem.

The analogy suggested by (1) and (2) leads us to formulate the following theorem.

Taylor's or the general mean value theorem. *If $f^{(n-1)}(x)$ is continuous for $a \leqq x \leqq b$, and $f^{(n)}(x)$ exists for $a < x < b$, then*

$$f(b) = f(a) + (b-a)f'(a) + \frac{(b-a)^2}{2!}f''(a) + \dots$$
$$+ \frac{(b-a)^{n-1}}{(n-1)!}f^{(n-1)}(a) + \frac{(b-a)^n}{n!}f^{(n)}(\xi),$$

where $a < \xi < b$; and if $b = a + h$, then

$$f(a+h) = f(a) + hf'(a) + \tfrac{1}{2}h^2 f''(a) + \dots$$
$$+ \frac{h^{n-1}}{(n-1)!}f^{(n-1)}(a) + \frac{h^n}{n!}f^{(n)}(a + \theta_n h),$$

where $0 < \theta_n < 1$.

The continuity of $f^{(n-1)}(x)$ naturally involves that of $f(x)$, $f'(x), \dots, f^{(n-2)}(x)$.

The proof proceeds on the same lines as in the special cases in which $n = 1$ and $n = 2$. We consider the function

$$F_n(x) - \left(\frac{b-x}{b-a}\right)^n F_n(a),$$

where

$$F_n(x) = f(b) - f(x) - (b-x)f'(x) - \dots - \frac{(b-x)^{n-1}}{(n-1)!}f^{(n-1)}(x).$$

This function vanishes for $x = a$ and $x = b$; its derivative is

$$\frac{n(b-x)^{n-1}}{(b-a)^n}\left\{F_n(a) - \frac{(b-a)^n}{n!}f^{(n)}(x)\right\};$$

and there must be some value of x between a and b for which the derivative vanishes. This leads at once to the result.

Examples LV. 1. Suppose that $f(x)$ is a polynomial of degree r. Then $f^{(n)}(x)$ is identically zero when $n > r$, and the theorem leads to the algebraical identity

$$f(a+h) = f(a) + hf'(a) + \frac{h^2}{2!}f''(a) + \dots + \frac{h^r}{r!}f^{(r)}(a).$$

2. By applying the theorem to $f(x) = 1/x$, and supposing x and $x+h$ positive, obtain the result

$$\frac{1}{x+h} = \frac{1}{x} - \frac{h}{x^2} + \frac{h^2}{x^3} - \dots + \frac{(-1)^{n-1}h^{n-1}}{x^n} + \frac{(-1)^n h^n}{(x+\theta_n h)^{n+1}}.$$

[Since $$\frac{1}{x+h} = \frac{1}{x} - \frac{h}{x^2} + \frac{h^2}{x^3} - \dots + \frac{(-1)^{n-1}h^{n-1}}{x^n} + \frac{(-1)^n h^n}{x^n(x+h)},$$

we can verify the result by showing that $x^n(x+h)$ can be put in the form $(x+\theta_n h)^{n+1}$, or that $x^n(x+h)$ lies between x^{n+1} and $(x+h)^{n+1}$.]

3. Obtain the formula

$$\sin(x+h) = \sin x + h\cos x - \frac{h^2}{2!}\sin x - \frac{h^3}{3!}\cos x + \dots$$
$$+ (-1)^{n-1}\frac{h^{2n-1}}{(2n-1)!}\cos x + (-1)^n\frac{h^{2n}}{2n!}\sin(x+\theta_{2n}h),$$

the corresponding formula for $\cos(x+h)$, and similar formulae involving powers of h extending up to h^{2n+1}.

4. Show that if m is a positive integer, and n a positive integer not greater than m, then

$$(x+h)^m = x^m + \binom{m}{1}x^{m-1}h + \dots + \binom{m}{n-1}x^{m-n+1}h^{n-1} + \binom{m}{n}(x+\theta_n h)^{m-n}h^n.$$

Show also that, if the interval $(x, x+h)$ does not include $x = 0$, the formula holds for all rational values of m and all positive integral values of n; and that, even if $x < 0 < x+h$ or $x+h < 0 < x$, the formula still holds if $m-n$ is positive.

5. The formula $f(x+h) = f(x) + hf'(x+\theta_1 h)$ is not true if $f(x) = 1/x$ and $x < 0 < x+h$. [For $f(x+h) - f(x) > 0$ and $hf'(x+\theta_1 h) = -h/(x+\theta_1 h)^2 < 0$; it is evident that the conditions for the truth of the mean value theorem are not satisfied.]

6. If $x = -a$, $h = 2a$, $f(x) = x^{\frac{1}{3}}$, then the equation

$$f(x+h) = f(x) + hf'(x+\theta_1 h)$$

is satisfied by $\theta_1 = \frac{1}{2} \pm \frac{1}{18}\sqrt{3}$. [This example shows that the result of the theorem may hold even if the conditions under which it was proved are not satisfied.]

7. **Newton's method of approximation to the roots of equations.** Let ξ be an approximation to a root of an algebraical equation $f(x) = 0$, the actual root being $\xi + h$. Then

$$0 = f(\xi+h) = f(\xi) + hf'(\xi) + \tfrac{1}{2}h^2 f''(\xi+\theta_2 h),$$

so that
$$h = -\frac{f(\xi)}{f'(\xi)} - \tfrac{1}{2}h^2 \frac{f''(\xi+\theta_2 h)}{f'(\xi)},$$

provided that $f'(\xi) \neq 0$.

If the root is a simple root, and h is sufficiently small, then there is a positive K such that $|f'(x)| > K$ for all the values of x which we are considering; and the root is

$$\xi + h = \xi - \frac{f(\xi)}{f'(\xi)} + O(h^2) = \xi_1 + O(h^2),$$

say. Thus ξ_1 is a better approximation to the root than ξ.

We can repeat the argument, taking ξ_1 in the place of ξ, and so obtain a series of still better approximations $\xi_2, \xi_3, \ldots$, with errors $O(h^4), O(h^8), \ldots$.

8. Apply this process to the equation $x^2 = 2$, taking $\xi = \frac{3}{2}$ as the first approximation. [We find $\xi_1 = \frac{17}{12} = 1{\cdot}417\ldots$, which is quite a good approximation, in spite of the roughness of the first. If now we repeat the process, we obtain $\xi_2 = \frac{577}{408} = 1{\cdot}414215\ldots$, which is correct to 5 places of decimals.]

9. By considering in this way the equation $x^2 - 1 - y = 0$, where y is small, show that

$$\sqrt{(1+y)} = 1 + \tfrac{1}{2}y - \frac{y^2}{4(2+y)} + O(y^4).$$

10. Show that the root of the equation in Ex. 7 is

$$\xi - \frac{f}{f'} - \frac{f^2 f''}{2f'^3} + O(|h|^3)$$

(the argument of every function being ξ).

11. The equation $\sin x = \alpha x$, where α is small, has a root nearly equal to π. Show that $(1-\alpha)\pi$ is a better approximation, and $(1-\alpha+\alpha^2)\pi$ a better still. [The method of Exs. 7–10 does not depend on $f(x) = 0$ being an algebraical equation, so long as f' and f'' are continuous and $f'(\xi) \neq 0$.]

12. Show that if $f^{(n+1)}(x)$ is continuous then the limit of the number θ_n which occurs in the general mean value theorem, when $h \to 0$, is $1/(n+1)$.

[For $f(x+h)$ is equal to each of

$$f(x)+\dots+\frac{h^n}{n!}f^{(n)}(x+\theta_n h),\quad f(x)+\dots+\frac{h^n}{n!}f^{(n)}(x)+\frac{h^{n+1}}{(n+1)!}f^{(n+1)}(x+\theta_{n+1}h),$$

where θ_{n+1} as well as θ_n lies between 0 and 1. Hence

$$f^{(n)}(x+\theta_n h)=f^{(n)}(x)+\frac{hf^{(n+1)}(x+\theta_{n+1}h)}{n+1}.$$

But if we apply the original mean value theorem to the function $f^{(n)}(x)$, taking $\theta_n h$ in place of h, we find

$$f^{(n)}(x+\theta_n h)=f^{(n)}(x)+\theta_n hf^{(n+1)}(x+\theta\theta_n h),$$

where θ also lies between 0 and 1. Hence

$$\theta_n f^{(n+1)}(x+\theta\theta_n h)=\frac{f^{(n+1)}(x+\theta_{n+1}h)}{n+1},$$

from which the result follows, since $f^{(n+1)}(x+\theta\theta_n h)$ and $f^{(n+1)}(x+\theta_{n+1}h)$ tend to the same limit $f^{(n+1)}(x)$ when $h\to 0$.]

151. Another form of Taylor's theorem. There is another form of Taylor's theorem in which we assume less than in § 150.

Suppose that $f(x)$ has n derivatives $f'(a)$, ..., $f^{(n)}(a)$ at $x=a$. The existence of $f^{(\nu)}(x)$ at any point involves the existence of $f^{(\nu-1)}(x)$ in an interval including the point, and its continuity at the point; so that the first $n-2$ derivatives are continuous in an interval including $x=a$, and the $(n-1)$th at $x=a$. But we do not assume even the existence of the nth derivative at any point except $x=a$.

Suppose first that $h \geqq 0$, and write

$$F_n(h)=f(a+h)-f(a)-hf'(a)-\dots-\frac{h^{n-1}}{(n-1)!}f^{(n-1)}(a).$$

Then $F_n(h)$ and its first $n-1$ derivatives vanish for $h=0$, and $F_n^{(n)}(0)=f^{(n)}(a)$. Hence, if we write

$$G(h)=F_n(h)-\frac{h^n}{n!}\{f^{(n)}(a)-\delta\},$$

where δ is positive, we have

$$G(0)=0,\quad G'(0)=0,\quad \dots,\quad G^{(n-1)}(0)=0,\quad G^{(n)}(0)=\delta>0.$$

It follows from the last two equations, and Theorem A of § 122, that $G^{(n-1)}(h)$ is increasing at $h=0$, and is positive for small positive h.

Next, $G^{(n-2)}(0) = 0$, and $G^{(n-1)}(h) > 0$ for small positive h; and therefore, by Cor. 1 of § 122, $G^{(n-2)}(h) > 0$ for small positive h.* Repeating the argument, we find successively that $G^{(n-3)}(h)$, $G^{(n-4)}(h)$, ..., and finally $G(h)$, are positive, i.e. that

$$F_n(h) > \frac{h^n}{n!}\{f^{(n)}(a) - \delta\}$$

for small positive h.

Similarly† we can prove that

$$F_n(h) < \frac{h^n}{n!}\{f^{(n)}(a) + \delta\}$$

for small positive h; and in these inequalities δ is an arbitrary positive number. It follows that

$$F_n(h) = \frac{h^n}{n!}\{f^{(n)}(a) + \eta\},$$

where $\eta \to 0$ when $h \to 0$ through positive values.

We can treat negative values of h similarly, and arrive at the following theorem.

If $f(x)$ has n derivatives at $x = a$, then

$$(1)\ \ f(a+h) = f(a) + hf'(a) + \dots + \frac{h^{n-1}}{(n-1)!}f^{(n-1)}(a) + \frac{h^n}{n!}\{f^{(n)}(a) + \eta\},$$

where $\eta \to 0$ with h.

We can also write (1), in the notation of § 98, as

$$(2)\ \ f(a+h) = f(a) + hf'(a) + \dots + \frac{h^{n-1}}{(n-1)!}f^{(n-1)}(a) + o(h^n).$$

We could also deduce this from the theorem of § 150, but only by assuming the continuity of $f^{(n)}(x)$ for $x = a$.

Examples LVI. 1. Show that if

$$a_0 + a_1 x + \dots + a_n x^n + o(x^n) = b_0 + b_1 x + \dots + b_n x^n + o(x^n)$$

when $x \to 0$, then $a_0 = b_0$, $a_1 = b_1$, ..., $a_n = b_n$.

[Making $x \to 0$, we see that $a_0 = b_0$. Now divide by x, and make $x \to 0$; repeating the process as often as is necessary.

* Or $G^{(n-2)}(h) = G^{(n-2)}(h) - G^{(n-2)}(0) = hG^{(n-1)}(\theta h) > 0$, by the mean value theorem.

† Changing the sign before δ in the definition of $G(h)$.

It follows that if $f(x)$ has n derivatives at $x = a$, and

$$f(a+h) = c_0 + c_1 h + \ldots + c_n h^n + o(h^n),$$

then $c_0, c_1, \ldots$ have the values in (2).]

2. Prove that
$$\frac{f(a+h)-f(a-h)}{2h} \to f'(a)$$
if the right-hand side exists.

3. Prove that
$$\frac{f(a+h)-2f(a)+f(a-h)}{h^2} \to f''(a)$$
if the right-hand side exists. (*Math. Trip.* 1925)

4. Prove that
$$\frac{3\sin 2\theta}{2(2+\cos 2\theta)} = \theta + \tfrac{4}{45}\theta^5 + o(\theta^5)$$
for small θ. (*Math. Trip.* 1935)

5. Prove that if $\sin x = xy^2$, and x and $y-1$ are small, then
$$y = 1 - \tfrac{1}{12}x^2 + \tfrac{1}{1440}x^4 + o(x^4), \quad x^2 = -12(y-1) + \tfrac{6}{5}(y-1)^2 + o\{(y-1)^2\}.$$
(*Math. Trip.* 1934)

152. Taylor's series. Suppose that $f(x)$ is a function which has differential coefficients of all orders in an interval $(a-\eta, a+\eta)$ surrounding the point $x = a$. Then, if h is numerically less than η, we have

$$f(a+h) = f(a) + hf'(a) + \ldots + \frac{h^{n-1}}{(n-1)!}f^{(n-1)}(a) + \frac{h^n}{n!}f^{(n)}(a+\theta_n h),$$

where $0 < \theta_n < 1$, for all values of n. Or, if

$$S_n = \sum_0^{n-1} \frac{h^\nu}{\nu!}f^{(\nu)}(a), \quad R_n = \frac{h^n}{n!}f^{(n)}(a+\theta_n h),$$

we have
$$f(a+h) - S_n = R_n.$$

Now let us suppose, in addition, that $R_n \to 0$ when $n \to \infty$. Then

$$f(a+h) = \lim_{n\to\infty} S_n = f(a) + hf'(a) + \frac{h^2}{2!}f''(a) + \ldots.$$

This expansion of $f(a+h)$ is known as *Taylor's series*. When $a = 0$ the formula reduces to

$$f(h) = f(0) + hf'(0) + \frac{h^2}{2!}f''(0) + \ldots,$$

which is known as *Maclaurin's series*. The function R_n is known as *Lagrange's form of the remainder*.

The reader should be careful to guard himself against supposing that the existence of all the derivatives of $f(x)$ is a sufficient condition for the validity of Taylor's series. A direct discussion of the behaviour of R_n is essential.

(1) **The cosine and sine series.** Let $f(x) = \sin x$. Then $f(x)$ has derivatives of all orders for all values of x. Also $|f^{(n)}(x)| \leqq 1$ for all values of x and n. Hence in this case $|R_n| \leqq h^n/n!$, which tends to zero as $n \to \infty$ (Ex. XXVII. 12) whatever the value of h. It follows that

$$\sin(x+h) = \sin x + h\cos x - \frac{h^2}{2!}\sin x - \frac{h^3}{3!}\cos x + \frac{h^4}{4!}\sin x + \ldots,$$

for all values of x and h. In particular

$$\sin h = h - \frac{h^3}{3!} + \frac{h^5}{5!} - \ldots,$$

for all values of h. Similarly we can prove that

$$\cos(x+h) = \cos x - h\sin x - \frac{h^2}{2!}\cos x + \frac{h^3}{3!}\sin x + \ldots, \quad \cos h = 1 - \frac{h^2}{2!} + \frac{h^4}{4!} - \ldots.$$

(2) **The binomial series.** Let $f(x) = (1+x)^m$, where m is any rational number, positive or negative. Then

$$f^{(n)}(x) = m(m-1)\ldots(m-n+1)(1+x)^{m-n},$$

and Maclaurin's series (with x for h) takes the form

$$(1+x)^m = 1 + \binom{m}{1}x + \binom{m}{2}x^2 + \ldots.$$

When m is a positive integer the series terminates, and we obtain the ordinary formula for the binomial theorem with a positive integral exponent. In the general case

$$R_n = \frac{x^n}{n!}f^{(n)}(\theta_n x) = \binom{m}{n}x^n(1+\theta_n x)^{m-n},$$

and in order to show that Maclaurin's series really represents $(1+x)^m$ for any range of values of x when m is not a positive integer, we must show that $R_n \to 0$ for every value of x in that range. This is in fact so if $-1 < x < 1$, and may be proved, when $0 \leqq x < 1$, by means of the expression given above for R_n, since $(1+\theta_n x)^{m-n} < 1$ if $n > m$, and $\binom{m}{n}x^n \to 0$ as $n \to \infty$ (Ex. XXVII. 13). But a difficulty arises if $-1 < x < 0$, since $1+\theta_n x < 1$ and $(1+\theta_n x)^{m-n} > 1$ if $n > m$; knowing only that $0 < \theta_n < 1$, we cannot be sure that $1+\theta_n x$ is not quite small and $(1+\theta_n x)^{m-n}$ quite large.

In fact, in order to prove the binomial theorem by means of Taylor's theorem, we need some different form for R_n, such as will be given later (§167).

153. Applications of Taylor's theorem. A. Maxima and minima. Taylor's theorem may be applied to give greater theoretical completeness to the tests of Ch. VI, §§ 123, 124, though the results are not of much practical importance. It will be remembered that, assuming that $\phi(x)$ has derivatives of the first two orders, we stated the following sufficient conditions for a maximum or minimum of $\phi(x)$ at $x = \xi$: *for a maximum*, $\phi'(\xi) = 0$, $\phi''(\xi) < 0$; *for a minimum*, $\phi'(\xi) = 0$, $\phi''(\xi) > 0$. It is evident that these tests fail if $\phi''(\xi)$ as well as $\phi'(\xi)$ is zero.

Let us suppose that $\phi(x)$ has n derivatives

$$\phi'(x), \quad \phi''(x), \quad \dots, \quad \phi^{(n)}(x),$$

and that all save the last vanish when $x = \xi$. Then

$$\phi(\xi+h) - \phi(\xi) = \frac{h^n}{n!}\phi^{(n)}(\xi) + o(h^n),$$

by (2) of § 151, and this has to be of constant sign for all small h, positive or negative. This evidently requires that n should be even; and if n is even there will be a maximum or a minimum according as $\phi^{(n)}(\xi)$ is negative or positive.

Thus we obtain the test: *if there is to be a maximum or minimum the first derivative which does not vanish must be an even derivative, and there will be a maximum if it is negative, a minimum if it is positive.*

Examples LVII. 1. Verify the result when $\phi(x) = (x-a)^m$, m being a positive integer, and $\xi = a$.

2. Test the function $(x-a)^m (x-b)^n$, where m and n are positive integers, for maxima and minima at the points $x = a$, $x = b$. Draw graphs of the different possible forms of the curve $y = (x-a)^m (x-b)^n$.

3. Test the functions $\sin x - x$, $\sin x - x + \dfrac{x^3}{3!}$, $\sin x - x + \dfrac{x^3}{3!} - \dfrac{x^5}{5!}$, ..., $\cos x - 1$, $\cos x - 1 + \dfrac{x^2}{2!}$, $\cos x - 1 + \dfrac{x^2}{2!} - \dfrac{x^4}{4!}$, ... for maxima or minima at $x = 0$.

154. B. The calculation of certain limits. It is often necessary to calculate the limiting values of ratios which assume the form '0/0' on substitution of particular values of a variable. We suppose that the value in question is $x = 0$. There are various methods.

(a) Suppose that $f(x)$ and $\phi(x)$ are differentiable at $x = 0$, and that $f(0) = \phi(0) = 0$, $\phi'(0) \neq 0$. Then

$$f(x) = xf'(0) + o(x), \quad \phi(x) = x\phi'(0) + o(x)$$

and therefore
$$\frac{f(x)}{\phi(x)} \to \frac{f'(0)}{\phi'(0)}.$$

More generally, if the functions have n derivatives at $x = 0$, and the first $n-1$ derivatives of each set vanish, while $\phi^{(n)}(0) \neq 0$, then, by the theorem of § 151,

$$f(x) = \frac{x^n}{n!} f^{(n)}(0) + o(x^n), \quad \phi(x) = \frac{x^n}{n!} \phi^{(n)}(0) + o(x^n),$$

and
$$\frac{f(x)}{\phi(x)} \to \frac{f^{(n)}(0)}{\phi^{(n)}(0)}.$$

(b) It is often better to use the theorem of § 128. If $f(x)$ and $\phi(x)$ are continuous for $0 \leqq x \leqq h$ and differentiable for $0 < x \leqq h$, $f(0) = 0$ and $\phi(0) = 0$, $\phi(h) \neq 0$, and $f'(x)$ and $\phi'(x)$ never vanish for the same x, then

$$\frac{f(h)}{\phi(h)} = \frac{f'(\xi)}{\phi'(\xi)} \tag{1}$$

for some ξ between 0 and h.

Suppose now that

$$\frac{f'(x)}{\phi'(x)} \to l \tag{2}$$

when $x \to 0$ through positive values. Then there is an interval $(0, k)$ inside which $\phi'(x)$ does not vanish.* It follows, by the theorem of § 129, that $\phi'(x)$ is of constant sign for $0 < x < k$; and therefore, by Cor. 2 of § 122, that $\phi(x)$ is of constant sign for $0 < x < k$. Hence (1) is true for every positive h less than k, and

$$\frac{f(h)}{\phi(h)} \to l.$$

* For otherwise the left-hand side of (2) would be meaningless for an infinity of small positive values of x.

That is to say

$$(3) \qquad \lim_{x\to+0}\frac{f(x)}{\phi(x)} = \lim_{x\to+0}\frac{f'(x)}{\phi'(x)}$$

whenever the second limit exists.

There are naturally similar theorems in which '$x\to+0$' is replaced by '$x\to-0$' or by '$x\to 0$'; and the argument may be repeated as often as is necessary. Thus

$$\lim_{x\to 0}\frac{f(x)}{\phi(x)} = \lim_{x\to 0}\frac{f^{(n)}(x)}{\phi^{(n)}(x)},$$

for any n, whenever $f^{(\nu)}(0)=0$ and $\phi^{(\nu)}(0)=0$ for $0 \leqq \nu < n$ and the limit on the right exists.

The same process shows that $f/\phi\to+\infty$ when $f'/\phi'\to+\infty$.

If we wish to deduce (3) from the mean value theorem of § 126, we must assume that $f'(x)$ and $\phi'(x)$ are continuous at $x=0$ (at any rate for approach to 0 from the right). Then

$$f(x) = xf'(\theta_1 x), \quad \phi(x) = x\phi'(\theta_2 x),$$

where θ_1 and θ_2 each lie between 0 and 1. The conclusion follows because $f'(\theta_1 x)\to f'(0)$ and $\phi'(\theta_2 x)\to\phi'(0)$.

The advantage of the procedure (2) is shown by Ex. LVIII. 3 below. Here

$$f(x) = \tan x - x, \quad \phi(x) = x - \sin x,$$

$$f(0) = f'(0) = f''(0) = 0, \ \phi(0) = \phi'(0) = \phi''(0) = 0, \ f'''(0) = 2, \ \phi'''(0) = 1,$$

and (*a*) gives the limit 2. This argument requires three differentiations of each function. But

$$\frac{f'(x)}{\phi'(x)} = \frac{\sec^2 x - 1}{1-\cos x} = \sec^2 x(1-\cos x)\to 2$$

and we can obtain the result much more quickly by method (*b*).

There are many variants of the theorems of this section. Thus x may tend to a or to infinity instead of to 0; and the meaningless form of f/ϕ may be '∞/∞' instead of '$0/0$'. Such variations may usually be reduced to the standard case by some simple substitution.

Examples LVIII. 1. If $f = x^2\sin\dfrac{1}{x}$, $\phi = x$, then $\dfrac{f}{\phi}\to 0$. Here

$$\frac{f'}{\phi'} = 2x\sin\frac{1}{x} - \cos\frac{1}{x}$$

which oscillates when $x\to 0$. Thus f/ϕ may tend to a limit when f'/ϕ' does not; our condition is sufficient but not necessary.

2. Find the limit of

$$\frac{x-(n+1)x^{n+1}+nx^{n+2}}{(1-x)^2},$$

as $x \to 1$.

3. Find the limits as $x \to 0$ of

$$\frac{\tan x - x}{x - \sin x}, \quad \frac{\tan nx - n\tan x}{n\sin x - \sin nx}.$$

4. $\dfrac{1-4\sin^2\frac{1}{6}\pi x}{1-x^2} \to \frac{1}{6}\pi\sqrt{3}$ when $x \to 1$. (*Math. Trip.* 1932)

5. Find the limit of $x\{\sqrt{(x^2+a^2)}-x\}$ as $x \to \infty$. [Put $x = 1/y$.]

6. Prove that

$$\lim_{x\to n}(x-n)\operatorname{cosec} x\pi = \frac{(-1)^n}{\pi}, \quad \lim_{x\to n}\frac{1}{x-n}\left\{\operatorname{cosec} x\pi - \frac{(-1)^n}{(x-n)\pi}\right\} = \frac{(-1)^n\pi}{6},$$

n being any integer; and evaluate the corresponding limits involving $\cot x\pi$.

7. Find the limits as $x \to 0$ of

$$\frac{1}{x^3}\left(\operatorname{cosec} x - \frac{1}{x} - \frac{x}{6}\right), \quad \frac{1}{x^3}\left(\cot x - \frac{1}{x} + \frac{x}{3}\right).$$

8. $(\sin x \arcsin x - x^2)/x^6 \to \frac{1}{18}$, $(\tan x \arctan x - x^2)/x^6 \to \frac{2}{9}$, when $x \to 0$.

155. C. The contact of plane curves. Two curves are said to *intersect* (or *cut*) at a point if the point lies on each of them. They are said to *touch* at the point if they have the same tangent at the point.

Let us suppose now that $f(x)$, $\phi(x)$ are two functions which possess derivatives of all orders at $x = \xi$, and let us consider the curves $y = f(x)$, $y = \phi(x)$. In general $f(\xi)$ and $\phi(\xi)$ will not be equal. In this case the abscissa $x = \xi$ does not correspond to a point of intersection of the curves. If however $f(\xi) = \phi(\xi)$, then the curves intersect in the point $x = \xi$, $y = f(\xi) = \phi(\xi)$. In order that the curves should touch at this point it is necessary and sufficient that the first derivatives $f'(x)$, $\phi'(x)$ should also have the same value when $x = \xi$.

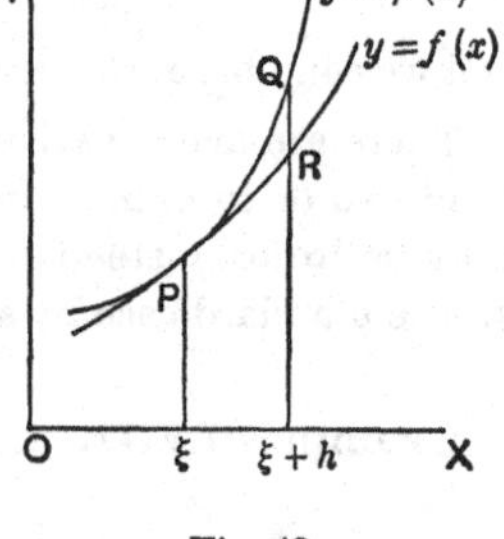

Fig. 43

The contact of the curves in this case may be regarded from a different point of view. In the figure the two curves are drawn touching at P, and QR is equal to $\phi(\xi+h)-f(\xi+h)$, or, since

$$\phi(\xi)=f(\xi), \quad \phi'(\xi)=f'(\xi),$$

to

$$\tfrac{1}{2}h^2\{\phi''(\xi+\theta h)-f''(\xi+\theta h)\},$$

where θ lies between 0 and 1. Hence

$$\lim\frac{QR}{h^2}=\tfrac{1}{2}\{\phi''(\xi)-f''(\xi)\},$$

when $h\to 0$. In other words, when the curves touch at the point whose abscissa is ξ, *the difference of their ordinates at the point whose abscissa is* $\xi+h$ *is at least of the second order of smallness in* h.

It is evident that the degree of smallness of QR may be taken as a kind of measure of the closeness of the contact of the curves. It is at once suggested that if the first $n-1$ derivatives of f and ϕ have equal values when $x=\xi$, then QR will be of the nth order of smallness; and the reader will have no difficulty in proving that this is so and that

$$\lim\frac{QR}{h^n}=\frac{1}{n!}\{\phi^{(n)}(\xi)-f^{(n)}(\xi)\}.$$

We are therefore led to frame the following definition:

Contact of the nth order. *If* $f(\xi)=\phi(\xi)$, $f'(\xi)=\phi'(\xi)$, ..., $f^{(n)}(\xi)=\phi^{(n)}(\xi)$, *but* $f^{(n+1)}(\xi)\neq\phi^{(n+1)}(\xi)$, *then the curves* $y=f(x)$, $y=\phi(x)$ *will be said to have contact of the nth order at the point whose abscissa is* ξ.

The preceding discussion makes the notion of contact of the nth order dependent on the choice of axes, and fails entirely when the tangent to the curves is parallel to the axis of y. We can deal with this case by taking y as the independent and x as the dependent variable. It is better, however, to consider x and y as functions of a parameter t. A good account of the theory will be found in Fowler's tract *The elementary differential geometry of plane curves*, or in de la Vallée Poussin's *Cours d'analyse*, 6th edition, vol. II, pp. 372 *et seq.*

Examples LIX. 1. Let $\phi(x) = ax+b$, so that $y = \phi(x)$ is a straight line. The conditions for contact at the point for which $x = \xi$ are $f(\xi) = a\xi+b$, $f'(\xi) = a$. If we determine a and b so as to satisfy these equations, we find $a = f'(\xi)$, $b = f(\xi)-\xi f'(\xi)$, and the equation of the tangent to $y = f(x)$ at the point $x = \xi$ is

$$y = xf'(\xi)+\{f(\xi)-\xi f'(\xi)\},$$

or $y-f(\xi) = (x-\xi)f'(\xi)$. Cf. Ex. XXXIX. 5.

2. The fact that the line is to have simple contact with the curve determines the line completely. In order that the tangent should have contact of the second order with the curve we must have $f''(\xi) = \phi''(\xi)$, i.e. $f''(\xi) = 0$. A point at which the tangent to a curve has contact of the second order is called a *point of inflexion*.

3. Find the points of inflexion on the graphs of the functions

$$3x^4-6x^3+1, \quad 2x/(1+x^2), \quad \sin x, \quad a\cos^2 x+b\sin^2 x, \quad \tan x, \quad \arctan x.$$

4. Show that the conic $ax^2+2hxy+by^2+2gx+2fy+c = 0$ cannot have a point of inflexion unless it is degenerate. [Here

$$ax+hy+g+(hx+by+f)\,y_1 = 0$$

and
$$a+2hy_1+by_1^2+(hx+by+f)\,y_2 = 0,$$

suffixes denoting differentiations. Thus at a point of inflexion

$$a+2hy_1+by_1^2 = 0,$$

or
$$a(hx+by+f)^2-2h(ax+hy+g)\,(hx+by+f)+b(ax+hy+g)^2 = 0,$$

or
$$(ab-h^2)\,\{ax^2+2hxy+by^2+2gx+2fy\}+af^2-2fgh+bg^2 = 0.$$

But this is inconsistent with the equation of the conic unless

$$af^2-2fgh+bg^2 = c(ab-h^2)$$

or $abc+2fgh-af^2-bg^2-ch^2 = 0$; and this is the condition that the conic should degenerate into two straight lines.]

5. The curve
$$y = \frac{ax^2+2bx+c}{\alpha x^2+2\beta x+\gamma}$$

has one or three points of inflexion according as the roots of

$$\alpha x^2+2\beta x+\gamma = 0$$

are real or complex.

[The equation of the curve can, by a change of origin (cf. Ex. XLVI. 15), be reduced to the form

$$\eta = \frac{\xi}{A\xi^2+2B\xi+C} = \frac{\xi}{A(\xi-p)\,(\xi-q)},$$

where p, q are real or conjugate. The condition for a point of inflexion will be found to be $\xi^3-3pq\xi+pq(p+q) = 0$, which has one or three real roots according as $\{pq(p-q)\}^2$ is positive or negative, i.e. according as p and q are real or conjugate.]

6. Show that when the curve of Ex. 5 has three points of inflexion, they lie on a straight line. [The equation $\xi^3-3pq\xi+pq(p+q)=0$ can be put in the form $(\xi-p)(\xi-q)(\xi+p+q)+(p-q)^2\xi=0$, so that the points of inflexion lie on the line $\xi+A(p-q)^2\eta+p+q=0$ or

$$A\xi-4(AC-B^2)\eta=2B.]$$

7. Find the inflexions of the curve

$$54y=(x+5)^2(x^3-10),$$

and draw a rough sketch of the curve for $-6<x<3$. (*Math. Trip.* 1936)

[See Ex. XLVI. 10.]

8. **Contact of a circle with a curve. Curvature*.** The circle

$$(x-a)^2+(y-b)^2=r^2 \quad \dots\dots\dots\dots\dots\dots(1)$$

will have second order contact with the curve $y=f(x)$ at (ξ,η) if y, y_1, and y_2 have the same values for the two curves when $x=\xi$.

Differentiating (1) twice, and putting $x=\xi$, we obtain

$$(\xi-a)^2+(\eta-b)^2=r^2,\quad (\xi-a)+(\eta-b)\eta_1=0,\quad 1+\eta_1^2+(\eta-b)\eta_2=0,$$

where η, η_1, η_2 mean $f(\xi)$, $f'(\xi)$, $f''(\xi)$. These equations give

$$a=\xi-\frac{\eta_1(1+\eta_1^2)}{\eta_2},\quad b=\eta+\frac{1+\eta_1^2}{\eta_2},\quad r=\frac{(1+\eta_1^2)^{\frac{3}{2}}}{\eta_2}.$$

The circle which has contact of the second order with the curve at the point (ξ,η) is called the *circle of curvature*, and its radius the *radius of curvature*. The *measure of curvature* (or simply the *curvature*) is the reciprocal of the radius: thus the measure of curvature is $\eta_2(1+\eta_1^2)^{-\frac{3}{2}}$.

9. Verify that the curvature of a circle is constant and equal to the reciprocal of the radius; and show that the circle is the only curve whose curvature is constant.

10. Find the centre and radius of curvature at any point of the conics $y^2=4ax$, $(x/a)^2+(y/b)^2=1$.

11. Show that in general one conic can be drawn to have contact of the fourth order with the curve $y=f(x)$ at a given point P.

12. An infinity of conics can be drawn having contact of the third order with the curve at P. Show that their centres all lie on a straight line.

* A much fuller discussion of the theory of curvature will be found in Fowler's tract referred to on p. 297.

[Take the tangent and normal as axes. Then the equation of the conic is of the form $2y = ax^2 + 2hxy + by^2$, and when x is small one value of y may be expressed (Ch. V, Misc. Ex. 24) in the form

$$y = \tfrac{1}{2}ax^2 + \tfrac{1}{2}ahx^3 + o(x^3).$$

This expression must be the same as

$$y = \tfrac{1}{2}f''(0)\,x^2 + \tfrac{1}{6}f'''(0)\,x^3 + o\,(x^3);$$

and so $a = f''(0)$, $h = f'''(0)/3f''(0)$ (Ex. LVI. 1). But the centre lies on the line $ax + hy = 0$.]

13. The locus of the centres of conics which have contact of the third order with the ellipse $(x/a)^2 + (y/b)^2 = 1$ at the point $(a\cos\alpha, b\sin\alpha)$ is the diameter $x/(a\cos\alpha) = y/(b\sin\alpha)$. [For the ellipse itself is one such conic.]

156. Differentiation of functions of several variables. So far we have been concerned exclusively with functions of a single variable x, but there is nothing to prevent us applying the notion of differentiation to functions of several variables $x, y, \ldots$.

Suppose then that $f(x, y)$ is a function of two* real variables x and y, and that the limits

$$\lim_{h\to 0}\frac{f(x+h, y) - f(x, y)}{h}, \quad \lim_{k\to 0}\frac{f(x, y+k) - f(x, y)}{k}$$

exist for all values of x and y in question, that is to say that $f(x, y)$ possesses a derivative df/dx or $D_x f(x, y)$ with respect to x and a derivative df/dy or $D_y f(x, y)$ with respect to y. It is usual to call these derivatives the *partial differential coefficients* of f, and to denote them by

$$\frac{\partial f}{\partial x}, \quad \frac{\partial f}{\partial y}$$

or

$$f'_x(x, y), \quad f'_y(x, y)$$

or simply f'_x, f'_y or f_x, f_y. The reader must not suppose, however, that these new notations imply any essential novelty of idea: 'partial differentiation' with respect to x is exactly the same process as ordinary differentiation, the only novelty lying in the presence in f of a second variable y independent of x.

* The new points which arise when we consider functions of several variables are illustrated sufficiently when there are two variables only. We take for granted the obvious generalisations of our theorems to three or more variables.

Our definitions presuppose the independence of x and y. If x and y are related, y is a function $\phi(x)$ of x, and

$$f(x, y) = f\{x, \phi(x)\}$$

is a function of the single variable x. And if $x = \phi(t)$, $y = \psi(t)$, then $f(x, y)$ is a function of t.

Examples LX. 1. Prove that if $x = r\cos\theta$, $y = r\sin\theta$, so that $r = \sqrt{(x^2+y^2)}$, $\theta = \arctan(y/x)$, then

$$\frac{\partial r}{\partial x} = \frac{x}{\sqrt{(x^2+y^2)}}, \quad \frac{\partial r}{\partial y} = \frac{y}{\sqrt{(x^2+y^2)}}, \quad \frac{\partial \theta}{\partial x} = -\frac{y}{x^2+y^2}, \quad \frac{\partial \theta}{\partial y} = \frac{x}{x^2+y^2},$$

$$\frac{\partial x}{\partial r} = \cos\theta, \quad \frac{\partial y}{\partial r} = \sin\theta, \quad \frac{\partial x}{\partial \theta} = -r\sin\theta, \quad \frac{\partial y}{\partial \theta} = r\cos\theta.$$

2. Account for the fact that $\dfrac{\partial r}{\partial x} \neq 1\Big/\left(\dfrac{\partial x}{\partial r}\right)$ and $\dfrac{\partial \theta}{\partial x} \neq 1\Big/\left(\dfrac{\partial x}{\partial \theta}\right)$. [When we were considering a function y of one variable x it followed from the definitions that dy/dx and dx/dy were reciprocals. This is no longer so when we are dealing with functions of two variables. Let P (Fig. 44) be the point (x, y) or (r, θ). To find $\partial r/\partial x$ we must increase x, say by an increment $MM_1 = \delta x$, while keeping y constant. This brings P to P_1. If we take $OP' = OP$ along OP_1, the increment of r is $P'P_1 = \delta r$, say; and $\partial r/\partial x = \lim(\delta r/\delta x)$. If on the other hand we want to calculate $\partial x/\partial r$, x and y being now regarded as functions of r and θ, we must increase r by Δr, say, keeping θ constant. Let us suppose that this brings P to P_2, and write $PP_2 = \Delta r$. The corresponding increment of x is $MM_1 = \Delta x$, say, and

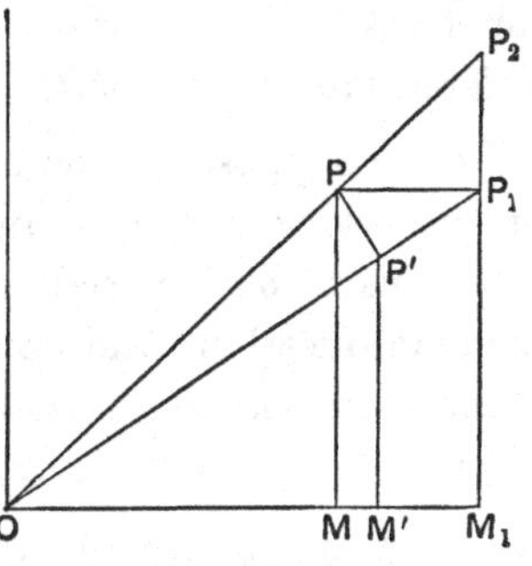

Fig. 44

$$\partial x/\partial r = \lim(\Delta x/\Delta r).$$

Now $\Delta x = \delta x$*: but $\Delta r \neq \delta r$. Indeed it is easy to see from the figure that

$$\lim(\delta r/\delta x) = \lim(P'P_1/PP_1) = \cos\theta,$$

but $$\lim(\Delta r/\Delta x) = \lim(PP_2/PP_1) = \sec\theta,$$

so that $$\lim(\delta r/\Delta r) = \cos^2\theta.]$$

3. Prove that if $z = f(ax+by)$ then $b\dfrac{\partial z}{\partial x} = a\dfrac{\partial z}{\partial y}$.

* Of course the fact that $\Delta x = \delta x$ is due merely to the particular value of Δr that we have chosen (viz. PP_2). Any other choice would give us values of Δx, Δr proportional to those used here.

4. Find $X_x, X_y, \dots$ when $X+Y=x$, $Y=xy$. Express x, y as functions of X, Y and find $x_X, x_Y, \dots$.

5. Find $X_x, \dots$ when $X+Y+Z=x$, $Y+Z=xy$, $Z=xyz$; express x, y, z in terms of X, Y, Z and find $x_X, \dots$.

[There is no difficulty in extending the ideas of the last section to functions of any number of variables. But the reader must be careful to impress on his mind that the notion of the partial derivative of a function of several variables is only determinate when *all* the independent variables are specified. Thus if $u=x+y+z$, x, y, and z being the independent variables, then $u_x=1$. But if we regard u as a function of the variables x, $x+y=\eta$, and $x+y+z=\zeta$, so that $u=\zeta$, then $u_x=0$.]

157. Differentiation of a function of two functions. There is a theorem concerning the differentiation of a function of one variable, known generally as the *theorem of the total differential coefficient*, which is very important and depends on the notions explained in the preceding section regarding functions of two variables. This theorem gives us a rule for differentiating $f\{\phi(t), \psi(t)\}$, with respect to t.

Let us suppose, in the first instance, that $f(x,y)$ is a function of the two variables x and y, and that f'_x, f'_y are continuous functions of both variables (§ 108) for all of their values which come in question. And now let us suppose that the variation of x and y is restricted in that (x,y) lies on a curve

$$x=\phi(t), \quad y=\psi(t),$$

where ϕ and ψ are functions of t with continuous differential coefficients $\phi'(t)$, $\psi'(t)$. Then $f(x,y)$ reduces to a function of the single variable t, say $F(t)$. The problem is to determine $F'(t)$.

Suppose that, when t changes to $t+\tau$, x and y change to $x+\xi$ and $y+\eta$. Then by definition

$$\begin{aligned}\frac{dF(t)}{dt} &= \lim_{\tau\to 0}\frac{1}{\tau}[f\{\phi(t+\tau), \psi(t+\tau)\} - f\{\phi(t), \psi(t)\}] \\ &= \lim\frac{1}{\tau}\{f(x+\xi, y+\eta) - f(x,y)\} \\ &= \lim\left[\frac{f(x+\xi, y+\eta) - f(x, y+\eta)}{\xi}\frac{\xi}{\tau} + \frac{f(x, y+\eta) - f(x,y)}{\eta}\frac{\eta}{\tau}\right].\end{aligned}$$

But, by the mean value theorem,

$$\frac{f(x+\xi, y+\eta)-f(x, y+\eta)}{\xi} = f'_x(x+\theta\xi, y+\eta),$$

$$\frac{f(x, y+\eta)-f(x, y)}{\eta} = f'_y(x, y+\theta'\eta),$$

where θ and θ' each lie between 0 and 1. When $\tau \to 0$, $\xi \to 0$ and $\eta \to 0$, and $\xi/\tau \to \phi'(t)$, $\eta/\tau \to \psi'(t)$: also

$$f'_x(x+\theta\xi, y+\eta) \to f'_x(x, y), \quad f'_y(x, y+\theta'\eta) \to f'_y(x, y).$$

Hence

$$F'(t) = D_t f\{\phi(t), \psi(t)\} = f'_x(x, y)\,\phi'(t) + f'_y(x, y)\,\psi'(t),$$

where we are to put $x = \phi(t)$, $y = \psi(t)$ after carrying out the differentiations with respect to x and y. This result may also be expressed in the form

$$\frac{df}{dt} = \frac{\partial f}{\partial x}\frac{dx}{dt} + \frac{\partial f}{\partial y}\frac{dy}{dt}.$$

Examples LXI. 1. Suppose

$$\phi(t) = \frac{1-t^2}{1+t^2}, \quad \psi(t) = \frac{2t}{1+t^2},$$

so that the locus of (x, y) is the circle $x^2+y^2 = 1$. Then

$$F'(t) = -\frac{4t}{(1+t^2)^2} f'_x + \frac{2(1-t^2)}{(1+t^2)^2} f'_y,$$

where x and y are to be put equal to $(1-t^2)/(1+t^2)$ and $2t/(1+t^2)$ after carrying out the differentiations.

It is instructive to verify this formula in particular cases. Suppose, e.g., that $f(x, y) = x^2+y^2$. Then $f'_x = 2x$, $f'_y = 2y$, and

$$F'(t) = 2x\phi'(t) + 2y\psi'(t) = 0,$$

which is correct because $F(t) = 1$.

2. Verify the theorem in the same way when (*a*) $x = t^m$, $y = 1-t^m$, $f(x, y) = x+y$; (*b*) $x = a\cos t$, $y = a\sin t$, $f(x, y) = x^2+y^2$.

3. One of the most important cases is that in which t is x itself. We then obtain

$$D_x f\{x, \psi(x)\} = D_x f(x, y) + D_y f(x, y)\,\psi'(x),$$

where y is to be replaced by $\psi(x)$ after differentiation.

It was this case which led to the introduction of the notation $\partial f/\partial x$, $\partial f/\partial y$. For it would seem natural to use the notation df/dx for *either* of the functions $D_x f\{x, \psi(x)\}$ and $D_x f(x, y)$, in one of which y is put equal to $\psi(x)$ before and in the other after differentiation. Suppose for example that $y = 1-x$ and $f(x,y) = x+y$. Then $D_x f(x, 1-x) = D_x 1 = 0$, but $D_x f(x,y) = 1$.

The distinction between the two functions is adequately shown by denoting the first by df/dx and the second by $\partial f/\partial x$, in which case the theorem takes the form

$$\frac{df}{dx} = \frac{\partial f}{\partial x} + \frac{\partial f}{\partial y}\frac{dy}{dx};$$

though this notation is also open to objection, in that it is a little misleading to denote the functions $f\{x, \psi(x)\}$ and $f(x,y)$, whose forms as functions of x are quite different from one another, by the same letter f in df/dx and $\partial f/\partial x$.

4. If the result of eliminating t between $x = \phi(t)$, $y = \psi(t)$ is $f(x,y) = 0$, then

$$\frac{\partial f}{\partial x}\frac{dx}{dt} + \frac{\partial f}{\partial y}\frac{dy}{dt} = 0.$$

5. If x and y are functions of t, and r and θ are the polar coordinates of (x, y), then $r' = (xx'+yy')/r$, $\theta' = (xy'-yx')/r^2$, dashes denoting differentiations with respect to t.

158. We have assumed that f'_x and f'_y are continuous functions of the two variables x and y in the sense of § 108. It is not sufficient to assume merely that they exist for all x and y.

In fact we can infer very little from the mere existence of f'_x and f'_y; we cannot even infer that f is continuous. Consider, for instance, the function used as an example in § 108, and defined by

$$f(x,y) = \frac{2xy}{x^2+y^2}$$

when $x \neq 0$, $y \neq 0$, and by $f = 0$ when one at least of x and y is zero. Then

$$f'_x(x,y) = -\frac{2y(x^2-y^2)}{(x^2+y^2)^2}, \quad f'_y(x,y) = \frac{2x(x^2-y^2)}{(x^2+y^2)^2}$$

at all points except the origin. Also

$$f'_x(0,0) = \lim_{h\to 0}\frac{f(h,0)-f(0,0)}{h} = \lim_{h\to 0}\frac{0}{h} = 0;$$

and similarly $f'_y(0,0) = 0$. Thus f'_x and f'_y exist for all x, y; but (as we saw in § 108) f is discontinuous at the origin.

The function defined by

$$f(x,y) = \frac{2xy}{x^2+y^2}(x+y)$$

when $x \neq 0$, $y \neq 0$, and by $f = 0$ when $x = 0$ or $y = 0$, is continuous everywhere, including the origin; and in this case also

$$f'_x(0,0) = f'_y(0,0) = 0.$$

Suppose now that $x = y = t$. Then $F(t) = f(t,t) = 2t$, and $F'(0) = 2$; but

$$f'_x \frac{dx}{dt} + f'_y \frac{dy}{dt} = 0.1 + 0.1 = 0$$

when $t = 0$, so that the result of the last section is false.

In what follows we shall assume the continuity of all derivatives which occur.

159. The mean value theorem for functions of two variables. Many of the results of the last chapter depended upon the mean value theorem

$$f(x+h) - f(x) = hf'(x+\theta h).$$

We may write this as

$$\delta y = f'(x + \theta\,\delta x)\,\delta x,$$

where $y = f(x)$. We shall now suppose that $z = f(x,y)$ is a function of the two independent variables x and y, and that x and y receive increments h, k or δx, δy respectively; and express the corresponding increment of z, viz.

$$\delta z = f(x+h, y+k) - f(x,y),$$

in terms of h, k and the derivatives of z with respect to x and y.

Let $f(x+ht, y+kt) = F(t)$. Then

$$f(x+h, y+k) - f(x,y) = F(1) - F(0) = F'(\theta),$$

where $0 < \theta < 1$. But, by § 157,

$$\begin{aligned} F'(t) &= D_t f(x+ht, y+kt) \\ &= hf'_x(x+ht, y+kt) + kf'_y(x+ht, y+kt). \end{aligned}$$

Hence finally

$$\delta z = f(x+h, y+k) - f(x,y) = hf'_x(x+\theta h, y+\theta k) + kf'_y(x+\theta h, y+\theta k),$$

which is the formula desired. Since f'_x, f'_y are continuous functions of x and y, we have

$$\begin{aligned} f'_x(x+\theta h, y+\theta k) &= f'_x(x,y) + \epsilon_{h,k}, \\ f'_y(x+\theta h, y+\theta k) &= f'_y(x,y) + \eta_{h,k}, \end{aligned}$$

where $\epsilon_{h,k}$ and $\eta_{h,k}$ tend to zero as h and k tend to zero. Hence the theorem may be written in the form

$$\delta z = (f'_x + \epsilon)\,\delta x + (f'_y + \eta)\,\delta y \quad \ldots\ldots\ldots\ldots\ldots(1),$$

where ϵ and η are small when δx and δy are small.

The result embodied in (1) may be expressed by saying that the equation

$$\delta z = f'_x\,\delta x + f'_y\,\delta y$$

is *approximately* true; i.e. that the difference between the two sides of the equation is small in comparison with the larger of δx and δy*. We must say '*the larger of δx and δy*' because one of them might be small in comparison with the other; we might indeed have $\delta x = 0$ or $\delta y = 0$.

If any equation of the form $\delta z = \lambda\,\delta x + \mu\,\delta y$ is 'approximately true', then $\lambda = f'_x$, $\mu = f'_y$. For

$$\delta z - f'_x\,\delta x - f'_y\,\delta y = \epsilon\,\delta x + \eta\,\delta y, \quad \delta z - \lambda\,\delta x - \mu\,\delta y = \epsilon'\delta x + \eta'\delta y,$$

where ϵ, η, ϵ', η' all tend to zero as δx and δy tend to zero; and so

$$(\lambda - f'_x)\,\delta x + (\mu - f'_y)\,\delta y = \rho\,\delta x + \sigma\delta y,$$

where ρ and σ tend to zero. Hence, if ζ is any assigned positive number, we can choose ω so that

$$|(\lambda - f'_x)\,\delta x + (\mu - f'_y)\,\delta y| < \zeta(|\delta x| + |\delta y|)$$

for all values of δx and δy numerically less than ω. Taking $\delta y = 0$ we obtain $|(\lambda - f'_x)\,\delta x| < \zeta\,|\delta x|$, or $|\lambda - f'_x| < \zeta$, and this can be true, for arbitrary ζ, only if $\lambda = f'_x$. Similarly $\mu = f'_y$.

We have proved that (1) is true if f'_x and f'_y are continuous, but this condition is not at all necessary. Suppose, for example, that $\phi(x, y)$ is any continuous function of x and y, and that

$$z = f(x, y) = (x + y)\,\phi(x, y).$$

Then

$$f'_x(0, 0) = \lim \frac{h\phi(h, 0)}{h} = \phi(0, 0),$$

and similarly $f'_y(0, 0) = \phi(0, 0)$; and plainly

$$z = \{\phi(0, 0) + \epsilon\}\,x + \{\phi(0, 0) + \eta\}\,y,$$

where ϵ and η tend to zero with x and y. This is equivalent to (1), with $x = y = 0$. But we have not assumed that $\phi(x, y)$ is differentiable with respect to x or y, and f'_x and f'_y need not exist anywhere except at the origin.

The equation (1) is sometimes taken as the definition of a 'differentiable

* Or with $|\delta x| + |\delta y|$ or $\sqrt{(\delta x^2 + \delta y^2)}$.

function of two variables'; $f(x, y)$ is said to be *differentiable at the point* (x, y) if

$$f(x+h, y+k)-f(x, y) = (A+\epsilon)h+(B+\eta)k,$$

where A and B depend only on x and y, and ϵ and η tend to zero when h and k tend to zero; and to be *differentiable in a region* if it is differentiable at all points of that region. In this case f'_x and f'_y exist and are equal to A and B, but need not be continuous. The hypothesis is intermediate between the weaker hypothesis 'f'_x and f'_y exist' and the stronger hypothesis 'f'_x and f'_y are continuous'. The definition has many advantages, but the hypothesis of continuity is sufficiently general for our purposes here. See W. H. Young, "The fundamental theorems of the differential calculus", *Cambridge Math. Tracts*, No. 11, and de la Vallée Poussin, *Cours d'analyse*, 6th edition, vol. II, Ch. III.

160. Differentials. In the applications of the calculus, especially in geometry, it is usually most convenient to work with equations expressed not, like equation (1) of § 159, in terms of the increments δx, δy, δz of the functions x, y, z, but in terms of what are called their *differentials* dx, dy, dz.

Let us return for a moment to a function $y = f(x)$ of a single variable x. If f is differentiable then

$$\delta y = \{f'(x)+\epsilon\}\,\delta x \quad \text{..................(1)},$$

where ϵ tends to zero with δx. The equation

$$\delta y = f'(x)\,\delta x \quad \text{......................(2)}$$

is 'approximately' true.

We have up to the present attributed no meaning to the isolated symbol dy. We now agree to *define* dy by the equation

$$dy = f'(x)\,\delta x \quad \text{......................(3)}.$$

If we choose for y the particular function x, we obtain

$$dx = \delta x \quad \text{.........................(4)},$$

so that

$$dy = f'(x)\,dx \quad \text{......................(5)};$$

and if we divide both sides of (5) by dx we obtain

$$\frac{dy}{dx} = f'(x) \quad \text{.........................(6)},$$

where dy/dx denotes not, as heretofore, the differential coefficient of y, but the quotient of the differentials dy, dx. The symbol

dy/dx thus acquires a double meaning; but there is no inconvenience in this, since (6) is true whichever meaning we choose.

We pass now to the corresponding definitions connected with a function z of two independent variables x and y. We define the differential dz by the equation

$$dz = f'_x \delta x + f'_y \delta y \quad \text{......................(7).}$$

Putting $z = x$ and $z = y$ in turn, we obtain

$$dx = \delta x, \quad dy = \delta y \quad \text{......................(8),}$$

so that

$$dz = f'_x dx + f'_y dy \quad \text{......................(9),}$$

which is the exact equation corresponding to the approximate equation (1) of § 159.

One property of the equation (9) deserves special remark. We saw in § 157 that if $z = f(x, y)$, x and y being not independent but functions of a single variable t, so that z is also a function of t alone, then

$$\frac{dz}{dt} = \frac{\partial f}{\partial x}\frac{dx}{dt} + \frac{\partial f}{\partial y}\frac{dy}{dt}.$$

Multiplying this equation by dt and observing that

$$dx = \frac{dx}{dt}dt, \quad dy = \frac{dy}{dt}dt, \quad dz = \frac{dz}{dt}dt,$$

we obtain

$$dz = f'_x dx + f'_y dy,$$

which is the same in form as (9). Thus *the formula which expresses dz in terms of dx and dy is the same whether the variables x and y are independent or not.* This remark is of great importance in applications.

It should also be observed that if z is a function of the two independent variables x and y, and

$$dz = \lambda dx + \mu dy,$$

then $\lambda = f'_x$, $\mu = f'_y$. This follows at once from § 159.

It is obvious that the theorems and definitions of the last three sections are capable of immediate extension to functions of any number of variables. The differential notation has great technical advantages, particularly in geometry.

Examples LXII. 1. The area of an ellipse is A, and a, b are the semi-axes. Prove that

$$\frac{dA}{A} = \frac{da}{a} + \frac{db}{b}.$$

2. Express Δ, the area of a triangle ABC, as a function of (i) a, B, C, (ii) A, b, c, and (iii) a, b, c, and establish the formulae

$$\frac{d\Delta}{\Delta} = 2\frac{da}{a} + \frac{c\,dB}{a\sin B} + \frac{b\,dC}{a\sin C}, \quad \frac{d\Delta}{\Delta} = \cot A\,dA + \frac{db}{b} + \frac{dc}{c},$$

$$d\Delta = R(\cos A\,da + \cos B\,db + \cos C\,dc),$$

where R is the radius of the circum-circle.

3. The sides of a triangle vary in such a way that the area remains constant, so that a may be regarded as a function of b and c. Prove that

$$\frac{\partial a}{\partial b} = -\frac{\cos B}{\cos A}, \quad \frac{\partial a}{\partial c} = -\frac{\cos C}{\cos A}.$$

[This follows from the equations

$$da = \frac{\partial a}{\partial b}db + \frac{\partial a}{\partial c}dc, \quad \cos A\,da + \cos B\,db + \cos C\,dc = 0.]$$

4. If a, b, c vary so that R remains constant, then

$$\frac{da}{\cos A} + \frac{db}{\cos B} + \frac{dc}{\cos C} = 0,$$

and so

$$\frac{\partial a}{\partial b} = -\frac{\cos A}{\cos B}, \quad \frac{\partial a}{\partial c} = -\frac{\cos A}{\cos C}.$$

[Use the formulae $a = 2R\sin A, \ldots$, and the facts that R and $A+B+C$ are constant.]

5. If z is a function of u and v, which are functions of x and y, then

$$\frac{\partial z}{\partial x} = \frac{\partial z}{\partial u}\frac{\partial u}{\partial x} + \frac{\partial z}{\partial v}\frac{\partial v}{\partial x}, \quad \frac{\partial z}{\partial y} = \frac{\partial z}{\partial u}\frac{\partial u}{\partial y} + \frac{\partial z}{\partial v}\frac{\partial v}{\partial y}.$$

[We have

$$dz = \frac{\partial z}{\partial u}du + \frac{\partial z}{\partial v}dv, \quad du = \frac{\partial u}{\partial x}dx + \frac{\partial u}{\partial y}dy, \quad dv = \frac{\partial v}{\partial x}dx + \frac{\partial v}{\partial y}dy.$$

Substitute for du and dv in the first equation and compare the result with the equation

$$dz = \frac{\partial z}{\partial x}dx + \frac{\partial z}{\partial y}dy.]$$

6. If $ur\cos\theta = 1$, $\tan\theta = v$, and $F(r,\theta) = G(u,v)$, then

$$rF_r = -uG_u, \quad F_\theta = uvG_u + (1+v^2)\,G_v.$$

(*Math. Trip.* 1932)

7. Let z be a function of x and y, and let X, Y, Z be defined by the equations

$$x = a_1X+b_1Y+c_1Z, \quad y = a_2X+b_2Y+c_2Z, \quad z = a_3X+b_3Y+c_3Z.$$

Then Z may be expressed as a function of X and Y. Express Z_X, Z_Y in terms of z_x, z_y. [Let these differential coefficients be denoted by P, Q and p, q. Then $dz-p\,dx-q\,dy = 0$, or

$$(c_1p+c_2q-c_3)\,dZ+(a_1p+a_2q-a_3)\,dX+(b_1p+b_2q-b_3)\,dY = 0.$$

Comparing this equation with $dZ-P\,dX-Q\,dY = 0$ we see that

$$P = -\frac{a_1p+a_2q-a_3}{c_1p+c_2q-c_3}, \quad Q = -\frac{b_1p+b_2q-b_3}{c_1p+c_2q-c_3}.]$$

8. If $\quad (a_1x+b_1y+c_1z)\,p+(a_2x+b_2y+c_2z)\,q = a_3x+b_3y+c_3z,$

then $\quad (a_1X+b_1Y+c_1Z)\,P+(a_2X+b_2Y+c_2Z)\,Q = a_3X+b_3Y+c_3Z.$

(*Math. Trip.* 1899)

9. **Differentiation of implicit functions.** Suppose that $f(x,y)$ and its derivatives f'_x and f'_y are continuous in the neighbourhood of the point (a,b), and that

$$f(a,b) = 0, \quad f'_b(a,b) \neq 0.$$

Then we can find a neighbourhood of (a,b) throughout which $f'_y(x,y)$ has always the same sign. Let us suppose, for example, that $f'_y(x,y)$ is positive near (a,b). Then $f(x,y)$ is, for any value of x sufficiently near to a, and for values of y sufficiently near to b, an increasing function of y in the stricter sense of § 95. It follows, by the theorem of § 109, that there is a unique continuous function y which is equal to b when $x = a$ and which satisfies the equation $f(x,y) = 0$ for all values of x sufficiently near to a.

If $f(x,y) = 0$, $x = a+h$, $y = b+k$, then

$$0 = f(x,y)-f(a,b) = (f'_a+\epsilon)\,h+(f'_b+\eta)\,k,$$

where ϵ and η tend to zero with h and k. Thus

$$\frac{k}{h} = -\frac{f'_a+\epsilon}{f'_b+\eta} \to -\frac{f'_a}{f'_b},$$

or

$$\frac{dy}{dx} = -\frac{f'_a}{f'_b}.$$

10. The equation of the tangent to the curve $f(x,y) = 0$, at the point (x_0, y_0), is

$$(x-x_0)f'_x(x_0,y_0)+(y-y_0)f'_y(x_0,y_0) = 0.$$

11. The result of eliminating u between the equations $y = f(x,u)$ and $z = \phi(x,u)$ is expressed as $z = F(x,y)$. Prove that

$$F_x = \frac{f_u\phi_x-f_x\phi_u}{f_u}, \quad F_y = \frac{\phi_u}{f_u}.$$

(*Math. Trip.* 1933)

12. **Maxima and minima.** We can define maximum and minimum values of a function of two variables by making the obvious changes in the definition of § 123. It is plain that if (a, b) gives a maximum of $f(x, y)$ then a gives a maximum of $f(x, b)$, so that f'_x vanishes at (a, b). Similarly f'_y vanishes, and

$$f'_x = 0, \quad f'_y = 0$$

or (what is the same thing) $\qquad df = 0$

are *necessary* conditions for a maximum or a minimum. The problem of finding *sufficient* conditions is more complex and we shall not consider it here.

13. If y is defined as a function of x by $g(x, y) = 0$, and $f(x, y)$ has a maximum at a point, then (since the formulae for differentials are the same whether the variables are independent or not) $df = 0$ at the maximum, while $dg = 0$ for all x, y. In other words, $f'_x dx + f'_y dy = 0$ provided $g'_x dx + g'_y dy = 0$; and so

$$\frac{f'_x}{g'_x} = \frac{f'_y}{g'_y} \quad \text{..............................(1).}$$

If g'_x or g'_y vanishes, then (1) is to be interpreted as meaning that the corresponding one of f'_x or f'_y vanishes.

Similarly, if z is defined by $g(x, y, z) = 0$, and $f(x, y, z)$ has a maximum, then

$$\frac{f'_x}{g'_x} = \frac{f'_y}{g'_y} = \frac{f'_z}{g'_z}$$

(with a similar gloss).

14. If α, β, γ are positive, A, B, C are the angles of a triangle, and $\sin^\alpha A \sin^\beta B \sin^\gamma C$ is a maximum, then

$$\tan^2 A = \frac{\alpha(\alpha+\beta+\gamma)}{\beta\gamma}, \quad \tan^2 B = \frac{\beta(\alpha+\beta+\gamma)}{\gamma\alpha}, \quad \tan^2 C = \frac{\gamma(\alpha+\beta+\gamma)}{\alpha\beta}.$$

(*Math. Trip.* 1935)

161. Definite integrals and areas. It will be remembered that in Ch. VI, § 148, we assumed that if $f(x)$ is a continuous function of x, and $P_1 P$ is an arc of the graph of $y = f(x)$, then the region bounded by $P_1 P$, the ordinates $P_1 N_1$ and PN, and the segment $N_1 N$ of the axis of x, has associated with it a number which we call its area. It is plain that, if $ON = x$, and we allow x to vary, this area is a function of x, which we denote by $F(x)$.

Making this assumption, we proved in § 148 that $F'(x) = f(x)$, and we showed how this result might be used in the calculation of the areas of particular curves. But we have still to justify the fundamental assumption that there is such a number as the area $F(x)$.

We know what is meant by the area of a *rectangle*, and that it is measured by the product of its sides. Also the properties of triangles, parallelograms, and polygons proved by Euclid enable us to attach a definite meaning to the areas of such figures. But nothing which we know so far provides us with a direct definition of the area of a figure bounded by curved lines. We shall now show how to give a definition of $F(x)$ which will enable us to prove its existence.

We suppose $f(x)$ continuous throughout the interval (a, b), and divide up the interval into a number of sub-intervals by means of the points of division $x_0, x_1, x_2, \ldots, x_n$, where

$$a = x_0 < x_1 < \ldots < x_{n-1} < x_n = b.$$

We denote by δ_ν the interval $(x_\nu, x_{\nu+1})$, and by m_ν the lower bound (§ 103) of $f(x)$ in δ_ν, and write

$$s = m_0\delta_0 + m_1\delta_1 + \ldots + m_{n-1}\delta_{n-1} = \Sigma m_\nu \delta_\nu,$$

say. It is evident that, if M is the upper bound of $f(x)$ in (a, b), then $s \leqq M(b-a)$. The aggregate of values of s is therefore, in the language of § 103, bounded above, and has an upper bound which we will denote by j. No value of s exceeds j, but there are values of s which exceed any number less than j.

In the same way, if M_ν is the upper bound of $f(x)$ in δ_ν, we can define the sum

$$S = \Sigma M_\nu \delta_\nu.$$

It is evident that, if m is the lower bound of $f(x)$ in (a, b), then $S \geqq m(b-a)$. The aggregate of values of S is therefore bounded below, and has a lower bound which we will denote by J. No value of S is less than J, but there are values of S less than any number greater than J.

It will help to make clear the significance of the sums s and S if we observe that, in the simple case in which $f(x)$ increases steadily from $x = a$ to $x = b$, m_ν is $f(x_\nu)$ and M_ν is $f(x_{\nu+1})$. In this case s is the total area of the rectangles shaded in Fig. 45, and S is the area bounded by a thick line. In the general case s and S will still be areas composed of rectangles, respectively included in and including the curvilinear region whose area we are trying to define.

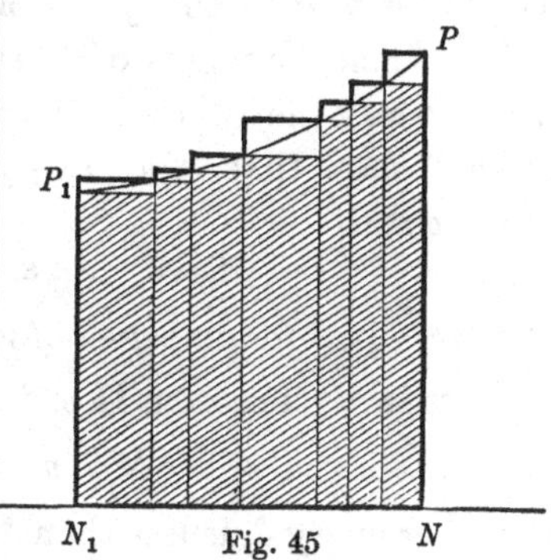

Fig. 45

We shall now show that *no sum such as s can exceed any sum such as S*. Let s, S be the sums corresponding to one mode of sub-division, and s', S' those corresponding to another. We have to show that $s \leqq S'$ and $s' \leqq S$.

We can form a third mode of sub-division by taking as dividing points all points which are such for either s, S or s', S'. Let $\mathbf{s}$, $\mathbf{S}$ be the sums corresponding to this third mode of sub-division. Then it is easy to see that

$$\mathbf{s} \geqq s, \quad \mathbf{s} \geqq s', \quad \mathbf{S} \leqq S, \quad \mathbf{S} \leqq S' \text{..............} (1).$$

For example, $\mathbf{s}$ differs from s in that at least one interval δ_ν which occurs in s is divided into a number of smaller intervals

$$\delta_{\nu,1},\ \delta_{\nu,2},\ \ldots,\ \delta_{\nu,p},$$

so that a term $m_\nu \delta_\nu$ of s is replaced in $\mathbf{s}$ by a sum

$$m_{\nu,1}\delta_{\nu,1} + m_{\nu,2}\delta_{\nu,2} + \ldots + m_{\nu,p}\delta_{\nu,p},$$

where $m_{\nu,1}$, $m_{\nu,2}$, ... are the lower bounds of $f(x)$ in $\delta_{\nu,1}$, $\delta_{\nu,2}$, ... But evidently $m_{\nu,1} \geqq m_\nu$, $m_{\nu,2} \geqq m_\nu$, ..., so that the sum just written is not less than $m_\nu \delta_\nu$. Hence $\mathbf{s} \geqq s$; and the other inequalities (1) can be established in the same way. But, since $\mathbf{s} \leqq \mathbf{S}$, it follows that

$$s \leqq \mathbf{s} \leqq \mathbf{S} \leqq S',$$

which is what we wanted to prove.

It also follows that $j \leqq J$. For we can find an s as near to j as we please and an S as near to J as we please*, and so $j > J$ would involve the existence of an s and an S for which $s > S$.

* The s and the S do not in general correspond to the same mode of sub-division.

So far we have made no use of the fact that $f(x)$ is continuous. We shall now show that $j = J$, and that the sums s, S tend to the limit J when the points of division x_ν are multiplied indefinitely in such a way that all the intervals δ_ν tend to zero. More precisely, we shall show that, *given any positive number* ϵ, *it is possible to find* δ *so that*

$$0 \leqslant J - s < \epsilon, \quad 0 \leqslant S - J < \epsilon$$

whenever $\delta_\nu < \delta$ *for all values of* ν.

There is, by Theorem II of § 107, a number δ such that

$$M_\nu - m_\nu < \epsilon/(b-a),$$

whenever every δ_ν is less than δ. Hence

$$S - s = \Sigma (M_\nu - m_\nu)\,\delta_\nu < \epsilon.$$

But
$$S - s = (S - J) + (J - j) + (j - s);$$

and all the three terms on the right-hand side are positive (or zero), and therefore all less than ϵ. Since $J - j$ is a constant, it must be zero. Hence $j = J$ and $0 \leqslant j - s < \epsilon$, $0 \leqslant S - J < \epsilon$, as was to be proved.

We define the area of $N_1 N P P_1$ as being *the common limit of* s *and* S, *that is to say* J. It is easy to give a more general form to this definition. Consider the sum

$$\sigma = \Sigma f_\nu \delta_\nu,$$

where f_ν denotes the value of $f(x)$ at any point in δ_ν. Then σ plainly lies between s and S, and so tends to the limit J when the intervals δ_ν tend to zero. We may therefore define the area as the limit of σ.

162. The definite integral. Let us now suppose that $f(x)$ is a continuous function, so that the region bounded by the curve $y = f(x)$, the ordinates $x = a$ and $x = b$, and the axis of x, has a definite area. We proved in Ch. VI, § 148, that if $F(x)$ is an 'integral function' of $f(x)$, i.e. if

$$F'(x) = f(x), \quad F(x) = \int f(x)\,dx,$$

then the area in question is $F(b) - F(a)$.

Since it is not always practicable to determine the form of $F(x)$, it is convenient to have a formula which represents the area N_1NPP_1 and contains no explicit reference to $F(x)$. We shall write

$$(N_1NPP_1) = \int_a^b f(x)\,dx.$$

The expression on the right-hand side of this equation may then be regarded as being defined in either of two ways. We may regard it as simply an abbreviation for $F(b) - F(a)$, where $F(x)$ is some integral function of $f(x)$, whether an actual formula expressing it is known or not; or we may regard it as the value of the area N_1NPP_1, defined directly in § 161.

The number $$\int_a^b f(x)\,dx$$

is called a *definite integral*; a and b are called its *lower and upper limits*; $f(x)$ is called the *subject of integration* or *integrand*; and the interval (a, b) the *range of integration*. The definite integral depends on a and b and the form of the function $f(x)$ only, and is not a function of x. On the other hand, the integral function

$$F(x) = \int f(x)\,dx$$

is sometimes called the *indefinite integral* of $f(x)$.

The distinction between the definite and the indefinite integral is merely one of point of view. The definite integral $\int_a^b f(x)\,dx = F(b) - F(a)$ is a function of b, and may be regarded as a particular integral function of $f(b)$. On the other hand the indefinite integral $F(x)$ can always be expressed by means of a definite integral, since

$$F(x) = F(a) + \int_a^x f(t)\,dt.$$

But when we are considering 'indefinite integrals' or 'integral functions' we are usually thinking of *a relation between two functions*, in virtue of which one is the derivative of the other; and when we are considering a 'definite integral' we are not as a rule concerned with any possible variation of the limits.

It should be observed that the integral $\int_a^x f(t)\,dt$, having a differential coefficient $f(x)$, is *a fortiori* a continuous function of x.

Since $1/x$ is continuous for all positive values of x, the investigations of the preceding paragraphs supply us with a proof of the existence of the function $\log x$, which we agreed to assume provisionally in § 131.

163. Area of a sector of a circle. The circular functions. The theory of the trigonometrical functions $\cos x$, $\sin x$, etc., as usually presented in text-books of elementary trigonometry, rests on an unproved assumption. An *angle* is the configuration formed by two straight lines OA, OP; and there is no difficulty in translating this 'geometrical' definition into purely analytical terms. The assumption comes at the next stage, when it is assumed that *angles are capable of numerical measurement*, that is to say that there is a real number x associated with the configuration, just as there is a real number associated with the region of § 148. This point once admitted, $\cos x$ and $\sin x$ may be defined in the ordinary way, and there is no further difficulty of principle in the elaboration of the theory. The whole difficulty lies in the question, *what is the x which occurs in $\cos x$ and $\sin x$?* To answer this question, we must define the measure of an angle, and we are now in a position to do so. The most natural definition would be this: suppose that AP is an arc of a circle whose centre is O and whose radius is unity, so that $OA = OP = 1$. Then x, the measure of the angle, is *the length of the arc AP*. This is, in substance, the definition adopted in the text-books, in the accounts which they give of the theory of 'circular measure'. It has however, for our present purpose, a fatal defect; for we have not proved that the arc of a curve, even of a circle, possesses a length. The notion of the length of a curve is capable of precise mathematical analysis just as much as that of an area; but the analysis, although of the same general character as that of the preceding sections, is decidedly more difficult, and it is

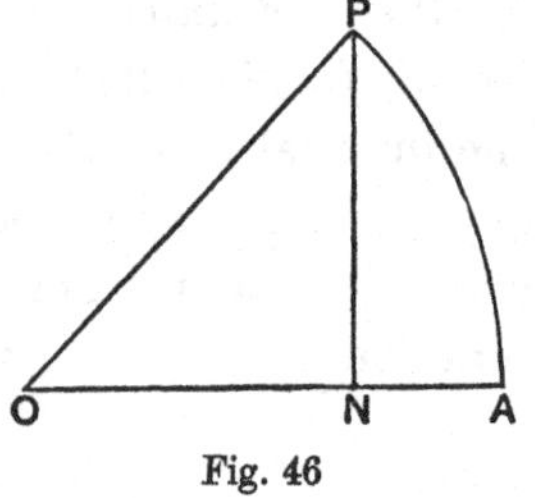

Fig. 46

impossible that we should give any general treatment of the subject here.

We must therefore found our definition on the notion not of length but of *area*. We define the measure of the angle AOP as *twice the area of the sector AOP of the unit circle.*

Suppose, in particular, that OA is $y = 0$ and that OP is $y = mx$, where $m > 0$. The area is a function of m, which we may denote by $\phi(m)$. The point P is $(\mu, m\mu)$, where

$$\mu = \frac{1}{\sqrt{(1+m^2)}}, \quad \sqrt{(1-\mu^2)} = \frac{m}{\sqrt{(1+m^2)}}, \quad m = \frac{\sqrt{(1-\mu^2)}}{\mu},$$

and

$$\phi(m) = \tfrac{1}{2}m\mu^2 + \int_\mu^1 \sqrt{(1-x^2)}\,dx = \tfrac{1}{2}\mu\sqrt{(1-\mu^2)} + \int_\mu^1 \sqrt{(1-x^2)}\,dx.$$

Hence

$$\frac{d\phi}{d\mu} = \tfrac{1}{2}\sqrt{(1-\mu^2)} - \frac{\mu^2}{2\sqrt{(1-\mu^2)}} - \sqrt{(1-\mu^2)} = -\frac{1}{2\sqrt{(1-\mu^2)}},$$

$$\frac{d\phi}{dm} = \frac{d\phi}{d\mu}\frac{d\mu}{dm} = \frac{1}{2\sqrt{(1-\mu^2)}}\frac{m}{(1+m^2)^{\frac{3}{2}}} = \frac{1}{2(1+m^2)},$$

and so
$$\phi(m) = \tfrac{1}{2}\int_0^m \frac{dt}{1+t^2}.$$

Thus the analytical equivalent of our definition would be to define $\arctan m$ by the equation

$$\arctan m = \int_0^m \frac{dt}{1+t^2}.$$

The theory of the circular functions will be worked out from this starting point in Ch. IX.

Examples LXIII. Calculation of the definite from the indefinite integral. 1. Show that, if $b > a \geqq 0$ and $n > -1$, then

$$\int_a^b x^n\,dx = \frac{b^{n+1} - a^{n+1}}{n+1}.$$

2. $\displaystyle\int_a^b \cos mx\,dx = \frac{\sin mb - \sin ma}{m}; \quad \int_a^b \sin mx\,dx = \frac{\cos ma - \cos mb}{m}.$

3. $\displaystyle\int_a^b \frac{dx}{1+x^2} = \arctan b - \arctan a; \quad \int_0^1 \frac{dx}{1+x^2} = \tfrac{1}{4}\pi.$

[There is an apparent difficulty here owing to the fact that $\arctan x$ is a many valued function. The difficulty may be avoided by observing that, in the equation

$$\int_0^x \frac{dt}{1+t^2} = \arctan x,$$

$\arctan x$ must denote an angle lying between $-\frac{1}{2}\pi$ and $\frac{1}{2}\pi$. For the integral vanishes when $x = 0$ and increases steadily and continuously as x increases. Thus the same is true of $\arctan x$, which therefore tends to $\frac{1}{2}\pi$ as $x \to \infty$. In the same way we can show that $\arctan x \to -\frac{1}{2}\pi$ as $x \to -\infty$. Similarly, in the equation

$$\int_0^x \frac{dt}{\sqrt{(1-t^2)}} = \arcsin x,$$

where $-1 < x < 1$, $\arcsin x$ denotes an angle lying between $-\frac{1}{2}\pi$ and $\frac{1}{2}\pi$. Thus, if a and b are both numerically less than unity, we have

$$\int_a^b \frac{dx}{\sqrt{(1-x^2)}} = \arcsin b - \arcsin a.]$$

4. $\displaystyle\int_0^1 \frac{dx}{1+2x\cos\alpha+x^2} = \frac{\alpha}{2\sin\alpha}$ if $-\pi < \alpha < \pi$, except when $\alpha = 0$, when the value of the integral is $\frac{1}{2}$, which is the limit of $\frac{1}{2}\alpha\operatorname{cosec}\alpha$ as $\alpha \to 0$.

5. $\displaystyle\int_0^1 \sqrt{(1-x^2)}\,dx = \tfrac{1}{4}\pi$; $\displaystyle\int_0^a \sqrt{(a^2-x^2)}\,dx = \tfrac{1}{4}\pi a^2$ if $a > 0$.

6. $\displaystyle\int_{-1}^1 \frac{dx}{\sqrt{(1-2\alpha x+\alpha^2)}}$ is equal to 2 if $-1 < \alpha < 1$ and to $2/\alpha$ if $|\alpha| > 1$.

(*Math. Trip.* 1933)

7. $\displaystyle\int_0^\pi \frac{dx}{a+b\cos x} = \frac{\pi}{\sqrt{(a^2-b^2)}}$ if $a > |b|$. [For the form of the indefinite integral see Exs. LIII. 3, 4. If $|a| < |b|$ then the subject of integration has an infinity between 0 and π. What is the value of the integral when a is negative and $-a > |b|$?]

8. $\displaystyle\int_0^{\frac{1}{2}\pi} \frac{dx}{a^2\cos^2 x+b^2\sin^2 x} = \frac{\pi}{2ab}$ if a and b are positive. What is the value of the integral when a and b have opposite signs, or when both are negative?

9. **Fourier's integrals.** Prove that if m and n are positive integers then

$$\int_0^{2\pi} \cos mx \sin nx\,dx$$

is always equal to zero, and

$$\int_0^{2\pi} \cos mx \cos nx\,dx, \quad \int_0^{2\pi} \sin mx \sin nx\,dx$$

are equal to zero unless $m = n$, when each is equal to π.

10. Prove that $\int_0^{\pi} \cos mx \cos nx\,dx$ and $\int_0^{\pi} \sin mx \sin nx\,dx$ are each equal to zero except when $m = n$, when each is equal to $\frac{1}{2}\pi$; and that

$$\int_0^{\pi} \cos mx \sin nx\,dx = \frac{2n}{n^2 - m^2}, \quad \int_0^{\pi} \cos mx \sin nx\,dx = 0,$$

according as $n - m$ is odd or even.

11. Prove that
$$\int_0^{\pi} \cos m\theta (\cos\theta)^n\,d\theta = 0$$
if m and n are positive integers and $m > n$. (*Math. Trip.* 1928)

12. Evaluate

$$\int_0^1 \frac{4x^2+3}{8x^2+4x+5}dx, \quad \int_0^c \frac{x\,dx}{\sqrt{(x+c)}}, \quad \int_0^{\pi} \frac{dx}{5+3\cos x}, \quad \int_0^{\frac{1}{2}\pi} \frac{dx}{1+2\cos x},$$

$$\int_0^{\alpha} \frac{dx}{\cos 2\alpha - \cos x} \quad (0 < \alpha < \tfrac{2}{3}\pi), \quad \int_0^1 \arctan x\,dx.$$

(*Math. Trip.* 1927, 1928, 1929, 1930, 1936)

164. Calculation of the definite integral from its definition as the limit of a sum. In a few cases we can evaluate a definite integral by direct calculation, starting from the definitions of §§ 161 and 162. As a rule it is much simpler to use the indefinite integral, but the reader will find it instructive to work through a few examples.

Examples LXIV. 1. Evaluate $\int_a^b x\,dx$ by dividing (a, b) into n equal parts by the points of division $a = x_0, x_1, x_2, \ldots, x_n = b$, and calculating the limit as $n \to \infty$ of

$$(x_1 - x_0)f(x_0) + (x_2 - x_1)f(x_1) + \ldots + (x_n - x_{n-1})f(x_{n-1}).$$

[This sum is

$$\frac{b-a}{n}\left[a + \left(a + \frac{b-a}{n}\right) + \left(a + 2\frac{b-a}{n}\right) + \ldots + \left\{a + (n-1)\frac{b-a}{n}\right\}\right]$$

$$= \frac{b-a}{n}\left[na + \frac{b-a}{n}\{1 + 2 + \ldots + (n-1)\}\right] = (b-a)\left\{a + (b-a)\frac{n(n-1)}{2n^2}\right\},$$

which tends to the limit $\frac{1}{2}(b^2-a^2)$ as $n\to\infty$. Verify the result by graphical reasoning.]

2. Calculate $\int_a^b x\,dx$, where $0<a<b$, by dividing (a,b) into n parts by the points of division $a, ar, ar^2, \dots, ar^{n-1}, ar^n$, where $r^n = b/a$. Apply the same method to the more general integral $\int_a^b x^m\,dx$.

3. Calculate $\int_a^b x^2\,dx$, $\int_a^b \cos mx\,dx$ and $\int_a^b \sin mx\,dx$ by the method of Ex. 1.

4. Prove that $n\sum_{r=0}^{n-1}\frac{1}{n^2+r^2}\to\frac{1}{4}\pi$ when $n\to\infty$.

[This follows from the fact that

$$\frac{n}{n^2}+\frac{n}{n^2+1^2}+\dots+\frac{n}{n^2+(n-1)^2}=\sum_{r=0}^{n-1}\frac{1/n}{1+(r/n)^2},$$

which tends to the limit $\int_0^1\frac{dx}{1+x^2}$ when $n\to\infty$, in virtue of the direct definition of the integral.]

5. Prove that $\frac{1}{n^2}\sum_{r=0}^{n-1}\sqrt{(n^2-r^2)}\to\frac{1}{4}\pi$. [The limit is $\int_0^1\sqrt{(1-x^2)}\,dx$.]

165. General properties of the definite integral. The definition of the definite integral as the limit of a sum presupposed (i) that f is continuous and (ii) that $a<b$. We define its value, when $a>b$, by

$$(1)\qquad \int_a^b f(x)\,dx = -\int_b^a f(x)\,dx,$$

and when $a=b$ by

$$(2)\qquad \int_a^a f(x)\,dx = 0.$$

These definitions become theorems if we define the integrals by means of the function $F(x)$; for $F(b)-F(a) = -\{F(a)-F(b)\}$, $F(a)-F(a)=0$.

We have then, for any a and b,

$$(3)\qquad \int_a^b f(x)\,dx+\int_b^c f(x)\,dx=\int_a^c f(x)\,dx.$$

$$(4)\qquad \int_a^b kf(x)\,dx = k\int_a^b f(x)\,dx.$$

$$(5) \qquad \int_a^b \{f(x)+\phi(x)\}\,dx = \int_a^b f(x)\,dx + \int_a^b \phi(x)\,dx.$$

The reader will find it an instructive exercise to write out formal proofs of these properties, in each case giving a proof starting from (α) the definition by means of the integral function and (β) the direct definition.

The following theorems are also important.

(6) *If* $f(x) \geqq 0$ *when* $a \leqq x \leqq b$, *then* $\int_a^b f(x)\,dx \geqq 0$.

We have only to observe that the sum s of § 156 cannot be negative. It will be shown later (Misc. Ex. 43, p. 340) that the value of the integral cannot be zero unless $f(x)$ is always equal to zero: this may also be inferred from the first corollary of § 122.

(7) *If* $H \leqq f(x) \leqq K$ *when* $a \leqq x \leqq b$, *then*

$$H(b-a) \leqq \int_a^b f(x)\,dx \leqq K(b-a).$$

This follows at once if we apply (6) to $f(x) - H$ and $K - f(x)$.

$$(8) \qquad \int_a^b f(x)\,dx = (b-a)f(\xi),$$

where ξ *lies between* a *and* b.

This follows from (7). For we can take H to be the least and K the greatest value of $f(x)$ in (a, b). Then the integral is equal to $\eta(b-a)$, where η lies between H and K. But, since $f(x)$ is continuous, there must be a value of ξ for which $f(\xi) = \eta$ (§ 101).

If $F(x)$ is the integral function, we can write the result of (8) in the form

$$F(b) - F(a) = (b-a)\,F'(\xi),$$

so that (8) appears now to be a special case of the mean value theorem of § 126. We may call (8) the *first mean value theorem for integrals.*

(9) **The generalised mean value theorem for integrals.** *If* $\phi(x)$ *is positive, and* H *and* K *are defined as in* (7), *then*

$$H\int_a^b \phi(x)\,dx \leqq \int_a^b f(x)\,\phi(x)\,dx \leqq K\int_a^b \phi(x)\,dx;$$

and $$\int_a^b f(x)\,\phi(x)\,dx = f(\xi)\int_a^b \phi(x)\,dx,$$

where ξ lies between a and b.

This follows at once by applying theorem (6) to the integrals

$$\int_a^b \{f(x)-H\}\,\phi(x)\,dx, \quad \int_a^b \{K-f(x)\}\,\phi(x)\,dx.$$

(10) The fundamental theorem of the integral calculus. *The function*

$$F(x) = \int_a^x f(t)\,dt$$

has a derivative equal to $f(x)$.

This has been proved already in § 148, but it is convenient to restate the result here as a formal theorem. It follows as a corollary, as was pointed out in § 162, that $F(x)$ *is a continuous function of x.*

Examples LXV. 1. Show, by means of the direct definition of the definite integral, and equations (1)–(5) above, that

$$\text{(i)}\quad \int_{-a}^{a} \phi(x^2)\,dx = 2\int_0^a \phi(x^2)\,dx, \quad \int_{-a}^{a} x\phi(x^2)\,dx = 0;$$

$$\text{(ii)}\quad \int_0^{\frac{1}{2}\pi} \phi(\cos x)\,dx = \int_0^{\frac{1}{2}\pi} \phi(\sin x)\,dx = \tfrac{1}{2}\int_0^{\pi} \phi(\sin x)\,dx;$$

$$\text{(iii)}\quad \int_0^{m\pi} \phi(\cos^2 x)\,dx = m\int_0^{\pi} \phi(\cos^2 x)\,dx,$$

m being an integer. [The truth of these equations will appear geometrically intuitive if the graphs of the functions under the sign of integration are sketched.]

2. Prove that $$\int_0^{\pi} \frac{\sin(n+\frac{1}{2})\theta}{\sin\frac{1}{2}\theta}\,d\theta = \pi$$

if n is a positive integer or zero. What is the value of the integral for negative integral n?

3. Prove that $\displaystyle\int_0^{\pi} \frac{\sin nx}{\sin x}\,dx$ is equal to π or to 0 according as n is odd or even. (*Math. Trip.* 1933)

4. Prove that $$\int_0^{\pi} \left(\frac{\sin nx}{\sin x}\right)^2 dx = n\pi$$

for all positive integral values of n. (*Math. Trip.* 1933)

[For Ex. 2 use the identity

$$\frac{\sin(n+\frac{1}{2})x}{\sin\frac{1}{2}x} = 1+2\cos x+2\cos 2x+\dots+2\cos nx,$$

and for Ex. 3 the identity

$$\frac{\sin nx}{\sin x} = 2\cos(n-1)x+2\cos(n-3)x+\dots,$$

where the last term is 1 or $2\cos x$. To prove Ex. 4, square the last identity and use Ex. LXIII. 10.]

5. If $\phi(x) = \frac{1}{2}a_0+a_1\cos x+b_1\sin x+a_2\cos 2x+\dots+a_n\cos nx+b_n\sin nx$, and k is a positive integer not greater than n, then

$$\int_0^{2\pi}\phi(x)\,dx = \pi a_0,\quad \int_0^{2\pi}\cos kx\,\phi(x)\,dx = \pi a_k,\quad \int_0^{2\pi}\sin kx\,\phi(x)\,dx = \pi b_k.$$

If $k>n$ then the value of each of the last two integrals is zero. [Use Ex. LXIII. 9.]

6. If $f(x)\leqq\phi(x)$ when $a\leqq x\leqq b$, then $\int_a^b f\,dx\leqq\int_a^b\phi\,dx$.

7. Prove that

$$0<\int_0^{\frac{1}{2}\pi}\sin^{n+1}x\,dx<\int_0^{\frac{1}{2}\pi}\sin^n x\,dx,\quad 0<\int_0^{\frac{1}{4}\pi}\tan^{n+1}x\,dx<\int_0^{\frac{1}{4}\pi}\tan^n x\,dx.$$

8*. If $n>1$ then $\cdot 5<\int_0^{\frac{1}{2}}\frac{dx}{\sqrt{(1-x^{2n})}}<\cdot 524$. [The first inequality follows from the fact that $\sqrt{(1-x^{2n})}<1$, the second from the fact that

$$\sqrt{(1-x^{2n})}>\sqrt{(1-x^2)}.]$$

9. Prove that $\frac{1}{2}<\int_0^1\frac{dx}{\sqrt{(4-x^2+x^3)}}<\frac{1}{6}\pi$.

10. Prove that $\cdot 573<\int_1^2\frac{dx}{\sqrt{(4-3x+x^3)}}<\cdot 595$. [Put $x = 1+u$: then replace $2+3u^2+u^3$ by $2+4u^2$ and by $2+3u^2$.]

11. If α and ϕ are positive acute angles then

$$\phi<\int_0^{\phi}\frac{dx}{\sqrt{(1-\sin^2\alpha\sin^2 x)}}<\frac{\phi}{\sqrt{(1-\sin^2\alpha\,\sin^2\phi)}}.$$

If $\alpha = \phi = \frac{1}{6}\pi$, then the integral lies between $\cdot 523$ and $\cdot 541$.

12. Prove that $\left|\int_a^b f(x)\,dx\right|\leqq\int_a^b|f(x)|\,dx$.

* Exs. 8–11 are taken from Gibson's *Elementary treatise on the calculus*.

[If σ is the sum considered at the end of § 161, and σ' the corresponding sum formed from the function $|f(x)|$, then $|\sigma| \leqq \sigma'$.]

13. If $|f(x)| \leqq M$, then $\left|\int_a^b f(x)\phi(x)\,dx\right| \leqq M\int_a^b |\phi(x)|\,dx$.

166. Integration by parts and by substitution. It follows from § 141 that, if $f'(x)$ and $\phi'(x)$ are continuous,

$$\int_a^b f(x)\phi'(x)\,dx = f(b)\phi(b) - f(a)\phi(a) - \int_a^b f'(x)\phi(x)\,dx.$$

This formula is known as the formula for *integration of a definite integral by parts.*

Again, we know (§ 136) that if $F(t)$ is the integral function of $f(t)$ then

$$\int f\{\phi(x)\}\phi'(x)\,dx = F\{\phi(x)\}.$$

Hence, if $\phi(a) = c$, $\phi(b) = d$, we have

$$\int_c^d f(t)\,dt = F(d) - F(c) = F\{\phi(b)\} - F\{\phi(a)\} = \int_a^b f\{\phi(x)\}\phi'(x)\,dx;$$

which is the formula for the transformation of a definite integral by *substitution.*

These formulae often enable us to determine the value of a definite integral without knowledge of the form of the integral function $F(x)$. The definite integral depends only upon the difference between two particular values of $F(x)$, which may often be found by some special device when the form of $F(x)$ is unknown.

Examples LXVI. 1. Prove that

$$\int_a^b xf''(x)\,dx = \{bf'(b) - f(b)\} - \{af'(a) - f(a)\}.$$

2. More generally, $\int_a^b x^m f^{(m+1)}(x)\,dx = F(b) - F(a)$, where

$$F(x) = x^m f^{(m)}(x) - mx^{m-1}f^{(m-1)}(x) + m(m-1)x^{m-2}f^{(m-2)}(x) - \dots + (-1)^m m!\,f(x).$$

3. Prove that

$$\int_0^1 \operatorname{arc\,sin} x\,dx = \tfrac{1}{2}\pi - 1, \quad \int_0^1 x \operatorname{arc\,tan} x\,dx = \tfrac{1}{4}\pi - \tfrac{1}{2}.$$

4. Prove that if a and b are positive then

$$\int_0^{\frac{1}{2}\pi} \frac{x\cos x \sin x\,dx}{(a^2\cos^2 x+b^2\sin^2 x)^2} = \frac{\pi}{4ab^2(a+b)}.$$

[Integrate by parts and use Ex. LXIII. 8.]

5. Evaluate

$$\int_1^2 \frac{dx}{x(1+x^4)},\quad \int_8^{15} \frac{dx}{(x-3)\sqrt{(x+1)}},\quad \int_0^1 \frac{x\,dx}{1+\sqrt{x}},\quad \int_0^{\frac{1}{4}\pi} \sec^3 x\,dx,\quad \int_0^{\frac{1}{4}\pi} \sqrt{(\tan x)}\,dx,$$

$$\int_{-\frac{1}{2}\pi}^{\frac{1}{2}\pi} \frac{dx}{5+7\cos x+\sin x},\quad \int_0^{\frac{1}{2}\pi} \frac{1+2\cos x}{(2+\cos x)^2}\,dx,\quad \int_0^{\frac{1}{2}\pi} \sin^{\frac{3}{2}} x \cos^3 x\,dx$$

by appropriate substitutions. (*Math. Trip.* 1924, 1925, 1926, 1931)

6. If $f_1(x) = \int_0^x f(t)\,dt,\ f_2(x) = \int_0^x f_1(t)\,dt,\ \dots,\ f_k(x) = \int_0^x f_{k-1}(t)\,dt,$

then
$$f_k(x) = \frac{1}{(k-1)!}\int_0^x f(t)\,(x-t)^{k-1}\,dt. \quad (\textit{Math. Trip. } 1933)$$

[Integrate repeatedly by parts.]

7. Prove by integration by parts that if $u_{m,n} = \int_0^1 x^m(1-x)^n\,dx$, where m and n are positive integers, then $(m+n+1)\,u_{m,n} = nu_{m,n-1}$, and deduce that

$$u_{m,n} = \frac{m!\,n!}{(m+n+1)!}.$$

8. Prove that if $u_n = \int_0^{\frac{1}{4}\pi} \tan^n x\,dx$ then $u_n + u_{n-2} = \dfrac{1}{n-1}$. Hence evaluate the integral for all positive integral values of n.

[Put $\tan^n x = \tan^{n-2} x\,(\sec^2 x - 1)$ and integrate by parts.]

9. Prove that if $u_n = \int_0^{\frac{1}{2}\pi} \sin^n x\,dx$ then $u_n = \dfrac{n-1}{n} u_{n-2}$. [Write $\sin^{n-1} x \sin x$ for $\sin^n x$ and integrate by parts.]

10. Deduce from Ex. 9 that u_n is equal to

$$\frac{2.4.6\ldots(n-1)}{3.5.7\ldots n},\quad \tfrac{1}{2}\pi\,\frac{1.3.5\ldots(n-1)}{2.4.6\ldots n},$$

according as n is odd or even. (*Math. Trip.* 1935)

11. **The second mean value theorem.** If $f(x)$ is a function of x which has a continuous differential coefficient of constant sign for all values of x from $x = a$ to $x = b$, then there is a number ξ between a and b such that

$$\int_a^b f(x)\,\phi(x)\,dx = f(a)\int_a^\xi \phi(x)\,dx + f(b)\int_\xi^b \phi(x)\,dx.$$

[Let $\int_a^x \phi(t)\,dt = \Phi(x)$. Then

$$\int_a^b f(x)\,\phi(x)\,dx = \int_a^b f(x)\,\Phi'(x)\,dx = f(b)\,\Phi(b) - \int_a^b f'(x)\,\Phi(x)\,dx$$
$$= f(b)\,\Phi(b) - \Phi(\xi)\int_a^b f'(x)\,dx,$$

by theorem (9) of § 165: i.e.

$$\int_a^b f(x)\,\phi(x)\,dx = f(b)\,\Phi(b) + \{f(a) - f(b)\}\,\Phi(\xi),$$

which is equivalent to the result stated.]

12. **Bonnet's form of the second mean value theorem.** If $f'(x)$ is continuous and of constant sign, and $f(b)$ and $f(a)-f(b)$ have the same sign, then

$$\int_a^b f(x)\,\phi(x)\,dx = f(a)\int_a^X \phi(x)\,dx,$$

where X lies between a and b. For

$$f(b)\,\Phi(b) + \{f(a) - f(b)\}\,\Phi(\xi) = \mu f(a),$$

where μ lies between $\Phi(\xi)$ and $\Phi(b)$, and so is the value of $\Phi(x)$ for a value of x such as X. The important case is that in which $0 \leqq f(b) \leqq f(x) \leqq f(a)$.

Prove similarly that if $f(a)$ and $f(b)-f(a)$ have the same sign, then

$$\int_a^b f(x)\,\phi(x)\,dx = f(b)\int_X^b \phi(x)\,dx,$$

where X lies between a and b.

13. Prove that $\left|\int_X^{X'} \frac{\sin x}{x}\,dx\right| < \frac{2}{X}$ if $X' > X > 0$. [Apply the first formula of Ex. 12, and note that the integral of $\sin x$ over any interval is numerically less than 2.]

14. Establish the results of Ex. LXV. 1 by means of the rule for substitution. [For example, in (iii), divide the range into m equal parts and use the substitutions $x = \pi + y$, $x = 2\pi + y$, ...]

15. Prove that $\int_a^b F(x)\,dx = \int_a^b F(a+b-x)\,dx$.

16. Prove that $\int_0^{\frac{1}{2}\pi} \cos^m x \sin^m x\,dx = 2^{-m}\int_0^{\frac{1}{2}\pi} \cos^m x\,dx$.

17. Prove that $\int_0^{\pi} x\phi(\sin x)\,dx = \frac{1}{2}\pi\int_0^{\pi} \phi(\sin x)\,dx$. [Put $x = \pi - y$.]

18. Prove that $\displaystyle\int_0^{\pi} \frac{x \sin x}{1+\cos^2 x}\,dx = \tfrac{1}{4}\pi^2$, $\displaystyle\int_0^{\pi} x \sin^6 x \cos^4 x\,dx = \tfrac{3}{512}\pi^2$.

(*Math. Trip.* 1927)

19. Show by means of the transformation $x = a\cos^2\theta + b\sin^2\theta$ that $\displaystyle\int_a^b \sqrt{\{(x-a)(b-x)\}}\,dx = \tfrac{1}{8}\pi(b-a)^2$.

20. Show by means of the substitution $(a+b\cos x)(a-b\cos y) = a^2-b^2$ that

$$\int_0^{\pi}(a+b\cos x)^{-n}\,dx = (a^2-b^2)^{-(n-\frac{1}{2})}\int_0^{\pi}(a-b\cos y)^{n-1}\,dy$$

when n is a positive integer and $a > |b|$, and evaluate the integral when $n = 1, 2, 3$.

21. If m and n are positive integers then

$$\int_a^b (x-a)^m (b-x)^n\,dx = (b-a)^{m+n+1}\frac{m!\,n!}{(m+n+1)!}.$$

[Put $x = a+(b-a)y$, and use Ex. 7.]

167. Proof of Taylor's theorem by integration by parts. We can use the method of partial integration to obtain another proof of Taylor's theorem.

Let $f(x)$ be a function whose first n derivatives are continuous, and let

$$F_n(x) = f(b) - f(x) - (b-x)f'(x) - \dots - \frac{(b-x)^{n-1}}{(n-1)!}f^{(n-1)}(x).$$

Then

$$F_n'(x) = -\frac{(b-x)^{n-1}}{(n-1)!}f^{(n)}(x),$$

and so

$$F_n(a) = F_n(b) - \int_a^b F_n'(x)\,dx = \frac{1}{(n-1)!}\int_a^b (b-x)^{n-1}f^{(n)}(x)\,dx.$$

If now we write $a+h$ for b, and transform the integral by putting $x = a+th$, we obtain

$$f(a+h) = f(a) + hf'(a) + \dots + \frac{h^{n-1}}{(n-1)!}f^{(n-1)}(a) + R_n \quad \dots(1),$$

where

$$R_n = \frac{h^n}{(n-1)!}\int_0^1 (1-t)^{n-1}f^{(n)}(a+th)\,dt \quad \dots\dots\dots(2).$$

Now, if p is any positive integer not greater than n, we have, by theorem (9) of § 165,

$$\int_0^1 (1-t)^{n-1} f^{(n)}(a+th)\,dt = \int_0^1 (1-t)^{n-p}(1-t)^{p-1} f^{(n)}(a+th)\,dt$$
$$= (1-\theta)^{n-p} f^{(n)}(a+\theta h)\int_0^1 (1-t)^{p-1}\,dt,$$

where $0<\theta<1$. Hence

$$R_n = \frac{(1-\theta)^{n-p} f^{(n)}(a+\theta h)\,h^n}{p(n-1)!} \quad \text{..............(3).}$$

If we take $p = n$ we obtain Lagrange's form of R_n (§ 152). If on the other hand we take $p = 1$ we obtain *Cauchy's form*, viz.

$$R_n = \frac{(1-\theta)^{n-1} f^{(n)}(a+\theta h)\,h^n}{(n-1)!} \quad \text{...............(4).}$$

This proof of Taylor's theorem has the advantage of leading to an exact formula for R_n, viz. (2), which does not involve an undetermined number θ. It is (considered simply as a proof of Lagrange's form of the theorem) less general than that of § 150, since we have assumed the continuity of $f^{(n)}(x)$. The argument of § 150 can be modified so as to give the formulae (3) and (4).

168. Application of Cauchy's form to the binomial series. If $f(x) = (1+x)^m$, where m is not a positive integer, then Cauchy's form of the remainder is

$$R_n = \frac{m(m-1)\dots(m-n+1)}{1.2\dots(n-1)} \frac{(1-\theta)^{n-1}x^n}{(1+\theta x)^{n-m}}.$$

Now $(1-\theta)/(1+\theta x)$ is less than unity, so long as $-1<x<1$, whether x is positive or negative; and $(1+\theta x)^{m-1}$ is less than $(1+|x|)^{m-1}$ if $m>1$ and than $(1-|x|)^{m-1}$ if $m<1$. Hence

$$|R_n| < |m|\,(1 \pm |x|)^{m-1}\left|\binom{m-1}{n-1}\right| |x|^n = \rho_n,$$

say. But $\rho_n \to 0$ as $n \to \infty$, by Ex. XXVII. 13, and so $R_n \to 0$. The truth of the binomial theorem is thus established for all rational values of m and all values of x between -1 and 1. It will be remembered that the difficulty in using Lagrange's form, in § 152 (2), arose in connection with negative values of x.

169. Approximate formulae for definite integrals. Simpson's rule. There are a number of approximations to the value of a definite integral which are important in numerical work. The simplest is

$$\int_a^b f(x)\,dx = \tfrac{1}{2}(b-a)\{f(a)+f(b)\} \quad \text{............(1)}.$$

Here we replace the area P_1N_1NP of §148 by the polygon P_1N_1NP, and the formula is exact when $f(x)$ is linear. It may be shown (see Ex. LXVII. 2, p. 330) that, if $f(x)$ has two derivatives $f'(x)$ and $f''(x)$, then the error in (1) is

$$-\tfrac{1}{12}(b-a)^3 f''(\xi)$$

for a value of ξ between a and b. In practice, of course, we should divide up the range of integration into a number of smaller pieces and apply the formula to each of them separately.

A much better formula is

$$\int_a^b f(x)\,dx = \tfrac{1}{6}(b-a)\left\{f(a)+4f\left(\frac{a+b}{2}\right)+f(b)\right\} \quad \text{......(2)},$$

which is generally known as *Simpson's rule*. We shall prove that, if $f(x)$ has four derivatives $f'(x), f''(x), f'''(x)$, and $f^{\text{iv}}(x)$, then the error in (2) is

$$-\tfrac{1}{2880}(b-a)^5 f^{\text{iv}}(\xi)$$

for a ξ between a and b. Incidentally this proves that Simpson's rule is exact for polynomials of the third or lower degree.

We write $c-h$, $c+h$ for a, b, and consider the function

$$\phi(t) = \psi(t) - \left(\frac{t}{h}\right)^5 \psi(h),$$

where $$\psi(t) = \int_{c-t}^{c+t} f(x)\,dx - \tfrac{1}{3}t\{f(c+t)+4f(c)+f(c-t)\}.$$

Differentiating three times, we find

$$\phi'(t) = \tfrac{2}{3}\{f(c+t)-2f(c)+f(c-t)\} - \tfrac{1}{3}t\{f'(c+t)-f'(c-t)\} - \frac{5t^4}{h^5}\psi(h),$$

$$\phi''(t) = \tfrac{1}{3}\{f'(c+t)-f'(c-t)\} - \tfrac{1}{3}t\{f''(c+t)+f''(c-t)\} - \frac{20t^3}{h^5}\psi(h),$$

$$\phi'''(t) = -\tfrac{1}{3}t\{f'''(c+t)-f'''(c-t)\} - \frac{60t^2}{h^5}\psi(h).$$

Hence, by the mean value theorem,

$$\phi'''(t) = -\tfrac{2}{3}t^2\left\{f^{\text{iv}}(\xi) + \frac{90}{h^5}\psi(h)\right\} \quad\text{.................(3)},$$

where ξ lies in $(c-t, c+t)$.

Now $\phi(0) = \phi(h) = 0$, and therefore, by Rolle's theorem, $\phi'(t_1) = 0$ for a t_1 between 0 and h. Also $\phi'(0) = 0$, and therefore $\phi''(t_2) = 0$ for a t_2 between 0 and t_1, and *a fortiori* between 0 and h. Finally $\phi''(0) = 0$, and therefore $\phi'''(t_3) = 0$ for a t_3 between 0 and h. Hence, by (3)

$$f^{\text{iv}}(\xi) = -\frac{90}{h^5}\psi(h)$$

for a ξ between $c-t_3$ and $c+t_3$, and therefore between $c-h$ and $c+h$. But this is

$$\int_{c-h}^{c+h} f(x)\,dx - \tfrac{1}{3}h\{f(c+h) + 4f(c) + f(c-h)\} = -\frac{h^5}{90}f^{\text{iv}}(\xi)$$

or

$$\int_a^b f(x)\,dx = \tfrac{1}{6}(b-a)\left\{f(a) + 4f\left(\frac{a+b}{2}\right) + f(b)\right\} - \frac{(b-a)^5}{2880}f^{\text{iv}}(\xi).$$

In practice again we divide the range of integration into pieces and apply the rule to each.

Examples LXVII. 1. Prove that, if $f(x)$ has two derivatives,

$$f(x+h) - 2f(x) + f(x-h) = h^2 f''(\xi),$$

where ξ lies between $x-h$ and $x+h$. (*Math. Trip.* 1925)

[Use the auxiliary function

$$\phi(t) = f(x+t) - 2f(x) + f(x-t) - \left(\frac{t}{h}\right)^2 \{f(x+h) - 2f(x) + f(x-h)\}.]$$

2. Prove that the error in (1) above is $-\frac{1}{12}(b-a)^3 f''(\xi)$, where $a < \xi < b$. [Use the auxiliary functions

$$\psi(t) = \int_{c-t}^{c+t} f(x)\,dx - t\{f(c+t) + f(c-t)\}, \quad \phi(t) = \psi(t) - \left(\frac{t}{h}\right)^3 \psi(h).]$$

3. Prove that

$$\int_a^b f(x)\,dx = (b-a)f\left(\frac{a+b}{2}\right) + \tfrac{1}{24}(b-a)^3 f''(\xi),$$

where $a < \xi < b$.

4. Apply Simpson's rule to the calculation of π from the formula $\frac{1}{4}\pi = \int_0^1 \frac{dx}{1+x^2}$. [The result is ·7833.... If we divide the integral into two, from 0 to $\frac{1}{2}$ and from $\frac{1}{2}$ to 1, and apply the rule to each part, we obtain ·7853916.... The correct value is ·7853921....]

5. Show that $8{\cdot}9 < \int_3^5 \sqrt{(4+x^2)}\,dx < 9$. (*Math. Trip.* 1903)

6. Apply Simpson's rule, with five ordinates, to calculate

$$\int_1^2 \sqrt{\left(x-\frac{1}{x}\right)}\,dx$$

to two decimal places. (*Math. Trip.* 1934)

7. Show that $\int_0^4 x^3\sqrt{(4x-x^2)}\,dx = 88$ approximately.

(*Math. Trip.* 1933)

170. Integrals of complex functions of a real variable. So far we have always supposed that the subject of integration in a definite integral is real. We define the integral of a complex function $f(x) = \phi(x)+i\psi(x)$ of the real variable x, between the limits a and b, by the equations

$$\int_a^b f(x)\,dx = \int_a^b \{\phi(x)+i\psi(x)\}\,dx = \int_a^b \phi(x)\,dx + i\int_a^b \psi(x)\,dx;$$

and it is evident that the properties of such integrals may be deduced from those of the real integrals already considered.

There is one of these properties that we shall make use of later on. It is expressed by the inequality

$$\left|\int_a^b f(x)\,dx\right| \leqq \int_a^b |f(x)|\,dx \quad \text{..............(1)*}.$$

This inequality may be deduced without difficulty from the definitions of §§161 and 162. If δ_ν has the same meaning as in §161, ϕ_ν and ψ_ν are the values of ϕ and ψ at a point of δ_ν, and $f_\nu = \phi_\nu + i\psi_\nu$, then we have

$$\begin{aligned}\int_a^b f\,dx = \int_a^b \phi\,dx + i\int_a^b \psi\,dx &= \lim \Sigma\phi_\nu\delta_\nu + i\lim\Sigma\psi_\nu\delta_\nu \\ &= \lim\Sigma(\phi_\nu+i\psi_\nu)\,\delta_\nu = \lim\Sigma f_\nu\delta_\nu,\end{aligned}$$

* The corresponding inequality for a real integral was proved in Ex. LXV. 12.

and so $$\left|\int_a^b f\,dx\right| = |\lim \Sigma f_\nu \delta_\nu| = \lim |\Sigma f_\nu \delta_\nu|;$$

while $$\int_a^b |f|\,dx = \lim \Sigma |f_\nu|\,\delta_\nu.$$

The result now follows from the inequality

$$|\Sigma f_\nu \delta_\nu| \leqq \Sigma |f_\nu|\,\delta_\nu.$$

It is evident that the formulae (1) and (2) of § 167 remain true when f is a complex function $\phi + i\psi$.

MISCELLANEOUS EXAMPLES ON CHAPTER VII

1. Verify the terms given of the following Taylor's series:

$$\begin{aligned}
&(1) & \tan x &= x + \tfrac{1}{3}x^3 + \tfrac{2}{15}x^5 + \dots,\\
&(2) & \sec x &= 1 + \tfrac{1}{2}x^2 + \tfrac{5}{24}x^4 + \dots,\\
&(3) & x \operatorname{cosec} x &= 1 + \tfrac{1}{6}x^2 + \tfrac{7}{360}x^4 + \dots,\\
&(4) & x \cot x &= 1 - \tfrac{1}{3}x^2 - \tfrac{1}{45}x^4 - \dots.
\end{aligned}$$

2. Show that if $f(x)$ and its first $n+2$ derivatives are continuous, and $f^{(n+1)}(0) \neq 0$, and θ_n is the value of θ which occurs in Lagrange's form of the remainder after n terms of Taylor's series, then

$$\theta_n = \frac{1}{n+1} + \frac{n}{2(n+1)^2(n+2)} \frac{f^{(n+2)}(0)}{f^{(n+1)}(0)} x + o(x).$$

[Follow the method of Ex. LV. 12.]

3. Establish the formulae

$$\text{(i)} \qquad \begin{vmatrix} f(a) & f(b) \\ g(a) & g(b) \end{vmatrix} = (b-a) \begin{vmatrix} f(a) & f'(\beta) \\ g(a) & g'(\beta) \end{vmatrix},$$

where β lies between a and b, and

$$\text{(ii)} \quad \begin{vmatrix} f(a) & f(b) & f(c) \\ g(a) & g(b) & g(c) \\ h(a) & h(b) & h(c) \end{vmatrix} = \tfrac{1}{2}(b-c)(c-a)(a-b) \begin{vmatrix} f(a) & f'(\beta) & f''(\gamma) \\ g(a) & g'(\beta) & g''(\gamma) \\ h(a) & h'(\beta) & h''(\gamma) \end{vmatrix},$$

where β and γ lie between the least and greatest of a, b, c. [To prove (ii) consider the function

$$\phi(x) = \begin{vmatrix} f(a) & f(b) & f(x) \\ g(a) & g(b) & g(x) \\ h(a) & h(b) & h(x) \end{vmatrix} - \frac{(x-a)(x-b)}{(c-a)(c-b)} \begin{vmatrix} f(a) & f(b) & f(c) \\ g(a) & g(b) & g(c) \\ h(a) & h(b) & h(c) \end{vmatrix},$$

which vanishes when $x = a$, $x = b$, and $x = c$. Its first derivative must vanish for two distinct values of x lying between the least and greatest of a, b, c, by Theorem B of § 122; and therefore its second derivative must vanish for a value γ of x satisfying the same condition. We thus obtain the formula

$$\begin{vmatrix} f(a) & f(b) & f(c) \\ g(a) & g(b) & g(c) \\ h(a) & h(b) & h(c) \end{vmatrix} = \tfrac{1}{2}(c-a)(c-b) \begin{vmatrix} f(a) & f(b) & f''(\gamma) \\ g(a) & g(b) & g''(\gamma) \\ h(a) & h(b) & h''(\gamma) \end{vmatrix}.$$

The reader will now complete the proof without difficulty.]

4. If $F(x)$ is a function which has continuous derivatives of the first n orders, of which the first $n-1$ vanish when $x = 0$, and $A \leqq F^{(n)}(x) \leqq B$ when $0 \leqq x \leqq h$, then

$$A\frac{x^n}{n!} \leqq F(x) \leqq B\frac{x^n}{n!}$$

when $0 \leqq x \leqq h$. Apply this result to

$$f(x) - f(0) - xf'(0) - \dots - \frac{x^{n-1}}{(n-1)!}f^{(n-1)}(0),$$

and deduce Taylor's theorem.

5. If $\Delta_h \phi(x) = \phi(x) - \phi(x+h)$, $\Delta_h^2 \phi(x) = \Delta_h\{\Delta_h \phi(x)\}$, and so on, and $\phi(x)$ has derivatives of the first n orders, then

$$\Delta_h^n \phi(x) = \sum_{r=0}^{n} (-1)^r \binom{n}{r} \phi(x+rh) = (-h)^n \phi^{(n)}(\xi),$$

where ξ lies between x and $x+nh$. [Use the auxiliary function

$$\psi(t) = \Delta_t^n \phi(x) - \left(\frac{t}{h}\right)^n \Delta_h^n \phi(x).$$

Ex. LXVII. 1 is substantially the case $n = 2$.]

6. Deduce from Ex. 5 that $x^{n-m}\Delta_h^n x^m \to m(m-1)\dots(m-n+1)h^n$ when $x \to \infty$, m being any rational number and n any positive integer. In particular prove that

$$x\sqrt{x}\{\sqrt{x} - 2\sqrt{(x+1)} + \sqrt{(x+2)}\} \to -\tfrac{1}{4}.$$

7. Suppose that $y = \phi(x)$ is a function of x with continuous derivatives of the first four orders, and that $\phi(0) = 0$, $\phi'(0) = 1$, so that

$$y = \phi(x) = x + a_2x^2 + a_3x^3 + a_4x^4 + o(x^4).$$

Prove that

$$x = \psi(y) = y - a_2y^2 + (2a_2^2 - a_3)y^3 - (5a_2^3 - 5a_2a_3 + a_4)y^4 + o(y^4),$$

and that

$$\frac{\phi(x)\psi(x) - x^2}{x^4} \to a_2^2$$

when $x \to 0$.

8. The coordinates (ξ, η) of the centre of curvature of the curve $x = f(t)$, $y = F(t)$, at the point (x, y), are given by

$$-\frac{\xi - x}{y'} = \frac{\eta - y}{x'} = \frac{x'^2 + y'^2}{x'y'' - x''y'};$$

and the radius of curvature of the curve is $(x'^2 + y'^2)^{\frac{3}{2}}/(x'y'' - x''y')$, dashes denoting differentiations with respect to t.

9. The coordinates (ξ, η) of the centre of curvature of the curve $27ay^2 = 4x^3$, at the point (x, y), are given by

$$3a(\xi + x) + 2x^2 = 0, \quad \eta = 4y + (9ay)/x.$$

(*Math. Trip.* 1899)

10. Prove that the circle of curvature at a point (x, y) will have contact of the third order with the curve if $(1 + y_1^2)\, y_3 = 3y_1 y_2^2$ at that point. Prove also that the circle is the only curve which possesses this property at every point; and that the only points on a conic which possess the property are the extremities of the axes. [Cf. Ch. VI, Misc. Ex. 13 (iv).]

11. The conic of closest contact with the curve

$$y = ax^2 + bx^3 + cx^4 + \dots + kx^n$$

at the origin is $\quad a^3y = a^4x^2 + a^2bxy + (ac - b^2)\, y^2.$

Deduce that the conic of closest contact at the point (ξ, η) of the curve $y = f(x)$ is

$$18\eta_2^3 T = 9\eta_2^4(x - \xi)^2 + 6\eta_2^2\eta_3(x - \xi)\, T + (3\eta_2\eta_4 - 4\eta_3^2)\, T^2,$$

where $T = (y - \eta) - \eta_1(x - \xi)$. (*Math. Trip.* 1907)

12. **Homogeneous functions*.** If $u = x^n f(y/x, z/x, \dots)$ then u is unaltered, save for a factor λ^n, when $x, y, z, \dots$ are all increased in the ratio $\lambda : 1$. In these circumstances u is called a *homogeneous function of degree n* in the variables $x, y, z, \dots$. Prove that if u is homogeneous and of degree n then

$$x\frac{\partial u}{\partial x} + y\frac{\partial u}{\partial y} + z\frac{\partial u}{\partial z} + \dots = nu.$$

This result is known as *Euler's theorem* on homogeneous functions.

13. If u is homogeneous and of degree n then $u_x, u_y, \dots$ are homogeneous and of degree $n - 1$.

14. Let $f(x, y) = 0$ be an equation in x and y (e.g. $x^n + y^n - x = 0$), and let $F(x, y, z) = 0$ be the form it assumes when made homogeneous by the introduction of a third variable z in place of unity (e.g. $x^n + y^n - xz^{n-1} = 0$).

* In this and the following examples the reader is to assume the continuity of all the derivatives which occur.

Show that the equation of the tangent at the point (ξ, η) of the curve $f(x, y) = 0$ is

$$xF_\xi + yF_\eta + F_\zeta = 0,$$

where F_ξ, F_η, F_ζ denote the values of F_x, F_y, F_z when $x = \xi$, $y = \eta$, $z = \zeta = 1$.

15. Dependent and independent functions. Jacobians or functional determinants. Suppose that u and v are functions of x and y connected by an identical relation

$$\phi(u, v) = 0 \quad \ldots\ldots\ldots\ldots\ldots\ldots\ldots\ldots(1).$$

Differentiating (1) with respect to x and y, we obtain

$$\frac{\partial\phi}{\partial u}\frac{\partial u}{\partial x} + \frac{\partial\phi}{\partial v}\frac{\partial v}{\partial x} = 0, \quad \frac{\partial\phi}{\partial u}\frac{\partial u}{\partial y} + \frac{\partial\phi}{\partial v}\frac{\partial v}{\partial y} = 0 \quad \ldots\ldots\ldots\ldots(2),$$

and, eliminating the derivatives of ϕ,

$$J = \begin{vmatrix} u_x & u_y \\ v_x & v_y \end{vmatrix} = u_x v_y - u_y v_x = 0 \quad \ldots\ldots\ldots\ldots(3),$$

where u_x, u_y, v_x, v_y are the derivatives of u and v with respect to x and y. This condition is therefore *necessary* for the existence of a relation such as (1). It can be proved that the condition is also *sufficient*; for this we may refer to Goursat's *Cours d'analyse*, 3rd edition, vol. I, pp. 126 *et seq.*

Two functions u and v are said to be *dependent* or *independent* according as they are or are not connected by such a relation as (1). It is usual to call J the *Jacobian* or *functional determinant* of u and v with respect to x and y, and to write

$$J = \frac{\partial(u, v)}{\partial(x, y)}.$$

Similar results hold for functions of any number of variables. Thus three functions u, v, w of three variables x, y, z are or are not connected by a relation $\phi(u, v, w) = 0$ according as

$$J = \begin{vmatrix} u_x & u_y & u_z \\ v_x & v_y & v_z \\ w_x & w_y & w_z \end{vmatrix} = \frac{\partial(u, v, w)}{\partial(x, y, z)}$$

does or does not vanish for all values of x, y, z.

16. Show that $ax^2 + by^2 + cz^2 + 2fyz + 2gzx + 2hxy$ can be expressed as a product of two linear functions of x, y, and z if and only if

$$abc + 2fgh - af^2 - bg^2 - ch^2 = 0.$$

[Write down the condition that $px + qy + rz$ and $p'x + q'y + r'z$ should be connected with the given function by a functional relation.]

17. If u and v are functions of ξ and η, which are themselves functions of x and y, then

$$\frac{\partial(u, v)}{\partial(x, y)} = \frac{\partial(u, v)}{\partial(\xi, \eta)} \frac{\partial(\xi, \eta)}{\partial(x, y)}.$$

Extend the result to any number of variables.

18. Let $f(x)$ be a function of x whose derivative is $1/x$ and which vanishes when $x = 1$. Show that if $u = f(x) + f(y)$, $v = xy$, then $u_x v_y - u_y v_x = 0$, and hence that u and v are connected by a functional relation. By putting $y = 1$, show that this relation must be $f(x) + f(y) = f(xy)$. Prove in a similar manner that if the derivative of $f(x)$ is $1/(1+x^2)$, and $f(0) = 0$, then $f(x)$ must satisfy the equation

$$f(x) + f(y) = f\left(\frac{x+y}{1-xy}\right).$$

19. Prove that if $f(x) = \int_0^x \frac{dt}{\sqrt{(1-t^4)}}$ then

$$f(x) + f(y) = f\left\{\frac{x\sqrt{(1-y^4)} + y\sqrt{(1-x^4)}}{1+x^2y^2}\right\}.$$

20. Show that if a functional relation exists between

$$u = f(x) + f(y) + f(z), \quad v = f(y)f(z) + f(z)f(x) + f(x)f(y), \quad w = f(x)f(y)f(z),$$

then f must be a constant. [The condition for a functional relation will be found to be

$$f'(x)f'(y)f'(z)\{f(y) - f(z)\}\{f(z) - f(x)\}\{f(x) - f(y)\} = 0.]$$

21. If $f(y, z)$, $f(z, x)$, and $f(x, y)$ are connected by a functional relation then $f(x, x)$ is independent of x. (*Math. Trip.* 1909)

22. If $u = 0$, $v = 0$, $w = 0$ are the equations of three circles, rendered homogeneous as in Ex. 14, then the equation

$$\frac{\partial(u, v, w)}{\partial(x, y, z)} = 0$$

represents the circle which cuts them all orthogonally.

(*Math. Trip.* 1900)

23. Calculate $\frac{\partial(\lambda, \mu)}{\partial(x, y)}$ when

$$\frac{x^2}{a^2+\lambda} + \frac{y^2}{b^2+\lambda} = \frac{x^2}{a^2+\mu} + \frac{y^2}{b^2+\mu} = 1.$$

(*Math. Trip.* 1936)

24. If A, B, C are three functions of x such that

$$\begin{vmatrix} A & A' & A'' \\ B & B' & B'' \\ C & C' & C'' \end{vmatrix}$$

vanishes identically, then we can find constants λ, μ, ν such that

$$\lambda A+\mu B+\nu C$$

vanishes identically; and conversely. [The converse is almost obvious. To prove the direct theorem let $\alpha = BC'-B'C, \ldots$. Then $\alpha' = BC''-B''C, \ldots$, and it follows from the vanishing of the determinant that $\beta\gamma'-\beta'\gamma = 0, \ldots$; and so that the ratios $\alpha:\beta:\gamma$ are constant. But $\alpha A+\beta B+\gamma C = 0$.]

25. Suppose that three variables x, y, z are connected by a relation in virtue of which (i) z is a function of x and y, with derivatives z_x, z_y, and (ii) x is a function of y and z, with derivatives x_y, x_z. Prove that

$$x_y = -z_y/z_x, \quad x_z = 1/z_x.$$

[We have $\qquad dz = z_x dx+z_y dy, \quad dx = x_y dy+x_z dz.$

The result of substituting for dx in the first equation is

$$dz = (z_x x_y+z_y)\,dy+z_x x_z dz,$$

which can be true only if $z_x x_y+z_y = 0$, $z_x x_z = 1$.]

26. Four variables x, y, z, u are connected by two relations in virtue of which any two can be expressed as functions of the others. Show that

$$x_y^z y_z^u+x_u^z u_z^y = 0, \quad y_z^u z_x^u x_y^u = -y_z^x z_x^y x_y^z = 1, \quad x_z^u z_x^y+y_z^u z_y^x = 1,$$

where y_z^u denotes the derivative of y, when expressed as a function of z and u, with respect to z. (*Math. Trip.* 1897, 1928)

27. The variables x, y, z are connected by

$$x^2+y^2+z^2-3xyz = 0,$$

and $\phi(x,y,z) = x^3y^2z$. Determine the value of ϕ_x at $(1, 1, 1)$ when the independent variables are (i) x and y, (ii) x and z; and explain geometrically the difference between the meanings of ϕ_x in the two cases.

(*Math. Trip.* 1936)

28. If $x^2 = vw$, $y^2 = wu$, $z^2 = uv$ and $f(x,y,z) = \phi(u,v,w)$, then

$$xf_x+yf_y+zf_z = u\phi_u+v\phi_v+w\phi_w.$$

(*Math. Trip.* 1933)

29. Find the value of $\phi_y(0,0)$ when

$$\phi(x,y) = \frac{(x+y)^2\,(x-y)}{x^2+y^2}$$

unless x and y are both zero, and $\phi(0,0) = 0$; and explain why $\phi(x,y) = 0$ does not define y as a single-valued function of x near the origin.

(*Math. Trip.* 1928)

30. The function $\phi(u,v,x,y)$ is homogeneous, of the second degree, in u, v; $\phi_u = p$, $\phi_v = q$; and $\phi(u,v,x,y)$, when expressed in terms of p, q, x, y,

is $\psi(p,q,x,y)$. Prove that

$$\psi_p = u, \quad \psi_q = v, \quad \psi_x = -\phi_x, \quad \psi_y = -\phi_y.$$

(*Math. Trip.* 1936)

[By Euler's theorem (Ex. 12), $u\phi_u + v\phi_v = 2\phi$, or $pu+qv = 2\psi$, when u and v are expressed as functions of p, q, x, y. Hence

$$u+pu_p+qv_p = 2\psi_p.$$

But
$$\psi_p = \phi_u u_p + \phi_v v_p = pu_p + qv_p,$$
and so $\psi_p = u$. The other results may be proved similarly.]

31. If $a>0$, $ac-b^2>0$, and $x_1>x_0$, then

$$\int_{x_0}^{x_1} \frac{dx}{ax^2+2bx+c} = \frac{1}{\sqrt{(ac-b^2)}} \arctan\left\{\frac{(x_1-x_0)\sqrt{(ac-b^2)}}{ax_1x_0+b(x_1+x_0)+c}\right\},$$

the inverse tangent lying between 0 and π*.

32. Evaluate the integral $\displaystyle\int_{-1}^{1} \frac{\sin\alpha\, dx}{1-2x\cos\alpha+x^2}$. For what values of α is the integral a discontinuous function of α? (*Math. Trip.* 1904)

[The value of the integral is $\frac{1}{2}\pi$ if $2n\pi<\alpha<(2n+1)\pi$, and $-\frac{1}{2}\pi$ if $(2n-1)\pi<\alpha<2n\pi$, n being any integer; and 0 if α is a multiple of π.]

33. If $ax^2+2bx+c>0$ when $x_0 \leqq x \leqq x_1$, $f(x) = \sqrt{(ax^2+2bx+c)}$, and

$$y=f(x), \quad y_0=f(x_0), \quad y_1=f(x_1), \quad X=(x_1-x_0)/(y_1+y_0),$$

then
$$\int_{x_0}^{x_1} \frac{dx}{y} = \frac{1}{\sqrt{a}}\log\frac{1+X\sqrt{a}}{1-X\sqrt{a}}, \quad \frac{2}{\sqrt{(-a)}}\arctan\{X\sqrt{(-a)}\},$$

according as a is positive or negative. In the latter case the inverse tangent lies between 0 and $\frac{1}{2}\pi$. [It will be found that the substitution $t=\dfrac{x-x_0}{y+y_0}$ reduces the integral to the form $2\displaystyle\int_0^X \frac{dt}{1-at^2}$.]

34. Prove that
$$\int_0^a \frac{dx}{x+\sqrt{(a^2-x^2)}} = \tfrac{1}{4}\pi. \qquad (\textit{Math. Trip.}\ 1913)$$

35. If $a>1$ then $\displaystyle\int_{-1}^{1} \frac{\sqrt{(1-x^2)}}{a-x}\,dx = \pi\{a-\sqrt{(a^2-1)}\}$.

36. If $p>1$, $0<q<1$, then

$$\int_0^1 \frac{dx}{\sqrt{[\{1+(p^2-1)x\}\{1-(1-q^2)x\}]}} = \frac{2\omega}{(p+q)\sin\omega},$$

where ω is the positive acute angle whose cosine is $(1+pq)/(p+q)$.

* In connection with Exs. 31, 33, 36, 38 see a paper by Bromwich in vol. xxxv of the *Messenger of mathematics*.

37. If $a>b>0$, then $\displaystyle\int_0^{2\pi}\frac{\sin^2\theta\,d\theta}{a-b\cos\theta}=\frac{2\pi}{b^2}\{a-\sqrt{(a^2-b^2)}\}$.

(*Math. Trip.* 1904)

38. Prove that if $a>\sqrt{(b^2+c^2)}$ then

$$\int_0^{\pi}\frac{d\theta}{a+b\cos\theta+c\sin\theta}=\frac{2}{\sqrt{(a^2-b^2-c^2)}}\arctan\left\{\frac{\sqrt{(a^2-b^2-c^2)}}{c}\right\},$$

the inverse tangent lying between 0 and π.

39. Prove that if $m\geqq 1$ and

$$I_{m,n}=\int_0^{\frac{1}{2}\pi}\sin^m x\cos nx\,dx,\quad J_{m,n}=\int_0^{\frac{1}{2}\pi}\sin^m x\sin nx\,dx,$$

then

$$(m+n)\,I_{m,n}=\sin\tfrac{1}{2}n\pi-mJ_{m-1,n-1};$$

and express $I_{m,n}$ in terms of $I_{m-2,n-2}$ when $m\geqq 2$. (*Math. Trip.* 1933)

40. Prove, by integrating the inequalities

$$\frac{1-\sin^{2n-1}x}{2n-1}>\frac{1-\sin^{2n}x}{2n}>\frac{1-\sin^{2n+1}x}{2n+1}$$

from 0 to $\frac{1}{2}\pi$, and using Ex. LXVI. 10, that

$$p_{n-1}\left(1+\frac{2n-1}{2n}p_{n-1}\right)>\frac{4n}{\pi}>p_n(p_n-1),$$

where

$$p_n=\frac{3.5\ldots(2n+1)}{2.4\ldots 2n}.$$

(*Math. Trip.* 1924)

41. Find a formula of reduction for $\displaystyle\int_0^x\sin^{2n-1}\theta\,d\theta$, and deduce that

$$1=\cos x+\tfrac{1}{2}\cos x\sin^2 x+\ldots+\frac{1.3\ldots(2n-3)}{2.4\ldots(2n-2)}\cos x\sin^{2n-2}x+r_n,$$

$$\alpha=\sin\alpha+\frac{1}{2}\frac{\sin^3\alpha}{3}+\ldots+\frac{1.3\ldots(2n-3)}{2.4\ldots(2n-2)}\frac{\sin^{2n-1}\alpha}{2n-1}+R_n,$$

where

$$r_n=\frac{3.5\ldots(2n-1)}{2.4\ldots(2n-2)}\int_0^x\sin^{2n-1}\theta\,d\theta,$$

and

$$R_n=\int_0^\alpha r_n\,dx=\frac{3.5\ldots(2n-1)}{2.4\ldots(2n-2)}\int_0^\alpha(\alpha-x)\sin^{2n-1}x\,dx.$$

Prove that $x+\alpha\cos x\geqq\alpha$ if $0\leqq x\leqq\alpha\leqq\frac{1}{2}\pi$, and hence that

$$R_n\leqq\frac{1.3\ldots(2n-1)}{2.4\ldots 2n}\alpha\sin^{2n}\alpha.$$

(*Math. Trip.* 1924)

42. Prove, by the substitution $\sqrt{(1+x^4)}=(1+x^2)\cos\phi$, or otherwise, that

$$\int_0^1\frac{1-x^2}{1+x^2}\frac{dx}{\sqrt{(1+x^4)}}=\frac{\pi}{4\sqrt{2}}.$$

(*Math. Trip.* 1923)

43. If $f(x)$ is continuous and never negative, and $\int_a^b f(x)\,dx = 0$, then $f(x) = 0$ for all values of x between a and b. [If $f(x)$ were equal to a positive number k when $x = \xi$, say, then we could, in virtue of the continuity of $f(x)$, find an interval $(\xi-\delta, \xi+\delta)$ throughout which $f(x) > \frac{1}{2}k$; and then the value of the integral would be greater than δk.]

44. **Schwarz's inequality.** Prove that

$$\left(\int_a^b \phi\psi\,dx\right)^2 \leqq \int_a^b \phi^2\,dx \int_a^b \psi^2\,dx.$$

[Observe that

$$\int_a^b (\lambda\phi+\mu\psi)^2\,dx = \lambda^2\int_a^b \phi^2\,dx + 2\lambda\mu\int_a^b \phi\psi\,dx + \mu^2\int_a^b \psi^2\,dx$$

cannot be negative. The inequality can also be deduced as a limiting case from Cauchy's inequality (Ch. I, Misc. Ex. 10).]

45. If $P_n(x) = \dfrac{1}{(\beta-\alpha)^n n!}\left(\dfrac{d}{dx}\right)^n \{(x-\alpha)(\beta-x)\}^n$, then $P_n(x)$ is a polynomial of degree n, which possesses the property that

$$\int_\alpha^\beta P_n(x)\,\theta(x)\,dx = 0$$

if $\theta(x)$ is any polynomial of degree less than n. [Integrate by parts $m+1$ times, where m is the degree of $\theta(x)$, and observe that $\theta^{(m+1)}(x) = 0$.]

46. Prove that $\int_\alpha^\beta P_m(x)\,P_n(x)\,dx = 0$ if $m \neq n$, but that if $m = n$ then the value of the integral is $(\beta-\alpha)/(2n+1)$.

47. If $Q_n(x)$ is a polynomial of degree n, which possesses the property that $\int_\alpha^\beta Q_n(x)\,\theta(x)\,dx = 0$ if $\theta(x)$ is any polynomial of degree less than n, then $Q_n(x)$ is a constant multiple of $P_n(x)$.

[We can choose κ so that $Q_n - \kappa P_n$ is of degree $n-1$: then

$$\int_\alpha^\beta Q_n(Q_n-\kappa P_n)\,dx = 0, \quad \int_\alpha^\beta P_n(Q_n-\kappa P_n)\,dx = 0,$$

and so

$$\int_\alpha^\beta (Q_n-\kappa P_n)^2\,dx = 0.$$

Now apply Ex. 43.]

48. If $\phi(x)$ is a polynomial of the fifth degree, then

$$\int_0^1 \phi(x)\,dx = \tfrac{1}{18}\{5\phi(\alpha)+8\phi(\tfrac{1}{2})+5\phi(\beta)\},$$

α and β being the roots of the equation $x^2 - x + \frac{1}{10} = 0$.

(*Math. Trip.* 1909)

CHAPTER VIII

THE CONVERGENCE OF INFINITE SERIES AND INFINITE INTEGRALS

171. In Ch. IV we explained what was meant by saying that an infinite series is *convergent*, *divergent*, or *oscillatory*, and illustrated our definitions by a few simple examples, mainly derived from the geometrical series

$$1+x+x^2+\dots$$

and other series connected with it. In this chapter we shall pursue the subject in a more systematic manner, and prove a number of theorems which enable us to determine when the simplest series which occur commonly in analysis are convergent.

We shall often use the notation

$$u_m+u_{m+1}+\dots+u_n=\sum_m^n u_\nu,$$

and write $\sum\limits_0^\infty u_n$, or simply Σu_n, for the infinite series

$$u_0+u_1+u_2+\dots*.$$

172. Series of positive terms. The theory of the convergence of series is comparatively simple when all the terms of the series considered are positive†. We shall consider such series first, not only because they are the easiest to deal with, but also

* It is immaterial whether we denote our series by $u_1+u_2+\dots$ (as in Ch. IV) or by $u_0+u_1+\dots$ (as here). Later in this chapter we shall be concerned with series of the type $a_0+a_1x+a_2x^2+\dots$: for these the latter notation is clearly more convenient. We shall therefore adopt this as our standard notation. But we shall not adhere to it systematically, and we shall suppose that u_1 is the first term whenever this course is more convenient. It is more convenient, for example, when dealing with the series $1+\frac{1}{2}+\frac{1}{3}+\dots$, to suppose that $u_n=1/n$ and that the series begins with u_1, than to suppose that $u_n=1/(n+1)$ and that the series begins with u_0. This remark applies, e.g., to Ex. LXVIII. 4.

† Here and in what follows 'positive' is to be regarded as including zero.

because the discussion of the convergence of a series containing negative or complex terms can often be made to depend upon a similar discussion of a series of positive terms only.

When we are discussing the convergence or divergence of a series we may disregard any finite number of terms. Thus, when a series contains a finite number only of negative or complex terms, we may omit them and apply the theorems which follow to the remainder.

173. It will be well to recall the following fundamental theorems established in § 77.

A. *A series of positive terms must be convergent or diverge to* ∞, *and cannot oscillate.*

B. *A necessary and sufficient condition that* Σu_n *should be convergent is that there should be a number* K *such that*

$$u_0 + u_1 + \dots + u_n < K$$

for all values of n.

C. **The comparison theorem.** *If* Σu_n *is convergent, and* $v_n \leqq u_n$ *for all values of* n, *then* Σv_n *is convergent, and* $\Sigma v_n \leqq \Sigma u_n$. *More generally, if* $v_n \leqq K u_n$, *where* K *is a constant, then* Σv_n *is convergent and* $\Sigma v_n \leqq K \Sigma u_n$. *And if* Σu_n *is divergent, and* $v_n \geqq K u_n$, *with a positive* K, *then* Σv_n *is divergent.*

Moreover, in inferring the convergence or divergence of Σv_n by means of one of these tests, it is sufficient to know that the test is satisfied for *sufficiently large* values of n, i.e. for all values of n greater than a definite value n_0. But of course the conclusion that $\Sigma v_n \leqq K \Sigma u_n$ does not necessarily hold in this case.

A particularly useful case of this theorem is

D. *If* Σu_n *is convergent (divergent) and* u_n/v_n *tends to a limit other than zero as* $n \to \infty$, *then* Σv_n *is convergent (divergent).*

174. First applications of these tests. The most important theorem we have proved about the convergence of any particular series is that Σr^n is convergent if $r<1$ and divergent if $r \geqq 1$*, and it is natural to take $u_n = r^n$ in Theorem C. We find

1. *The series Σv_n is convergent if $v_n \leqq Kr^n$, where $r<1$, for all sufficiently large values of n.*

When $K=1$, this condition may be written in the form $v_n^{1/n} \leqq r$. Hence we obtain what is known as *Cauchy's test* for the convergence of a series of positive terms; viz.

2. *The series Σv_n is convergent if $v_n^{1/n} \leqq r$, where $r<1$, for all sufficiently large values of n.*

On the other hand

3. *The series Σv_n is divergent if $v_n^{1/n} \geqq 1$ for an infinity of values of n.*

This is obvious because $v_n^{1/n} \geqq 1$ involves $v_n \geqq 1$.

175. Ratio tests. There are also very useful tests involving the ratio v_{n+1}/v_n of two successive terms of a series. In these we must suppose u_n and v_n *strictly* positive.

Suppose that $u_n > 0$, $v_n > 0$, and that

$$\frac{v_{n+1}}{v_n} \leqq \frac{u_{n+1}}{u_n} \quad \text{.........................(1)}$$

for sufficiently large n, say for $n \geqq n_0$. Then

$$v_n = \frac{v_{n_0+1}}{v_{n_0}} \frac{v_{n_0+2}}{v_{n_0+1}} \cdots \frac{v_n}{v_{n-1}} v_{n_0} \leqq \frac{u_{n_0+1}}{u_{n_0}} \frac{u_{n_0+2}}{u_{n_0+1}} \cdots \frac{u_n}{u_{n-1}} v_{n_0} = \frac{v_{n_0}}{u_{n_0}} u_n,$$

so that $v_n \leqq Ku_n$, where K is independent of n. Similarly

$$\frac{v_{n+1}}{v_n} \geqq \frac{u_{n+1}}{u_n} \quad \text{.........................(2)}$$

for $n \geqq n_0$ implies $v_n \geqq Ku_n$, with a positive K. Hence

4. *If* (1) *is true for sufficiently large n, and Σu_n is convergent, then Σv_n is convergent.*

5. *If* (2) *is true for sufficiently large n, and Σu_n is divergent, then Σv_n is divergent.*

* Throughout this chapter r is positive, in the wider sense which includes zero.

Taking $u_n = r^n$ in Theorem 4 we find

6. *The series Σv_n is convergent if $v_{n+1}/v_n \leqq r$, where $r < 1$, for all sufficiently large n.*

This test is known as *d'Alembert's test*. The corresponding divergence test, that Σv_n is divergent if $v_{n+1}/v_n \geqq r$, where $r \geqq 1$, for all sufficiently large n, is trivial.

We shall see later that d'Alembert's test is theoretically less general than Cauchy's, in that Cauchy's test can always be applied when d'Alembert's test can, and often when it cannot: see, for example, Ex. LXVIII. 9 below. Ratio tests are useless for 'irregular' series, such as $0+\frac{1}{2}+0+\frac{1}{4}+0+\frac{1}{8}+\dots$. None the less d'Alembert's test is very useful in practice, because when v_n is a complicated function v_{n+1}/v_n is often much less complicated and so easier to work with.

It often happens that v_{n+1}/v_n or $v_n^{1/n}$ tends to a limit as $n \to \infty$*. When this limit is less than 1, it is evident that the conditions of Theorems 2 or 6 above are satisfied. Thus

7. *If $v_n^{1/n}$ or v_{n+1}/v_n tends to a limit less than unity when $n \to \infty$, then Σv_n is convergent.*

It is almost obvious that if either function tends to a limit greater than unity, then Σv_n is divergent. We leave the formal proof of this as an exercise to the reader. But when $v_n^{1/n}$ or v_{n+1}/v_n tends to 1 these tests fail. They fail also when $v_n^{1/n}$ or v_{n+1}/v_n oscillates in such a way that, while always less than 1, it assumes for an infinity of values of n values approaching indefinitely near to 1; and the tests which involve v_{n+1}/v_n fail also when that ratio oscillates so as to be sometimes less than and sometimes greater than 1. When $v_n^{1/n}$ behaves in this way Theorem 3 is sufficient to prove the divergence of the series. But it is clear that there is a wide margin of cases in which more subtle tests will be needed.

* It will be proved in Ch. IX (Ex. LXXXVII. 36) that $v_n^{1/n} \to l$ whenever $v_{n+1}/v_n \to l$. That the converse is not true may be seen by supposing that $v_n = 1$ when n is odd and $v_n = 2$ when n is even.

Examples LXVIII. 1. Apply Cauchy's and d'Alembert's tests (as specialised in 7 above) to the series $\Sigma n^k r^n$, where k is a positive integer.

[Here
$$\frac{v_{n+1}}{v_n} = \left(\frac{n+1}{n}\right)^k r \to r,$$
and d'Alembert's test shows that the series is convergent if $r<1$ and divergent if $r>1$. The test fails if $r=1$; but the series is then obviously divergent. Since $\lim n^{1/n} = 1$ (Ex. XXVII. 11), Cauchy's test leads to the same conclusions.]

2. Consider the series $\Sigma(An^k + Bn^{k-1} + \dots + K)r^n$. [We may suppose A positive. If the coefficient of r^n is denoted by $P(n)$, then $P(n) \sim An^k$ and, by D of § 173, the series behaves like $\Sigma n^k r^n$.]

3. Consider $\Sigma \dfrac{An^k + Bn^{k-1} + \dots + K}{\alpha n^l + \beta n^{l-1} + \dots + \kappa} r^n \quad (A>0, \alpha>0).$

[The series behaves like $\Sigma n^{k-l} r^n$. The case in which $r=1$, $k<l$ requires further consideration.]

4. We have seen (Ch. IV, Misc. Ex. 25) that the series
$$\Sigma \frac{1}{n(n+1)}, \quad \Sigma \frac{1}{n(n+1)\dots(n+p)}$$
are convergent. Show that Cauchy's and d'Alembert's tests both fail when applied to them. [For $\lim u_n^{1/n} = \lim (u_{n+1}/u_n) = 1$.]

5. Show that the series Σn^{-p}, where p is an integer not less than 2, is convergent. [Since $n(n+1)\dots(n+p-1) \sim n^p$, this follows from the convergence of the series considered in Ex. 4. We proved in § 77 (7) that the series is divergent if $p=1$, and it is obviously divergent if $p \leqq 0$.]

6. Show that the series of Ex. 3 is convergent if $r=1$, $l>k+1$, and divergent if $r=1$, $l \leqq k+1$.

7. If m_n is a positive integer, and $m_{n+1} > m_n$, then the series Σr^{m_n} is convergent if $r<1$ and divergent if $r \geqq 1$. For example the series $1 + r + r^4 + r^9 + \dots$ is convergent if $r<1$ and divergent if $r \geqq 1$.

8. Sum the series $1 + 2r + 2r^4 + \dots$ to 24 places of decimals when $r = \cdot 1$ and to 2 places when $r = \cdot 9$. [If $r = \cdot 1$, then the first 5 terms give the sum $1 \cdot 2002000020000002$, and the error is
$$2r^{25} + 2r^{36} + \dots < 2r^{25} + 2r^{36} + 2r^{47} + \dots = 2r^{25}/(1-r^{11}) < 3 \,.\, 10^{-25}.$$
If $r = \cdot 9$, then the first 8 terms give the sum $5 \cdot 458\dots$, and the error is less than $2r^{64}/(1-r^{17}) < \cdot 003$.]

9. If $0<a<b<1$, then the series $a+b+a^2+b^2+a^3+\dots$ is convergent. Show that Cauchy's test may be applied to this series, but that d'Alembert's test fails. [For $v_{2n+1}/v_{2n} = (b/a)^{n+1}\to\infty$, $v_{2n+2}/v_{2n+1} = b(a/b)^{n+2}\to 0$.]

10. The series $\Sigma\dfrac{r^n}{n!}$ and $\Sigma\dfrac{r^n}{n^n}$ are convergent for all r, and $\Sigma n!\,r^n$ and $\Sigma n^n r^n$ for no r except $r=0$. (*Math. Trip.* 1935, 1936)

11. The series
$$\Sigma\left(\frac{nr}{n+1}\right)^n,\quad \Sigma\frac{\{(n+1)r\}^n}{n^{n+1}}$$
are convergent if $r<1$ and divergent if $r\geqq 1$. (*Math. Trip.* 1927, 1928)

[Use § 73 and § 77 (7) when $r=1$.]

12. If Σu_n is convergent then so are Σu_n^2 and $\Sigma\dfrac{u_n}{1+u_n}$.

13. If Σu_n^2 is convergent then so is $\Sigma n^{-1}u_n$. [For $2n^{-1}u_n\leqq u_n^2+n^{-2}$ and Σn^{-2} is convergent.]

14. Show that $1+\dfrac{1}{3^2}+\dfrac{1}{5^2}+\dots=\dfrac{3}{4}\left(1+\dfrac{1}{2^2}+\dfrac{1}{3^2}+\dots\right)$ and
$$1+\frac{1}{2^2}+\frac{1}{3^2}+\frac{1}{5^2}+\frac{1}{6^2}+\frac{1}{7^2}+\frac{1}{9^2}+\dots=\frac{15}{16}\left(1+\frac{1}{2^2}+\frac{1}{3^2}+\dots\right).$$

[To prove the first result we note that
$$1+\frac{1}{2^2}+\frac{1}{3^2}+\dots=\left(1+\frac{1}{2^2}\right)+\left(\frac{1}{3^2}+\frac{1}{4^2}\right)+\dots$$
$$=1+\frac{1}{3^2}+\frac{1}{5^2}+\dots+\frac{1}{2^2}\left(1+\frac{1}{2^2}+\frac{1}{3^2}+\dots\right),$$
by Theorems (8), (6) and (4) of § 77.]

15. Prove by a *reductio ad absurdum* that Σn^{-1} is divergent. [If the series were convergent we should have, by the argument used in Ex. 14,
$$1+\tfrac{1}{2}+\tfrac{1}{3}+\dots=(1+\tfrac{1}{3}+\tfrac{1}{5}+\dots)+\tfrac{1}{2}(1+\tfrac{1}{2}+\tfrac{1}{3}+\dots),$$
or
$$\tfrac{1}{2}+\tfrac{1}{4}+\tfrac{1}{6}+\dots=1+\tfrac{1}{3}+\tfrac{1}{5}+\dots$$
which is absurd, since every term of the first series is less than the corresponding term of the second.]

176. Before proceeding further in the investigation of tests of convergence and divergence, we shall prove an important general theorem concerning series of positive terms.

Dirichlet's theorem*. *The sum of a series of positive terms is the same in whatever order the terms are taken.*

This theorem asserts that if we have a convergent series of positive terms, $u_0 + u_1 + u_2 + \dots$ say, and form any other series

$$v_0 + v_1 + v_2 + \dots$$

out of the same terms, by taking them in any new order, then the second series is convergent and has the same sum as the first. Of course no terms must be omitted: every u must come somewhere among the v's, and *vice versa*.

The proof is extremely simple. Let s be the sum of the series of u's. Then the sum of any number of terms, selected from the u's, is not greater than s. But every v is a u, and therefore the sum of any number of terms selected from the v's is not greater than s. Hence Σv_n is convergent, and its sum t is not greater than s. But we can show in exactly the same way that $s \leqq t$. Thus $s = t$.

177. Multiplication of series of positive terms. An immediate corollary from Dirichlet's theorem is the following theorem: *if* $u_0 + u_1 + u_2 + \dots$ *and* $v_0 + v_1 + v_2 + \dots$ *are two convergent series of positive terms, and* s *and* t *are their respective sums, then the series*

$$u_0 v_0 + (u_1 v_0 + u_0 v_1) + (u_2 v_0 + u_1 v_1 + u_0 v_2) + \dots$$

is convergent and has the sum st.

Arrange all the possible products of pairs $u_m v_n$ in the form of a doubly infinite array

$$\begin{array}{ccccc}
u_0 v_0 & u_1 v_0 & u_2 v_0 & u_3 v_0 & \dots \\
u_0 v_1 & u_1 v_1 & u_2 v_1 & u_3 v_1 & \dots \\
u_0 v_2 & u_1 v_2 & u_2 v_2 & u_3 v_2 & \dots \\
u_0 v_3 & u_1 v_3 & u_2 v_3 & u_3 v_3 & \dots \\
\dots & \dots & \dots & \dots & \dots
\end{array}$$

* This theorem seems to have first been stated explicitly by Dirichlet in 1837. It was no doubt known to earlier writers, and in particular to Cauchy.

We can rearrange these terms in the form of a simply infinite series in a variety of ways. Among these are the following.

(1) We begin with the single term $u_0 v_0$ for which $m+n=0$; then we take the two terms $u_1 v_0$, $u_0 v_1$ for which $m+n=1$; then the three terms $u_2 v_0$, $u_1 v_1$, $u_0 v_2$ for which $m+n=2$; and so on. We thus obtain the series

$$u_0 v_0 + (u_1 v_0 + u_0 v_1) + (u_2 v_0 + u_1 v_1 + u_0 v_2) + \dots$$

of the theorem.

(2) We begin with the single term $u_0 v_0$ for which both suffixes are zero; then we take the terms $u_1 v_0$, $u_1 v_1$, $u_0 v_1$ which involve a suffix 1 but no higher suffix; then the terms $u_2 v_0$, $u_2 v_1$, $u_2 v_2$, $u_1 v_2$, $u_0 v_2$ which involve a suffix 2 but no higher suffix; and so on. The sums of these groups of terms are respectively equal to

$$u_0 v_0, \quad (u_0+u_1)(v_0+v_1) - u_0 v_0,$$
$$(u_0+u_1+u_2)(v_0+v_1+v_2) - (u_0+u_1)(v_0+v_1), \quad \dots$$

and the sum of the first $n+1$ groups is

$$(u_0+u_1+\dots+u_n)(v_0+v_1+\dots+v_n),$$

and tends to st as $n \to \infty$. When the sum of the series is formed in this manner the sum of the first one, two, three, ... groups comprises all the terms in the first, second, third, ... rectangles indicated in the diagram on p. 347.

The sum of the series formed in the second manner is st. But the first series is (when the brackets are removed) a rearrangement of the second; and therefore, by Dirichlet's theorem, it converges to the sum st. Thus the theorem is proved.

Examples LXIX. 1. Verify that if $r<1$ then

$$1+r^2+r+r^4+r^6+r^3+\dots = 1+r+r^3+r^2+r^5+r^7+\dots = 1/(1-r).$$

2*. If either of the series $u_0+u_1+\dots$, $v_0+v_1+\dots$ is divergent, then so is the series $u_0 v_0 + (u_1 v_0 + u_0 v_1) + (u_2 v_0 + u_1 v_1 + u_0 v_2) + \dots$, except in the trivial case in which every term of one series is zero.

* In Exs. 2–4 the series considered are of course series of positive terms.

3. If the series $u_0+u_1+\dots$, $v_0+v_1+\dots$, $w_0+w_1+\dots$ converge to sums r, s, t, then the series $\Sigma\lambda_k$, where $\lambda_k = \Sigma u_m v_n w_p$, the summation being extended to all sets of values of m, n, p such that $m+n+p = k$, converges to the sum rst.

4. If Σu_n and Σv_n converge to sums s and t, then the series Σw_n, where $w_n = \Sigma u_l v_m$, the summation extending to all pairs l, m for which $lm = n$, converges to the sum st.

178. Further tests for convergence and divergence. The examples on pp. 345–6 suffice to show that there are simple and interesting types of series of positive terms which cannot be dealt with by the general tests of §§ 174–5. In fact, if we consider the simple type of series in which u_{n+1}/u_n tends to a limit as $n\to\infty$, *the tests of* §§ 174–5 *generally fail when this limit is* 1. Thus in Ex. LXVIII. 5 these tests failed, and we had to fall back upon a special device, which was in essence that of using the series of Ex. LXVIII. 4 as our comparison series, instead of the geometric series.

The fact is that the geometric series, by comparison with which the tests of §§ 174–5 were obtained, is not only convergent but *very rapidly* convergent. The tests derived from comparison with it are therefore naturally very crude, and much more delicate tests are often wanted.

We proved in Ex. XXVII. 7 that $n^k r^n \to 0$ when $n\to\infty$, provided $r<1$, whatever the value of k; and in Ex. LXVIII. 1 we proved more than this, viz. that the series $\Sigma n^k r^n$ is convergent. It follows that the sequence $r, r^2, r^3, \dots, r^n, \dots$, where $r<1$, diminishes more rapidly than the sequence $1^{-k}, 2^{-k}, 3^{-k}, \dots, n^{-k}, \dots$. This seems at first paradoxical if r is not much less than unity and k is large. Thus of the two sequences

$$\tfrac{2}{3}, \tfrac{4}{9}, \tfrac{8}{27}, \dots;\quad 1, \tfrac{1}{4096}, \tfrac{1}{531441}, \dots$$

whose general terms are $(\frac{2}{3})^n$ and n^{-12}, the second seems at first sight to decrease much more rapidly. But if only we go far enough into the sequences we shall find the terms of the first sequence very much the smaller. For example,

$$(\tfrac{2}{3})^4 = \tfrac{16}{81} < \tfrac{1}{5},\quad (\tfrac{2}{3})^{12} < (\tfrac{1}{5})^3 < (\tfrac{1}{10})^2,\quad (\tfrac{2}{3})^{1000} < (\tfrac{1}{10})^{166},$$

while

$$1000^{-12} = 10^{-36};$$

so that the 1000th term of the first sequence is less than the 10^{130}th part of the corresponding term of the second sequence. Thus the series $\Sigma(\frac{2}{3})^n$ is very much more rapidly convergent than the series Σn^{-12}, and even this series is very much more rapidly convergent than Σn^{-2}*.

There are two tests, *Maclaurin's* (or Cauchy's) *integral test* and *Cauchy's condensation test*, which are particularly useful when those of §§ 174–5 fail. In these we make an additional assumption about u_n, viz. that it decreases steadily with n. This condition is satisfied by the most important series.

But before we proceed to these two tests, we prove a simple and useful theorem which we shall call *Abel's theorem*†. It states a *necessary* condition for the convergence of series of this particular type.

179. Abel's (or Pringsheim's) theorem. *If Σu_n is a convergent series of positive and decreasing terms, then* $\lim nu_n = 0$.

For $u_{n+1} + u_{n+2} + \dots \to 0$, and *a fortiori*

$$u_{n+1} + u_{n+2} + \dots + u_{2n} \to 0,$$

and the left-hand side is at least nu_{2n}. Hence $2nu_{2n} = 2(nu_{2n}) \to 0$. Also

$$(2n+1)\,u_{2n+1} \leqq \frac{2n+1}{2n}\,2nu_{2n} \to 0;$$

and therefore $nu_n \to 0$.

Examples LXX. 1. Use Abel's theorem to show that Σn^{-1} and $\Sigma(an+b)^{-1}$ are divergent. [Here $nu_n \to 1$ or $nu_n \to 1/a$.]

2. Show that Abel's theorem is not true if we omit the condition that u_n decreases as n increases. [The series

$$1 + \frac{1}{2^2} + \frac{1}{3^2} + \frac{1}{4} + \frac{1}{5^2} + \frac{1}{6^2} + \frac{1}{7^2} + \frac{1}{8^2} + \frac{1}{9} + \frac{1}{10^2} + \dots,$$

in which $u_n = 1/n$ or $1/n^2$, according as n is or is not a perfect square, is convergent, since it may be rearranged in the form

$$\frac{1}{2^2} + \frac{1}{3^2} + \frac{1}{5^2} + \frac{1}{6^2} + \frac{1}{7^2} + \frac{1}{8^2} + \frac{1}{10^2} + \dots + \left(1 + \frac{1}{4} + \frac{1}{9} + \dots\right),$$

and each of these series is convergent. But, since $nu_n = 1$ whenever n is a perfect square, it is not true that $nu_n \to 0$.]

* Five terms suffice to give the sum of Σn^{-12} correctly to 7 places of decimals, whereas some 10,000,000 are needed to give an equally good approximation to Σn^{-2}. A large number of numerical results of this character will be found in the appendix (compiled by Mr J. Jackson) to the author's tract "Orders of infinity" (*Cambridge math. tracts*, No. 12).

† The theorem was discovered by Abel but forgotten, and rediscovered by Pringsheim.

3. *The converse of Abel's theorem is not true*, i.e. it is not true that, if u_n decreases with n and $\lim nu_n = 0$, then Σu_n is convergent.

[Take the series Σn^{-1} and multiply the first term by 1, the second by $\frac{1}{2}$, the next two by $\frac{1}{3}$, and next four by $\frac{1}{4}$, the next eight by $\frac{1}{5}$, and so on. On grouping in brackets the terms of the new series thus formed we obtain

$$1+\tfrac{1}{2}\cdot\tfrac{1}{2}+\tfrac{1}{3}(\tfrac{1}{3}+\tfrac{1}{4})+\tfrac{1}{4}(\tfrac{1}{5}+\tfrac{1}{6}+\tfrac{1}{7}+\tfrac{1}{8})+\dots;$$

and this series is divergent, since its terms are greater than those of

$$1+\tfrac{1}{2}\cdot\tfrac{1}{2}+\tfrac{1}{3}\cdot\tfrac{1}{2}+\tfrac{1}{4}\cdot\tfrac{1}{2}+\dots,$$

which is divergent. But it is easy to see that the terms of the series

$$1+\tfrac{1}{2}\cdot\tfrac{1}{2}+\tfrac{1}{3}\cdot\tfrac{1}{3}+\tfrac{1}{3}\cdot\tfrac{1}{4}+\tfrac{1}{4}\cdot\tfrac{1}{5}+\tfrac{1}{4}\cdot\tfrac{1}{6}+\dots$$

satisfy the condition that $nu_n \to 0$. In fact $nu_n = 1/\nu$ if $2^{\nu-2} < n \leqq 2^{\nu-1}$, and $\nu \to \infty$ as $n \to \infty$.]

180. Maclaurin's (or Cauchy's) integral test*. If u_n decreases steadily as n increases, we can write $u_n = \phi(n)$ and suppose that $\phi(n)$ is the value assumed, when $x = n$, by a continuous and steadily decreasing function $\phi(x)$ of the continuous variable x. Then, if ν is any positive integer, we have

$$\phi(\nu-1) \geqq \phi(x) \geqq \phi(\nu)$$

when $\nu - 1 \leqq x \leqq \nu$. Let

$$v_\nu = \phi(\nu-1) - \int_{\nu-1}^{\nu} \phi(x)\,dx = \int_{\nu-1}^{\nu} \{\phi(\nu-1)-\phi(x)\}\,dx,$$

so that
$$0 \leqq v_\nu \leqq \phi(\nu-1)-\phi(\nu).$$

Then Σv_ν is a series of positive terms, and

$$v_2+v_3+\dots+v_n \leqq \phi(1)-\phi(n) \leqq \phi(1).$$

Hence Σv_ν is convergent, and so $v_2+v_3+\dots+v_n$ or

$$\sum_{1}^{n-1} \phi(\nu) - \int_1^n \phi(x)\,dx$$

tends to a positive limit, not exceeding $\phi(1)$, when $n \to \infty$.

Let us write
$$\Phi(\xi) = \int_1^{\xi} \phi(x)\,dx,$$

* The test was discovered by Maclaurin and rediscovered by Cauchy, to whom it is often attributed.

so that $\Phi(\xi)$ is a continuous and steadily increasing function of ξ. Then

$$u_1 + u_2 + \dots + u_{n-1} - \Phi(n)$$

tends to a positive limit, not greater than $\phi(1)$, when $n \to \infty$. Hence Σu_ν is convergent or divergent according as $\Phi(n)$ tends to a limit or to infinity as $n \to \infty$, and therefore, since $\Phi(n)$ increases steadily, according as $\Phi(\xi)$ tends to a limit or to infinity as $\xi \to \infty$. Hence *if $\phi(x)$ is a function of x which is positive and continuous for all values of x greater than unity, and decreases steadily as x increases, then the series*

$$\phi(1) + \phi(2) + \dots$$

does or does not converge according as

$$\Phi(\xi) = \int_1^\xi \phi(x)\,dx$$

does or does not tend to a limit l as $\xi \to \infty$; and, in the first case, the sum of the series is not greater than $\phi(1) + l$.

The sum must in fact be less than $\phi(1) + l$. For it follows from (6) of § 165, and Ch. VII, Misc. Ex. 43, that $v_\nu < \phi(\nu - 1) - \phi(\nu)$ unless $\phi(x) = \phi(\nu)$ throughout the interval $(\nu - 1, \nu)$; and this cannot be true for all values of ν.

181. The series Σn^{-s}. Much the most important application of the integral test is to the series

$$1^{-s} + 2^{-s} + 3^{-s} + \dots + n^{-s} + \dots,$$

where s is any rational number. We have seen already (§ 77 and Exs. LXVIII. 15, LXX. 1) that the series is divergent when $s = 1$.

If $s \leqq 0$ then it is obvious that the series is divergent. If $s > 0$ then u_n decreases as n increases, and we can apply the test. Here

$$\Phi(\xi) = \int_1^\xi \frac{dx}{x^s} = \frac{\xi^{1-s} - 1}{1 - s},$$

unless $s = 1$. If $s > 1$ then $\xi^{1-s} \to 0$ as $\xi \to \infty$, and

$$\Phi(\xi) \to \frac{1}{s-1} = l,$$

say. And if $s < 1$ then $\xi^{1-s} \to \infty$ as $\xi \to \infty$, and so $\Phi(\xi) \to \infty$. Thus *the series Σn^{-s} is convergent if $s > 1$, divergent if $s \leqq 1$, and in the first case its sum is less than $s/(s-1)$.*

We could of course prove that the series is divergent when $s<1$ by comparing it with the divergent series Σn^{-1}.

It is however interesting to see how the integral test may be applied to the series Σn^{-1}, when the preceding analysis fails. In this case

$$\Phi(\xi)=\int_1^{\xi}\frac{dx}{x},$$

and it is easy to see that $\Phi(\xi)\to\infty$ when $\xi\to\infty$. For if $\xi>2^n$ then

$$\Phi(\xi)>\int_1^{2^n}\frac{dx}{x}=\int_1^2\frac{dx}{x}+\int_2^4\frac{dx}{x}+\dots+\int_{2^{n-1}}^{2^n}\frac{dx}{x}.$$

But by putting $x=2^r u$ we obtain

$$\int_{2^r}^{2^{r+1}}\frac{dx}{x}=\int_1^2\frac{du}{u},$$

and so $\Phi(\xi)>n\int_1^2\frac{du}{u}$, which shows that $\Phi(\xi)\to\infty$ when $\xi\to\infty$.

Examples LXXI. 1. Prove by an argument similar to that used above, and without integration, that $\Phi(\xi)=\int_1^{\xi}\frac{dx}{x^s}$, where $s<1$, tends to infinity with ξ.

2. The series Σn^{-2}, $\Sigma n^{-\frac{3}{2}}$, $\Sigma n^{-\frac{11}{10}}$ are convergent, and their sums are not greater than 2, 3, 11 respectively. The series $\Sigma n^{-\frac{1}{2}}$, $\Sigma n^{-\frac{10}{11}}$ are divergent.

3. The series $\Sigma\frac{n^s}{n^t+a}$, where $a>0$, is convergent or divergent according as $t>1+s$ or $t\leqq 1+s$. [Compare with Σn^{s-t}.]

4. Discuss the convergence or divergence of the series

$$\Sigma\frac{a_1 n^{s_1}+a_2 n^{s_2}+\dots+a_k n^{s_k}}{b_1 n^{t_1}+b_2 n^{t_2}+\dots+b_l n^{t_l}},$$

where all the letters denote positive numbers and the s's and t's are rational and arranged in descending order of magnitude.

5. Prove that if $m>0$ then

$$\frac{1}{m^2}+\frac{1}{(m+1)^2}+\frac{1}{(m+2)^2}+\dots<\frac{m+1}{m^2}.$$

6. Prove that $$\sum_1^{\infty}\frac{1}{n^2+1}<\tfrac{1}{2}+\tfrac{1}{4}\pi.$$

7. Prove that $$-\tfrac{1}{2}\pi<\sum_1^{\infty}\frac{a}{a^2+n^2}<\tfrac{1}{2}\pi.$$ (*Math. Trip.* 1909)

8. Prove that

$$2\sqrt{n}-2<\frac{1}{\sqrt{1}}+\frac{1}{\sqrt{2}}+\dots+\frac{1}{\sqrt{n}}<2\sqrt{n}-1,$$

$$\tfrac{1}{2}\pi<\frac{1}{2\sqrt{1}}+\frac{1}{3\sqrt{2}}+\frac{1}{4\sqrt{3}}+\dots<\tfrac{1}{2}(\pi+1).$$

(*Math. Trip.* 1911)

9. If $\phi(n)\to l>1$ then the series $\Sigma n^{-\phi(n)}$ is convergent. If $\phi(n)\to l<1$ then it is divergent.

10. Prove that if $a>0$, $b>0$, and $0<s<1$, then

$$\psi(n)=(a+b)^{-s}+(a+2b)^{-s}+\dots+(a+nb)^{-s}-\frac{(a+nb)^{1-s}}{b(1-s)}$$

tends to a limit A when $n\to\infty$. Prove also that $\psi(n)-\psi(n-1)=O(n^{-s-1})$, and deduce that $\psi(n)=A+O(n^{-s})$. (*Math. Trip.* 1926)

182. Cauchy's condensation test. The second of the two tests mentioned in § 178 is as follows: *if $u_n=\phi(n)$ is a decreasing function of n, then the series $\Sigma\phi(n)$ is convergent or divergent according as $\Sigma 2^n\phi(2^n)$ is convergent or divergent.*

We can prove this by an argument which we have used already (§ 77) for the special series Σn^{-1}. In the first place

$$\phi(3)+\phi(4)\geqq 2\phi(4),$$

$$\phi(5)+\phi(6)+\phi(7)+\phi(8)\geqq 4\phi(8),$$

....

$$\phi(2^n+1)+\phi(2^n+2)+\dots+\phi(2^{n+1})\geqq 2^n\phi(2^{n+1}).$$

If $\Sigma 2^n\phi(2^n)$ diverges then so do $\Sigma 2^{n+1}\phi(2^{n+1})$ and $\Sigma 2^n\phi(2^{n+1})$, and then the inequalities just obtained show that $\Sigma\phi(n)$ diverges.

On the other hand

$$\phi(2)+\phi(3)\leqq 2\phi(2),\quad \phi(4)+\phi(5)+\phi(6)+\phi(7)\leqq 4\phi(4),$$

and so on; and from this set of inequalities it follows that if $\Sigma 2^n\phi(2^n)$ converges then so does $\Sigma\phi(n)$. Thus the theorem is established.

For our present purposes the field of application of this test is practically the same as that of the integral test. It enables us to discuss the series Σn^{-s} with equal ease. For Σn^{-s} will converge or diverge according as $\Sigma 2^n 2^{-ns}$ converges or diverges, i.e. according as $s > 1$ or $s \leqq 1$.

Examples LXXII. 1. Show that if a is any positive integer greater than 1 then $\Sigma\phi(n)$ is convergent or divergent according as $\Sigma a^n \phi(a^n)$ is convergent or divergent. [Use the same arguments as above, taking groups of a, a^2, a^3, ... terms.]

2. If $\Sigma 2^n \phi(2^n)$ converges then $\lim 2^n \phi(2^n) = 0$. Hence deduce Abel's theorem of § 179.

183. Further ratio tests. If $u_n = n^{-s}$ then, by Taylor's theorem,

$$\frac{u_{n+1}}{u_n} = \left(1+\frac{1}{n}\right)^{-s} = 1 - \frac{s}{n} + \frac{s(s-1)}{2n^2}\left(1+\frac{\theta}{n}\right)^{-s-2},$$

where $0 < \theta < 1$, and so

$$\frac{u_{n+1}}{u_n} = 1 - \frac{s}{n} + O\left(\frac{1}{n^2}\right).$$

Suppose now that

$$\frac{v_{n+1}}{v_n} = 1 - \frac{a}{n} + O\left(\frac{1}{n^2}\right) \qquad \text{.................(1).}$$

If $a > 1$, we can choose s so that $1 < s < a$, and then $v_{n+1}/v_n < u_{n+1}/u_n$ for sufficiently large n. But Σu_n is convergent, and therefore, by 4 of § 175, Σv_n is convergent. Similarly, if $a < 1$, we can choose s so that $a < s < 1$, and prove the divergence of Σv_n by comparison with the divergent series Σu_n. It follows that *if* v_n *satisfies* (1), *then* Σv_n *is convergent if* $a > 1$ *and divergent if* $a < 1$. We must leave the case $a = 1$ to the next chapter (Ex. XC. 5).

We can prove similarly that if (1) is true, with any positive a, and $0 < s < a$, then $v_n \leqq Kn^{-s}$, and so $v_n \to 0$.

Let us consider in particular the 'hypergeometric' series

$$\Sigma v_n = 1 + \frac{\alpha.\beta}{1.\gamma} + \frac{\alpha(\alpha+1).\beta(\beta+1)}{1.2.\gamma(\gamma+1)} + \dots \qquad \text{.........(2),}$$

where α, β, γ are real, and none of them is zero or a negative integer. Then the terms are ultimately of constant sign, and

$$\frac{v_{n+1}}{v_n} = \frac{(\alpha+n)(\beta+n)}{(1+n)(\gamma+n)} = 1 - \frac{\gamma+1-\alpha-\beta}{n} + O\left(\frac{1}{n^2}\right).$$

Hence *the series* (2) *is convergent when* $\gamma > \alpha + \beta$ *and divergent when* $\gamma < \alpha + \beta$. *In particular the series*

$$1 + \frac{n}{1} + \frac{n(n+1)}{1.2} + \dots$$

is convergent when $n < 0$ *and divergent when* $n > 0$. Also $v_n \to 0$ if $\gamma > \alpha + \beta - 1$.

184. Infinite integrals. The integral test of § 180 shows that, if $\phi(x)$ is a positive and decreasing function of x, then the series $\Sigma\phi(n)$ is convergent or divergent according as the integral function $\Phi(x)$ does or does not tend to a limit as $x \to \infty$. Let us suppose that it does tend to a limit, and that

$$\lim_{x\to\infty} \int_1^x \phi(t)\,dt = l.$$

Then we shall say that *the integral*

$$\int_1^\infty \phi(t)\,dt$$

is convergent, and has the value l; and we shall call the integral an *infinite integral.*

So far we have supposed $\phi(t)$ positive and decreasing. But it is natural to extend our definition to other cases. Nor is there any special point in supposing the lower limit to be unity. We are accordingly led to formulate the following definition.

If $\phi(t)$ *is a function of* t *continuous for* $t \geqq a$, *and*

$$\lim_{x\to\infty} \int_a^x \phi(t)\,dt = l,$$

then we shall say that the infinite integral

$$\int_a^\infty \phi(t)\,dt \quad \dots\dots\dots\dots\dots\dots\dots\dots(1)$$

is convergent and has the value l.

The ordinary integral between limits a and A, as defined in Ch. VII, we shall sometimes call in contrast a *finite* integral.

On the other hand, when

$$\int_a^x \phi(t)\,dt \to \infty,$$

we shall say that the integral *diverges* to ∞, and we can give a similar definition of divergence to $-\infty$. Finally, when none of these alternatives occur, we shall say that the integral *oscillates*, *finitely* or *infinitely*, as $x \to \infty$.

These definitions suggest the following remarks.

(i) If we write $$\int_a^x \phi(t)\,dt = \Phi(x),$$

then the integral converges, diverges, or oscillates according as $\Phi(x)$ tends to a limit, tends to ∞ (or to $-\infty$), or oscillates, as $x \to \infty$. If $\Phi(x)$ tends to a limit, which we may denote by $\Phi(\infty)$, then the value of the integral is $\Phi(\infty)$. More generally, if $\Phi(x)$ is any integral function of $\phi(x)$, then the value of the integral is $\Phi(\infty) - \Phi(a)$.

(ii) In the special case in which $\phi(t)$ is always positive it is clear that $\Phi(x)$ is an increasing function of x. Hence the only alternatives are convergence and divergence to ∞.

(iii) The general principle of convergence, corresponding to that of §96, is: *a necessary and sufficient condition for the convergence of the integral* (1) *is that*

$$\left| \int_{x_1}^{x_2} \phi(x)\,dx \right| < \delta$$

for $x_2 > x_1 \geqq X(\delta)$.

(iv) The reader should not be puzzled by the use of the term *infinite integral* to denote something which has a definite value such as 2 or $\frac{1}{2}\pi$. The distinction between an infinite integral and a finite integral is similar to that between an infinite series and a finite series, and no one supposes that an infinite series is necessarily divergent.

(v) The integral $\int_a^x \phi(t)\,dt$ was defined in §§161–2 as a *simple* limit, i.e. as the limit of a certain finite sum. The infinite integral is therefore *the limit of a limit*, or what is known as a *repeated* limit. The notion of the infinite integral is essentially more complex than that of the finite integral, of which it is a development.

(vi) The integral test of §180 may now be stated in the form: *if* $\phi(x)$

is positive and decreases steadily as x increases, then the infinite series $\Sigma\phi(n)$ and the infinite integral $\int_1^\infty \phi(x)\,dx$ converge or diverge together.

(vii) The reader will find no difficulty in formulating and proving theorems for infinite integrals analogous to those stated in (1)–(6) of § 77. Thus the result analogous to (2) is that *if $\int_a^\infty \phi(x)\,dx$ is convergent, and $b > a$, then $\int_b^\infty \phi(x)\,dx$ is convergent and*

$$\int_a^\infty \phi(x)\,dx = \int_a^b \phi(x)\,dx + \int_b^\infty \phi(x)\,dx.$$

185. The case in which $\phi(x)$ is positive. It is natural to consider what are the general theorems, concerning the convergence or divergence of the infinite integral (1) of § 184, analogous to theorems A–D of § 173. That A is true of integrals as well as of series we have already seen in § 184 (ii). Corresponding to B we have the theorem that *a necessary and sufficient condition for the convergence of the integral* (1) *is that there should be a constant K such that*

$$\int_a^x \phi(t)\,dt < K$$

for all values of x greater than a. Similarly, corresponding to C, we have the theorem: *if $\int_a^\infty \phi(x)\,dx$ is convergent, and $\psi(x) \leqq K\phi(x)$ for all values of x greater than a, then $\int_a^\infty \psi(x)\,dx$ is convergent and*

$$\int_a^\infty \psi(x)\,dx \leqq K \int_a^\infty \phi(x)\,dx.$$

We leave it to the reader to formulate the corresponding test for divergence.

We may observe that d'Alembert's test (§ 175), depending as it does on the notion of successive terms, has no analogue for integrals; and that the analogue of Cauchy's test is not of much importance, and in any case could only be formulated when we have investigated in greater detail the theory of the function

$\phi(x) = r^x$, as we shall do in Ch. IX. The most important special tests are obtained by comparison with the integral

$$\int_a^\infty \frac{dx}{x^s} \qquad (a > 0),$$

whose convergence or divergence we have investigated in § 181, and are as follows: *if* $\phi(x) < Kx^{-s}$, *where* $s > 1$, *when* $x \geqq a$, *then* $\int_a^\infty \phi(x)\,dx$ *is convergent; and if* $\phi(x) > Kx^{-s}$, *where* $K > 0$ *and* $s \leqq 1$, *when* $x \geqq a$, *then the integral is divergent; and in particular, if* $\lim x^s\phi(x) = l$, *where* $l > 0$, *then the integral is convergent or divergent according as* $s > 1$ *or* $s \leqq 1$.

There is one fundamental property of a convergent infinite series in regard to which the analogy between infinite series and infinite integrals breaks down. If $\Sigma\phi(n)$ is convergent then $\phi(n) \to 0$; but it is *not* always true, even when $\phi(x)$ is always positive, that if $\int_a^\infty \phi(x)\,dx$ is convergent then $\phi(x) \to 0$.

Consider for example the function $\phi(x)$ whose graph is indicated by the thick line in the figure. Here the height of the peaks corresponding to the points $x = 1, 2, 3, \ldots$ is in each case unity, and the breadth of the peak

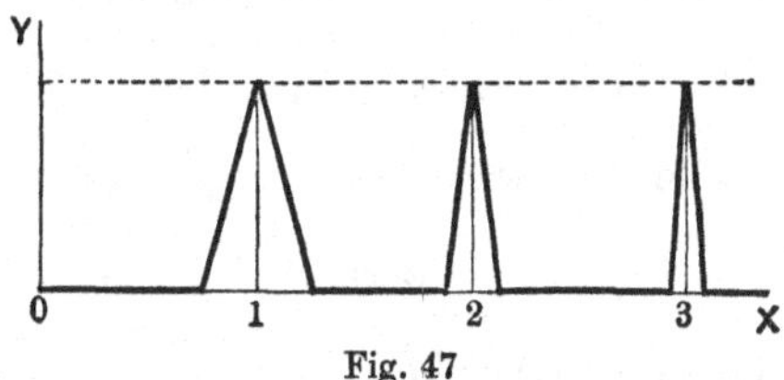

Fig. 47

corresponding to $x = n$ is $2/(n+1)^2$. The area of the peak is $1/(n+1)^2$, and it is evident that, for any value of ξ,

$$\int_0^\xi \phi(x)\,dx < \sum_0^\infty \frac{1}{(n+1)^2},$$

so that $\int_0^\infty \phi(x)\,dx$ is convergent; but it is not true that $\phi(x) \to 0$.

Examples LXXIII. 1. The integral

$$\int_a^\infty \frac{\alpha x^r + \beta x^{r-1} + \ldots + \lambda}{Ax^s + Bx^{s-1} + \ldots + L}\,dx,$$

where α and A are positive and a is greater than the greatest root (if any) of the denominator, is convergent if $s > r+1$ and otherwise divergent.

2. Which of the integrals

$$\int_a^\infty \frac{dx}{\sqrt{x}},\quad \int_a^\infty \frac{dx}{x^{\frac{4}{3}}},\quad \int_a^\infty \frac{dx}{c^2+x^2},\quad \int_a^\infty \frac{x\,dx}{c^2+x^2},\quad \int_a^\infty \frac{x^2\,dx}{c^2+x^2},\quad \int_a^\infty \frac{x^2\,dx}{\alpha+2\beta x^2+\gamma x^4}$$

are convergent? In the first two integrals it is supposed that $a > 0$, and in the last that a is greater than the greatest root (if any) of the denominator.

3. The integrals $\int_a^\xi \cos x\,dx$, $\int_a^\xi \cos(\alpha x+\beta)\,dx$ oscillate finitely as $\xi \to \infty$.

4. The integrals $\int_a^\xi x\cos x\,dx$, $\int_a^\xi x^n\cos(\alpha x+\beta)\,dx$, where n is any positive integer, oscillate infinitely as $\xi \to \infty$.

5. **Integrals to $-\infty$.** If $\int_\xi^a \phi(x)\,dx$ tends to a limit l as $\xi \to -\infty$, then we say that $\int_{-\infty}^a \phi(x)\,dx$ is convergent and equal to l. Such integrals possess properties in every respect analogous to those of the integrals discussed in the preceding sections: the reader will find no difficulty in formulating them.

6. **Integrals from $-\infty$ to $+\infty$.** If the integrals

$$\int_{-\infty}^a \phi(x)\,dx,\quad \int_a^\infty \phi(x)\,dx$$

are both convergent, and have the values k, l respectively, then we say that

$$\int_{-\infty}^\infty \phi(x)\,dx$$

is convergent and has the value $k+l$.

7. Prove that

$$\int_{-\infty}^0 \frac{dx}{1+x^2} = \int_0^\infty \frac{dx}{1+x^2} = \tfrac{1}{2}\int_{-\infty}^\infty \frac{dx}{1+x^2} = \tfrac{1}{2}\pi.$$

8. Prove that $\int_{-\infty}^\infty \phi(x^2)\,dx = 2\int_0^\infty \phi(x^2)\,dx$, provided that the integral $\int_0^\infty \phi(x^2)\,dx$ is convergent.

9. Prove that if $\int_0^\infty x\phi(x^2)\,dx$ is convergent then $\int_{-\infty}^\infty x\phi(x^2)\,dx = 0$.

10. **Analogue of Abel's theorem of § 179.** *If $\phi(x)$ is positive and steadily decreases, and $\int_a^\infty \phi(x)\,dx$ is convergent, then $x\phi(x) \to 0$.* Prove this (*a*) by means of Abel's theorem and the integral test and (*b*) directly, by arguments analogous to those of § 179.

11. If $a = x_0 < x_1 < x_2 < \dots$ and $x_n \to \infty$, and $u_n = \int_{x_n}^{x_{n+1}} \phi(x)\,dx$, then the convergence of $\int_a^\infty \phi(x)\,dx$ involves that of Σu_n. If $\phi(x)$ is always positive the converse statement is also true. [That the converse is not true in general is shown by the example in which $\phi(x) = \cos x$, $x_n = n\pi$.]

186. Application to infinite integrals of the rules for substitution and integration by parts. The rules for the transformation of a definite integral which were discussed in § 166 may be extended so as to apply to infinite integrals.

(1) **Transformation by substitution.** Suppose that

$$\int_a^\infty \phi(x)\,dx \quad \text{..........................(1)}$$

is convergent. Further suppose that, for any value of ξ greater than a, we have, as in § 166,*

$$\int_a^\xi \phi(x)\,dx = \int_b^\tau \phi\{f(t)\}f'(t)\,dt \quad \text{..............(2),}$$

where $a = f(b)$, $\xi = f(\tau)$. Finally suppose that the functional relation $x = f(t)$ is such that $x \to \infty$ when $t \to \infty$. Then, making τ and so ξ tend to ∞ in (2), we see that the integral

$$\int_b^\infty \phi\{f(t)\}f'(t)\,dt \quad \text{.......................(3)}$$

is convergent and equal to the integral (1).

On the other hand it may happen that $\xi \to \infty$ when $\tau \to -\infty$ or when $\tau \to c$. In the first case we obtain

$$\int_a^\infty \phi(x)\,dx = \lim_{\tau \to -\infty} \int_b^\tau \phi\{f(t)\}f'(t)\,dt$$

$$= -\lim_{\tau \to -\infty} \int_\tau^b \phi\{f(t)\}f'(t)\,dt = -\int_{-\infty}^b \phi\{f(t)\}f'(t)\,dt.$$

* f and ϕ are now interchanged.

In the second case we obtain

$$\int_a^\infty \phi(x)\,dx = \lim_{\tau\to c}\int_b^\tau \phi\{f(t)\}f'(t)\,dt \quad \text{...........(4).}$$

We shall return to this equation in § 188.

There are of course corresponding results for integrals over $(-\infty, a)$ or $(-\infty, \infty)$ which the reader will be able to formulate for himself.

Examples LXXIV. 1. Show, by means of the substitution $x = t^\alpha$, that if $s > 1$ and $\alpha > 0$ then

$$\int_1^\infty x^{-s}\,dx = \alpha\int_1^\infty t^{\alpha(1-s)-1}\,dt;$$

and verify the result by calculating the value of each integral directly.

2. If $\int_a^\infty \phi(x)\,dx$ is convergent then it is equal to one or other of

$$\alpha\int_{(a-\beta)/\alpha}^\infty \phi(\alpha t+\beta)\,dt, \quad -\alpha\int_{-\infty}^{(a-\beta)/\alpha} \phi(\alpha t+\beta)\,dt,$$

according as α is positive or negative.

3. If $\phi(x)$ is a positive and steadily decreasing function of x, and α and β are any positive numbers, then the convergence of the series $\Sigma\phi(n)$ implies and is implied by that of the series $\Sigma\phi(\alpha n+\beta)$. [It follows at once, on making the substitution $x = \alpha t+\beta$, that the integrals

$$\int_a^\infty \phi(x)\,dx, \quad \int_{(a-\beta)/\alpha}^\infty \phi(\alpha t+\beta)\,dt$$

converge or diverge together. Now use the integral test.]

4. Show that $\int_1^\infty \dfrac{dx}{(1+x)\sqrt{x}}$ [Put $x = t^2$.]

5. Evaluate $\int_0^\infty \dfrac{dx}{(1+x^2)^n}$ and $\int_0^\infty \dfrac{dx}{(1+x^2)^{n+\frac{1}{2}}}$, n being a positive integer.

(*Math. Trip.* 1929, 1935)

[The substitution $x = \cot\theta$ reduces the integrals to $\int_0^{\frac{1}{2}\pi} \sin^{2n-2}\theta\,d\theta$ and $\int_0^{\frac{1}{2}\pi} \sin^{2n-1}\theta\,d\theta$. Now use Ex. LXVI. 10.]

6. If $\phi(x)\to h$ as $x\to\infty$, and $\phi(x)\to k$ as $x\to-\infty$, then

$$\int_{-\infty}^\infty \{\phi(x-a)-\phi(x-b)\}\,dx = -(a-b)(h-k).$$

[For $\int_{-\xi'}^{\xi}\{\phi(x-a)-\phi(x-b)\}\,dx = \int_{-\xi'}^{\xi}\phi(x-a)\,dx - \int_{-\xi'}^{\xi}\phi(x-b)\,dx$

$$= \int_{-\xi'-a}^{\xi-a}\phi(t)\,dt - \int_{-\xi'-b}^{\xi-b}\phi(t)\,dt = \int_{-\xi'-a}^{-\xi'-b}\phi(t)\,dt - \int_{\xi-a}^{\xi-b}\phi(t)\,dt.$$

The first of these two integrals may be expressed in the form

$$(a-b)\,k + \int_{-\xi'-a}^{-\xi'-b}\rho\,dt,$$

where $\rho \to 0$ as $\xi' \to \infty$, and the modulus of the last integral does not exceed $|a-b|\,\kappa$, where κ is the greatest value of ρ throughout the interval $(-\xi'-a, -\xi'-b)$. Hence

$$\int_{-\xi'-a}^{-\xi'-b}\phi(t)\,dt \to (a-b)\,k.$$

The second integral may be discussed similarly.]

(2) **Integration by parts.** The formula for integration by parts (§ 166) is

$$\int_a^{\xi} f(x)\,\phi'(x)\,dx = f(\xi)\,\phi(\xi) - f(a)\,\phi(a) - \int_a^{\xi} f'(x)\,\phi(x)\,dx.$$

Suppose now that $\xi \to \infty$. Then if any two of the three terms in the above equation which involve ξ tend to limits, so does the third, and we obtain the result

$$\int_a^{\infty} f(x)\,\phi'(x)\,dx = \lim_{\xi\to\infty} f(\xi)\,\phi(\xi) - f(a)\,\phi(a) - \int_a^{\infty} f'(x)\,\phi(x)\,dx.$$

There are of course similar results for integrals to $-\infty$, or from $-\infty$ to ∞.

Examples LXXV. 1. Show that $\int_0^{\infty}\frac{x}{(1+x)^3}\,dx = \frac{1}{2}\int_0^{\infty}\frac{dx}{(1+x)^2} = \frac{1}{2}$.

2. If m and $n-1$ are positive integers, and $I_{m,n} = \int_0^{\infty}\frac{x^m\,dx}{(1+x)^{m+n}}$, then $(m+n-1)\,I_{m,n} = mI_{m-1,n}$. Hence prove that

$$I_{m,n} = \frac{m!\,(n-2)!}{(m+n-1)!}.$$

3. Prove that $\int_1^{\infty}\frac{\sqrt{x}}{(1+x)^2}\,dx = \frac{1}{2}+\frac{1}{4}\pi$. [Put $x = t^2$, when we obtain

$$2\int_1^{\infty}\frac{t^2\,dt}{(1+t^2)^2} = -\int_1^{\infty} t\frac{d}{dt}\left(\frac{1}{1+t^2}\right)dt.$$

Now integrate by parts.]

4. Prove by partial integration that, if u_n is the first integral of Ex. LXXIV. 5, and $n > 1$, then $(2n-2)u_n = (2n-3)u_{n-1}$; and so evaluate u_n. (*Math. Trip.* 1935)

[Observe that

$$u_{n-1} - u_n = \int_0^\infty \frac{x^2\,dx}{(1+x^2)^n} = -\frac{1}{2(n-1)}\int_0^\infty x\frac{d}{dx}\left\{\frac{1}{(1+x^2)^{n-1}}\right\}dx.]$$

187. Other types of infinite integrals. It was assumed, in the definition of the ordinary or finite integral given in Ch. VII, that (1) the range of integration is finite and (2) the subject of integration is continuous.

It is possible, however, to extend the notion of the 'definite integral' so as to apply to many cases in which these conditions are not satisfied. The 'infinite' integrals which we have discussed in the preceding sections, for example, differ from those of Ch. VII in that the range of integration is infinite. We shall now suppose that it is the second condition (2) which is not satisfied. The most important case is that in which $\phi(x)$ is continuous throughout the range of integration (a, A) except for a finite number of values of x, say $x = \xi_1, \xi_2, \ldots$, while $\phi(x) \to \infty$ or $\phi(x) \to -\infty$ as x tends to any of these exceptional values from either side.

It is evident that we need only consider the case in which (a, A) contains *one* such point ξ. When there is more than one such point we can divide up (a, A) into a finite number of sub-intervals each of which contains only one; and, if the value of the integral over each of these sub-intervals has been defined, we can then define the integral over the whole interval as being the sum of the integrals over each sub-interval. Further, we can suppose that the one point ξ in (a, A) comes at one of the limits a, A. For, if it comes between a and A, we can define $\int_a^A \phi(x)\,dx$ as

$$\int_a^\xi \phi(x)\,dx + \int_\xi^A \phi(x)\,dx,$$

assuming each of these integrals to have been satisfactorily defined. We shall suppose, then, that $\xi = a$; it is evident that the definitions to which we are led will apply, with trifling changes, to the case in which $\xi = A$.

Let us then suppose $\phi(x)$ to be continuous throughout (a, A) except for $x = a$, while $\phi(x) \to \infty$ as $x \to a$ through values greater than a. A typical example of such a function is given by

$$\phi(x) = (x-a)^{-s},$$

where $s > 0$; or, in particular, if $a = 0$, by $\phi(x) = x^{-s}$. Let us therefore consider how we can define

$$\int_0^A \frac{dx}{x^s} \quad \text{..........................(1),}$$

when $s > 0$.

The integral $\int_{1/A}^{\infty} y^{s-2}\,dy$ is convergent if $s < 1$ (§ 185) and means $\lim_{\eta\to\infty} \int_{1/A}^{\eta} y^{s-2}\,dy$. But if we make the substitution $y = 1/x$, we obtain

$$\int_{1/A}^{\eta} y^{s-2}\,dy = \int_{1/\eta}^{A} x^{-s}\,dx.$$

Thus $\lim_{\eta\to\infty} \int_{1/\eta}^{A} x^{-s}\,dx$, or, what is the same thing, $\lim_{\epsilon\to+0} \int_{\epsilon}^{A} x^{-s}\,dx$, exists provided that $s < 1$; and it is natural to define the value of the integral (1) as being equal to this limit. Similar considerations lead us to define $\int_a^A (x-a)^{-s}\,dx$ by the equation

$$\int_a^A (x-a)^{-s}\,dx = \lim_{\epsilon\to+0} \int_{a+\epsilon}^{A} (x-a)^{-s}\,dx.$$

We are thus led to the following general definition: *if the integral*

$$\int_{a+\epsilon}^{A} \phi(x)\,dx$$

tends to a limit l as $\epsilon \to +0$, we shall say that the integral

$$\int_a^A \phi(x)\,dx$$

is convergent and has the value l.

Similarly, when $\phi(x) \to \infty$ as x tends to the upper limit A, we define $\int_a^A \phi(x)\,dx$ as being

$$\lim_{\epsilon\to+0} \int_a^{A-\epsilon} \phi(x)\,dx:$$

and then, as we explained above, we can extend our definitions to cover the case in which the interval (a, A) contains any finite number of infinities of $\phi(x)$.

An integral in which the subject of integration tends to ∞ or to $-\infty$ as x tends to some value or values included in the range of integration will be called an *infinite integral of the second kind*: the *first kind* of infinite integrals being the class discussed in §§ 184 *et seq.* Nearly all the remarks (i)–(vii) made at the end of § 184 apply to infinite integrals of the second kind as well as to those of the first.

We framed our definitions with functions in mind which tend to infinity for special values of x, but they can be used also when the integrand has discontinuities of other types. Thus if $f(x) = -1$ for $-1 \leqq x < 0$, $f(0) = 0$, and $f(x) = 1$ for $0 < x \leqq 1$, then $\int_{-1}^{1} f(x)\,dx$ means

$$\lim_{\eta\to+0}\int_{-1}^{-\eta} f(x)\,dx + \lim_{\epsilon\to+0}\int_{\epsilon}^{1} f(x)\,dx = \lim_{\eta\to+0}(-1+\eta) + \lim_{\epsilon\to+0}(1-\epsilon) = 0.$$

The definition may also be used when $f(x)$ has oscillatory discontinuities, e.g. if $f(x) = \sin(1/x)$.

188. We may now write the equation (4) of § 186 in the form

$$\int_{a}^{\infty} \phi(x)\,dx = \int_{b}^{c} \phi\{f(t)\}f'(t)\,dt \;\ldots\ldots\ldots\ldots\ldots\ldots(1).$$

The integral on the right-hand side is defined as the limit, as $\tau \to c$, of the corresponding integral over the range (b, τ), i.e. as an infinite integral of the second kind. And when $\phi\{f(t)\}f'(t)$ has an infinity at $t = c$ the integral is essentially an infinite integral. Suppose, for example, that $\phi(x) = (1+x)^{-m}$, where $1 < m < 2$, and $a = 0$, and that $f(t) = t/(1-t)$. Then $b = 0$, $c = 1$, and (1) becomes

$$\int_{0}^{\infty} \frac{dx}{(1+x)^m} = \int_{0}^{1} (1-t)^{m-2}\,dt \;\ldots\ldots\ldots\ldots\ldots\ldots(2);$$

and the integral on the right-hand side is an infinite integral of the second kind.

On the other hand it may happen that $\phi\{f(t)\}f'(t)$ is continuous for $t = c$. In this case

$$\int_{b}^{c} \phi\{f(t)\}f'(t)\,dt$$

is a finite integral, and

$$\lim_{T\to c}\int_b^T \phi\{f(t)\}f'(t)\,dt = \int_b^c \phi\{f(t)\}f'(t)\,dt,$$

in virtue of the corollary to Theorem (10) of § 165. The substitution $x = f(t)$ then transforms an infinite into a finite integral. This case arises if $m \geqq 2$ in the example considered a moment ago.

Examples LXXVI. 1. If $\phi(x)$ is continuous except for $x = a$, while $\phi(x) \to \infty$ as $x \to a$, then a necessary and sufficient condition that $\int_a^A \phi(x)\,dx$ should be convergent is that we can find a constant K such that

$$\int_{a+\epsilon}^{A} \phi(x)\,dx < K$$

for all positive values of ϵ.

It is clear that we can choose a number A' between a and A so that $\phi(x)$ is positive throughout (a, A'). If $\phi(x)$ is positive throughout the whole interval (a, A) then we can identify A' and A. Now

$$\int_{a-\epsilon}^{A} \phi(x)\,dx = \int_{a-\epsilon}^{A'} \phi(x)\,dx + \int_{A'}^{A} \phi(x)\,dx.$$

The first integral on the right-hand side of the above equation increases as ϵ decreases, and therefore tends to a limit or to ∞; and the truth of the result stated becomes evident.

If the condition is not satisfied then $\int_{a-\epsilon}^{A} \phi(x)\,dx \to \infty$. We shall then say that the integral $\int_a^A \phi(x)\,dx$ *diverges* to ∞. It is clear that if $\phi(x) \to \infty$ as $x \to a+0$ then convergence and divergence to ∞ are the only alternatives for the integral. We may discuss similarly the case in which $\phi(x) \to -\infty$.

2. Prove that
$$\int_a^A (x-a)^{-s}\,dx = \frac{(A-a)^{1-s}}{1-s}$$
if $s < 1$, while the integral is divergent if $s \geqq 1$.

3. If $\phi(x)$ is continuous for $a < x \leqq A$, and $0 \leqq \phi(x) < K(x-a)^{-s}$, where $s < 1$, then $\int_a^A \phi(x)\,dx$ is convergent; and if $\phi(x) > K(x-a)^{-s}$, where $s \geqq 1$, then the integral is divergent. [This is a particular case of a general comparison theorem analogous to that stated in § 185.]

4. Are the integrals

$$\int_a^A \frac{dx}{\sqrt{\{(x-a)(A-x)\}}},\quad \int_a^A \frac{dx}{(A-x)\sqrt[3]{(x-a)}},\quad \int_a^A \frac{dx}{(A-x)\sqrt[3]{(A-x)}},$$

$$\int_a^A \frac{dx}{\sqrt{(x^2-a^2)}},\quad \int_a^A \frac{dx}{\sqrt[3]{(A^3-x^3)}},\quad \int_a^A \frac{dx}{x^2-a^2},\quad \int_a^A \frac{dx}{A^3-x^3}$$

convergent or divergent?

5. The integrals $\int_{-1}^{1} \frac{dx}{\sqrt[3]{x}}$, $\int_{a-1}^{a+1} \frac{dx}{\sqrt[3]{(x-a)}}$ are convergent, and the value of each is zero.

6. The integral $\int_0^\pi \frac{dx}{\sqrt{(\sin x)}}$ is convergent. [The subject of integration tends to ∞ as x tends to either limit.]

7. The integral $\int_0^\pi \frac{dx}{(\sin x)^s}$ is convergent if and only if $s<1$.

8. Show that $\int_0^h \frac{\sin x}{x^p}\,dx$, where $h>0$, is convergent if $p<2$. Show also that, if $0<p<2$, the integrals

$$\int_0^\pi \frac{\sin x}{x^p}\,dx,\quad \int_\pi^{2\pi} \frac{\sin x}{x^p}\,dx,\quad \int_{2\pi}^{3\pi} \frac{\sin x}{x^p}\,dx,\ \ldots$$

alternate in sign and steadily decrease in absolute value. [Transform the integral whose limits are $k\pi$ and $(k+1)\pi$ by the substitution $x = k\pi + y$.]

9. Show that $\int_0^h \frac{\sin x}{x^p}\,dx$, where $0<p<2$, attains its greatest value when $h=\pi$. (*Math. Trip.* 1911)

10. $\int_0^{\frac{1}{2}\pi} (\cos x)^l (\sin x)^m\,dx$ is convergent if and only if $l>-1$, $m>-1$.

11. Such an integral as $\int_0^\infty \frac{x^{s-1}\,dx}{1+x}$, where $s<1$, does not fall directly under any of our previous definitions. For the range of integration is infinite and the subject of integration tends to ∞ as $x \to +0$. It is natural to define this integral as being equal to the sum

$$\int_0^1 \frac{x^{s-1}\,dx}{1+x} + \int_1^\infty \frac{x^{s-1}\,dx}{1+x},$$

provided that these two integrals are both convergent.

The first integral is convergent if $s>0$. The second is convergent if $s<1$. Thus the integral from 0 to ∞ is convergent if and only if $0<s<1$.

12. Prove that $\int_0^\infty \frac{x^{s-1}}{1+x^t}\,dx$ is convergent if and only if $0<s<t$.

13. The integral $\int_0^\infty \frac{x^{s-1}-x^{t-1}}{1-x}\,dx$ is convergent if and only if $0<s<1$, $0<t<1$. [It should be noticed that the subject of integration is undefined when $x=1$; but $(x^{s-1}-x^{t-1})/(1-x)\to t-s$ as $x\to 1$ from either side; so that the subject of integration becomes a continuous function of x if we assign to it the value $t-s$ when $x=1$.

It often happens that the subject of integration has a discontinuity which is due simply to a failure in its definition at a particular point in the range of integration, and can be removed by attaching a particular value to it at that point. In this case it is usual to suppose the definition of the subject of integration completed in this way. Thus the integrals

$$\int_0^{\frac{1}{2}\pi} \frac{\sin mx}{x}\,dx, \quad \int_0^{\frac{1}{2}\pi} \frac{\sin mx}{\sin x}\,dx$$

are ordinary finite integrals, if the subjects of integration are regarded as having the value m when $x=0$.]

14. **Substitution and integration by parts.** The formulae for transformation by substitution and integration by parts may of course be extended to infinite integrals of the second as well as of the first kind. The reader should formulate the general theorems for himself, on the lines of §186.

15. Prove by integration by parts that if $s>0$, $t>1$, then

$$\int_0^1 x^{s-1}(1-x)^{t-1}\,dx = \frac{t-1}{s}\int_0^1 x^s(1-x)^{t-2}\,dx.$$

16. If $s>0$ then $\int_0^1 \frac{x^{s-1}\,dx}{1+x} = \int_1^\infty \frac{t^{-s}\,dt}{1+t}$. [Put $x=1/t$.]

17. If $0<s<1$ then $\int_0^1 \frac{x^{s-1}+x^{-s}}{1+x}\,dx = \int_0^\infty \frac{t^{-s}\,dt}{1+t} = \int_0^\infty \frac{t^{s-1}\,dt}{1+t}$.

18. If $a+b>0$ then

$$\int_b^\infty \frac{dx}{(x+a)\sqrt{(x-b)}} = \frac{\pi}{\sqrt{(a+b)}}. \qquad \textit{(Math. Trip. 1909)}$$

[Put $x-b=t^2$.]

19. If $I_n = \int_0^a (a^2-x^2)^n\,dx$, and $n>0$, then $(2n+1)\,I_n = 2na^2 I_{n-1}$.

(*Math. Trip.* 1934)

[Observe that

$$I_n = \int_0^a (a^2-x^2)^n \frac{d}{dx} x\,dx = 2n\int_0^a x^2(a^2-x^2)^{n-1}\,dx = 2n(a^2 I_{n-1} - I_n).$$

The result may be used to evaluate I_n when n is a positive integer. The substitution $x = a\cos\theta$ reduces I_n to the integral of Ex. LXVI. 10.]

20. Show, by means of the substitution $x = t/(1-t)$, that if l and m are both positive then

$$\int_0^\infty \frac{x^{l-1}}{(1+x)^{l+m}}\,dx = \int_0^1 t^{l-1}(1-t)^{m-1}\,dt.$$

21. Show, by means of the substitution $x = pt/(p+1-t)$, that if l, m, and p are all positive then

$$\int_0^1 x^{l-1}(1-x)^{m-1}\frac{dx}{(x+p)^{l+m}} = \frac{1}{(1+p)^l p^m}\int_0^1 t^{l-1}(1-t)^{m-1}\,dt.$$

22. Prove that

$$\int_a^b \frac{dx}{\sqrt{\{(x-a)(b-x)\}}} = \pi, \quad \int_a^b \frac{x\,dx}{\sqrt{\{(x-a)(b-x)\}}} = \tfrac{1}{2}\pi(a+b),$$

(i) by means of the substitution $x = a+(b-a)t^2$, (ii) by means of the substitution $(b-x)/(x-a) = t$, and (iii) by means of the substitution $x = a\cos^2 t + b\sin^2 t$.

23. Prove that if p and q are positive and $f(p,q) = \int_0^1 x^{p-1}(1-x)^{q-1}\,dx$,

then $f(p+1,q)+f(p,q+1) = f(p,q)$, $qf(p+1,q) = pf(p,q+1)$.

Express $f(p+1,q)$ and $f(p,q+1)$ in terms of $f(p,q)$; and prove that

$$f(p,n) = \frac{(n-1)!}{p(p+1)\dots(p+n-1)},$$

where n is a positive integer. (*Math. Trip.* 1926)

24. Establish the formulae

$$\int_0^1 \frac{f(x)\,dx}{\sqrt{(1-x^2)}} = \int_0^{\frac{1}{2}\pi} f(\sin\theta)\,d\theta,$$

$$\int_a^b \frac{f(x)\,dx}{\sqrt{\{(x-a)(b-x)\}}} = 2\int_0^{\frac{1}{2}\pi} f(a\cos^2\theta + b\sin^2\theta)\,d\theta.$$

25. Prove that $\displaystyle\int_1^2 \frac{dx}{(x+1)\sqrt{(x^2-1)}} = \frac{1}{\sqrt{3}}$. (*Math. Trip.* 1930)

26. Prove that

$$\int_0^1 \frac{dx}{(1+x)(2+x)\sqrt{\{x(1-x)\}}} = \pi\left(\frac{1}{\sqrt{2}} - \frac{1}{\sqrt{6}}\right).$$

[Put $x = \sin^2\theta$ and use Ex. LXIII. 7.] (*Math. Trip.* 1912)

189. Some care has occasionally to be exercised in applying the rule for transformation by substitution. Suppose for example that

$$J = \int_1^7 (x^2 - 6x + 13)\, dx.$$

We find by direct integration that $J = 48$. Now let us apply the substitution

$$y = x^2 - 6x + 13,$$

which gives $x = 3 \pm \sqrt{(y-4)}$. Since $y = 8$ when $x = 1$ and $y = 20$ when $x = 7$, we appear to be led to the result

$$J = \int_8^{20} y \frac{dx}{dy}\, dy = \pm \tfrac{1}{2} \int_8^{20} \frac{y\, dy}{\sqrt{(y-4)}}.$$

The indefinite integral is

$$\tfrac{1}{3}(y-4)^{\frac{3}{2}} + 4(y-4)^{\frac{1}{2}},$$

and so we obtain the value $\pm \frac{80}{3}$, which is wrong whichever sign we choose.

The explanation is to be found in a closer consideration of the relation between x and y. The function $x^2 - 6x + 13$ has a minimum for $x = 3$, when $y = 4$. As x increases from 1 to 3, y decreases from 8 to 4, and dx/dy is negative, so that

$$\frac{dx}{dy} = -\frac{1}{2\sqrt{(y-4)}}.$$

As x increases from 3 to 7, y increases from 4 to 20, and the other sign must be chosen. Thus

$$J = \int_1^7 y\, dx = \int_8^4 \left\{-\frac{y}{2\sqrt{(y-4)}}\right\} dy + \int_4^{20} \frac{y}{2\sqrt{(y-4)}}\, dy,$$

a formula which will be found to lead to the correct result.

Similarly, if we transform the integral $\int_0^\pi dx = \pi$ by the substitution $x = \arcsin y$, we must observe that dx/dy is $(1-y^2)^{-\frac{1}{2}}$ or $-(1-y^2)^{-\frac{1}{2}}$ according as $0 \leqq x < \frac{1}{2}\pi$ or $\frac{1}{2}\pi < x \leqq \pi$.

Example. Verify the results of transforming the integrals

$$\int_0^1 (4x^2 - x + \tfrac{1}{16})\, dx, \quad \int_0^\pi \cos^2 x\, dx$$

by the substitutions $4x^2 - x + \frac{1}{16} = y$, $x = \arcsin y$ respectively.

190. Series of positive and negative terms. Our definitions of the sum of an infinite series, and the value of an infinite integral, whether of the first or the second kind, apply to series of terms or integrals of functions whose values may be of either sign. But the special tests for convergence or divergence which we have established in this chapter, and the examples by which we have illustrated them, have had reference almost entirely to cases in which these values are all positive or all negative.

In the case of a series it has always been assumed, explicitly or tacitly, that any conditions imposed upon u_n may be violated for a finite number of terms: all that is necessary is that such a condition (e.g. that the terms are positive) should be satisfied from some term onwards. Similarly in the case of an infinite integral the conditions have been supposed to be satisfied for all values of x greater than some value x_0, or for all values of x within some interval $(a, a+\delta)$ which includes the value a near which the subject of integration tends to infinity. Thus our tests apply to such a series as

$$\Sigma \frac{n^2-10}{n^4},$$

since $n^2 - 10 > 0$ when $n \geqq 4$, and to such integrals as

$$\int_1^\infty \frac{3x-7}{(x+1)^3}\,dx, \quad \int_0^1 \frac{1-2x}{\sqrt{x}}\,dx,$$

since $3x - 7 > 0$ when $x > \frac{7}{3}$, and $1 - 2x > 0$ when $0 < x < \frac{1}{2}$.

But when the changes of sign of u_n *persist throughout the series*, i.e. when the number of both positive and negative terms is infinite, as in the series $1 - \frac{1}{2} + \frac{1}{3} - \frac{1}{4} + \dots$; or when $\phi(x)$ continually changes sign when $x \to \infty$, as in the integral

$$\int_1^\infty \frac{\sin x}{x^s}\,dx,$$

or when $x \to a$, where a is a point of discontinuity of $\phi(x)$, as in the integral

$$\int_a^A \sin\left(\frac{1}{x-a}\right)\frac{dx}{x-a};$$

then the problem of discussing convergence or divergence becomes more difficult. For now we have to consider the possibility of oscillation as well as of convergence or divergence.

191. Absolutely convergent series. Let us then consider a series Σu_n in which any term may be either positive or negative. Let

$$|u_n| = \alpha_n,$$

so that $\alpha_n = u_n$ if u_n is positive and $\alpha_n = -u_n$ if u_n is negative. Further, let $v_n = u_n$ or $v_n = 0$, according as u_n is positive or negative, and $w_n = -u_n$ or $w_n = 0$, according as u_n is negative or positive; or, what is the same thing, let v_n or w_n be equal to α_n according as u_n is positive or negative, the other being in either case equal to zero. Then it is evident that v_n and w_n are always positive, and that

$$u_n = v_n - w_n, \quad \alpha_n = v_n + w_n.$$

If, for example, our series is $1-(\frac{1}{2})^2+(\frac{1}{3})^2-\dots$, then $u_n = (-1)^{n-1}/n^2$ and $\alpha_n = 1/n^2$, while $v_n = 1/n^2$ or $v_n = 0$ according as n is odd or even, and $w_n = 1/n^2$ or $w_n = 0$ according as n is even or odd.

We can now distinguish two cases.

A. Suppose that the series $\Sigma\alpha_n$ is convergent. This is the case, for instance, in the example above, where $\Sigma\alpha_n$ is

$$1+(\tfrac{1}{2})^2+(\tfrac{1}{3})^2+\dots.$$

Then both Σv_n and Σw_n are convergent: for (Ex. XXX. 18) any series selected from the terms of a convergent series of positive terms is convergent. And hence, by Theorem (6) of § 77, Σu_n or $\Sigma(v_n - w_n)$ is convergent and equal to $\Sigma v_n - \Sigma w_n$.

We are thus led to formulate the following definition.

DEFINITION. *When $\Sigma\alpha_n$ or $\Sigma|u_n|$ is convergent, the series Σu_n is said to be* **absolutely convergent.**

And what we have proved above amounts to this: *if Σu_n is absolutely convergent then it is convergent; so are the series formed by its positive and negative terms taken separately; and the sum of the series is equal to the sum of the positive terms plus the sum of the negative terms.*

The reader should guard himself against supposing that the statement 'an absolutely convergent series is convergent' is a tautology. When we say that Σu_n is 'absolutely convergent' we do *not* assert directly that Σu_n

is convergent: we assert the convergence of another series $\Sigma|u_n|$, and it is not evident that this precludes oscillation on the part of Σu_n.

Examples LXXVII. 1. Employ the 'general principle of convergence' (§ 84, Theorem 2) to prove that an absolutely convergent series is convergent. [Since $\Sigma|u_n|$ is convergent, we can, when any positive number δ is assigned, choose n_0 so that

$$|u_{n_1+1}|+|u_{n_1+2}|+\dots+|u_{n_2}|<\delta$$

when $n_2>n_1\geqq n_0$. *A fortiori*

$$|u_{n_1+1}+u_{n_1+2}+\dots+u_{n_2}|<\delta,$$

and therefore Σu_n is convergent.]

2. If Σa_n is a convergent series of positive terms, and $|b_n|\leqq Ka_n$, then Σb_n is absolutely convergent.

3. If Σa_n is a convergent series of positive terms, then the series $\Sigma a_n x^n$ is absolutely convergent when $-1\leqq x\leqq 1$.

4. If Σa_n is a convergent series of positive terms, then the series $\Sigma a_n \cos n\theta$, $\Sigma a_n \sin n\theta$ are absolutely convergent for all values of θ. [Examples are afforded by the series $\Sigma r^n \cos n\theta$, $\Sigma r^n \sin n\theta$ of § 88.]

5. Any series selected from the terms of an absolutely convergent series is absolutely convergent. [For the series of the moduli of its terms is a selection from the series of the moduli of the terms of the original series.]

6. Prove that if $\Sigma|u_n|$ is convergent then $|\Sigma u_n|\leqq\Sigma|u_n|$, and that the only case of equality is that in which every term has the same sign.

192. Extension of Dirichlet's theorem to absolutely convergent series. Dirichlet's theorem (§ 176) shows that the terms of a series of positive terms may be rearranged in any way without affecting its sum. It is now easy to see that any absolutely convergent series has the same property. For let Σu_n be so rearranged as to become $\Sigma u'_n$, and let α'_n, v'_n, w'_n be formed from u'_n as α_n, v_n, w_n were formed from u_n. Then $\Sigma\alpha'_n$ is convergent, since it is a rearrangement of $\Sigma\alpha_n$, and so are $\Sigma v'_n$, $\Sigma w'_n$, which are rearrangements of Σv_n, Σw_n. Also, by Dirichlet's theorem, $\Sigma v'_n=\Sigma v_n$ and $\Sigma w'_n=\Sigma w_n$, and so

$$\Sigma u'_n=\Sigma v'_n-\Sigma w'_n=\Sigma v_n-\Sigma w_n=\Sigma u_n.$$

193. Conditionally convergent series. B. We have now to consider the second case indicated above, viz. that in which the series of moduli $\Sigma\alpha_n$ diverges to ∞.

DEFINITION. *If Σu_n is convergent, but $\Sigma|u_n|$ divergent, the original series is said to be* **conditionally convergent.**

In the first place we note that, if Σu_n is conditionally convergent, then the series Σv_n, Σw_n of § 191 must both diverge to ∞. They cannot both converge, since this would involve the convergence of $\Sigma(v_n+w_n)$ or $\Sigma\alpha_n$. And if one of them, say Σw_n, is convergent, and Σv_n divergent, then

$$\sum_0^N u_n = \sum_0^N v_n - \sum_0^N w_n \quad \text{..................(1)},$$

and therefore tends to ∞ with N, which is contrary to the hypothesis that Σu_n is convergent.

Hence Σv_n, Σw_n are both divergent. It is clear from equation (1) above that the sum of a conditionally convergent series is the limit of the difference of two functions each of which tends to ∞ with n. It is obvious too that Σu_n no longer possesses the property of convergent series of positive terms (Ex. XXX. 18), and all absolutely convergent series (Ex. LXXVII. 5), that any selection from the terms itself forms a convergent series. And it seems more than likely that the property prescribed by Dirichlet's theorem will not be possessed by conditionally convergent series; at any rate the proof of § 192 fails completely, since it depends essentially on the convergence of Σv_n and Σw_n separately. We shall see in a moment that this conjecture is well founded, and that the theorem is not true for series such as we are now considering.

194. Tests of convergence for conditionally convergent series. It is not to be expected that we should be able to find tests for conditional convergence as simple and general as those of §§ 173 *et seq.* It is naturally more difficult to formulate tests of convergence for series whose convergence, as is shown by equation (1) above, depends essentially on the cancelling of the positive by the negative terms. In the first place *there are no comparison tests for convergence of conditionally convergent series.*

For suppose we wish to infer the convergence of Σv_n from that of Σu_n. We have to compare

$$v_0 + v_1 + \dots + v_n, \quad u_0 + u_1 + \dots + u_n.$$

If every u and every v were positive, and (a) every v less than the corresponding u, we could at once infer that

$$v_0 + v_1 + \dots + v_n < u_0 + u_1 + \dots + u_n,$$

and so that Σv_n is convergent. If the u's only were positive and (b) every v *numerically* less than the corresponding u, we could infer that

$$|v_0| + |v_1| + \dots + |v_n| < u_0 + u_1 + \dots + u_n,$$

and so that Σv_n is absolutely convergent. But in the general case, when the u's and v's are both unrestricted as to sign, all that we can infer from (b) is that

$$|v_0| + |v_1| + \dots + |v_n| < |u_0| + |u_1| + \dots + |u_n|.$$

This would enable us to infer the absolute convergence of Σv_n from the absolute convergence of Σu_n; but if Σu_n is only conditionally convergent we can draw no inference at all.

Example. We shall see shortly that the series $1 - \frac{1}{2} + \frac{1}{3} - \frac{1}{4} + \dots$ is convergent. But the series $\frac{1}{2} + \frac{1}{3} + \frac{1}{4} + \frac{1}{5} + \dots$ is divergent, although each of its terms is numerically less than the corresponding term of the former series.

It is therefore only natural that such tests as we can obtain should be of a much more special character than those given in the early part of this chapter.

195. Alternating series. The simplest conditionally convergent series are *alternating series*, series whose terms are alternately positive and negative. The convergence of the most important series of this type is established by the following theorem.

If $\phi(n)$ is a positive function of n which tends **steadily** *to zero as $n \to \infty$, then the series*

$$\phi(0)-\phi(1)+\phi(2)-\dots$$

is convergent, and its sum lies between $\phi(0)$ and $\phi(0)-\phi(1)$.

Let us write $\phi_0, \phi_1, \dots$ for $\phi(0), \phi(1), \dots$; and let

$$s_n = \phi_0-\phi_1+\phi_2-\dots+(-1)^n\phi_n.$$

Then

$$s_{2n+1}-s_{2n-1} = \phi_{2n}-\phi_{2n+1} \geqq 0, \quad s_{2n}-s_{2n-2} = -(\phi_{2n-1}-\phi_{2n}) \leqq 0.$$

Hence $s_0, s_2, s_4, \dots, s_{2n}, \dots$ is a decreasing sequence, and therefore tends to a limit or to $-\infty$, and $s_1, s_3, s_5, \dots, s_{2n+1}, \dots$ is an increasing sequence, and therefore tends to a limit or to ∞. But $\lim(s_{2n+1}-s_{2n}) = \lim(-1)^{2n+1}\phi_{2n+1} = 0$, from which it follows that both sequences must tend to limits, and that the two limits must be the same. That is to say, the sequence $s_0, s_1, \dots, s_n, \dots$ tends to a limit. Since $s_0 = \phi_0$, $s_1 = \phi_0-\phi_1$, it is clear that this limit lies between ϕ_0 and $\phi_0-\phi_1$.

Examples LXXVIII. 1. The series

$$1-\frac{1}{2}+\frac{1}{3}-\frac{1}{4}+\dots, \quad 1-\frac{1}{\sqrt{2}}+\frac{1}{\sqrt{3}}-\frac{1}{\sqrt{4}}+\dots,$$

$$\Sigma\frac{(-1)^n}{n+a}, \quad \Sigma\frac{(-1)^n}{\sqrt{(n+a)}}, \quad \Sigma\frac{(-1)^n}{\sqrt{n}+\sqrt{a}}, \quad \Sigma\frac{(-1)^n}{(\sqrt{n}+\sqrt{a})^2},$$

where $a>0$, are conditionally convergent.

2. The series $\Sigma(-1)^n(n+a)^{-s}$, where $a>0$, is absolutely convergent if $s>1$, conditionally convergent if $0<s\leqq 1$, and oscillatory if $s\leqq 0$.

3. The sum of the series of § 195 lies between s_n and s_{n+1} for all values of n; and the error committed by taking the sum of the first n terms instead of the sum of the whole series is not greater numerically than the modulus of the $(n+1)$th term.

4. Consider the series $\qquad \Sigma\dfrac{(-1)^n}{\sqrt{n}+(-1)^n},$

which we suppose to begin with the term for which $n = 2$, to avoid any difficulty as to the definitions of the first few terms. This series may be written in the form

$$\Sigma\left[\left\{\frac{(-1)^n}{\sqrt{n}+(-1)^n}-\frac{(-1)^n}{\sqrt{n}}\right\}+\frac{(-1)^n}{\sqrt{n}}\right]$$

or

$$\Sigma\left\{\frac{(-1)^n}{\sqrt{n}}-\frac{1}{n+(-1)^n\sqrt{n}}\right\} = \Sigma(\psi_n-\chi_n),$$

say. The series $\Sigma\psi_n$ is convergent; but $\Sigma\chi_n$ is divergent, since all its terms are positive and $\lim n\chi_n = 1$. Hence the original series is divergent, although it is of the form $\phi_2 - \phi_3 + \phi_4 - \dots$, where $\phi_n \to 0$. This example shows that the condition that ϕ_n should tend *steadily* to zero is essential to the truth of the theorem. The reader will easily verify that

$$\sqrt{(2n+1)} - 1 < \sqrt{(2n)} + 1,$$

so that this condition is not satisfied here.

5. If the conditions of § 195 are satisfied except that ϕ_n tends steadily to a positive limit l, then the series $\Sigma(-1)^n \phi_n$ oscillates finitely.

6. The series $\Sigma(-1)^n \dfrac{a(a+1)\dots(a+n+1)}{b(b+1)\dots(b+n+1)}$, where neither a nor b is 0 or a negative integer, converges if and only if $a<b$. (*Math. Trip.* 1927)

[Call the series $\Sigma(-1)^n\phi_n$, and suppose first that a and b are positive. If $a \geqq b$ then $\phi_{n+1} \geqq \phi_n$, and ϕ_n does not tend to 0. If $a<b$ then $\phi_{n+1} < \phi_n$ and $\phi_n \to 0$ (§ 183), so that the conditions of the general theorem are satisfied.

In the general case we can choose N so that $a' = a+N$ and $b' = b+N$ are both positive, and ϕ_n is a multiple of ψ_{n-N}, where

$$\psi_n = \frac{a'(a'+1)\dots(a'+n+1)}{b'(b'+1)\dots(b'+n+1)}.]$$

7. **Alteration of the sum of a conditionally convergent series by rearrangement of the terms.** Let s be the sum of the series

$$1 - \tfrac{1}{2} + \tfrac{1}{3} - \tfrac{1}{4} + \dots,$$

and s_{2n} the sum of its first $2n$ terms, so that $\lim s_{2n} = s$; and rearrange the series as

$$1 + \tfrac{1}{3} - \tfrac{1}{2} + \tfrac{1}{5} + \tfrac{1}{7} - \tfrac{1}{4} + \dots \qquad \dots\dots\dots(1),$$

two positive terms being followed by one negative term. If t_{3n} is the sum of the first $3n$ terms of the new series, then

$$t_{3n} = 1 + \frac{1}{3} + \dots + \frac{1}{4n-1} - \frac{1}{2} - \frac{1}{4} - \dots - \frac{1}{2n}$$

$$= s_{2n} + \frac{1}{2n+1} + \frac{1}{2n+3} + \dots + \frac{1}{4n-1}.$$

Now
$$\lim\left[\frac{1}{2n+1} - \frac{1}{2n+2} + \frac{1}{2n+3} - \dots + \frac{1}{4n-1} - \frac{1}{4n}\right] = 0,$$

since the sum of the terms inside the bracket is less than $n/(2n+1)(2n+2)$; and

$$\lim\left(\frac{1}{2n+2} + \frac{1}{2n+4} + \dots + \frac{1}{4n}\right) = \tfrac{1}{2}\lim\frac{1}{n}\sum_{r=1}^{n}\frac{1}{1+(r/n)} = \tfrac{1}{2}\int_1^2 \frac{dx}{x},$$

after §§ 161 and 164. Hence

$$\lim t_{3n} = s + \tfrac{1}{2}\int_1^2 \frac{dx}{x},$$

and it follows that the sum of the series (1) is not s, but the right-hand side of the last equation. Later on we shall give the actual values of the sums of the two series: see § 220, Ex. xc. 7, and Ch. IX, Misc. Ex. 19.

It can indeed be proved that a conditionally convergent series can always be so rearranged as to converge to any sum whatever, or to diverge to ∞ or to $-\infty$. For a proof we may refer to Bromwich's *Infinite series*, 2nd edition, p. 74.

8. The series $1+\frac{1}{\sqrt{3}}-\frac{1}{\sqrt{2}}+\frac{1}{\sqrt{5}}+\frac{1}{\sqrt{7}}-\frac{1}{\sqrt{4}}+\dots$ diverges to ∞. [Here

$$t_{3n} = s_{2n}+\frac{1}{\sqrt{(2n+1)}}+\frac{1}{\sqrt{(2n+3)}}+\dots+\frac{1}{\sqrt{(4n-1)}} > s_{2n}+\frac{n}{\sqrt{(4n-1)}},$$

where $s_{2n} = 1-\frac{1}{\sqrt{2}}+\dots-\frac{1}{\sqrt{2n}}$, which tends to a limit when $n\to\infty$.]

196. Abel's and Dirichlet's tests of convergence. A more general test, which includes the test of § 195 as a particular case, is the following.

Dirichlet's test. *If ϕ_n satisfies the same conditions as in* § 195, *and Σa_n is any series which converges or oscillates finitely, then the series*

$$a_0\phi_0+a_1\phi_1+a_2\phi_2+\dots$$

is convergent.

The reader will easily verify the identity

$$\begin{aligned} &a_0\phi_0+a_1\phi_1+\dots+a_n\phi_n \\ &\qquad = s_0(\phi_0-\phi_1)+s_1(\phi_1-\phi_2)+\dots+s_{n-1}(\phi_{n-1}-\phi_n)+s_n\phi_n, \end{aligned}$$

where $s_n = a_0+a_1+\dots+a_n$. Now the series $(\phi_0-\phi_1)+(\phi_1-\phi_2)+\dots$ is convergent, since the sum to n terms is $\phi_0-\phi_n$ and $\lim\phi_n = 0$; and all its terms are positive. Also since Σa_n, if not actually convergent, at any rate oscillates finitely, we can determine a constant K so that $|s_\nu|<K$ for all values of ν. Hence the series

$$\Sigma s_\nu(\phi_\nu-\phi_{\nu+1})$$

is absolutely convergent, and so

$$s_0(\phi_0-\phi_1)+s_1(\phi_1-\phi_2)+\dots+s_{n-1}(\phi_{n-1}-\phi_n)$$

tends to a limit as $n\to\infty$. Finally ϕ_n, and therefore $s_n\phi_n$, tends to zero; and therefore

$$a_0\phi_0+a_1\phi_1+\dots+a_n\phi_n$$

tends to a limit, i.e. the series $\Sigma a_\nu\phi_\nu$ is convergent.

Abel's test. There is another test, due to Abel, which, though of less frequent application than Dirichlet's, is sometimes useful.

Suppose that ϕ_n, as in Dirichlet's test, is a positive and decreasing function of n, but that its limit as $n \to \infty$ is not necessarily zero. Thus we postulate less about ϕ_n, but to make up for this we postulate more about Σa_n, viz. that it is *convergent*. Then we have the theorem: *if ϕ_n is a positive and decreasing function of n, and Σa_n is convergent, then $\Sigma a_n \phi_n$ is convergent.*

For ϕ_n has a limit as $n \to \infty$, say l: and $\lim(\phi_n - l) = 0$. Hence, by Dirichlet's test, $\Sigma a_n(\phi_n - l)$ is convergent; and since Σa_n is convergent it follows that $\Sigma a_n \phi_n$ is convergent.

This theorem may be stated as follows: *a convergent series remains convergent if we multiply its terms by any sequence of positive and decreasing factors.*

Examples LXXIX. 1. Dirichlet's and Abel's tests may also be established by means of the general principle of convergence (§ 84). Let us suppose, for example, that the conditions of Abel's test are satisfied. We have identically

$$a_m\phi_m + a_{m+1}\phi_{m+1} + \dots + a_n\phi_n = s_{m,m}(\phi_m - \phi_{m+1}) + s_{m,m+1}(\phi_{m+1} - \phi_{m+2})$$
$$+ \dots + s_{m,n-1}(\phi_{n-1} - \phi_n) + s_{m,n}\phi_n \dots \quad \dots\dots(1),$$

where $$s_{m,\nu} = a_m + a_{m+1} + \dots + a_\nu.$$

The left-hand side of (1) therefore lies between $h\phi_m$ and $H\phi_m$, where h and H are the algebraically least and greatest of $s_{m,m}, s_{m,m+1}, \dots, s_{m,n}$. But, given any positive δ, we can choose m_0 so that $|s_{m,\nu}| < \delta$ when $m \geqq m_0$, and so

$$|a_m\phi_m + a_{m+1}\phi_{m+1} + \dots + a_n\phi_n| < \delta\phi_m \leqq \delta\phi_1$$

when $n > m \geqq m_0$. Thus the series $\Sigma a_n\phi_n$ is convergent.

2. The series $\Sigma \cos n\theta$ and $\Sigma \sin n\theta$ oscillate finitely when θ is not a multiple of π. For, if we denote the sums of the first n terms of the two series by s_n and t_n, and write $z = \operatorname{Cis}\theta$, so that $|z| = 1$ and $z \neq 1$, we have

$$|s_n + it_n| = \left|\frac{1-z^n}{1-z}\right| \leqq \frac{1+|z^n|}{|1-z|} \leqq \frac{2}{|1-z|};$$

and so $|s_n|$ and $|t_n|$ are also not greater than $2/|1-z|$. That the series are not actually convergent follows from the fact that their nth terms do not tend to zero (Ex. XXIV. 7).

The sine series converges to zero if θ is a multiple of π. The cosine series oscillates finitely if θ is an odd multiple of π and diverges if θ is an even multiple of π.

It follows that *if ϕ_n is a positive function of n which tends steadily to zero as $n \to \infty$, then the series*

$$\Sigma\phi_n \cos n\theta, \quad \Sigma\phi_n \sin n\theta$$

are convergent, except perhaps the first series when θ is a multiple of 2π. In this case the first series reduces to $\Sigma\phi_n$, which may or may not be convergent: the second series vanishes identically. If $\Sigma\phi_n$ is convergent then both series are absolutely convergent (Ex. LXXVII. 4) for all values of θ, and the whole interest of the result lies in its application to the case in which $\Sigma\phi_n$ is divergent. And in this case the series above written are conditionally and *not* absolutely convergent, as will be proved in Ex. 6 below. If we put $\theta = \pi$ in the cosine series we are led back to the result of § 195, since $\cos n\pi = (-1)^n$.

3. The series $\Sigma n^{-s}\cos n\theta$, $\Sigma n^{-s}\sin n\theta$ are convergent if $s > 0$, unless (in the case of the first series) θ is a multiple of 2π and $0 < s \leqq 1$.

4. The series of Ex. 3 are in general absolutely convergent if $s > 1$, conditionally convergent if $0 < s \leqq 1$, and oscillatory if $s \leqq 0$ (finitely if $s = 0$ and infinitely if $s < 0$). Mention any exceptional cases.

5. If $\Sigma a_n n^{-s}$ is convergent or oscillates finitely, then $\Sigma a_n n^{-t}$ is convergent when $t > s$.

6. If ϕ_n is a positive function of n which tends steadily to 0 as $n \to \infty$, and $\Sigma\phi_n$ is divergent, then the series $\Sigma\phi_n \cos n\theta$, $\Sigma\phi_n \sin n\theta$ are *not* absolutely convergent, except the sine series when θ is a multiple of π. [For suppose, e.g., that $\Sigma\phi_n |\cos n\theta|$ is convergent. Since $\cos^2 n\theta \leqq |\cos n\theta|$, it follows that $\Sigma\phi_n \cos^2 n\theta$ or

$$\tfrac{1}{2}\Sigma\phi_n(1+\cos 2n\theta)$$

is convergent. But this is impossible, since $\Sigma\phi_n$ is divergent and $\Sigma\phi_n \cos 2n\theta$, by Dirichlet's test, convergent, unless θ is a multiple of π, in which case it is obvious that $\Sigma\phi_n |\cos n\theta|$ is divergent. The reader should write out the corresponding argument for the sine series, noting where it fails when θ is a multiple of π.]

197. Series of complex terms. So far we have confined ourselves to series all of whose terms are real. We shall now consider the series

$$\Sigma u_n = \Sigma(v_n + iw_n),$$

where v_n and w_n are real. The consideration of such series does not introduce any really new difficulties. The series is convergent if, and only if, the series

$$\Sigma v_n, \quad \Sigma w_n$$

are separately convergent. There is however one class of such series so important as to require special treatment. Accordingly we give the following definition, which is an obvious extension of that of § 191.

DEFINITION. *The series* Σu_n, *where* $u_n = v_n + iw_n$, *is said to be* **absolutely convergent** *if the series* Σv_n *and* Σw_n *are absolutely convergent.*

THEOREM. *A necessary and sufficient condition for the absolute convergence of* Σu_n *is the convergence of* $\Sigma |u_n|$ *or* $\Sigma \sqrt{(v_n^2 + w_n^2)}$.

For if Σu_n is absolutely convergent, then both of the series $\Sigma |v_n|$ and $\Sigma |w_n|$ are convergent, and so $\Sigma\{|v_n| + |w_n|\}$ is convergent: but

$$|u_n| = \sqrt{(v_n^2 + w_n^2)} \leqq |v_n| + |w_n|,$$

and therefore $\Sigma |u_n|$ is convergent. On the other hand

$$|v_n| \leqq \sqrt{(v_n^2 + w_n^2)}, \quad |w_n| \leqq \sqrt{(v_n^2 + w_n^2)},$$

so that $\Sigma |v_n|$ and $\Sigma |w_n|$ are convergent whenever $\Sigma |u_n|$ is convergent.

It is obvious that *an absolutely convergent series is convergent,* since its real and imaginary parts converge separately. And Dirichlet's theorem (§§ 176, 192) may be extended at once to absolutely convergent complex series by applying it to the separate series Σv_n and Σw_n.

The convergence of an absolutely convergent series may also be deduced directly from the general principle of convergence (cf. Ex. LXXVII. 1). We leave this as an exercise to the reader.

198. Power series. One of the most important parts of the theory of the ordinary functions of elementary analysis (such as the sine and cosine, and the logarithm and exponential, which will be discussed in the next chapter) is that which is concerned with their expansion in series of the form $\Sigma a_n x^n$. Such a series is called a *power series in x*. We have already come across some cases of expansion in series of this kind in connection with Taylor's and Maclaurin's series (§ 152). There, however, we were concerned only with a real variable x. We shall now consider a few general properties of power series in z, where z is a complex variable.

A. *A power series $\Sigma a_n z^n$ may be convergent for all values of z, for a certain region of values, or for no values except $z = 0$.*

It is sufficient to give an example of each possibility.

1. *The series $\Sigma \dfrac{z^n}{n!}$ is convergent for all values of z.* For if $u_n = \dfrac{z^n}{n!}$ then

$$\frac{|u_{n+1}|}{|u_n|} = \frac{|z|}{n+1} \to 0$$

whatever value z may have. Hence, by d'Alembert's test, $\Sigma |u_n|$ is convergent for all values of z, and the original series is absolutely convergent for all values of z. We shall see later on that a power series, when convergent, is *generally* absolutely convergent.

2. *The series $\Sigma n!z^n$ is not convergent for any value of z except $z = 0$.* For if $u_n = n!z^n$ then $|u_{n+1}|/|u_n| = (n+1)|z|$, which tends to ∞ with n, unless $z = 0$. Hence (cf. Exs. XXVII. 1, 2, 5) the modulus of the nth term tends to ∞ with n; and so the series cannot converge, except when $z = 0$. It is obvious that any power series converges when $z = 0$.

3. *The series Σz^n is always convergent when $|z| < 1$, and never convergent when $|z| \geqq 1$.* This was proved in § 88. Thus we have an example of each of the three possibilities.

199. B. *If a power series $\Sigma a_n z^n$ is convergent for a particular value of z, say $z_1 = r_1(\cos\theta_1 + i\sin\theta_1)$, then it is absolutely convergent for all values of z such that $|z| < r_1$.*

For $\lim a_n z_1^n = 0$, since $\Sigma a_n z_1^n$ is convergent, and therefore we can find a number K such that $|a_n z_1^n| < K$ for all values of n. But, if $|z| = r < r_1$, we have

$$|a_n z^n| = |a_n z_1^n| \left(\frac{r}{r_1}\right)^n < K\left(\frac{r}{r_1}\right)^n,$$

and the result follows by comparison with the convergent geometrical series $\Sigma (r/r_1)^n$.

In other words, if the series converges at P *then it converges absolutely at all points nearer to the origin than P.*

Example. Show that the result is true even if the series oscillates finitely when $z = z_1$. [If $s_n = a_0 + a_1 z_1 + \dots + a_n z_1^n$ then we can find K so that $|s_n| < K$ for all values of n. But

$$|a_n z_1^n| = |s_n - s_{n-1}| \leqq |s_{n-1}| + |s_n| < 2K,$$

and the argument can be completed as before.]

200. The region of convergence of a power series. The circle of convergence. Let $z = r$ be any point on the positive real axis. If the power series converges when $z = r$ then it converges absolutely at all points inside the circle $|z| = r$. In particular it converges for all real values of z less than r.

Now let us divide the points r of the positive real axis into two classes, the class at which the series converges and the class at which it does not. The first class must contain at least the one point $z = 0$. The second class, on the other hand, need not exist, since the series may converge for all values of z. Suppose however that it does exist, and that the first class of points includes points besides $z = 0$. Then it is clear that every point of the first class lies to the left of every point of the second class. Hence there is a point, say the point $z = R$, which divides the two classes, and may itself belong to either one or the other. *Then the series is absolutely convergent at all points inside the circle* $|z| = R$.

For suppose that this circle cuts OX in A (Fig. 48), and that P is a point inside it. We can draw a concentric circle, of radius less than R, so as to include P inside it. Let this circle cut OX in Q. Then the series is convergent at Q, and therefore, by Theorem B, absolutely convergent at P.

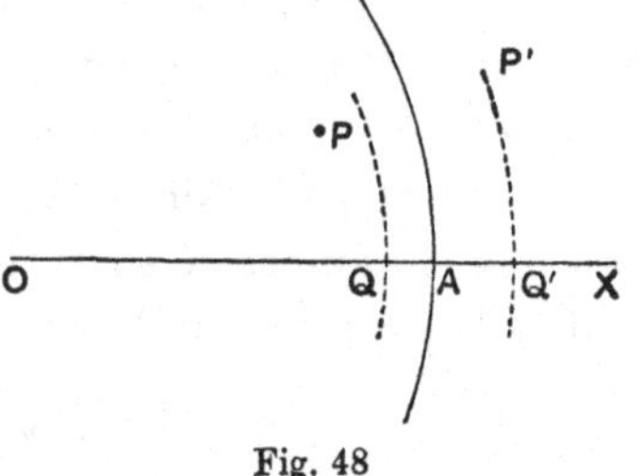

Fig. 48

On the other hand the series cannot converge at any point P' *outside* the circle. For if it converged at P' it would converge absolutely at all points nearer to O than P; and this is absurd, as it does not converge at any point between A and Q'.

So far we have excepted the cases in which the power series (1) does not converge at any point on the positive real axis except $z = 0$ or (2) converges at all points on the positive real axis. It is clear that in case (1) the power series converges nowhere except when $z = 0$, and that in case (2) it is absolutely convergent everywhere. Thus we obtain the following result: *a power series either*

(1) *converges for $z = 0$ and for no other values of z; or*

(2) *converges absolutely for all values of z; or*

(3) *converges absolutely for all values of z within a certain circle of radius R, and does not converge for any value of z outside this circle.*

In case (3) the circle is called the *circle of convergence* and its radius the *radius of convergence* of the power series.

It should be observed that this general result gives no information at all about the behaviour of the series *on* the circle of convergence. The examples which follow show that as a matter of fact there are very diverse possibilities.

Examples LXXX. 1. The series $1 + az + a^2z^2 + \dots$, where $a > 0$, has a radius of convergence equal to $1/a$. It does not converge anywhere on its circle of convergence, diverging when $z = 1/a$ and oscillating finitely at all other points on the circle.

2. The series $\dfrac{z}{1^2} + \dfrac{z^2}{2^2} + \dfrac{z^3}{3^2} + \dots$ has its radius of convergence equal to 1; it converges absolutely at all points on its circle of convergence.

3. More generally, if $|a_{n+1}|/|a_n| \to \lambda$, or $|a_n|^{1/n} \to \lambda$, as $n \to \infty$, then the series $a_0 + a_1z + a_2z^2 + \dots$ has $1/\lambda$ as its radius of convergence. In the first case

$$\lim |a_{n+1}z^{n+1}|/|a_nz^n| = \lambda |z|,$$

which is less or greater than unity according as $|z|$ is less or greater than $1/\lambda$, so that we can use d'Alembert's test (§ 175, 6). In the second case we can use Cauchy's test (§ 174, 2) similarly.

4. **The logarithmic series.** The series

$$z - \tfrac{1}{2}z^2 + \tfrac{1}{3}z^3 - \dots$$

is called (for reasons which will appear later) the 'logarithmic' series. It follows from Ex. 3 that its radius of convergence is unity.

When z is on the circle of convergence we may write $z = \cos\theta + i\sin\theta$, and the series assumes the form

$$\cos\theta - \tfrac{1}{2}\cos 2\theta + \tfrac{1}{3}\cos 3\theta - \dots + i(\sin\theta - \tfrac{1}{2}\sin 2\theta + \tfrac{1}{3}\sin 3\theta - \dots).$$

The real and imaginary parts are both convergent, though not absolutely convergent, unless θ is an odd multiple of π (Exs. LXXIX. 3, 4, with $\theta + \pi$ for θ). If θ is an odd multiple of π then $z = -1$, and the series is $-1 - \frac{1}{2} - \frac{1}{3} - \dots$ and diverges to $-\infty$. Thus the logarithmic series converges at all points of its circle of convergence except the point $z = -1$.

5. **The binomial series.** Consider the series

$$1+mz+\frac{m(m-1)}{2!}z^2+\frac{m(m-1)(m-2)}{3!}z^3+\dots.$$

If m is a positive integer then the series terminates. In general

$$\frac{|a_{n+1}|}{|a_n|}=\frac{|m-n|}{n+1}\to 1,$$

so that the radius of convergence is unity. We shall not discuss the question of its convergence on the circle, which is a little more difficult, here*.

201. Uniqueness of a power series. If $\Sigma a_n z^n$ is a power series which is convergent for some values of z at any rate besides $z=0$, and $f(z)$ is its sum, then

$$f(z)=a_0+a_1z+\dots+a_mz^m+o(z^m)$$

for every m, when $z\to 0$. For, if μ is any positive number less than the radius of convergence of the series, then $|a_n|\mu^n<K$, where K is independent of n (cf. § 199); and so, if $|z|<\mu$,

$$\left|f(z)-\sum_0^m a_\nu z^\nu\right|\leqq|a_{m+1}||z|^{m+1}+|a_{m+2}||z|^{m+2}+\dots$$

$$<K\left(\frac{|z|}{\mu}\right)^{m+1}\left(1+\frac{|z|}{\mu}+\frac{|z|^2}{\mu^2}+\dots\right)=\frac{K|z|^{m+1}}{\mu^m(\mu-|z|)},$$

which is $O(|z|^{m+1})$ and *a fortiori* $o(|z|^m)$. In particular this is true for real positive z.

It now follows from Ex. LVI. 1 that, if $\Sigma a_n z^n=\Sigma b_n z^n$ for all z whose modulus is less than μ, then $a_n=b_n$ for all n. *The same function $f(z)$ cannot be represented by two different power series.*

202. Multiplication of series. We saw in § 177 that, if Σu_n and Σv_n are two convergent series of positive terms, then $\Sigma u_n\times\Sigma v_n=\Sigma w_n$, where

$$w_n=u_0v_n+u_1v_{n-1}+\dots+u_nv_0.$$

We can now extend this result to all cases in which Σu_n and Σv_n are *absolutely* convergent; for our proof was merely a simple application of Dirichlet's theorem, which we have already extended to all absolutely convergent series.

* The cases $z=1$ and $z=-1$ are discussed in § 222. For a complete discussion, see Bromwich, *Infinite series*, 2nd edition, pp. 287 *et seq.*; Hobson, *Plane trigonometry*, 5th edition, pp. 268 *et seq.*

Examples LXXXI. 1. If $|z|$ is less than the radius of convergence of either of the series $\Sigma a_n z^n$, $\Sigma b_n z^n$, then the product of the two series is $\Sigma c_n z^n$, where $c_n = a_0 b_n + a_1 b_{n-1} + \dots + a_n b_0$.

2. If the radius of convergence of $\Sigma a_n z^n$ is R, and $f(z)$ is the sum of the series when $|z| < R$, and $|z|$ is less than either R or unity, then $f(z)/(1-z) = \Sigma s_n z^n$, where $s_n = a_0 + a_1 + \dots + a_n$.

3. Prove, by squaring the series for $(1-z)^{-1}$, that

$$(1-z)^{-2} = 1 + 2z + 3z^2 + \dots$$

if $|z| < 1$.

4. Prove similarly that $(1-z)^{-3} = 1 + 3z + 6z^2 + \dots$, the general term being $\frac{1}{2}(n+1)(n+2)z^n$.

5. **The binomial theorem for a negative integral exponent.** If $|z| < 1$, and m is a positive integer, then

$$\frac{1}{(1-z)^m} = 1 + mz + \frac{m(m+1)}{1.2} z^2 + \dots + \frac{m(m+1)\dots(m+n-1)}{1.2\dots n} z^n + \dots.$$

[Assume the truth of the theorem for all indices up to m. Then, by Ex. 2, $1/(1-z)^{m+1} = \Sigma s_n z^n$, where

$$s_n = 1 + m + \frac{m(m+1)}{1.2} + \dots + \frac{m(m+1)\dots(m+n-1)}{1.2\dots n}$$

$$= \frac{(m+1)(m+2)\dots(m+n)}{1.2\dots n},$$

as is easily proved by induction (whether m be an integer or not).]

6. Prove by multiplication of series that if

$$f(m,z) = 1 + \binom{m}{1} z + \binom{m}{2} z^2 + \dots,$$

and $|z| < 1$, then $f(m,z)f(m',z) = f(m+m',z)$. [This equation forms the basis of Euler's proof of the binomial theorem. The coefficient of z^n in the product series is

$$\binom{m'}{n} + \binom{m}{1}\binom{m'}{n-1} + \binom{m}{2}\binom{m'}{n-2} + \dots + \binom{m}{n-1}\binom{m'}{1} + \binom{m}{n},$$

a polynomial in m and m'. When m and m' are positive integers this polynomial must reduce to $\binom{m+m'}{k}$, in virtue of the binomial theorem for a positive integral exponent; and if two such polynomials are equal for all positive integral values of m and m' then they must be equal identically.]

7. If $f(z) = 1 + z + \frac{z^2}{2!} + \dots$ then $f(z)f(z') = f(z+z')$. [For the series for

$f(z)$ is absolutely convergent for all values of z: and it is easy to see that if $u_n = \dfrac{z^n}{n!}$, $v_n = \dfrac{z'^n}{n!}$, then $w_n = \dfrac{(z+z')^n}{n!}$.]

8. If $$C(z) = 1 - \frac{z^2}{2!} + \frac{z^4}{4!} - \dots, \quad S(z) = z - \frac{z^3}{3!} + \frac{z^5}{5!} - \dots,$$
then
$$C(z+z') = C(z)\,C(z') - S(z)\,S(z'), \quad S(z+z') = S(z)\,C(z') + C(z)\,S(z'),$$
and
$$\{C(z)\}^2 + \{S(z)\}^2 = 1.$$

9. **Failure of the multiplication theorem.** That the theorem is not always true when Σu_n and Σv_n are not absolutely convergent may be seen by considering the case in which
$$u_n = v_n = \frac{(-1)^n}{\sqrt{(n+1)}}.$$
Then
$$w_n = (-1)^n \sum_{r=0}^{n} \frac{1}{\sqrt{\{(r+1)(n+1-r)\}}}.$$
But $\sqrt{\{(r+1)(n+1-r)\}} \leqq \frac{1}{2}(n+2)$, and so $|w_n| > (2n+2)/(n+2)$, which tends to 2; so that Σw_n is certainly not convergent.

203. Absolutely and conditionally convergent infinite integrals. There is a theory for integrals analogous to that developed for series in §§ 191 *et seq.*

The infinite integral
$$\int_a^\infty f(x)\,dx \quad \dots\dots\dots(1)$$
is said to be *absolutely convergent* if
$$\int_a^\infty |f(x)|\,dx \quad \dots\dots\dots(2)$$
is convergent. We may define $g(x)$ and $h(x)$ by
$$f(x) = g(x) - h(x), \quad |f(x)| = g(x) + h(x).$$
Then $g(x)$ is $f(x)$ when $f(x)$ is positive, and zero when $f(x)$ is negative, and $h(x)$ is zero when $f(x)$ is positive and $-f(x)$ when $f(x)$ is negative, so that $g(x)$ and $h(x)$ correspond to the v_n and w_n of § 191. It is plain that $g(x) \geqq 0$, $h(x) \geqq 0$, and that $g(x)$ and $h(x)$ are continuous when $f(x)$ is continuous.

It then follows as in §§ 191 and 193 that the integrals
$$\int_a^\infty g(x)\,dx, \quad \int_a^\infty h(x)\,dx$$
are both convergent when (2) is convergent, but are both divergent when (1) is convergent and (2) is not; and that *an absolutely convergent integral is convergent.*

It is plain that, if $|f(x)| \leqq \phi(x)$, and $\int_a^\infty \phi(x)\,dx$ is convergent, then the integral (1) is absolutely convergent.

When (1) is convergent and (2) is not, we shall say that (1) is *conditionally convergent.* We shall not have much to do with conditionally convergent integrals here, but there is one special type of integral which is particularly important.

Suppose that $\phi'(x)$ is continuous; that $\phi(x) \geqq 0$, $\phi'(x) \leqq 0$; and that $\phi(x) \to 0$ when $x \to \infty$. Then $|\phi'(x)| = -\phi'(x)$, and

$$\int_a^\infty |\phi'(x)|\,dx = -\int_a^\infty \phi'(x)\,dx = -\lim_{X\to\infty} \int_a^X \phi'(x)\,dx$$
$$= \lim_{X\to\infty} \{\phi(a) - \phi(X)\} = \phi(a),$$

so that $\int_a^\infty \phi'(x)\,dx$ is absolutely convergent.

Now consider the integral

$$\int_a^\infty \phi(x) \cos tx\,dx \quad \ldots\ldots\ldots\ldots\ldots\ldots\ldots\ldots(3);$$

we may suppose t positive. We have

$$\int_a^X \phi(x) \cos tx\,dx = \frac{1}{t}\int_a^X \phi(x) \frac{d}{dx}\sin tx\,dx$$
$$= \frac{\sin tX}{t}\phi(X) - \frac{\sin ta}{t}\phi(a) - \frac{1}{t}\int_a^X \phi'(x) \sin tx\,dx \quad \ldots(4).$$

The first term tends to zero when $X \to \infty$. Also $|\sin tx| \leqq 1$, so that $|\phi'(x) \sin tx| \leqq |\phi'(x)|$. Hence $\int_a^\infty \phi'(x) \sin tx\,dx$ is absolutely convergent, and therefore convergent; and the last integral in (4) tends to a limit when $X \to \infty$. It follows that the left-hand side of (4) tends to a limit, and therefore that (3) is convergent. Similarly

$$\int_a^\infty \phi(x) \sin tx\,dx$$

is convergent.

The most important case is that in which $a > 0$ and $\phi(x) = x^{-s}$, where $s > 0$. The integrals are then absolutely convergent when $s > 1$, and conditionally convergent when $0 < s \leqq 1$.

Examples LXXXII. 1. The integral $\int_0^\infty \frac{\sin tx}{x^s}\,dx$ is convergent if $0 < s < 2$, and absolutely convergent if $1 < s < 2$. [Consider the ranges $(0, 1)$, $(1, \infty)$ separately.]

2. $\int_0^\infty \frac{x+1}{x^{\frac{3}{2}}} \sin x\,dx$ is convergent. (*Math. Trip.* 1930)

3. $\int_0^\infty \frac{1-\cos tx}{x^s}\,dx$ is convergent if $1<s<3$, and then absolutely.

4. $\int_0^\infty \frac{\sin x\,(1-\cos x)}{x^s}\,dx$ is convergent if $0<s<4$, and absolutely convergent if $1<s<4$. (*Math. Trip.* 1934)

5. $\int_0^\infty x^{-\alpha} \sin x^{1-\beta}\,dx$ is convergent if α lies between β and $2-\beta$. (*Math. Trip.* 1936)

[Put $x^{1-\beta} = y$, and consider the cases $\beta<1$ and $\beta>1$ separately.]

MISCELLANEOUS EXAMPLES ON CHAPTER VIII

1. Discuss the convergence of the series $\Sigma n^k\{\sqrt{(n+1)}-2\sqrt{n}+\sqrt{(n-1)}\}$, where k is real. (*Math. Trip.* 1890)

2. Show that
$$\Sigma n^r \Delta^k(n^s),$$
where
$$\Delta u_n = u_n - u_{n+1}, \quad \Delta^2 u_n = \Delta(\Delta u_n),$$
and so on, is convergent if and only if $k>r+s+1$, except when s is a positive integer less than k, when every term of the series is zero. [The result of Ch. VII, Misc. Ex. 6, shows that $\Delta^k(n^s)$ is in general of order n^{s-k}.]

3. Show that
$$\sum_1^\infty \frac{n^2+9n+5}{(n+1)(2n+3)(2n+5)(n+4)} = \frac{5}{36}.$$
(*Math. Trip.* 1912)

[Resolve the general term into partial fractions.]

4. If Σa_n is a divergent series of positive terms, and
$$a_{n-1} > \frac{a_n}{1+a_n}, \quad b_n = \frac{a_n}{1+na_n},$$
then Σb_n is divergent. (*Math. Trip.* 1931)

[It is easy to verify that $b_{n-1}>b_n$. Hence the convergence of Σb_n would imply that $nb_n \to 0$, and therefore that $na_n \to 0$. This gives $b_n \sim a_n$ and a contradiction.]

5. Show that the series
$$1-\frac{1}{1+z}+\frac{1}{2}-\frac{1}{2+z}+\frac{1}{3}-\frac{1}{3+z}+\dots$$
is convergent provided only that z is not a negative integer.

6. Investigate the convergence or divergence of the series

$$\Sigma \sin\frac{a}{n},\quad \Sigma\frac{1}{n}\sin\frac{a}{n},\quad \Sigma(-1)^n \sin\frac{a}{n},\quad \Sigma\left(1-\cos\frac{a}{n}\right),\quad \Sigma(-1)^n n\left(1-\cos\frac{a}{n}\right),$$

where a is real.

7. Discuss the convergence of the series

$$\sum_1^\infty \left(1+\frac{1}{2}+\frac{1}{3}+\dots+\frac{1}{n}\right)\frac{\sin(n\theta+\alpha)}{n},$$

where θ and α are real. (*Math. Trip.* 1899)

8. Prove that the series

$$1-\tfrac{1}{2}-\tfrac{1}{3}+\tfrac{1}{4}+\tfrac{1}{5}+\tfrac{1}{6}-\tfrac{1}{7}-\tfrac{1}{8}-\tfrac{1}{9}-\tfrac{1}{10}+\dots,$$

in which successive terms of the same sign form groups of 1, 2, 3, 4, ... terms, is convergent; but that the corresponding series in which the groups contain 1, 2, 4, 8, ... terms oscillates finitely. (*Math. Trip.* 1908)

9. If $u_1, u_2, u_3, \dots$ is a decreasing sequence of positive numbers whose limit is zero, then the series

$$u_1-\tfrac{1}{2}(u_1+u_2)+\tfrac{1}{3}(u_1+u_2+u_3)-\dots,\qquad u_1-\tfrac{1}{3}(u_1+u_3)+\tfrac{1}{5}(u_1+u_3+u_5)-\dots$$

are convergent. [For if $(u_1+u_2+\dots+u_n)/n = v_n$ then $v_1, v_2, v_3, \dots$ is also a decreasing sequence whose limit is zero (Ch. IV, Misc. Exs. 8, 16). This shows that the first series is convergent; the second we leave to the reader. In particular the series

$$1-\tfrac{1}{2}(1+\tfrac{1}{2})+\tfrac{1}{3}(1+\tfrac{1}{2}+\tfrac{1}{3})-\dots,\qquad 1-\tfrac{1}{3}(1+\tfrac{1}{3})+\tfrac{1}{5}(1+\tfrac{1}{3}+\tfrac{1}{5})-\dots$$

are convergent.]

10. If $u_0+u_1+u_2+\dots$ is a divergent series of positive and decreasing terms, then

$$(u_0+u_2+\dots+u_{2n})/(u_1+u_3+\dots+u_{2n+1}) \to 1.$$

11. Prove that $\lim\limits_{\alpha\to+0} \alpha\sum\limits_1^\infty n^{-1-\alpha} = 1$. [It follows from § 180 that

$$0 < 1^{-1-\alpha}+2^{-1-\alpha}+\dots+(n-1)^{-1-\alpha}-\int_1^n x^{-1-\alpha}\,dx \leqq 1,$$

and it is easy to deduce that $\Sigma n^{-1-\alpha}$ lies between $1/\alpha$ and $(\alpha+1)/\alpha$.]

12. Find the sum of the series $\sum\limits_1^\infty u_n$, where

$$u_n = \frac{x^n-x^{-n-1}}{(x^n+x^{-n})(x^{n+1}+x^{-n-1})} = \frac{1}{x-1}\left(\frac{1}{x^n+x^{-n}}-\frac{1}{x^{n+1}+x^{-n-1}}\right),$$

for all real values of x for which the series is convergent.

(*Math. Trip.* 1901)

[If $|x|$ is not equal to unity then the series has the sum $x/\{(x-1)(x^2+1)\}$. If $x = 1$ then $u_n = 0$ and the sum is 0. If $x = -1$ then $u_n = \frac{1}{2}(-1)^{n+1}$ and the series oscillates finitely.]

13. Find the sums of the series

$$\frac{z}{1+z}+\frac{2z^2}{1+z^2}+\frac{4z^4}{1+z^4}+\dots, \quad \frac{z}{1-z^2}+\frac{z^2}{1-z^4}+\frac{z^4}{1-z^8}+\dots$$

(in which all the indices are powers of 2), whenever they are convergent.

[The first series converges only if $|z|<1$, its sum then being $z/(1-z)$; the second series converges to $z/(1-z)$ if $|z|<1$ and to $1/(1-z)$ if $|z|>1$.]

14. If $|a_n| \leqq 1$ for all values of n then the equation

$$0 = 1+a_1 z+a_2 z^2+\dots$$

cannot have a root whose modulus is less than $\frac{1}{2}$, and the only case in which it can have a root whose modulus is equal to $\frac{1}{2}$ is that in which $a_n = -\operatorname{Cis}(n\theta)$, when $z = \frac{1}{2}\operatorname{Cis}(-\theta)$ is a root.

15. **Recurring series.** A power series $\Sigma a_n z^n$ is said to be a *recurring series* if its coefficients satisfy a relation of the type

$$a_n+p_1 a_{n-1}+p_2 a_{n-2}+\dots+p_k a_{n-k} = 0 \quad \dots\dots\dots\dots(1),$$

where $n \geqq k$ and $p_1, p_2, \dots, p_k$ are independent of n. Any recurring series is the expansion of a rational function of z. To prove this we observe in the first place that the series is certainly convergent for values of z whose modulus is sufficiently small. For it follows from (1) that $|a_n| \leqq G\alpha_n$, where α_n is the modulus of the numerically greatest of the preceding coefficients, and $G = |p_1|+|p_2|+\dots+|p_k|$; and from this that $|a_n| < KG^n$, where K is independent of n. Thus the recurring series is certainly convergent for values of z whose modulus is less than $1/G$.

But if we multiply the series $f(z) = \Sigma a_n z^n$ by $p_1 z, p_2 z^2, \dots, p_k z^k$, and add the results, we obtain a new series in which all the coefficients after the $(k-1)$th vanish in virtue of (1), so that

$$(1+p_1 z+p_2 z^2+\dots+p_k z^k) f(z) = P_0+P_1 z+\dots+P_{k-1} z^{k-1},$$

where $P_0, P_1, \dots, P_{k-1}$ are constants. The polynomial

$$1+p_1 z+p_2 z^2+\dots+p_k z^k$$

is called the *scale of relation* of the series.

Conversely, it follows from the known results as to the expression of any rational function as the sum of a polynomial and certain partial fractions of the type $A/(z-\alpha)^p$, and from the binomial theorem for a negative integral exponent, that any rational function whose denominator is not divisible by z can be expanded in a power series convergent for values of z whose modulus is sufficiently small, in fact if $|z|<\rho$, where ρ is the least of the moduli of the roots of the denominator (cf. Ch. IV, Misc. Exs. 26

et seq.). And it is easy to see, by reversing the argument above, that the series is a recurring series. Thus *a necessary and sufficient condition that a power series should be a recurring series is that it should be the expansion of such a rational function of z.*

16. Solution of difference equations. A relation of the type of (1) in Ex. 15 is called a *linear difference equation in a_n with constant coefficients.* Such equations may be solved by a method which will be sufficiently explained by an example. Suppose that the equation is

$$a_n - a_{n-1} - 8a_{n-2} + 12a_{n-3} = 0.$$

Consider the recurring power series $\Sigma a_n z^n$. We find, as in Ex. 15, that its sum is

$$\frac{a_0 + (a_1 - a_0) z + (a_2 - a_1 - 8a_0) z^2}{1 - z - 8z^2 + 12z^3} = \frac{A_1}{1-2z} + \frac{A_2}{(1-2z)^2} + \frac{B}{1+3z},$$

where A_1, A_2, and B are numbers easily expressible in terms of a_0, a_1, and a_2. Expanding each fraction separately we see that the coefficient of z^n is

$$a_n = 2^n\{A_1 + (n+1) A_2\} + (-3)^n B.$$

The values of A_1, A_2, B depend upon the first three coefficients a_0, a_1, a_2, which may of course be chosen arbitrarily.

17. The solution of the difference equation $u_n - 2\cos\theta\, u_{n-1} + u_{n-2} = 0$ is $u_n = A\cos n\theta + B\sin n\theta$, where A and B are arbitrary constants.

18. If u_n is a polynomial in n of degree k, then $\Sigma u_n z^n$ is a recurring series whose scale of relation is $(1-z)^{k+1}$. (*Math. Trip.* 1904)

19. Expand $9/\{(z-1)(z+2)^2\}$ in ascending powers of z. (*Math. Trip.* 1913)

20. A player tossing a coin is to score one point for every head he turns up and two for every tail, and is to play on until his score reaches or passes a total n. Show that his chance of making exactly the total n is $\frac{1}{3}\{2 + (-\frac{1}{2})^n\}$. (*Math. Trip.* 1898)

[If p_n is the probability then $p_n = \frac{1}{2}(p_{n-1} + p_{n-2})$; also $p_0 = 1$, $p_1 = \frac{1}{2}$.]

21. Prove that

$$\frac{1}{a+1} + \frac{1}{a+2} + \dots + \frac{1}{a+n} = \binom{n}{1}\frac{1}{a+1} - \binom{n}{2}\frac{1!}{(a+1)(a+2)} + \dots$$

if n is a positive integer and a is not one of the numbers $-1, -2, \dots, -n$.

[This follows from splitting up each term on the right-hand side into partial fractions. When $a > -1$, the result may be deduced very simply from the equation

$$\int_0^1 x^a \frac{1-x^n}{1-x}\,dx = \int_0^1 (1-x)^a \{1-(1-x)^n\} \frac{dx}{x}$$

by expanding $(1-x^n)/(1-x)$ and $1-(1-x)^n$ in powers of x and integrating each term separately. The result, being an algebraical identity, must be true for all values of a save $-1, -2, \ldots, -n$.]

22. Prove by multiplication of series that

$$\sum_0^\infty \frac{z^n}{n!} \sum_1^\infty \frac{(-1)^{n-1} z^n}{n.n!} = \sum_1^\infty \left(1+\frac{1}{2}+\frac{1}{3}+\ldots+\frac{1}{n}\right)\frac{z^n}{n!}.$$

[The coefficient of z^n will be found to be

$$\frac{1}{n!}\left\{\binom{n}{1}-\frac{1}{2}\binom{n}{2}+\frac{1}{3}\binom{n}{3}-\ldots\right\}.$$

Now use Ex. 21, taking $a=0$.]

23. Discuss as completely as you can the convergence of

$$\Sigma \frac{2n!}{n!\,n!} z^n$$

for real or complex z. (*Math. Trip.* 1924)

24. If $A_n \to A$ and $B_n \to B$ as $n \to \infty$, then

$$D_n = \frac{1}{n}(A_1 B_n + A_2 B_{n-1} + \ldots + A_n B_1) \to AB.$$

If, further, A_n and B_n are positive and steadily decreasing, so also is D_n.

[Let $A_n = A + \epsilon_n$. Then the expression given is equal to

$$A\frac{B_1+B_2+\ldots+B_n}{n} + \frac{\epsilon_1 B_n + \epsilon_2 B_{n-1} + \ldots + \epsilon_n B_1}{n}.$$

The first term tends to AB (Ch. IV, Misc. Ex. 16). The modulus of the second is less than $\beta\{|\epsilon_1|+|\epsilon_2|+\ldots+|\epsilon_n|\}/n$, where β is any number greater than the greatest value of $|B_\nu|$: and this expression tends to zero.]

25. Prove that if $c_n = a_1 b_n + a_2 b_{n-1} + \ldots + a_n b_1$ and

$$A_n = a_1+a_2+\ldots+a_n, \quad B_n = b_1+b_2+\ldots+b_n, \quad C_n = c_1+c_2+\ldots+c_n,$$

then

$$C_n = a_1 B_n + a_2 B_{n-1} + \ldots + a_n B_1 = b_1 A_n + b_2 A_{n-1} + \ldots + b_n A_1$$

and

$$C_1 + C_2 + \ldots + C_n = A_1 B_n + A_2 B_{n-1} + \ldots + A_n B_1.$$

Hence prove that if the series Σa_n, Σb_n are convergent and have the sums A, B, so that $A_n \to A$, $B_n \to B$, then

$$(C_1 + C_2 + \ldots + C_n)/n \to AB.$$

Deduce that if Σc_n is convergent then its sum is AB. This result is known as *Abel's theorem on the multiplication of series*. We have already seen that we can multiply the series Σa_n, Σb_n in this way if both series are *absolutely* convergent: Abel's theorem shows that we can do so even if one or both are not absolutely convergent, provided only that the product series is convergent.

26. If $$a_n = \frac{(-1)^n}{\sqrt{(n+1)}}, \quad A_n = a_0 + a_1 + \ldots + a_n,$$
$$b_n = a_0 a_n + a_1 a_{n-1} + \ldots + a_n a_0, \quad B_n = b_0 + b_1 + \ldots + b_n,$$
then (i) Σa_n converges to a sum A, (ii) $A_n = A + O(n^{-\frac{1}{2}})$, (iii) b_n oscillates finitely, (iv) $B_n = a_0 A_n + a_1 A_{n-1} + \ldots + a_n A_0$, and (v) B_n oscillates finitely. (*Math. Trip.* 1933)

27. Prove that
$$\tfrac{1}{2}(1 - \tfrac{1}{2} + \tfrac{1}{3} - \ldots)^2 = \tfrac{1}{2} - \tfrac{1}{3}(1 + \tfrac{1}{2}) + \tfrac{1}{4}(1 + \tfrac{1}{2} + \tfrac{1}{3}) - \ldots,$$
$$\tfrac{1}{2}(1 - \tfrac{1}{3} + \tfrac{1}{5} - \ldots)^2 = \tfrac{1}{2} - \tfrac{1}{4}(1 + \tfrac{1}{3}) + \tfrac{1}{6}(1 + \tfrac{1}{3} + \tfrac{1}{5}) - \ldots.$$
[Use Ex. 9 to establish the convergence of the series.]

28. Prove that if $m > -1$, $p > 0$, $n > 0$, and $U_{m,n} = \int_0^1 x^m (1 - x^p)^n \, dx$, then $(m + np + 1)\, U_{m,n} = np U_{m,n-1}$. Deduce that
$$\int_0^1 x^{-\frac{1}{4}} (1 - x^{\frac{1}{2}})^{\frac{5}{2}} \, dx = \frac{5}{16} \int_0^1 x^{-\frac{1}{4}} (1 - x^{\frac{1}{2}})^{\frac{1}{2}} \, dx,$$
and evaluate the integrals by a suitable transformation.

(*Math. Trip.* 1932)

29. Prove that
$$\int_a^\infty \frac{dx}{x^4 \sqrt{(a^2 + x^2)}} = \frac{2 - \sqrt{2}}{3a^4}, \quad \int_0^1 \frac{x^3 \arcsin x}{\sqrt{(1 - x^2)}} \, dx = \frac{7}{9}.$$
(*Math. Trip.* 1932)

30. Establish the formulae
$$\int_0^\infty F\{\sqrt{(x^2 + 1)} + x\} \, dx = \tfrac{1}{2} \int_1^\infty \left(1 + \frac{1}{y^2}\right) F(y) \, dy,$$
$$\int_0^\infty F\{\sqrt{(x^2 + 1)} - x\} \, dx = \tfrac{1}{2} \int_0^1 \left(1 + \frac{1}{y^2}\right) F(y) \, dy.$$
In particular, prove that if $n > 1$ then
$$\int_0^\infty \frac{dx}{\{\sqrt{(x^2 + 1)} + x\}^n} = \int_0^\infty \{\sqrt{(x^2 + 1)} - x\}^n \, dx = \frac{n}{n^2 - 1}.$$
[In this and the succeeding examples it is assumed that the integrals considered have a meaning in accordance with the definitions of §§ 184 *et seq.*]

31. Show that if $2y = ax - bx^{-1}$, where a and b are positive, then y increases steadily from $-\infty$ to ∞ as x increases from 0 to ∞. Hence show that
$$\int_0^\infty f\left\{\frac{1}{2}\left(ax - \frac{b}{x}\right)\right\} dx = \frac{1}{a} \int_{-\infty}^\infty f(y) \left\{1 + \frac{y}{\sqrt{(y^2 + ab)}}\right\} dy.$$
If $f(y)$ is even, this is $$\frac{2}{a} \int_0^\infty f(y) \, dy.$$

32. Show that if $2y = ax + bx^{-1}$, where a and b are positive, then two values of x correspond to any value of y greater than $\sqrt{(ab)}$. Denoting the greater of these by x_1 and the less by x_2, show that, as y increases from $\sqrt{(ab)}$ towards ∞, x_1 increases from $\sqrt{(b/a)}$ towards ∞, and x_2 decreases from $\sqrt{(b/a)}$ to 0. Hence show that

$$\int_{\sqrt{(b/a)}}^{\infty} f(y)\,dx_1 = \frac{1}{a}\int_{\sqrt{(ab)}}^{\infty} f(y)\left\{\frac{y}{\sqrt{(y^2-ab)}}+1\right\}dy,$$

$$\int_{0}^{\sqrt{(b/a)}} f(y)\,dx_2 = \frac{1}{a}\int_{\sqrt{(ab)}}^{\infty} f(y)\left\{\frac{y}{\sqrt{(y^2-ab)}}-1\right\}dy;$$

and that

$$\int_0^{\infty} f\left\{\frac{1}{2}\left(ax+\frac{b}{x}\right)\right\}dx = \frac{2}{a}\int_{\sqrt{(ab)}}^{\infty}\frac{yf(y)}{\sqrt{(y^2-ab)}}\,dy = \frac{2}{a}\int_0^{\infty} f\{\sqrt{(z^2+ab)}\}\,dz.$$

33. Prove the formula

$$\int_0^{\pi} f(\sec\tfrac{1}{2}x+\tan\tfrac{1}{2}x)\frac{dx}{\sqrt{(\sin x)}} = \int_0^{\pi} f(\operatorname{cosec} x)\frac{dx}{\sqrt{(\sin x)}}.$$

34. If a and b are positive, then

$$\int_0^{\infty}\frac{dx}{(x^2+a^2)(x^2+b^2)} = \frac{\pi}{2ab(a+b)},\quad \int_0^{\infty}\frac{x^2\,dx}{(x^2+a^2)(x^2+b^2)} = \frac{\pi}{2(a+b)}.$$

Deduce that if α, β, and γ are positive, and $\beta^2 \geqq \alpha\gamma$, then

$$\int_0^{\infty}\frac{dx}{\alpha x^4+2\beta x^2+\gamma} = \frac{\pi}{2\sqrt{(2\gamma A)}},\quad \int_0^{\infty}\frac{x^2\,dx}{\alpha x^4+2\beta x^2+\gamma} = \frac{\pi}{2\sqrt{(2\alpha A)}},$$

where $A = B+\sqrt{(\alpha\gamma)}$. Also deduce the last result from Ex. 31, by putting $f(y) = 1/(c^2+y^2)$. [The last two results remain true when $\beta^2<\alpha\gamma$, but their proof is then not quite so simple.]

35. Prove that if b is positive then

$$\int_0^{\infty}\frac{x^2\,dx}{(x^2-a^2)^2+b^2x^2} = \frac{\pi}{2b},\quad \int_0^{\infty}\frac{x^4\,dx}{\{(x^2-a^2)^2+b^2x^2\}^2} = \frac{\pi}{4b^3}.$$

36. If $\phi'(x)$ is continuous for $x>1$, then

$$\sum_{1\leqq n\leqq x}\phi(n) = [x]\,\phi(x) - \int_1^x [t]\,\phi'(t)\,dt,$$

where $[x]$ is the greatest integer contained in x. (*Math. Trip.* 1932)

37. If $\phi''(x) = O(x^{-\alpha})$, where $\alpha>1$, for large x, then

$$\int_n^{n+1}\{\phi(x)-\phi(n+\tfrac{1}{2})\}\,dx = O(n^{-\alpha})$$

and

$$\sum_1^n \phi(m+\tfrac{1}{2}) = \int_1^{n+1}\phi(x)\,dx + C + O(n^{1-\alpha}),$$

where C is independent of n. (*Math. Trip.* 1923)

[Observe that

$$\int_n^{n+1}\{\phi(x)-\phi(n+\tfrac{1}{2})\}\,dx = \int_0^{\frac{1}{2}}\{\phi(n+\tfrac{1}{2}+t)+\phi(n+\tfrac{1}{2}-t)-2\phi(n+\tfrac{1}{2})\}\,dt.]$$

38. If $$J_m = \int_0^x \sin^m\theta \sin a(x-\theta)\,d\theta$$

and m is an integer not less than 2, then

$$m(m-1)\,J_{m-2} = a\sin^m x + (m^2-a^2)\,J_m.$$

Deduce that

$$\cos ax = 1-\frac{a^2}{2!}\sin^2 x-\frac{a^2(2^2-a^2)}{4!}\sin^4 x-\frac{a^2(2^2-a^2)\,(4^2-a^2)}{6!}\sin^6 x-\dots.$$

(*Math. Trip.* 1923)

39. Prove that if

$$u_n = \int_0^{\frac{1}{2}\pi}\sin 2nx \cot x\,dx,\quad v_n = \int_0^{\frac{1}{2}\pi}\frac{\sin 2nx}{x}\,dx,$$

then $u_n = \frac{1}{2}\pi$ and $$v_n \to \int_0^\infty \frac{\sin x}{x}\,dx = v,$$

say. Prove also, by partial integration or otherwise, that $u_n - v_n \to 0$; and deduce that $v = \frac{1}{2}\pi$. (*Math. Trip.* 1924)

40. If a is positive, $f(x)$ is continuous except at the origin,

$$\int_0^a f(x)\,dx = \lim_{\epsilon\to 0}\int_\epsilon^a f(x)\,dx$$

exists, and $$g(x) = \int_x^a \frac{f(t)}{t}\,dt,$$

then $$\int_0^a g(x)\,dx = \int_0^a f(x)\,dx.$$ (*Math. Trip.* 1934)

CHAPTER IX

THE LOGARITHMIC, EXPONENTIAL, AND CIRCULAR FUNCTIONS OF A REAL VARIABLE

204. The number of essentially different types of functions with which we have been concerned in the foregoing chapters is not very large, the most important being polynomials, rational functions, algebraical functions, explicit or implicit, and trigonometrical functions, direct or inverse.

The gradual expansion of mathematical knowledge has been accompanied by the introduction into analysis of one new class of function after another. These new functions have generally been introduced because it appeared that some problem which was occupying the attention of mathematicians was incapable of solution by means of the functions already known. The process may fairly be compared with that by which the irrational and complex numbers were first introduced, when it was found that certain algebraical equations could not be solved by means of the numbers already recognised. One of the most fruitful sources of new functions has been the problem of integration. Attempts have been made to integrate some function $f(x)$ in terms of functions already known. These attempts have failed; and after a certain number of failures it has begun to appear probable that the problem is insoluble. Sometimes it has been *proved* that this is so; but as a rule such a strict proof has not been forthcoming until later on. Generally it has happened that mathematicians have taken the impossibility for granted as soon as they have become reasonably convinced of it, and have introduced a new function $F(x)$ *defined* by its possessing the required property, viz. that $F'(x) = f(x)$. Starting from this definition, they have investigated the properties of $F(x)$; and it has then appeared that $F(x)$ has properties which no finite combination of the functions previously known could possibly have; and thus the correctness of the assumption that the original problem could not possibly be solved has been established. One such case occurred in the preceding pages, when in Ch. VI we defined the function $\log x$ by means of the equation

$$\log x = \int \frac{dx}{x}.$$

Let us consider what grounds we have for supposing $\log x$ to be a really new function. We have seen already (Ex. XLII. 4) that it cannot be a rational function, since the derivative of a rational function is a rational function whose denominator contains only repeated factors. The question whether it can be an algebraical or trigonometrical function is more difficult. But it is very easy to become convinced by a few experiments that differentiation will never get rid of algebraical irrationalities. For example, the result of differentiating $\sqrt{(1+x)}$ any number of times is always the product of $\sqrt{(1+x)}$ by a rational function, and so generally. Similarly, if we differentiate a function which involves $\sin x$ or $\cos x$, one or other of these functions persists in the result.

We have, therefore, not indeed a strict proof that $\log x$ is a new function —that we do not profess to give*—but a reasonable presumption that it is. We shall therefore treat it as such, and we shall find on examination that its properties are quite unlike those of any function which we have as yet encountered.

205. Definition of $\log x$. We define $\log x$, the logarithm of x, by the equation

$$\log x = \int_1^x \frac{dt}{t}.$$

We must suppose that x is positive, since (Ex. LXXVI. 2) the integral has no meaning if the range of integration includes the point $x = 0$. We might have chosen a lower limit other than 1; but 1 proves to be the most convenient. With this definition $\log 1 = 0$.

We shall now consider how $\log x$ behaves as x varies from 0 towards ∞. It follows at once from the definition that $\log x$ is a

* For such a proof see the author's tract quoted on p. 254.

continuous function of x which increases steadily with x and has a derivative

$$\frac{d}{dx}\log x = \frac{1}{x};$$

and it follows from § 181 that $\log x$ tends to ∞ as $x \to \infty$.

If x is positive but less than 1, then $\log x$ is negative. For

$$\log x = \int_1^x \frac{dt}{t} = -\int_x^1 \frac{dt}{t} < 0.$$

Moreover, if we make the substitution $t = 1/u$ in the integral, we obtain

$$\log x = \int_1^x \frac{dt}{t} = -\int_1^{1/x} \frac{du}{u} = -\log\frac{1}{x}.$$

Thus $\log x$ tends steadily to $-\infty$ as x decreases from 1 to 0.

The general form of the graph of the logarithmic function is shown in Fig. 49. Since the derivative of $\log x$ is $1/x$, the slope of

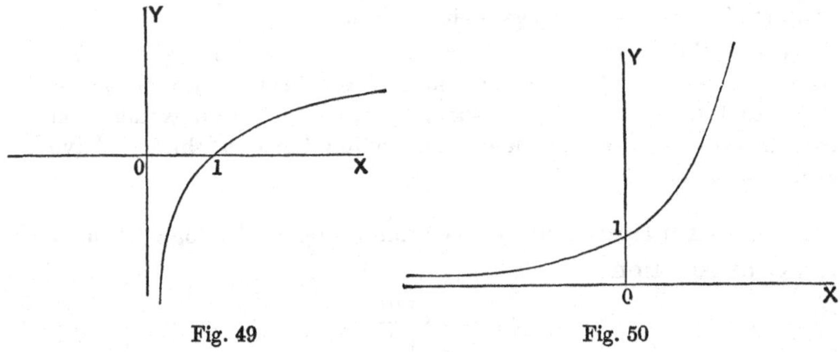

Fig. 49 Fig. 50

the curve is very gentle when x is very large, and very steep when x is very small.

Examples LXXXIII. 1. Prove from the definition that

(a) $\dfrac{x}{1+x} < \log(1+x) < x \quad (x>0)$, (b) $x < -\log(1-x) < \dfrac{x}{1-x} \quad (0<x<1)$.

[Thus for (a) observe that $\log(1+x) = \int_1^{1+x} \frac{dt}{t}$ and that the integrand lies between 1 and $1/(1+x)$.]

2. Prove the inequalities

(i) $x-\frac{1}{2}x^2<\log(1+x)\quad(x>0)$, (ii) $\dfrac{x-1}{x}<\log x<x-1\quad(x>1)$,

(iii) $4(x-1)-2\log x<2x\log x<x^2-1\quad(x>1)$,

(iv) $0<\dfrac{1}{x}-\log\dfrac{x+1}{x}<\dfrac{1}{2x^2}\quad(x>0)$,

(v) $\dfrac{2}{2x+1}<\log\dfrac{x+1}{x}<\dfrac{2x+1}{2x(x+1)}\quad(x>0)$.

(*Math. Trip.* 1931, 1933, 1936)

3. Prove that $\lim\limits_{x\to1}\dfrac{\log x}{x-1}=\lim\limits_{y\to0}\dfrac{\log(1+y)}{y}=1$. [Use Ex. 1.]

206. The functional equation satisfied by $\log x$. *The function* $\log x$ *satisfies the functional equation*

$$f(xy)=f(x)+f(y)\quad\text{....................(1).}$$

For, making the substitution $t=yu$, we see that

$$\log xy=\int_1^{xy}\frac{dt}{t}=\int_{1/y}^{x}\frac{du}{u}=\int_1^x\frac{du}{u}-\int_1^{1/y}\frac{du}{u}$$
$$=\log x-\log(1/y)=\log x+\log y,$$

which proves the theorem.

Examples LXXXIV. 1. It can be shown that there is no solution of the equation (1) which possesses a differential coefficient and is fundamentally distinct from $\log x$. For when we differentiate the functional equation, first with respect to x and then with respect to y, we obtain the two equations

$$yf'(xy)=f'(x),\quad xf'(xy)=f'(y);$$

and so, eliminating $f'(xy)$, $xf'(x)=yf'(y)$. But if this is true for every pair of values of x and y, then we must have $xf'(x)=C$, or $f'(x)=C/x$, where C is a constant. Hence

$$f(x)=\int\frac{C}{x}\,dx+C'=C\log x+C',$$

and substitution into (1) shows that $C'=0$. Thus there is no solution fundamentally distinct from $\log x$, except the trivial solution $f(x)=0$, obtained by taking $C=0$.

2. Show in the same way that there is no solution of the equation

$$f(x)+f(y)=f\left(\frac{x+y}{1-xy}\right)$$

which possesses a differential coefficient and is fundamentally distinct from arc tan x.

3. Prove that if $m+1>0$ then $\binom{m}{n} \to 0$ when $n \to \infty$.

[If m is integral then $u_n = \binom{m}{n} = 0$ for $n>m$, and there is nothing to prove. We suppose therefore that $p<m<p+1$, where p is an integer not less than -1. In this case $\dfrac{u_{\nu+1}}{u_\nu} = \dfrac{m-\nu}{\nu+1}$ is negative for $\nu \geqq p+1$, and less than 1 in absolute value, so that u_ν alternates in sign and $|u_\nu|$ decreases steadily. Also

$$\log \frac{|u_{\nu+1}|}{|u_\nu|} = \log \frac{\nu-m}{\nu+1} = \log\left(1-\frac{m+1}{\nu+1}\right) < -\frac{m+1}{\nu+1};$$

so that $\log |u_{n+1}| - \log |u_{p+1}| < -(m+1) \sum_{\nu=p+1}^{n} \frac{1}{\nu+1} \to -\infty$

when $n \to \infty$. Hence $u_{n+1} \to 0$.

If $m=-1$ then $u_n = (-1)^n$. If $m+1<0$ then $|u_n|$ increases with n. Prove that then $|u_n| \to \infty$.]

207. The manner in which $\log x$ **tends to infinity with** x. It will be remembered that in § 98 we defined functions which are of the first, second, third, ... orders of greatness for large x. A function $f(x)$ was said to be of the kth order of greatness when $f(x)/x^k$ tends to a limit different from zero as x tends to infinity.

It is easy to define a whole series of functions which tend to infinity with x with progressively diminishing rapidity. Thus $\sqrt{x}$, $\sqrt[3]{x}$, $\sqrt[4]{x}$, ... is such a series of functions. We may say generally that x^α, where α is any positive rational number, is of the αth order of greatness when x is large. We may suppose α as small as we please, e.g. less than ·0000001. And it might be thought that by giving α all possible values we should exhaust the possible 'orders of infinity' of $f(x)$. At any rate it might be supposed that if $f(x)$ tends to infinity with x, however slowly, we could always find a value of α so small that x^α would tend to infinity still more slowly; and similarly that, if $f(x)$ tends to infinity with x, however rapidly, we could always find a value of α so great that x^α would tend to infinity still more rapidly.

The behaviour of $\log x$ refutes any such expectation. *The* *logarithm of x tends to infinity with x, but more slowly than any positive power of x, integral or fractional.* In other words $\log x \to \infty$ but

$$\frac{\log x}{x^\alpha} \to 0$$

for *all* positive rational values of α.

208. Proof that $x^{-\alpha}\log x \to 0$ when $x \to \infty$. Let β be any positive rational number. Then $t^{-1} < t^{\beta-1}$ when $t > 1$, and so

$$\log x = \int_1^x \frac{dt}{t} < \int_1^x \frac{dt}{t^{1-\beta}};$$

so that

$$\log x < \frac{x^\beta - 1}{\beta} < \frac{x^\beta}{\beta}$$

for $x > 1$. Now if α is positive we can choose a smaller positive β, and then

$$0 < \frac{\log x}{x^\alpha} < \frac{x^{\beta-\alpha}}{\beta}.$$

But $x^{\beta-\alpha} \to 0$ when $x \to \infty$, since $\beta < \alpha$, and therefore $x^{-\alpha}\log x \to 0$.

209. The behaviour of $\log x$ when $x \to +0$. Since

$$x^{-\alpha}\log x = -y^\alpha \log y$$

if $x = 1/y$, it follows from the theorem proved above that

$$\lim_{y\to+0} y^\alpha \log y = -\lim_{x\to+\infty} x^{-\alpha}\log x = 0.$$

Thus $\log x$ tends to $-\infty$ and $\log(1/x) = -\log x$ to ∞ when x tends to zero by positive values, but $\log(1/x)$ tends to ∞ more slowly than any positive power of $1/x$, integral or fractional.

210. Scales of infinity. The logarithmic scale. Let us consider once more the series of functions

$$x,\quad \sqrt{x},\quad \sqrt[3]{x},\quad \dots,\quad \sqrt[n]{x},\quad \dots,$$

which possesses the property that, if $f(x)$ and $\phi(x)$ are any two of the functions contained in it, then $f(x)$ and $\phi(x)$ both tend to ∞ when $x \to \infty$, while $f(x)/\phi(x)$ tends to 0 or to ∞ according as $f(x)$ occurs to the right or the left of $\phi(x)$ in the series. We can now continue this series by the insertion of new terms to the right of all those already written down. We can begin with $\log x$, which tends to infinity more slowly than any of the old terms. Then $\sqrt{(\log x)}$ tends to ∞ more slowly than $\log x$, $\sqrt[3]{(\log x)}$ than $\sqrt{(\log x)}$, and so on. Thus we obtain a series

$$x,\ \sqrt{x},\ \sqrt[3]{x},\ \dots,\ \sqrt[n]{x},\ \dots,\ \log x,\ \sqrt{(\log x)},\ \sqrt[3]{(\log x)},\ \dots,\ \sqrt[n]{(\log x)},\ \dots$$

formed of two simply infinite series arranged one after the other. We can continue the series further by considering the function $\log\log x$, the logarithm of $\log x$. Since $x^{-\alpha}\log x \to 0$, for all positive values of α, it follows on putting $x = \log y$ that

$$(\log y)^{-\alpha}\log\log y = x^{-\alpha}\log x \to 0.$$

Thus $\log\log y$ tends to ∞ with y, but more slowly than any power of $\log y$. Hence we may continue our series in the form

$$x,\ \sqrt{x},\ \sqrt[3]{x},\ \dots,\ \log x,\ \sqrt{(\log x)},\ \sqrt[3]{(\log x)},\ \dots,$$
$$\log\log x,\ \sqrt{(\log\log x)},\ \dots,\ \sqrt[n]{(\log\log x)},\ \dots;$$

and it will now be obvious that by introducing the functions $\log\log\log x$, $\log\log\log\log x$, ... we can prolong the series to any extent we like. By putting $x = 1/y$ we obtain a similar scale of infinity for functions of y which tend to ∞ as y tends to 0 by positive values*.

Examples LXXXV. 1. Between any two terms $f(x)$, $F(x)$ of the series we can insert a new term $\phi(x)$ such that $\phi(x)$ tends to ∞ more slowly than $f(x)$ and more rapidly than $F(x)$. [Thus between $\sqrt{x}$ and $\sqrt[3]{x}$ we could insert $x^{\frac{5}{12}}$: between $\sqrt{(\log x)}$ and $\sqrt[3]{(\log x)}$ we could insert $(\log x)^{\frac{5}{12}}$. And generally $\phi(x) = \sqrt{\{f(x)\,F(x)\}}$ satisfies the conditions stated.]

2. Find a function which tends to ∞ more slowly than $\sqrt{x}$, but more rapidly than x^{α}, where α is any rational number less than $\frac{1}{2}$. [$x^{\frac{1}{2}}(\log x)^{-\beta}$, where β is any positive rational number, is such a function.]

3. Find a function which tends to ∞ more slowly than $\sqrt{x}$, but more rapidly than $\sqrt{x}(\log x)^{-\alpha}$, where α is any positive rational number. [The function $\sqrt{x}(\log\log x)^{-1}$ is such a function. It will be gathered from these examples that *incompleteness* is an inherent characteristic of the logarithmetic scale of infinity.]

4. How does the function

$$f(x) = \{x^{\alpha}(\log x)^{\alpha'}(\log\log x)^{\alpha''}\}/\{x^{\beta}(\log x)^{\beta'}(\log\log x)^{\beta''}\}$$

behave as x tends to ∞? [If $\alpha \neq \beta$ then the behaviour of

$$f(x) = x^{\alpha-\beta}(\log x)^{\alpha'-\beta'}(\log\log x)^{\alpha''-\beta''}$$

is dominated by that of $x^{\alpha-\beta}$. If $\alpha = \beta$ then the power of x disappears and the behaviour of $f(x)$ is dominated by that of $(\log x)^{\alpha'-\beta'}$, unless $\alpha' = \beta'$,

* For fuller information about 'scales of infinity' see the author's tract referred to on p. 350.

when it is dominated by that of $(\log\log x)^{\alpha''-\beta''}$. Thus $f(x)\to\infty$ if $\alpha>\beta$, or $\alpha=\beta$, $\alpha'>\beta'$, or $\alpha=\beta$, $\alpha'=\beta'$, $\alpha''>\beta''$, and $f(x)\to 0$ if $\alpha<\beta$, or $\alpha=\beta$, $\alpha'<\beta'$, or $\alpha=\beta$, $\alpha'=\beta'$, $\alpha''<\beta''$.]

5. Arrange the functions

$$\frac{x}{\sqrt{(\log x)}},\quad \frac{x\sqrt{(\log x)}}{\log\log x},\quad \frac{x\log\log x}{\sqrt{(\log x)}},\quad \frac{x\log\log\log x}{\sqrt{(\log\log x)}}$$

according to their order of magnitude for large x.

6. Prove that

$$\log(x+1)=\log x+O\left(\frac{1}{x}\right),\quad \tfrac{1}{2}\log\frac{x+1}{x-1}=\frac{1}{x}+O\left(\frac{1}{x^2}\right),$$

$$\log\log\frac{x+1}{x-1}=-\log x+\log 2+O\left(\frac{1}{x}\right),\quad \log(x\log x)\sim\log x$$

for large x.

7. Prove that

$$\frac{d}{dx}(\log x)^\alpha=\frac{\alpha}{x(\log x)^{1-\alpha}},\quad \frac{d}{dx}(\log\log x)^\alpha=\frac{\alpha}{x\log x\,(\log\log x)^{1-\alpha}},\quad \dots,$$

$$\int\frac{dx}{x\log x}=\log\log x,\quad \int\frac{dx}{x\log x\,\log\log x}=\log\log\log x,\quad \dots.$$

8. Prove that the curve $y=x^m(\log x)^n$, where x is positive and m and n are integers greater than 1, has at least two inflexions, and may have more. Sketch the curve when n is odd. (*Math. Trip.* 1927)

211. The number *e*. We shall now introduce a number, usually denoted by e, which is, like π, one of the fundamental constants of analysis.

We define e as *the number whose logarithm is* 1. In other words e is defined by the equation

$$1=\int_1^e\frac{dt}{t}.$$

Since $\log x$ is an increasing function of x, in the stricter sense of § 95, it can only pass once through the value 1. Hence our definition is unambiguous.

Now $\log xy=\log x+\log y$ and so

$$\log x^2=2\log x,\quad \log x^3=3\log x,\quad \dots,\quad \log x^n=n\log x,$$

where n is any positive integer. Hence

$$\log e^n=n\log e=n.$$

Again, if p and q are any positive integers, and $e^{p/q}$ denotes the positive qth root of e^p, we have

$$p = \log e^p = \log (e^{p/q})^q = q \log e^{p/q},$$

so that $\log e^{p/q} = p/q$. Thus, if y has any positive rational value, and e^y denotes the positive yth power of e, we have

$$\log e^y = y \quad \text{.........................(1)},$$

and $\log e^{-y} = -\log e^y = -y$. Hence the equation (1) is true for all rational values of y, positive or negative. In other words the equations

$$y = \log x, \quad x = e^y \quad \text{....................(2)}$$

are consequences of one another so long as y is rational and e^y has its positive value. At present we have not given any definition of a power such as e^y in which the index is irrational, and the function e^y is defined for rational values of y only.

Example. Prove that $2 < e < 3$. [In the first place it is evident that

$$\int_1^2 \frac{dt}{t} < 1,$$

and so $2 < e$. Also

$$\int_1^3 \frac{dt}{t} = \int_1^2 \frac{dt}{t} + \int_2^3 \frac{dt}{t} = \int_0^1 \frac{du}{2-u} + \int_0^1 \frac{du}{2+u} = 4\int_0^1 \frac{du}{4-u^2} > 1,$$

so that $e < 3$.]

212. The exponential function. We now define the *exponential function* e^y for all real values of y as the inverse of the logarithmic function. In other words we write

$$x = e^y$$

if $y = \log x$.

We saw that, as x varies from 0 towards ∞, y increases steadily, in the stricter sense, from $-\infty$ towards ∞. Thus to one value of x corresponds one value of y, and conversely. Also y is a continuous function of x, and it follows from § 110 that x is likewise a continuous function of y.

It is easy to give a direct proof of the continuity of the exponential function. For if $x = e^y$ and $x + \xi = e^{y+\eta}$ then

$$\eta = \int_x^{x+\xi} \frac{dt}{t}.$$

Thus $|\eta|$ is greater than $\xi/(x+\xi)$ if $\xi>0$, and than $|\xi|/x$ if $\xi<0$; and if η is small ξ must also be small.

Thus e^y is a positive and continuous function of y which increases steadily from 0 towards ∞ as y increases from $-\infty$ towards ∞. Moreover e^y is the positive yth power of the number e, in accordance with the elementary definitions, whenever y is a rational number. In particular $e^y = 1$ when $y = 0$. The general form of the graph of e^y is as shown in Fig. 50 (p. 400).

213. The principal properties of the exponential function. (1) If $x = e^y$, so that $y = \log x$, then

$$\frac{dy}{dx} = \frac{1}{x}, \quad \frac{dx}{dy} = x = e^y.$$

Thus *the derivative of the exponential function is equal to the function itself.* More generally

$$\frac{d}{dy}e^{ay} = ae^{ay}.$$

(2) *The exponential function satisfies the functional equation*

$$f(y+z) = f(y)f(z).$$

This follows, when y and z are rational, from the ordinary rules of indices. If y or z, or both, are irrational then we can choose two sequences $y_1, y_2, \dots, y_n, \dots$ and $z_1, z_2, \dots, z_n, \dots$ of rational numbers such that $\lim y_n = y$, $\lim z_n = z$. Then, since the exponential function is continuous, we have

$$e^y \times e^z = \lim e^{y_n} \times \lim e^{z_n} = \lim e^{y_n+z_n} = e^{y+z}.$$

In particular $e^y \times e^{-y} = e^0 = 1$, or $e^{-y} = 1/e^y$.

We may also deduce the functional equation satisfied by e^y from that satisfied by $\log x$. For if $y_1 = \log x_1$, $y_2 = \log x_2$, so that $x_1 = e^{y_1}$, $x_2 = e^{y_2}$, then $y_1 + y_2 = \log x_1 + \log x_2 = \log x_1 x_2$ and

$$e^{y_1+y_2} = e^{\log x_1 x_2} = x_1 x_2 = e^{y_1} \times e^{y_2}.$$

(3) *The function e^y tends to infinity with y more rapidly than any power of y, or*

$$\lim \frac{y^\alpha}{e^y} = \lim e^{-y}y^\alpha = 0$$

when $y \to \infty$, for all values of α however great.

We saw that $x^{-\beta}\log x \to 0$ when $x \to \infty$, for any positive value of β. Writing α for $1/\beta$, we see that $x^{-1}(\log x)^{\alpha} \to 0$ for any value of α. The result follows on putting $x = e^y$. It is clear also that $e^{\gamma y}$ tends to ∞ if $\gamma > 0$, and to 0 if $\gamma < 0$, and in each case more rapidly than any power of y.

From this result it follows that we can construct a 'scale of infinity' similar to that constructed in § 210, but extending in the opposite direction; i.e. a scale of functions which tend to ∞ more and more rapidly when $x \to \infty$*. The scale is

$$x,\quad x^2,\quad x^3,\quad \ldots,\quad e^x,\quad e^{2x},\quad \ldots,\quad e^{x^2},\quad \ldots,\quad e^{x^3},\quad \ldots,\quad e^{e^x},\quad \ldots,$$

where of course $e^{x^2}, \ldots, e^{e^x}, \ldots$ denote $e^{(x^2)}, \ldots, e^{(e^x)}, \ldots$.

The reader should try to apply the remarks about the logarithmic scale, made in § 210 and Exs. LXXXV, to this 'exponential scale' also. The two scales may of course (if the order of one is reversed) be combined into one scale

$$\ldots,\quad \log\log x,\quad \ldots,\quad \log x,\quad \ldots,\quad x,\quad \ldots,\quad e^x,\quad \ldots,\quad e^{e^x},\quad \ldots.$$

Examples LXXXVI. 1. If $D_y x = ax$ then $x = Ke^{ay}$, where K is a constant.

2. There is no solution of the equation $f(y+z) = f(y)f(z)$ fundamentally distinct from the exponential function. [We assume that $f(y)$ has a differential coefficient. Differentiating the equation with respect to y and z in turn, we obtain

$$f'(y+z) = f'(y)f(z),\quad f'(y+z) = f(y)f'(z).$$

Hence $f'(y)/f(y) = f'(z)/f(z)$, and therefore each is constant. Thus if $x = f(y)$ then $D_y x = ax$, where a is a constant, so that $x = Ke^{ay}$ (Ex. 1).]

3. Prove that $(e^{ay}-1)/y \to a$ when $y \to 0$. [Applying the mean value theorem, we obtain $e^{ay} - 1 = aye^{a\eta}$, where $0 < |\eta| < |y|$.]

4. Prove that $e^x - 1 - x$, $e^{-x} - 1 + x$, and $1 - \frac{1}{2}x^2 + \frac{1}{3}x^3 - (1+x)e^{-x}$ are positive and increase steadily for positive x. (*Math. Trip.* 1924)

5. Prove that
$$\left(\frac{d}{dx}\right)^m (x^n e^{-\sqrt{x}}) \to 0$$
when $x \to \infty$, for all integral m and n. (*Math. Trip.* 1936)

* The exponential function was introduced by inverting the equation $y = \log x$ into $x = e^y$; and we have accordingly, up to the present, used y as the independent and x as the dependent variable in discussing its properties. We shall now revert to the more natural plan of taking x as the independent variable, except when it is necessary to consider a pair of equations of the type $y = \log x$, $x = e^y$ simultaneously, or when there is some other special reason to the contrary.

214. The general power a^x. The function a^x has been defined only for rational values of x, except in the particular case when $a = e$. We shall now consider the case in which a is any positive number. Suppose that x is a positive rational number p/q. Then the positive value y of the power $a^{p/q}$ is given by $y^q = a^p$; from which it follows that

$$q \log y = p \log a, \quad \log y = (p/q) \log a = x \log a,$$

and so

$$y = e^{x \log a}.$$

We take this as our *definition* of a^x when x is irrational. Thus $10^{\sqrt{2}} = e^{\sqrt{2} \log 10}$. It is to be observed that a^x, when x is irrational, is defined only for positive values of a, and is itself essentially positive; and that $\log a^x = x \log a$. The most important properties of the function a^x are as follows.

(1) Whatever value a may have, $a^x \times a^y = a^{x+y}$ and $(a^x)^y = a^{xy}$. In other words the laws of indices hold for irrational no less than for rational indices. For, in the first place,

$$a^x \times a^y = e^{x \log a} \times e^{y \log a} = e^{(x+y) \log a} = a^{x+y};$$

and in the second

$$(a^x)^y = e^{y \log a^x} = e^{xy \log a} = a^{xy}.$$

(2) If $a > 1$ then $a^x = e^{x \log a} = e^{\alpha x}$, where α is positive. The graph of a^x is in this case similar to that of e^x, and $a^x \to \infty$ when $x \to \infty$, more rapidly than any power of x.

If $a < 1$ then $a^x = e^{x \log a} = e^{-\beta x}$, where β is positive. The graph of a^x is then similar in shape to that of e^x, but reversed as regards right and left, and $a^x \to 0$ when $x \to \infty$, more rapidly than any power of $1/x$.

(3) a^x is a differentiable function of x, and

$$D_x a^x = D_x e^{x \log a} = e^{x \log a} \log a = a^x \log a.$$

(4) a^x is also a differentiable function of a, and

$$D_a a^x = D_a e^{x \log a} = e^{x \log a}(x/a) = x a^{x-1}.$$

(5) It follows from (3) that

$$\lim \frac{a^x - 1}{x} = \log a;$$

for the left-hand side is the value of $D_x a^x$ for $x = 0$. The result is equivalent to that of Ex. LXXXVI. 3.

In the course of the preceding chapters a great many results involving the function a^x have been stated with the limitation that x is rational. The definition and theorems given in this section enable us to remove this restriction.

215. The representation of e^x as a limit. In Ch. IV, § 73, we proved that $\{1+(1/n)\}^n$ tends, when $n \to \infty$, to a limit which we denoted provisionally by e. We shall now identify this limit with the number e of the preceding sections. We can however establish a more general result, viz. that expressed by the equations

$$\lim_{n\to\infty} \left(1 + \frac{x}{n}\right)^n = \lim_{n\to\infty} \left(1 - \frac{x}{n}\right)^{-n} = e^x \quad \text{............(1).}$$

The result is very important, and we shall indicate alternative lines of proof.

(1) Since $$\frac{d}{dt} \log(1+xt) = \frac{x}{1+xt},$$

it follows that $$\lim_{h\to 0} \frac{\log(1+xh)}{h} = x.$$

If we put $h = 1/\xi$, we see that

$$\lim \xi \log\left(1 + \frac{x}{\xi}\right) = x$$

when $\xi \to \infty$ or $\xi \to -\infty$. Since the exponential function is continuous it follows that

$$\left(1 + \frac{x}{\xi}\right)^{\xi} = e^{\xi \log\{1+(x/\xi)\}} \to e^x$$

when $\xi \to \infty$ or $\xi \to -\infty$; i.e. that

$$\lim_{\xi\to\infty} \left(1 + \frac{x}{\xi}\right)^{\xi} = \lim_{\xi\to-\infty} \left(1 + \frac{x}{\xi}\right)^{\xi} = e^x \quad \text{............(2).}$$

If we suppose that $\xi \to \infty$ or $\xi \to -\infty$ through integral values only, we obtain the result expressed by the equations (1).

(2) If n is any positive integer and $x > 1$, we have

$$\int_1^x \frac{dt}{t^{1+(1/n)}} < \int_1^x \frac{dt}{t} < \int_1^x \frac{dt}{t^{1-(1/n)}},$$

or

$$n(1-x^{-1/n}) < \log x < n(x^{1/n}-1) \quad \ldots\ldots\ldots\ldots\ldots\ldots(3).$$

We write $y = \log x$, $x = e^y$. Then it follows from (3), after some simple transformations, that

$$\left(1+\frac{y}{n}\right)^n < e^y < \left(1-\frac{y}{n}\right)^{-n} \quad \ldots\ldots\ldots\ldots\ldots\ldots(4).$$

If $0 < n\xi < 1$ then $$1-(1-\xi)^n < n\{1-(1-\xi)\} = n\xi,$$

by (4) of § 74, and so $$(1-\xi)^{-n} < (1-n\xi)^{-1} \quad \ldots\ldots\ldots\ldots\ldots\ldots(5).$$

In particular (5) is true if $\xi = y^2/n^2$ and $n > y^2$. Hence

$$\left(1-\frac{y}{n}\right)^{-n} - \left(1+\frac{y}{n}\right)^n = \left(1+\frac{y}{n}\right)^n \left\{\left(1-\frac{y^2}{n^2}\right)^{-n} - 1\right\}$$

$$< e^y \left\{\left(1-\frac{y^2}{n}\right)^{-1} - 1\right\} = \frac{y^2 e^y}{n-y^2},$$

which tends to zero when $n \to \infty$; and (1) now follows from (4).

We leave it to the reader (i) to make the changes in the argument necessary when $0 < x < 1$ and (ii) to deduce the result for negative x.

216. The representation of $\log x$ as a limit. We can also prove that

$$\lim n(1-x^{-1/n}) = \lim n(x^{1/n}-1) = \log x.$$

For $$n(x^{1/n}-1) - n(1-x^{-1/n}) = n(x^{1/n}-1)(1-x^{-1/n}),$$

which tends to zero as $n \to \infty$, since $n(x^{1/n}-1)$ tends to a limit (§ 75) and $x^{-1/n}$ to 1 (Ex. XXVII. 10). The result now follows from the inequalities (3) of § 215.

Examples LXXXVII. 1. Prove, by taking $y = 1$ and $n = 6$ in the inequalities (4) of § 215, that $2{\cdot}5 < e < 2{\cdot}9$.

2. If $n\xi_n \to l$ as $n \to \infty$, then $(1+\xi_n)^n \to e^l$. [Writing $n\log(1+\xi_n)$ in the form

$$l\left(\frac{n\xi_n}{l}\right)\frac{\log(1+\xi_n)}{\xi_n},$$

and using Ex. LXXXIII. 3, we see that $n\log(1+\xi_n) \to l$.]

3. If $n\xi_n \to \infty$, then $(1+\xi_n)^n \to \infty$; and if $1+\xi_n > 0$ and $n\xi_n \to -\infty$, then

$$(1+\xi_n)^n \to 0.$$

4. Deduce from (1) of § 215 the theorem that e^y tends to infinity more rapidly than any power of y.

217. Common logarithms. The reader is probably familiar with the idea of a logarithm and its use in numerical calculation. He will remember that in elementary algebra $\log_a x$, the logarithm of x to the base a, is defined by the equations

$$x = a^y, \quad y = \log_a x.$$

This definition is of course applicable only when y is rational.

Our logarithms are therefore logarithms to the base e. For numerical work logarithms to the base 10 are used. If

$$y = \log x = \log_e x, \quad z = \log_{10} x,$$

then $x = e^y$ and also $x = 10^z = e^{z \log 10}$, so that

$$\log_{10} x = (\log_e x)/(\log_e 10).$$

Thus it is easy to pass from one system to the other when once $\log_e 10$ has been calculated.

It is not part of our purpose in this book to go into details concerning the practical uses of logarithms. If the reader is not familiar with them he should consult some text-book on algebra or trigonometry*.

Examples LXXXVIII. 1. Show that

$$D_x e^{ax} \cos bx = re^{ax} \cos(bx+\theta), \quad D_x e^{ax} \sin bx = re^{ax} \sin(bx+\theta),$$

where $r = \sqrt{(a^2+b^2)}$, $\cos\theta = a/r$, $\sin\theta = b/r$. Hence determine the nth derivatives of the functions $e^{ax}\cos bx$, $e^{ax}\sin bx$, and show in particular that

$$\left(\frac{d}{dx}\right)^n e^{ax} \sin bx = (a \sec\theta)^n e^{ax} \sin(bx+n\theta).$$

(*Math. Trip.* 1932)

2. If y_n is the nth derivative of $e^{ax}\sin bx$, then

$$y_{n+1} - 2ay_n + (a^2+b^2)\, y_{n-1} = 0. \quad (\textit{Math. Trip.}\ 1932)$$

* See for example Chrystal's *Algebra*, 2nd edition, vol. I, ch. XXI. The value of $\log_e 10$ is 2·302... and that of its reciprocal ·434....

3. If y_n is the nth derivative of x^2e^x, then

$$y_n = \tfrac{1}{2}n(n-1)\,y_2 - n(n-2)\,y_1 + \tfrac{1}{2}(n-1)(n-2)\,y.$$

(*Math. Trip.* 1934)

4. Trace the curve $y = e^{-ax}\sin bx$, where a and b are positive. Show that y has an infinity of maxima whose values form a geometrical progression and which lie on the curve

$$y = \frac{b}{\sqrt{(a^2+b^2)}}e^{-ax}.$$ (*Math. Trip.* 1912, 1935)

5. **Integrals containing the exponential function.** Prove that

$$\int e^{ax}\cos bx\,dx = \frac{a\cos bx + b\sin bx}{a^2+b^2}e^{ax}, \quad \int e^{ax}\sin bx\,dx = \frac{a\sin bx - b\cos bx}{a^2+b^2}e^{ax}.$$

[Denoting the two integrals by I, J, and integrating by parts, we obtain

$$aI = e^{ax}\cos bx + bJ, \quad aJ = e^{ax}\sin bx - bI.$$

Solve these equations for I and J.]

6. Prove that if $a > 0$ then

$$\int_0^\infty e^{-ax}\cos bx\,dx = \frac{a}{a^2+b^2}, \quad \int_0^\infty e^{-ax}\sin bx\,dx = \frac{b}{a^2+b^2}.$$

7. If $I_n = \int e^{ax}x^n\,dx$ then $aI_n = e^{ax}x^n - nI_{n-1}$. [Integrate by parts. It follows that I_n can be calculated for all positive integral values of n.]

8. Prove that, if n is a positive integer, then

$$\int_0^x e^{-t}t^n\,dt = n!\,e^{-x}\left(e^x - 1 - x - \frac{x^2}{2!} - \dots - \frac{x^n}{n!}\right)$$

and

$$\int_0^\infty e^{-x}x^n\,dx = n!.$$ (*Math. Trip.* 1935)

9. Prove that

$$\int_0^x e^t t^n\,dt = (-1)^{n-1}n!\,e^x\left\{e^{-x} - 1 + x - \frac{x^2}{2!} + \dots + (-1)^{n-1}\frac{x^n}{n!}\right\};$$

and deduce that, when $x > 0$, e^{-x} is greater than or less than the sum of the first $n+1$ terms of the series $1 - x + \dfrac{x^2}{2!} - \dots$, according as n is odd or even.

(*Math. Trip.* 1934)

10. If $u_n = \int_0^x e^{-t}t^n\,dt$ then $u_n - (n+x)\,u_{n-1} + (n-1)\,xu_{n-2} = 0$.

(*Math. Trip.* 1930)

11. Express $I_m = \int_0^\infty x^m e^{-x} \cos x\, dx$ and $J_m = \int_0^\infty x^m e^{-x} \sin x\, dx$ in terms of I_{m-1} and J_{m-1}; and show that

$$I_m - mI_{m-1} + \tfrac{1}{2}m(m-1)\,I_{m-2} = 0$$

if m is an integer greater than 1. Determine I_m by putting $I_m = m!\,u_m$ in the last relation. (*Math. Trip.* 1936)

12. Show how to find the integral of any rational function of e^x. [Put $x = \log u$, when $e^x = u$, $dx/du = 1/u$, and the integral is transformed into that of a rational function of u.]

13. Prove that we can integrate any function of the form

$$P(x, e^{ax}, e^{bx}, \ldots, \cos lx, \cos mx, \ldots, \sin lx, \sin mx, \ldots),$$

where P denotes a polynomial.

14. Prove that $\int_a^\infty e^{-\lambda x} R(x)\, dx$, where $\lambda > 0$ and a is greater than the greatest root of the denominator of $R(x)$, is convergent. [This follows from the fact that $e^{\lambda x}$ tends to infinity more rapidly than any power of x.]

15. Prove that $\int_{-\infty}^\infty e^{-\lambda x^2 + \mu x}\, dx$, where $\lambda > 0$, is convergent for all values of μ, and that the same is true of $\int_{-\infty}^\infty e^{-\lambda x^2+\mu x} x^n\, dx$, where n is any positive integer.

16. Draw the graphs of e^{x^2}, e^{-x^2}, xe^x, xe^{-x}, xe^{x^2}, xe^{-x^2}, and $x \log x$, determining any maxima and minima of the functions and any points of inflexion on their graphs.

17. Show that the equation $e^{ax} = bx$, where a and b are positive, has two real roots, one, or none, according as $b > ae$, $b = ae$, or $b < ae$. [The tangent to the curve $y = e^{ax}$ at the point $(\xi, e^{a\xi})$ is

$$y - e^{a\xi} = ae^{a\xi}(x - \xi),$$

which passes through the origin if $a\xi = 1$, so that the line $y = aex$ touches the curve at the point $(1/a, e)$. The result now becomes obvious when we draw the line $y = bx$. The reader should discuss the cases in which a or b or both are negative.]

18. Show that the equation $e^x = 1 + x$ has no real root except $x = 0$, and that $e^x = 1 + x + \frac{1}{2}x^2$ has three real roots.

19. Prove that $\dfrac{x^5}{e^x - 1}$ has two stationary values, one at the origin, and the other at $x = 5(1 - e^{-5})$ approximately. (*Math. Trip.* 1932)

20. **The hyperbolic functions.** The hyperbolic functions $\cosh x$*, $\sinh x$, ... are defined by the equations

$$\cosh x = \tfrac{1}{2}(e^x + e^{-x}), \quad \sinh x = \tfrac{1}{2}(e^x - e^{-x}),$$

$$\tanh x = \frac{\sinh x}{\cosh x}, \quad \coth x = \frac{\cosh x}{\sinh x}, \quad \operatorname{sech} x = \frac{1}{\cosh x}, \quad \operatorname{cosech} x = \frac{1}{\sinh x}.$$

Draw the graphs of these functions.

21. Establish the formulae

$$\cosh(-x) = \cosh x, \quad \sinh(-x) = -\sinh x, \quad \tanh(-x) = -\tanh x,$$

$$\cosh^2 x - \sinh^2 x = 1, \quad \operatorname{sech}^2 x + \tanh^2 x = 1, \quad \coth^2 x - \operatorname{cosech}^2 x = 1,$$

$$\cosh 2x = \cosh^2 x + \sinh^2 x, \quad \sinh 2x = 2 \sinh x \cosh x,$$

$$\cosh(x+y) = \cosh x \cosh y + \sinh x \sinh y,$$

$$\sinh(x+y) = \sinh x \cosh y + \cosh x \sinh y.$$

22. Verify that these formulae may be deduced from the corresponding formulae in $\cos x$ and $\sin x$, by writing $\cosh x$ for $\cos x$ and $i \sinh x$ for $\sin x$.

[It follows that the same is true of all the formulae involving $\cos nx$ and $\sin nx$ which are deduced from the corresponding elementary properties of $\cos x$ and $\sin x$. The reason for this analogy will appear in Ch. X.]

23. Express $\cosh x$ and $\sinh x$ in terms (*a*) of $\cosh 2x$, (*b*) of $\sinh 2x$. Discuss any ambiguities of sign that may occur. (*Math. Trip.* 1908)

24. Prove that

$$D_x \cosh x = \sinh x, \quad D_x \sinh x = \cosh x,$$

$$D_x \tanh x = \operatorname{sech}^2 x, \quad D_x \coth x = -\operatorname{cosech}^2 x,$$

$$D_x \operatorname{sech} x = -\operatorname{sech} x \tanh x, \quad D_x \operatorname{cosech} x = -\operatorname{cosech} x \coth x,$$

$$D_x \log \cosh x = \tanh x, \quad D_x \log |\sinh x| = \coth x,$$

$$D_x \arctan e^x = \tfrac{1}{2} \operatorname{sech} x, \quad D_x \log |\tanh \tfrac{1}{2}x| = \operatorname{cosech} x.$$

[All these formulae may of course be transformed into formulae in integration.]

25. Prove that $\cosh x \geqq 1$ and $-1 < \tanh x < 1$.

26. Show that if $-\tfrac{1}{2}\pi < x < \tfrac{1}{2}\pi$ and y is positive, and $\cos x \cosh y = 1$, then

$$y = \log(\sec x + \tan x), \quad D_x y = \sec x, \quad D_y x = \operatorname{sech} y.$$

27. **The inverse hyperbolic functions.** We write

$$s = \sinh x, \quad t = \tanh x, \quad c = \cosh x,$$

and suppose that x increases through all real values.

* 'Hyperbolic cosine': for an explanation of this phrase see Hobson's *Trigonometry*, ch. XVI.

(i) The function s increases steadily and assumes every real value once. The equation $\sinh x = s$ has a unique solution

$$x = \log\{s+\sqrt{(s^2+1)}\},$$

which we write as $\arg\sinh s$.

(ii) The function t increases steadily with x, and has the limits 1 and -1 when $x\to\infty$ and $x\to-\infty$. The equation $\tanh x = t$ has a unique solution

$$x = \tfrac{1}{2}\log\frac{1+t}{1-t},$$

which we write as $\arg\tanh t$.

(iii) The function c is even, and greater than 1 except when $x = 0$. It increases steadily when x is positive, and tends to ∞ when $x\to\infty$. The equation $\cosh x = c$ has two solutions

$$x = \log\{c+\sqrt{(c^2-1)}\}, \quad x = \log\{c-\sqrt{(c^2-1)}\},$$

equal numerically and opposite in sign. We call the first solution, which is positive, $\arg\cosh c$.

Thus $\arg\sinh x$, $\arg\tanh x$ are the one valued inverses of $\sinh x$ and $\tanh x$, while $\arg\cosh x$ may be regarded as one value of the two valued inverse of $\cosh x$. Verify that

$$\int\frac{dx}{\sqrt{(x^2+a^2)}} = \arg\sinh\frac{x}{a}, \quad \int\frac{dx}{\sqrt{(x^2-a^2)}} = \arg\cosh\frac{x}{a}, \quad \int\frac{dx}{x^2-a^2} = -\frac{1}{a}\arg\tanh\frac{x}{a}$$

if $a>0$, $x>a$ in the second formula, and $-a<x<a$ in the third. These formulae give us alternative methods of writing a good many of the formulae of Ch. VI.

28. Prove that

$$\int\frac{dx}{\sqrt{\{(x-a)(x-b)\}}} = 2\log\{\sqrt{(x-a)}+\sqrt{(x-b)}\} \qquad (a<b<x),$$

$$\int\frac{dx}{\sqrt{\{(a-x)(b-x)\}}} = -2\log\{\sqrt{(a-x)}+\sqrt{(b-x)}\} \quad (x<a<b),$$

$$\int\frac{dx}{\sqrt{\{(x-a)(b-x)\}}} = 2\arctan\sqrt{\left(\frac{x-a}{b-x}\right)} \qquad (a<x<b).$$

29. Solve the equation $a\cosh x + b\sinh x = c$, where $c>0$, showing that it has no real roots if $b^2+c^2-a^2<0$, while if $b^2+c^2-a^2>0$ it has two, one, or no real roots according as $a+b$ and $a-b$ are both positive, of opposite signs, or both negative. Discuss the case in which $b^2+c^2-a^2 = 0$.

30. Solve the simultaneous equations

$$\cosh x\cosh y = a, \quad \sinh x\sinh y = b.$$

31. $x^{1/x} \to 1$ as $x \to \infty$. [For $x^{1/x} = e^{(\log x)/x}$, and $(\log x)/x \to 0$. Cf. Ex. XXVII. 11.] Show also that the function $x^{1/x}$ has a maximum when $x = e$, and draw the graph of the function for positive values of x.

32. $x^x \to 1$ as $x \to +0$.

33. If $u_{n+1}/u_n \to l$, where $l > 0$, when $n \to \infty$, then $\sqrt[n]{u_n} \to l$.

[For
$$\log u_{n+1} - \log u_n \to \log l$$
and so
$$\log u_n \sim n \log l.$$
See Ch. IV, Misc. Ex. 17.]

34. $\sqrt[n]{(n!)} \sim e^{-1} n$ when $n \to \infty$. [Take $u_n = n^{-n} n!$ in Ex. 33.]

35. $\sqrt[n]{\left(\dfrac{2n!}{n!\,n!}\right)} \to 4.$

36. Discuss the approximate solution of the equation $e^x = x^{1000000}$.

[It is easy to see by general graphical considerations that the equation has two positive roots, one a little greater than 1 and one very large*, and one negative root a little greater than -1. To determine roughly the size of the large positive root we may proceed as follows. If $e^x = x^{1000000}$ then

$$x = 10^6 \log x,\quad \log x = 13{\cdot}82 + \log\log x,\quad \log\log x = 2{\cdot}63 + \log\left(1 + \frac{\log\log x}{13{\cdot}82}\right),$$

roughly, since $13{\cdot}82$ and $2{\cdot}63$ are approximate values of $\log 10^6$ and $\log\log 10^6$ respectively. It is easy to see from these equations that the ratios $\log x : 13{\cdot}82$ and $\log\log x : 2{\cdot}63$ do not differ greatly from unity, and that
$$x = 10^6(13{\cdot}82 + \log\log x) = 10^6(13{\cdot}82 + 2{\cdot}63) = 16450000$$
gives a tolerable approximation to the root, the error involved being roughly measured by $10^6(\log\log x - 2{\cdot}63)$ or $(10^6 \log\log x)/13{\cdot}82$ or $(10^6 \times 2{\cdot}63)/13{\cdot}82$, which is less than 200,000. The approximations are of course very rough, but suffice to give us a good idea of the scale of magnitude of the root.

Discuss similarly the equations $e^x = 1000000x^{1000000}$, $e^{x^2} = x^{1000000000}$.]

218. Logarithmic tests of convergence for series and integrals. We showed in Ch. VIII (§§ 181, 185) that

$$\sum_1^\infty \frac{1}{n^s},\quad \int_a^\infty \frac{dx}{x^s} \qquad (a > 0)$$

* The phrase 'very large' is of course not used here in the technical sense explained in Ch. IV. It means 'a good deal larger than the roots of such equations as usually occur in elementary mathematics'. The phrase 'a little greater than' must be interpreted similarly.

are convergent if $s>1$ and divergent if $s \leqq 1$. Thus Σn^{-1} is divergent, but $\Sigma n^{-1-\alpha}$ is convergent for all positive values of α.

We saw however in § 210 that with the aid of logarithms we can construct functions which tend to zero more rapidly than n^{-1} but less rapidly than any power $n^{-1-\alpha}$. For example $n^{-1}(\log n)^{-1}$ is such a function, and the question whether the series

$$\Sigma \frac{1}{n \log n}$$

is convergent or divergent cannot be settled by comparison with any series of the type Σn^{-s}.

The same is true of such series as

$$\Sigma \frac{1}{n \sqrt{(\log n)}}, \quad \Sigma \frac{\log \log n}{n(\log n)^2}.$$

It is important to find tests which shall enable us to decide whether series such as these are convergent or divergent; and such tests are easily deduced from the integral test of § 180.

For since

$$D_x(\log x)^{1-s} = \frac{1-s}{x(\log x)^s}, \quad D_x \log \log x = \frac{1}{x \log x},$$

we have

$$\int_a^\xi \frac{dx}{x(\log x)^s} = \frac{(\log \xi)^{1-s} - (\log a)^{1-s}}{1-s}, \quad \int_a^\xi \frac{dx}{x \log x} = \log \log \xi - \log \log a,$$

if $a>1$. The first integral tends to the limit $(\log a)^{1-s}/(s-1)$, when $\xi \to \infty$, if $s>1$, and to ∞ if $s<1$. The second integral tends to ∞. Hence *the series and integral*

$$\sum_{n_0}^{\infty} \frac{1}{n(\log n)^s}, \quad \int_a^\infty \frac{dx}{x(\log x)^s},$$

where n_0 and a are greater than unity, are convergent if $s>1$, divergent if $s \leqq 1$.

It follows that $\Sigma \phi(n)$ is convergent if

$$\phi(n) = O\left\{\frac{1}{n(\log n)^s}\right\},$$

where $s > 1$, for large n, and divergent if $\phi(n)$ is positive and

$$\frac{1}{\phi(n)} = O(n \log n).$$

We leave the statement of the corresponding theorem for integrals to the reader.

Examples LXXXIX. 1. The series

$$\Sigma \frac{(\log n)^p}{n^{1+s}}, \quad \Sigma \frac{(\log n)^p (\log \log n)^q}{n^{1+s}}, \quad \Sigma \frac{(\log \log n)^p}{n(\log n)^{1+s}},$$

where $s > 0$, are convergent for all values of p and q; and

$$\Sigma \frac{1}{n^{1-s}(\log n)^p}, \quad \Sigma \frac{1}{n^{1-s}(\log n)^p (\log \log n)^q}, \quad \Sigma \frac{1}{n(\log n)^{1-s} (\log \log n)^p}$$

divergent. For $(\log n)^p = O(n^\delta)$ for every p (however large) and every positive δ (however small), and $(\log \log n)^p = O\{(\log n)^\delta\}$. The factors involving $\log n$ and $\log \log n$ in the first two series of each set, and that involving $\log \log n$ in the third of each set, are negligible from the point of view of convergence.

2. The convergence or divergence of such series as

$$\Sigma \frac{1}{n \log n \log \log n}, \quad \Sigma \frac{\log \log \log n}{n \log n \sqrt{(\log \log n)}}$$

cannot be settled by the theorem of p. 418, since in each case the function under the sign of summation tends to zero more rapidly than $n^{-1}(\log n)^{-1}$ yet less rapidly than $n^{-1}(\log n)^{-1-\alpha}$, where α is any positive number. For such series we need a still more delicate test. The reader should be able, starting from the equations

$$\frac{d}{dx}(\log_k x)^{1-s} = \frac{1-s}{x \log x \log_2 x \ldots \log_{k-1} x (\log_k x)^s},$$

$$\frac{d}{dx}\log_{k+1} x = \frac{1}{x \log x \log_2 x \ldots \log_{k-1} x \log_k x},$$

where $\log_2 x = \log \log x$, $\log_3 x = \log \log \log x$, ...,* to prove the following theorem: *the series and integral*

$$\sum_{n_0}^{\infty} \frac{1}{n \log n \log_2 n \ldots \log_{k-1} n (\log_k n)^s}, \quad \int_a^{\infty} \frac{dx}{x \log x \log_2 x \ldots \log_{k-1} x (\log_k x)^s}$$

are convergent if $s > 1$ and divergent if $s \leqq 1$, n_0 and a being any numbers sufficiently great to ensure that $\log_k n$ and $\log_k x$ are positive when $n \geqq n_0$ or $x \geqq a$. These values of n_0 and a increase very rapidly as k increases:

* This notation must not be confused with that of p. 412.

thus $\log x > 0$ requires $x > 1$, $\log_2 x > 0$ requires $x > e$, $\log\log x > 0$ requires $x > e^e$, and so on; and it is easy to see that $e^e > 10$, $e^{e^e} > e^{10} > 20{,}000$, $e^{e^{e^e}} > e^{20000} > 10^{8000}$.

The reader should observe the extreme rapidity with which the higher exponential functions, such as e^{e^x} and $e^{e^{e^x}}$, increase with x. The same remark of course applies to such functions as a^{a^x} and $a^{a^{a^x}}$, where a has any value greater than unity. It has been computed that 9^{9^9} has about 369,693,100 figures, while $10^{10^{10}}$ has of course 10,000,000,001. Conversely, the rate of increase of the higher logarithmic functions is extremely slow. Thus to make $\log\log\log\log x > 1$ we have to suppose x a number with over 8000 figures*.

The number of protons in the universe has been estimated at 10^{80}, and the number of possible games of chess at $10^{10^{50}}$.

3. Prove that the integral $\int_0^a \frac{1}{x}\left\{\log\left(\frac{1}{x}\right)\right\}^s dx$, where $0<a<1$, is convergent if $s<-1$, divergent if $s \geqq -1$. [Consider the behaviour of

$$\int_\epsilon^a \frac{1}{x}\left\{\log\left(\frac{1}{x}\right)\right\}^s dx$$

when $\epsilon \to +0$. This result also may be sharpened by the introduction of higher logarithmic factors.]

4. Prove that $\int_0^1 \frac{1}{x}\left\{\log\left(\frac{1}{x}\right)\right\}^s dx$ is divergent for all values of s. [The last example shows that $s<-1$ is a necessary condition for convergence at the lower limit: but $\{\log(1/x)\}^s$ tends to ∞ like $(1-x)^s$, as $x \to 1-0$, if s is negative, and so the integral diverges at the upper limit when $s<-1$.]

5. The necessary and sufficient conditions for the convergence of $\int_0^1 x^{a-1}\left\{\log\left(\frac{1}{x}\right)\right\}^s dx$ are $a>0$, $s>-1$.

6. Investigate the convergence of $\int_0^\infty \frac{x^a\,dx}{(1+x)^b\{1+(\log x)^2\}}$.

(*Math. Trip.* 1934)

Examples XC. 1. **Euler's limit.** Show that

$$\phi(n) = 1+\frac{1}{2}+\frac{1}{3}+\dots+\frac{1}{n-1}-\log n$$

tends to a limit γ as $n\to\infty$, and that $0<\gamma\leqq 1$. [This follows at once from § 180. The value of γ is $\cdot 577\ldots$, and γ is usually called *Euler's constant*.]

* See the footnotes to pp. 350 and 404.

2. If a and b are positive then

$$\frac{1}{a}+\frac{1}{a+b}+\frac{1}{a+2b}+\dots+\frac{1}{a+(n-1)b}-\frac{1}{b}\log(a+nb)$$

tends to a limit as $n\to\infty$.

3. If $0<s<1$ then

$$\phi(n) = 1+2^{-s}+3^{-s}+\dots+(n-1)^{-s}-\frac{n^{1-s}}{1-s}$$

tends to a limit as $n\to\infty$.

4. Show that the series

$$\frac{1}{1}+\frac{1}{2(1+\frac{1}{2})}+\frac{1}{3(1+\frac{1}{2}+\frac{1}{3})}+\dots$$

is divergent. [Compare the general term of the series with $(n\log n)^{-1}$.

5. *The case $a = 1$ in* § 183. When $a = 1$ in equation (1) of § 183, we take $u_n = (n\log n)^{-1}$, when Σu_n is divergent. Since

$$\frac{\log(n+1)}{\log n} = \frac{1}{\log n}\left\{\log n+\frac{1}{n}+O\left(\frac{1}{n^2}\right)\right\} = 1+\frac{1}{n\log n}+O\left(\frac{1}{n^2}\right),$$

we have

$$\frac{u_{n+1}}{u_n} = \frac{n\log n}{(n+1)\log(n+1)} = 1-\frac{1}{n}-\frac{1}{n\log n}+O\left(\frac{1}{n^2}\right).$$

Hence $v_{n+1}/v_n > u_{n+1}/u_n$ for large n, and Σv_n is divergent.

6. Prove generally that if Σu_n is a series of positive terms, and

$$s_n = u_1+u_2+\dots+u_n,$$

then $\Sigma(u_n/s_{n-1})$ is convergent or divergent according as Σu_n is convergent or divergent. [If Σu_n is convergent then s_{n-1} tends to a positive limit l, and so $\Sigma(u_n/s_{n-1})$ is convergent. If Σu_n is divergent then $s_{n-1}\to\infty$, and

$$\frac{u_n}{s_{n-1}} > \log\left(1+\frac{u_n}{s_{n-1}}\right) = \log\frac{s_n}{s_{n-1}}$$

(Ex. LXXXIII. 1); and it is evident that

$$\log\frac{s_2}{s_1}+\log\frac{s_3}{s_2}+\dots+\log\frac{s_n}{s_{n-1}} = \log\frac{s_n}{s_1}$$

tends to ∞ as $n\to\infty$.]

7. Sum the series $1-\frac{1}{2}+\frac{1}{3}-\dots$. [We have

$$1+\frac{1}{2}+\dots+\frac{1}{2n} = \log(2n+1)+\gamma+o(1),$$

$$2\left(\frac{1}{2}+\frac{1}{4}+\dots+\frac{1}{2n}\right) = \log(n+1)+\gamma+o(1),$$

by Ex. 1, γ denoting Euler's constant. Subtracting and making $n \to \infty$ we see that the sum of the given series is $\log 2$. See also § 220.]

8. Prove that the series

$$\sum_{0}^{\infty}(-1)^n\left(1+\frac{1}{2}+\dots+\frac{1}{n+1}-\log n-C\right)$$

oscillates finitely except when $C=\gamma$, when it converges.

219. Series connected with the exponential and logarithmic functions. Expansion of e^x by Taylor's theorem.

Since all the derivatives of the exponential function are equal to the function itself, we have

$$e^x = 1+x+\frac{x^2}{2!}+\dots+\frac{x^{n-1}}{(n-1)!}+\frac{x^n}{n!}e^{\theta x},$$

where $0<\theta<1$. But $x^n/n!\to 0$ as $n\to\infty$, whatever be the value of x (Ex. XXVII. 12); and $e^{\theta x}<e^x$. Hence, making n tend to ∞, we have

$$e^x = 1+x+\frac{x^2}{2!}+\dots+\frac{x^n}{n!}+\dots \quad \dots\dots\dots\dots\dots(1).$$

The series on the right-hand side of this equation is known as the *exponential series*. In particular we have

$$e = 1+1+\frac{1}{2!}+\dots+\frac{1}{n!}+\dots \quad \dots\dots\dots\dots\dots(2);$$

and so

$$\left(1+1+\frac{1}{2!}+\dots+\frac{1}{n!}+\dots\right)^x = 1+x+\frac{x^2}{2!}+\dots+\frac{x^n}{n!}+\dots \quad \dots(3),$$

a result known as the *exponential theorem*. Also

$$a^x = e^{x\log a} = 1+(x\log a)+\frac{(x\log a)^2}{2!}+\dots \quad \dots\dots\dots(4)$$

for all positive values of a.

The reader will observe that the exponential series has the property of reproducing itself when every term is differentiated, and that no other series of powers of x would possess this property: for some further remarks in this connection see Appendix II.

The power series for e^x is so important that it is worth while to investigate it by an alternative method which does not depend upon Taylor's theorem. Let

$$E_n(x) = 1+x+\frac{x^2}{2!}+\dots+\frac{x^n}{n!},$$

and suppose that $x>0$. Then

$$\left(1+\frac{x}{n}\right)^n = 1+n\left(\frac{x}{n}\right)+\frac{n(n-1)}{1.2}\left(\frac{x}{n}\right)^2+\dots+\frac{n(n-1)\dots 1}{1.2\dots n}\left(\frac{x}{n}\right)^n,$$

which is less than $E_n(x)$ if $n>1$. And, provided $n>x$, we have also, by the binomial theorem for a negative integral exponent,

$$\left(1-\frac{x}{n}\right)^{-n} = 1+n\left(\frac{x}{n}\right)+\frac{n(n+1)}{1.2}\left(\frac{x}{n}\right)^2+\dots > E_n(x).$$

Thus
$$\left(1+\frac{x}{n}\right)^n < E_n(x) < \left(1-\frac{x}{n}\right)^{-n} \quad (n>x).$$

But (§ 215) the first and last functions tend to the limit e^x as $n\to\infty$, and therefore $E_n(x)$ must do the same. This proves (1) when x is positive. That it is true also when x is negative follows from the functional equation $f(x)f(y) = f(x+y)$ satisfied by the exponential series (Ex. LXXXI. 7).

Examples XCI. 1. Show that

$$\cosh x = 1+\frac{x^2}{2!}+\frac{x^4}{4!}+\dots, \quad \sinh x = x+\frac{x^3}{3!}+\frac{x^5}{5!}+\dots.$$

2. If x is positive then the greatest term in the exponential series is the $([x]+1)$th, unless x is an integer, when the preceding term is equal to it.

3. Show that $n!>(n/e)^n$. [For $n^n/n!$ is one term in the series for e^n.]

4. Prove that $e^n = \dfrac{n^n}{n!}(2+S_1+S_2)$, where

$$S_1 = \frac{1}{1+\nu}+\frac{1}{(1+\nu)(1+2\nu)}+\dots, \quad S_2 = (1-\nu)+(1-\nu)(1-2\nu)+\dots,$$

and $\nu = 1/n$; and deduce that $n!$ lies between $2(n/e)^n$ and $2(n+1)(n/e)^n$.

5. Employ the exponential series to prove that e^x tends to infinity more rapidly than any power of x. [Use the inequality $e^x>x^n/n!$.]

6. Show that e is not a rational number. [If $e = p/q$, where p and q are integers, we must have

$$\frac{p}{q} = 1+\frac{1}{2!}+\frac{1}{3!}+\dots+\frac{1}{q!}+\dots$$

or, multiplying by $q!$,

$$q!\left(\frac{p}{q}-1-1-\frac{1}{2!}-\dots-\frac{1}{q!}\right)=\frac{1}{q+1}+\frac{1}{(q+1)(q+2)}+\dots;$$

and this is absurd, since the left-hand side is integral, and the right-hand side less than $(q+1)^{-1}+(q+1)^{-2}+\dots=q^{-1}$.]

7. Sum the series $\sum_0^\infty P_r(n)\frac{x^n}{n!}$, where $P_r(n)$ is a polynomial of degree r in n. [We can express $P_r(n)$ in the form

$$A_0+A_1n+A_2n(n-1)+\dots+A_rn(n-1)\dots(n-r+1),$$

and
$$\sum_0^\infty P_r(n)\frac{x^n}{n!}=A_0\sum_0^\infty\frac{x^n}{n!}+A_1\sum_1^\infty\frac{x^n}{(n-1)!}+\dots+A_r\sum_r^\infty\frac{x^n}{(n-r)!}$$
$$=(A_0+A_1x+A_2x^2+\dots+A_rx^r)e^x.]$$

8. Show that

$$\sum_1^\infty\frac{n^3}{n!}x^n=(x+3x^2+x^3)e^x,\quad \sum_1^\infty\frac{n^4}{n!}x^n=(x+7x^2+6x^3+x^4)e^x;$$

and that if $S_n=1^3+2^3+\dots+n^3$ then

$$\sum_1^\infty S_n\frac{x^n}{n!}=\tfrac{1}{4}(4x+14x^2+8x^3+x^4)e^x.$$

In particular the last series is equal to zero when $x=-2$.

(*Math. Trip.* 1904)

9. Prove that $\Sigma(n/n!)=e$, $\Sigma(n^2/n!)=2e$, $\Sigma(n^3/n!)=5e$, and that $\Sigma(n^k/n!)$, where k is any positive integer, is a positive integral multiple of e.

10. Prove that $\sum_1^\infty\frac{(n-1)x^n}{(n+2)n!}=x^{-2}\{(x^2-3x+3)e^x+\tfrac{1}{2}x^2-3\}$.

[Multiply numerator and denominator by $n+1$, and proceed as in Ex. 7.]

11. Evaluate $\lim_{x\to 0}\frac{1-ae^{-x}-be^{-2x}-ce^{-3x}}{1-ae^x-be^{2x}-ce^{3x}}$ in the three cases (i) $a=3$, $b=-5$, $c=4$; (ii) $a=3$, $b=-4$, $c=2$; (iii) $a=3$, $b=-3$, $c=1$.

(*Math. Trip.* 1923)

12. Evaluate
$$\lim_{x\to 0}\frac{a^x-b^x}{c^x-d^x}$$
when a, b, c, d are positive and $c\neq d$. (*Math. Trip.* 1934)

13. Deduce the exponential series from the result of Ex. LXXXVIII. 9.

14. If

$$X_0=e^x,\quad X_1=e^x-1,\quad X_2=e^x-1-x,\quad X_3=e^x-1-x-\frac{x^2}{2!},\quad \dots,$$

then the derivative of X_ν is $X_{\nu-1}$. Hence prove that if $t>0$ then

$$X_1(t)=\int_0^t X_0\,dx<te^t,\quad X_2(t)=\int_0^t X_1\,dx<\int_0^t xe^x\,dx<e^t\int_0^t x\,dx=\frac{t^2}{2!}e^t,$$

and generally $X_\nu(t)<\dfrac{t^\nu}{\nu!}e^t$. Deduce the exponential theorem

15. Show that the expansion in powers of p of the positive root of $x^{2+p}=a^2$ begins with the terms

$$a\{1-\tfrac{1}{2}p\log a+\tfrac{1}{8}p^2\log a\,(2+\log a)\}.$$

(*Math. Trip.* 1909)

220. The logarithmic series. Another very important expansion in powers of x is that for $\log(1+x)$. Since

$$\log(1+x)=\int_0^x\frac{dt}{1+t},$$

and $1/(1+t)=1-t+t^2-\dots$ if t is numerically less than unity, it is natural to expect* that $\log(1+x)$ will be equal, when $-1<x<1$, to the series obtained by integrating each term of the series $1-t+t^2-\dots$ from $t=0$ to $t=x$, i.e. to the series $x-\frac{1}{2}x^2+\frac{1}{3}x^3-\dots$; and this is in fact true. For

$$\frac{1}{1+t}=1-t+t^2-\dots+(-1)^{m-1}t^{m-1}+\frac{(-1)^m t^m}{1+t},$$

and so, if $x>-1$,

$$\log(1+x)=\int_0^x\frac{dt}{1+t}=x-\frac{x^2}{2}+\dots+(-1)^{m-1}\frac{x^m}{m}+(-1)^m R_m,$$

where

$$R_m=\int_0^x\frac{t^m\,dt}{1+t}.$$

If $0\leqq x\leqq 1$ then

$$0\leqq R_m\leqq\int_0^x t^m\,dt=\frac{x^{m+1}}{m+1}\leqq\frac{1}{m+1}\to 0$$

when $m\to\infty$. If $-1<x<0$, and $x=-\xi$, so that $0<\xi<1$, then

$$R_m=(-1)^{m-1}\int_0^\xi\frac{u^m}{1-u}\,du$$

and

$$|R_m|\leqq\frac{1}{1-\xi}\int_0^\xi u^m\,du=\frac{\xi^{m+1}}{(m+1)(1-\xi)}\to 0;$$

* See Appendix II for some further remarks on this subject.

so that again $R_m \to 0$. Hence

$$\log(1+x) = x - \tfrac{1}{2}x^2 + \tfrac{1}{3}x^3 - \dots,$$

provided that $-1 < x \leqq 1$. If x lies outside these limits the series is not convergent. If $x = 1$ we obtain

$$\log 2 = 1 - \tfrac{1}{2} + \tfrac{1}{3} - \dots,$$

a result already proved otherwise (Ex. XC. 7).

221. The series for the inverse tangent. It is easy to prove in a similar manner that

$$\begin{aligned}\operatorname{arc}\tan x = \int_0^x \frac{dt}{1+t^2} &= \int_0^x (1 - t^2 + t^4 - \dots)\,dt \\ &= x - \tfrac{1}{3}x^3 + \tfrac{1}{5}x^5 - \dots\end{aligned}$$

when $-1 \leqq x \leqq 1$. The only differences are that the proof is a little simpler, since arc tan x is an odd function of x and we need only consider positive values of x; and that the series is convergent when $x = -1$ as well as when $x = 1$. We leave the discussion to the reader. The value of $\operatorname{arc}\tan x$ which is represented by the series is of course that which lies between $-\tfrac{1}{4}\pi$ and $\tfrac{1}{4}\pi$ when $-1 \leqq x \leqq 1$, and which we saw in Ch. VII (Ex. LXIII. 3) to be the value represented by the integral. If $x = 1$, we obtain the formula

$$\tfrac{1}{4}\pi = 1 - \tfrac{1}{3} + \tfrac{1}{5} - \dots.$$

Examples XCII. 1. $\log\left(\dfrac{1}{1-x}\right) = x + \tfrac{1}{2}x^2 + \tfrac{1}{3}x^3 + \dots$ if $-1 \leqq x < 1$.

2. $\operatorname{arg}\tanh x = \tfrac{1}{2}\log\left(\dfrac{1+x}{1-x}\right) = x + \tfrac{1}{3}x^3 + \tfrac{1}{5}x^5 + \dots$ if $-1 < x < 1$.

3. Prove that if x is positive then

$$\log(1+x) = \frac{x}{1+x} + \frac{1}{2}\left(\frac{x}{1+x}\right)^2 + \frac{1}{3}\left(\frac{x}{1+x}\right)^3 + \dots.$$

(*Math. Trip.* 1911)

4. Obtain the series for $\log(1+x)$ and $\operatorname{arc}\tan x$ by means of Taylor's theorem.

[A difficulty presents itself in the discussion of the remainder in the first series when x is negative, if Lagrange's form is used; Cauchy's form, viz.

$$R_n = \frac{(-1)^{n-1}(1-\theta)^{n-1}x^n}{(1+\theta x)^n},$$

should be used (cf. the corresponding discussion for the binomial series, §§ 152 (2) and 168).

In the case of the second series we have

$$\begin{aligned} D_x^n \arctan x &= D_x^{n-1}\{1+x^2)^{-1} \\ &= (-1)^{n-1}(n-1)!\,(x^2+1)^{-\frac{1}{2}n}\sin\{n\arctan(1/x)\} \end{aligned}$$

(Ex. XLV. 15), and there is no difficulty about the remainder, which is obviously not greater in absolute value than $1/n$*.]

5. Prove that $\log 2$ lies between the sums of the first $2n$ and the first $2n+1$ terms of the series $1-\frac{1}{2}+\frac{1}{3}-\ldots$. (*Math. Trip.* 1930)

6. Evaluate $$\lim_{x\to 1}\frac{1-x+\log x}{1-\sqrt{(2x-x^2)}}.$$ (*Math. Trip.* 1934)

7. If $y>0$ then

$$\log y = 2\left\{\frac{y-1}{y+1}+\frac{1}{3}\left(\frac{y-1}{y+1}\right)^3+\frac{1}{5}\left(\frac{y-1}{y+1}\right)^5+\ldots\right\}.$$

[Use the identity $y=\left(1+\dfrac{y-1}{y+1}\right)\Big/\left(1-\dfrac{y-1}{y+1}\right)$. This series may be used to calculate $\log 2$, a purpose for which the series $1-\frac{1}{2}+\frac{1}{3}-\ldots$, owing to the slowness of its convergence, is practically useless. Put $y=2$ and find $\log 2$ to three places of decimals.]

8. Find $\log 10$ to three places of decimals from the formula

$$\log 10 = 3\log 2+\log(1+\tfrac{1}{4}).$$

9. Prove that

$$\log\left(\frac{x+1}{x}\right) = 2\left\{\frac{1}{2x+1}+\frac{1}{3(2x+1)^3}+\frac{1}{5(2x+1)^5}+\ldots\right\}$$

if $x>0$, and that

$$\log\frac{(x-1)^2(x+2)}{(x+1)^2(x-2)} = 2\left\{\frac{2}{x^3-3x}+\frac{1}{3}\left(\frac{2}{x^3-3x}\right)^3+\frac{1}{5}\left(\frac{2}{x^3-3x}\right)^5+\ldots\right\}$$

if $x>2$. Given that $\log 2 = \cdot 6931471\ldots$ and $\log 3 = 1\cdot 0986123\ldots$, show, by putting $x=10$ in the second formula, that $\log 11 = 2\cdot 397895\ldots$.

(*Math. Trip.* 1912)

* The formula for $D_x^n \arctan x$ fails when $x=0$, since $\arctan(1/x)$ is then undefined. It is easy to see (cf. Ex. XLV. 15) that $\arctan(1/x)$ must then be interpreted as meaning $\frac{1}{2}\pi$.

10. Show that if log 2, log 5, and log 11 are known, then the formula

$$\log 13 = 3\log 11 + \log 5 - 9\log 2$$

gives log 13 with an error practically equal to ·00015. (*Math. Trip.* 1910)

11. Show that

$\frac{1}{2}\log 2 = 7a+5b+3c,\quad \frac{1}{2}\log 3 = 11a+8b+5c,\quad \frac{1}{2}\log 5 = 16a+12b+7c,$

where $a = \operatorname{arg\,tanh}\frac{1}{31}$, $b = \operatorname{arg\,tanh}\frac{1}{49}$, $c = \operatorname{arg\,tanh}\frac{1}{161}$.

[These formulae enable us to find log 2, log 3, and log 5 rapidly and with any degree of accuracy.]

12. Show that

$$\tfrac{1}{4}\pi = \arctan\tfrac{1}{2} + \arctan\tfrac{1}{3} = 4\arctan\tfrac{1}{5} - \arctan\tfrac{1}{239},$$

and calculate π to six places of decimals.

13. Expand $\log\{1-\log(1-x)\}$ up to the term in x^3; and deduce the corresponding expansion of $\log\{1+\log(1+x)\}$ by substituting $x/(1+x)$ for x. (*Math. Trip.* 1923)

14. Show that the expansion of $(1+x)^{1+x}$ in powers of x begins with the terms $1+x+x^2+\frac{1}{2}x^3$. (*Math. Trip.* 1910)

15. Show that

$$\log_{10} e - \sqrt{\{x(x+1)\}}\log_{10}\left(\frac{1+x}{x}\right) = \frac{\log_{10} e}{24x^2},$$

approximately, for large values of x. Apply the formula, when $x = 10$, to obtain an approximate value of $\log_{10} e$, and estimate the accuracy of the result. (*Math. Trip.* 1910)

16. If

$$2x = \log\frac{y-1}{2} + \sum_1^\infty \frac{(-1)^n}{n}\left(\frac{y-1}{2}\right)^n$$

and $1 < y \leqq 3$, then $y = -\coth x$. Find a similar expansion for $2x$ valid for $-3 \leqq y < -1$. (*Math. Trip.* 1927)

17. Using the logarithmic series and the facts that

$$\log_{10} 2{\cdot}3758 = {\cdot}3758099\ldots,\quad \log_{10} e = {\cdot}4343\ldots,$$

show that an approximate solution of the equation $x = 100\log_{10} x$ is 237·58121. (*Math. Trip.* 1910)

18. Expand $\log\cos x$ and $\log\sin x - \log x$ in powers of x as far as x^4, and verify that, to this order,

$$\log\sin x = \log x - \tfrac{1}{45}\log\cos x + \tfrac{64}{45}\log\cos\tfrac{1}{2}x.$$

(*Math. Trip.* 1908)

19. Show that $\int_0^x \frac{dt}{1+t^4} = x - \frac{1}{5}x^5 + \frac{1}{9}x^9 - \dots$ if $-1 \leqq x \leqq 1$. Deduce that

$$1 - \tfrac{1}{5} + \tfrac{1}{9} - \dots = \frac{\pi + 2\log(\sqrt{2}+1)}{4\sqrt{2}}. \quad (\textit{Math. Trip.}\ 1896)$$

[Proceed as in § 221 and use the result of Ex. XLVIII. 8. Sum $\frac{1}{3} - \frac{1}{7} + \frac{1}{11} - \dots$ similarly.]

20. Prove generally that if a and b are positive integers then

$$\frac{1}{a} - \frac{1}{a+b} + \frac{1}{a+2b} - \dots = \int_0^1 \frac{t^{a-1}\,dt}{1+t^b},$$

and so that the sum of the series can be found. Calculate in this way the sums of $1 - \frac{1}{4} + \frac{1}{7} - \dots$ and $\frac{1}{2} - \frac{1}{5} + \frac{1}{8} - \dots$.

222. The binomial series. We have already (§ 168) investigated the binomial theorem

$$(1+x)^m = 1 + \binom{m}{1}x + \binom{m}{2}x^2 + \dots,$$

assuming that $-1 < x < 1$ and that m is rational. When m is irrational we have

$$(1+x)^m = e^{m\log(1+x)},$$

$$\frac{d}{dx}(1+x)^m = \frac{m}{1+x}e^{m\log(1+x)} = m(1+x)^{m-1},$$

so that the rule for the differentiation of $(1+x)^m$ remains the same, and the proof of the theorem given in § 168 retains its validity. There remain the cases $x = 1$ and $x = -1$.

(1) When $x = 1$ the series is

$$1 + m + \frac{m(m-1)}{2!} + \frac{m(m-1)(m-2)}{3!} + \dots.$$

If $m+1 \leqq 0$ the general term u_n does not tend to zero (Ex. LXXXIV. 3). If $m+1 > 0$ then u_n ultimately alternates in sign and decreases steadily to zero, so that the series is convergent.

To sum the series, take $f(x) = (1+x)^m$ and write 0 for a and 1 for h in (1) of § 167. We obtain

$$2^m = u_0 + u_1 + \dots + u_{n-1} + R_n,$$

where

$$R_n = \frac{m(m-1)\dots(m-n+1)}{(n-1)!}\int_0^1 (1-t)^{n-1}(1+t)^{m-n}\,dt.$$

The integral here is less than n^{-1} for large n (since $m-n<0$ and $1+t \geqq 1$). Hence

$$|R_n| \leqq |u_n| \to 0.$$

Thus *the binomial series is convergent for $x = 1$ if and only if $m > -1$, and its sum is then 2^m.*

(2) When $x = -1$ we can sum the first $n+1$ terms of the series. The sum is 1 if $m = 0$. Otherwise, if we put $x = -1$ and $m = -\mu$, it is

$$1+\mu+\ldots+\frac{\mu(\mu+1)\ldots(\mu+n-1)}{n!} = \frac{(\mu+1)(\mu+2)\ldots(\mu+n)}{n!}$$

$$= (-1)^n \binom{m-1}{n}$$

(Ex. LXXXI. 5). This tends to zero when $m > 0$, and does not tend to a limit when $m < 0$ (Ex. LXXXIV. 3). Hence *the series is convergent for $x = -1$ if and only if $m \geqq 0$, and its sum is 1 when $m = 0$ and 0 when $m > 0$.*

Examples XCIII. 1. Prove that if $-1<x<1$ then

$$\frac{1}{\sqrt{(1+x^2)}} = 1-\frac{1}{2}x^2+\frac{1.3}{2.4}x^4-\ldots, \quad \frac{1}{\sqrt{(1-x^2)}} = 1+\frac{1}{2}x^2+\frac{1.3}{2.4}x^4+\ldots.$$

2. **Approximation to quadratic and other surds.** Let $\sqrt{M}$ be a quadratic surd whose numerical value is required. Let N^2 be the square nearest to M; and let $M = N^2+x$ or $M = N^2-x$, x being positive. Since x cannot be greater than N, x/N^2 is comparatively small and the surd $\sqrt{M} = N\sqrt{\{1 \pm (x/N^2)\}}$ can be expressed in a series

$$N\left\{1 \pm \frac{1}{2}\left(\frac{x}{N^2}\right) - \frac{1.1}{2.4}\left(\frac{x}{N^2}\right)^2 \pm \ldots\right\},$$

which is at any rate fairly rapidly convergent. Thus

$$\sqrt{67} = \sqrt{(64+3)} = 8\left\{1+\frac{1}{2}\left(\frac{3}{64}\right) - \frac{1.1}{2.4}\left(\frac{3}{64}\right)^2 + \ldots\right\}.$$

Verify that the error committed in taking $8\frac{3}{16}$ (the value given by the first two terms) as an approximate value is one of excess, and is less than $3^2/64^2$, which is less than $\cdot 003$.

3. If x is small compared with N^2 then

$$\sqrt{(N^2+x)} = N + \frac{x}{4N} + \frac{Nx}{2(2N^2+x)},$$

the error being of the order x^4/N^7. Apply the process to $\sqrt{997}$.

4. If M differs from N^3 by less than 1 per cent. of either, then $\sqrt[3]{M}$ differs from $\frac{2}{3}N+\frac{1}{3}MN^{-2}$ by less than $N/90000$. (*Math. Trip.* 1882)

5. If $M = N^4+x$, and x is small compared with N, then a good approximation for $\sqrt[4]{M}$ is

$$\frac{51}{56}N+\frac{5}{56}\frac{M}{N^3}+\frac{27Nx}{14(7M+5N^4)}.$$

Show that when $N = 10$, $x = 1$, this approximation is accurate to 16 places of decimals. (*Math. Trip.* 1886)

6. Show how to sum the series

$$\sum_0^\infty P_r(n)\binom{m}{n}x^n,$$

where $P_r(n)$ is a polynomial of degree r in n.

[Express $P_r(n)$ in the form $A_0+A_1n+A_2n(n-1)+\dots$ as in Ex. XCI. 7.]

223. An alternative method of development of the theory of the exponential and logarithmic functions. We shall now give an outline of a method of investigation of the properties of e^x and $\log x$ entirely different in logical order from that followed in the preceding pages. This method starts from the exponential series $1+x+\frac{x^2}{2!}+\dots$. We know that this series is convergent for all values of x, and we may therefore define the function $\exp x$ by the equation

$$\exp x = 1+x+\frac{x^2}{2!}+\dots \quad \text{.........................(1).}$$

We then prove, as in Ex. LXXXI. 7, that

$$\exp x \times \exp y = \exp(x+y) \quad \text{.....................(2).}$$

Also
$$\frac{\exp h-1}{h} = 1+\frac{h}{2!}+\frac{h^2}{3!}+\dots = 1+\rho(h),$$

where $\rho(h)$ is numerically less than

$$|\tfrac{1}{2}h|+|\tfrac{1}{2}h|^2+|\tfrac{1}{2}h|^3+\dots = \frac{|\frac{1}{2}h|}{1-|\frac{1}{2}h|},$$

so that $\rho(h)\to 0$ as $h\to 0$; and so

$$\frac{\exp(x+h)-\exp x}{h} = \exp x\left(\frac{\exp h-1}{h}\right)\to \exp x$$

as $h\to 0$, or
$$\frac{d}{dx}\exp x = \exp x \quad \text{..........................(3).}$$

Incidentally we have proved that $\exp x$ is a continuous function.

We have now a choice of procedure. Writing $y = \exp x$ and observing that $\exp 0 = 1$, we have

$$\frac{dy}{dx} = y, \quad x = \int_1^y \frac{dt}{t},$$

and, if we define the logarithmic function as the function inverse to the exponential function, we are brought back to the point of view adopted earlier in this chapter.

But we may proceed differently. From (2) it follows that if n is a positive integer then

$$(\exp x)^n = \exp nx, \quad (\exp 1)^n = \exp n.$$

If x is a positive rational fraction m/n, then

$$\{\exp(m/n)\}^n = \exp m = (\exp 1)^m,$$

and so $\exp(m/n)$ is equal to the positive value of $(\exp 1)^{m/n}$. This result may be extended to negative rational values of x by means of the equation

$$\exp x \exp(-x) = 1;$$

and so we have

$$\exp x = (\exp 1)^x = e^x,$$

say, where

$$e = \exp 1 = 1 + 1 + \frac{1}{2!} + \frac{1}{3!} + \dots,$$

for all rational values of x. Finally we define e^x, when x is irrational, as being equal to $\exp x$. The logarithm is then defined as the inverse function.

Example. Develop the theory of the binomial series

$$1 + \binom{m}{1}x + \binom{m}{2}x^2 + \dots = f(m, x),$$

where $-1 < x < 1$, in a similar manner, starting from the equation

$$f(m, x) f(m', x) = f(m + m', x)$$

(Ex. LXXXI. 6).

224. The analytical theory of the circular functions. We return now to a subject which we have already discussed briefly in § 163.

We have, throughout the body of this book, taken the elements of plane trigonometry for granted, and have used the trigonometrical or 'circular' functions $\cos x$, $\sin x$, $\tan x$, ... freely for purposes of illustration. We pointed out however in § 163 that the foundations of trigonometry are not quite so simple as a beginner might suppose, and that the ordinary presentation of the theory rests on certain presuppositions which demand careful analysis.

There are at least four obvious methods by which we may construct an analytical theory of the circular functions.

(i) *The geometrical method.* The most natural method is to follow as closely as we can the procedure of the ordinary text-books, translating the geometrical language which they employ into the language of analysis. We discussed this problem in § 163, and concluded that it involves one and only one serious difficulty. We have to show either that with any arc of a circle is associated a number which we call its *length*, or that with any sector of a circle is associated a number which we call its *area*. These demands are alternative, and when either of them has been satisfied our trigonometry will rest on a secure foundation. It is usual to adopt the first alternative, and to base trigonometry on the notion of length; but Ch. VII contains an accurate discussion of areas and not of lengths, so that we were naturally led to prefer the second alternative.

(ii) *The method of infinite series.* The second method, which is adopted in many treatises on analysis, is to define the trigonometrical functions as the exponential function was defined in § 223, namely by infinite series. We define $\cos x$ and $\sin x$ by the equations

$$(1)\qquad \cos x = 1 - \frac{x^2}{2!} + \frac{x^4}{4!} - \dots, \quad \sin x = x - \frac{x^3}{3!} + \frac{x^5}{5!} - \dots.$$

These series are absolutely convergent for all real values of x, and may be multiplied together as in § 223. We thus obtain the formula

$$\cos(x+y) = \cos x \cos y - \sin x \sin y$$

and the other addition formulae of trigonometry. The property of periodicity is a little more troublesome. We can prove from (1) that $\cos x$, which is positive for small values of x, changes its sign just once in the interval $(0, 2)$, say for $x = \xi$; and we define π by the equation $\frac{1}{2}\pi = \xi$. It is then easy to prove that $\sin\frac{1}{2}\pi = 1$, $\cos\pi = -1$, $\sin\pi = 0$; and the equations

$$\cos(x+\pi) = -\cos x, \quad \sin(x+\pi) = -\sin x$$

then follow from the addition formulae. A careful account of the theory, as based on these definitions, will be found in Whittaker and Watson's *Modern analysis*, Appendix A.

This theory is satisfactory enough, but it is more natural when we are considering $\cos z$ and $\sin z$ as functions of a complex variable z than when, as here, we are concerned with real variables and functions only.

(iii) *Definition of the sine by an infinite product.* A third method is to define $\sin x$ by the equation

$$\sin x = x\left(1-\frac{x^2}{\pi^2}\right)\left(1-\frac{x^2}{2^2\pi^2}\right)\left(1-\frac{x^2}{3^2\pi^2}\right)\ldots.$$

This method has many advantages, but naturally demands a knowledge of the theory of infinite products.

(iv) *Definition of the inverse functions by integrals.* There is a fourth method which is preferable here, since it follows the same lines as our treatment of the logarithmic function earlier in this chapter. We begin by defining the inverse tangent of x by the equation

(1) $$y = y(x) = \arctan x = \int_0^x \frac{dt}{1+t^2}.$$

This equation defines a unique value of y corresponding to every real value of x. Since the subject of integration is even, $y(x)$ is an odd function of x. Also, since y is continuous and strictly increasing, there is, by § 110, an inverse function $x = x(y)$, also continuous and strictly increasing. We write

(2) $$x = x(y) = \tan y.$$

If we define π by the equation

(3) $$\tfrac{1}{2}\pi = \int_0^\infty \frac{dt}{1+t^2},$$

then $x(y)$ is defined for $-\frac{1}{2}\pi < y < \frac{1}{2}\pi$.

We now write

(4) $$\cos y = \frac{1}{\sqrt{(1+x^2)}}, \quad \sin y = \frac{x}{\sqrt{(1+x^2)}},$$

where the square root is positive. Thus $\cos y$ and $\sin y$ are defined for $-\frac{1}{2}\pi<y<\frac{1}{2}\pi$. When $y\rightarrow\frac{1}{2}\pi$, $x\rightarrow\infty$, and so $\cos y\rightarrow 0$ and $\sin y\rightarrow 1$. We define $\cos\frac{1}{2}\pi$ and $\sin\frac{1}{2}\pi$ by the equations

$$(5) \qquad \cos\tfrac{1}{2}\pi = 0, \quad \sin\tfrac{1}{2}\pi = 1.$$

Then $\cos y$ and $\sin y$ are defined for $-\frac{1}{2}\pi<y\leqq\frac{1}{2}\pi$ and $\tan y$ for $-\frac{1}{2}\pi<y<\frac{1}{2}\pi$.

Finally, we define $\tan y$, $\cos y$, and $\sin y$, for values of y outside the interval $(-\frac{1}{2}\pi, \frac{1}{2}\pi)$, by the equations

$$(6) \quad \tan(y+\pi)=\tan y, \ \cos(y+\pi)=-\cos y, \ \sin(y+\pi)=-\sin y,$$

which extend our definitions successively to the intervals $(\frac{1}{2}\pi, \frac{3}{2}\pi)$, $(\frac{3}{2}\pi, \frac{5}{2}\pi)$, ..., $(-\frac{3}{2}\pi, -\frac{1}{2}\pi)$, $(-\frac{5}{2}\pi, -\frac{3}{2}\pi)$, ... The tangent is then defined for all values of π except the values $(k+\frac{1}{2})\pi$, where k is an integer. For these values the definition fails; and $\tan y$ tends to $+\infty$ or to $-\infty$ when y tends to one of them, the sign depending on whether y tends to the value in question from below or from above. On the other hand $\cos y$ and $\sin y$ are defined, and continuous, for all values of y.

Thus $\tan y\rightarrow+\infty$ when $y\rightarrow(k+\frac{1}{2})\pi-0$. The sign is reversed by a change of -0 to $+0$.

To see that $\cos y$ is continuous for $y=\frac{1}{2}\pi$ we observe (i) that $\cos\frac{1}{2}\pi=0$ by definition, (ii) that $\cos y\rightarrow 0$ when $y\rightarrow\frac{1}{2}\pi-0$, by (4), (iii) that $\cos y\rightarrow 0$ when $y\rightarrow-\frac{1}{2}\pi+0$, by (4), and therefore when $y\rightarrow\frac{1}{2}\pi+0$, by (6).

We have begun by defining $\operatorname{arc}\tan x$ and $\tan y$, and then defined $\cos y$ and $\sin y$ in terms of $\tan y$. We might have treated $\operatorname{arc}\sin x$ and $\sin y$ as our fundamental functions. In this case we should have defined $\operatorname{arc}\sin x$, in the range $(-1, 1)$ of values of x, by the equation

$$y = y(x) = \operatorname{arc}\sin x = \int_0^x \frac{dt}{\sqrt{(1-t^2)}},$$

where the square root is positive; $\sin y$ by inversion; π by $\frac{1}{2}\pi = \int_0^1 \frac{dt}{\sqrt{(1-t^2)}}$; and $\cos y$ and $\tan y$ by

$$\cos y = \sqrt{(1-x^2)}, \quad \tan y = \frac{x}{\sqrt{(1-x^2)}} \ (-1<x<1).$$

The procedure we have adopted is slightly more convenient.

225. We have now given all the definitions necessary, namely those expressed in § 224 by numbered equations. The further development of the theory depends upon the addition formulae.

We observe first that

$$(1+x^2)(1+y^2) = (1-xy)^2+(x+y)^2$$

and so

$$\frac{dx}{1+x^2}+\frac{dy}{1+y^2} = \frac{(1+y^2)\,dx+(1+x^2)\,dy}{(1-xy)^2+(x+y)^2}$$

$$= \frac{(1-xy)\,d(x+y)-(x+y)\,d(1-xy)}{(1-xy)^2+(x+y)^2} = \frac{dz}{1+z^2},$$

where
$$z = \frac{x+y}{1-xy}.$$

This suggests that

$$\operatorname{arc}\tan x+\operatorname{arc}\tan y = \operatorname{arc}\tan z;$$

but the functions are many valued, and the formula requires more careful examination.

We write
$$t = \frac{x_1+u}{1-x_1u}, \quad u = \frac{t-x_1}{1+x_1t},$$

so that
$$\frac{dt}{du} = \frac{1}{1-x_1u}+\frac{x_1(x_1+u)}{(1-x_1u)^2} = \frac{1+x_1^2}{(1-x_1u)^2} > 0.$$

Then t and u vary always in the same sense. As t increases from $-\infty$ to $-1/x_1$, u increases from $1/x_1$ to ∞, and as t increases from $-1/x_1$ to ∞, u increases from $-\infty$ to $1/x_1$. Also $u = 0$ when $t = x_1$, and $u = -x_1$ when $t = 0$*.

Suppose now that x_2 has any value such that the interval $(-x_1, x_2)$ of values of u does not include the point $u = 1/x_1$, for which t has an infinity. If $x_1 > 0$, x_2 must be less than $1/x_1$, and if $x_1 < 0$, x_2 must be greater than $1/x_1$. In these circumstances t increases or decreases steadily from 0 to

$$x = \frac{x_1+x_2}{1-x_1x_2}$$

when u increases or decreases from $-x_1$ to x_2. Since

$$\frac{1}{1+t^2} = \frac{(1-x_1u)^2}{(1+x_1^2)(1+u^2)},$$

* The reader should draw the graph of each variable as a function of the other.

we have

$$\arctan x = \arctan\frac{x_1+x_2}{1-x_1x_2} = \int_0^x \frac{dt}{1+t^2} = \int_{-x_1}^{x_2} \frac{du}{1+u^2}$$

$$= \int_0^{x_2} \frac{du}{1+u^2} + \int_{-x_1}^{0} \frac{du}{1+u^2} = \int_0^{x_2} \frac{du}{1+u^2} + \int_0^{x_1} \frac{du}{1+u^2}$$

$$= \arctan x_1 + \arctan x_2.$$

If now we write

$$y = \arctan x, \quad y_1 = \arctan x_1, \quad y_2 = \arctan x_2,$$

we have $y = y_1+y_2$ and

$$(1) \qquad \tan(y_1+y_2) = x = \frac{x_1+x_2}{1-x_1x_2} = \frac{\tan y_1 + \tan y_2}{1-\tan y_1 \tan y_2},$$

which is the addition formula for the tangent.

The formula is at present proved only under certain restrictions on the values of the variables, viz. that $x_2 < 1/x_1$ if $x_1 > 0$ and $x_2 > 1/x_1$ if $x_1 < 0$. When $x_1 > 0$ and $x_2 \to 1/x_1$ from below, $x \to \infty$ and $y \to \frac{1}{2}\pi$; and when $x_1 < 0$ and $x_2 \to 1/x_1$ from above, $x \to -\infty$ and $y \to -\frac{1}{2}\pi$. Our restrictions therefore amount to this, that *y_1, y_2, and y_1+y_2 must all lie in the range $(-\frac{1}{2}\pi, \frac{1}{2}\pi)$.*

These restrictions are however unnecessary.

The restriction on y_1+y_2 arose from our supposing that the interval $(-x_1, x_2)$ does not include $1/x_1$. Suppose that this condition is violated, e.g., to fix our ideas, that $x_1 > 0$ and $x_2 > 1/x_1$. Then, when u increases from $-x_2$ to x_1, t increases from 0 to ∞, and then changes sign and increases from $-\infty$ to x. We have thus

$$\int_{-x_1}^{x_2} \frac{du}{1+u^2} = \int_0^{\infty} \frac{dt}{1+t^2} + \int_{-\infty}^{x} \frac{dt}{1+t^2}$$

$$= \int_0^{\infty} \frac{dt}{1+t^2} + \int_{-\infty}^{0} \frac{dt}{1+t^2} + \int_0^{x} \frac{dt}{1+t^2} = \pi + \arctan x.$$

Hence $\qquad \arctan x = \arctan x_1 + \arctan x_2 - \pi,$

and so, by (6),

$$\tan(y_1+y_2) = \tan(y_1+y_2-\pi) = \tan y$$

$$= \frac{x_1+x_2}{1-x_1x_2} = \frac{\tan y_1 + \tan y_2}{1-\tan y_1 \tan y_2}.$$

We may deal similarly with the case in which $x_1 < 0$; and it follows that (1) is valid whenever y_1 and y_2 lie in $(-\frac{1}{2}\pi, \frac{1}{2}\pi)$.

Finally, since each side of (1) is, by (6), a periodic function of y_1 or of y_2, (1) holds without reservation, except when y_1, y_2, or y_1+y_2 is an odd multiple of $\frac{1}{2}\pi$, in which case it ceases to have a meaning.

226. From (1) of § 225 and (4) of § 224 we deduce

$$\cos^2(y_1+y_2) = \frac{(1-\tan y_1 \tan y_2)^2}{(1+\tan^2 y_1)(1+\tan^2 y_2)}$$

$$= (\cos y_1 \cos y_2 - \sin y_1 \sin y_2)^2,$$

and $$\cos(y_1+y_2) = \pm(\cos y_1 \cos y_2 - \sin y_1 \sin y_2).$$

To determine the sign put $y_2 = 0$. The equation reduces to $\cos y_1 = \pm \cos y_1$, so that the positive sign must be chosen when $y_2 = 0$. Since both sides change sign when y_2 is increased by π, the formula holds, with the positive sign, when y_2 is any multiple of π. Further, both sides are continuous functions of y_2, so that the sign could change only when each side vanishes, that is to say for the values $\dots, -\frac{1}{2}\pi - y_1, \frac{1}{2}\pi - y_1, \frac{3}{2}\pi - y_1, \dots$, one each in every interval of length π. Since we have seen that there is, in each such interval, a value of y_2 for which the sign is positive, it follows that it must be always positive. Hence

(2) $$\cos(y_1+y_2) = \cos y_1 \cos y_2 - \sin y_1 \sin y_2;$$

and the formula for $\sin(y_1+y_2)$ can be proved similarly.

MISCELLANEOUS EXAMPLES ON CHAPTER IX*

1. Given that $\log_{10} e = \cdot 4343$ and that 2^{10} and 3^{21} are nearly equal to powers of 10, calculate $\log_{10} 2$ and $\log_{10} 3$ to four places of decimals. (*Math. Trip.* 1905)

2. Show that $\log_{10} n$ cannot be a rational number if n is any positive integer not a power of 10. [If n is not divisible by 10, and $\log_{10} n = p/q$, we have $10^p = n^q$, which is impossible, since 10^p ends with 0 and n^q does

* A number of these examples are taken from Bromwich's *Infinite series*.

not. If $n = 10^a N$, where N is not divisible by 10, then $\log_{10} N$, and therefore

$$\log_{10} n = a + \log_{10} N,$$

cannot be rational.]

3. For what values of x are the functions $\log x$, $\log\log x$, $\log\log\log x$, ... (*a*) equal to 0, (*b*) equal to 1, (*c*) not defined? Consider also the same question for the functions lx, llx, $lllx$, ..., where $lx = \log|x|$.

4. Show that

$$\log x - \binom{n}{1}\log(x+1) + \binom{n}{2}\log(x+2) - \ldots + (-1)^n \log(x+n)$$

is negative and increases steadily towards 0 as x increases from 0 towards ∞.

[The derivative of the function is

$$\sum_1^n (-1)^r \binom{n}{r} \frac{1}{x+r} = \frac{n!}{x(x+1)\ldots(x+n)},$$

as is easily seen by splitting up the right-hand side into partial fractions. This expression is positive, and the function itself tends to zero as $x \to \infty$, since $\log(x+r) = \log x + o(1)$ and $1 - \binom{n}{1} + \binom{n}{2} - \ldots = 0$.]

5. Prove that

$$\left(\frac{d}{dx}\right)^n \frac{\log x}{x} = \frac{(-1)^n n!}{x^{n+1}}\left(\log x - 1 - \frac{1}{2} - \ldots - \frac{1}{n}\right).$$

(*Math. Trip.* 1909)

6. If $x > -1$ then $x^2 > (1+x)\{\log(1+x)\}^2$. (*Math. Trip.* 1906)

[Put $1 + x = e^\xi$, and use the fact that $\sinh\xi > \xi$ when $\xi > 0$.]

7. Show that $\dfrac{\log(1+x)}{x}$ and $\dfrac{x}{(1+x)\log(1+x)}$ both decrease steadily as x increases from 0 towards ∞.

8. Show that, as x increases from -1 towards ∞, the function $(1+x)^{-1/x}$ assumes once and only once every value between 0 and 1.

(*Math. Trip.* 1910)

9. Show that $\dfrac{1}{\log(1+x)} - \dfrac{1}{x} \to \dfrac{1}{2}$ as $x \to 0$.

10. Show that $\dfrac{1}{\log(1+x)} - \dfrac{1}{x}$ decreases steadily from 1 to 0 as x increases from -1 towards ∞. [The function is undefined when $x = 0$, but if we attribute to it the value $\frac{1}{2}$ when $x = 0$ it becomes continuous for $x = 0$. Use Ex. 6 to show that the derivative is negative.]

11. Prove that $\psi(x) = \frac{1}{2}\sin x \tan x - \log\sec x$

is positive and increasing for $0<x<\frac{1}{2}\pi$, and that $\psi(x) = O(x^6)$ for small x. (*Math. Trip.* 1930)

12. If
$$\phi(x) = \frac{3\int_0^x (1+\sec t)\log\sec t\,dt}{\log\sec x\{x+\log(\sec x+\tan x)\}}$$
then (i) $\phi(x)$ is even; (ii) $\phi(x)=1+\frac{1}{120}x^4$, approximately, for small x; and (iii) $\phi(x)\to\frac{3}{2}$ when $x\to\frac{1}{2}\pi$ through values less than $\frac{1}{2}\pi$. (*Math. Trip.* 1930)

13. Show that $e^x>Mx^N$, where M and N are large positive numbers, if x is greater than the greater of $2\log M$ and $16N^2$.

[It is easy to prove that $\log x<2\sqrt{x}$; and so the inequality given is certainly satisfied if
$$x>\log M+2N\sqrt{x},$$
and therefore certainly satisfied if $\frac{1}{2}x>\log M$, $\frac{1}{2}x>2N\sqrt{x}$.]

14. Show that the sequence
$$a_1 = e,\quad a_2 = e^{e^2},\quad a_3 = e^{e^{e^3}},\quad \dots$$
tends to infinity more rapidly than any member of the exponential scale.

[Let $e_1(x) = e^x$, $e_2(x) = e^{e_1(x)}$, and so on. Then, if $e_k(x)$ is any member of the exponential scale, $a_n>e_k(n)$ when $n>k$.]

15. If p and q are positive integers then
$$\frac{1}{pn+1}+\frac{1}{pn+2}+\dots+\frac{1}{qn}\to\log\left(\frac{q}{p}\right)$$
as $n\to\infty$. [Cf. Ex. LXXVIII. 7.]

16. Prove that if x is positive then $n\log\{\frac{1}{2}(1+x^{1/n})\}\to-\frac{1}{2}\log x$ as $n\to\infty$. [We have
$$n\log\{\tfrac{1}{2}(1+x^{1/n})\} = n\log\{1-\tfrac{1}{2}(1-x^{1/n})\} = \tfrac{1}{2}n(1-x^{1/n})\frac{\log(1-u)}{u}$$
where $u = \frac{1}{2}(1-x^{1/n})$. Now use § 216 and Ex. LXXXIII. 3.]

17. Prove that if a and b are positive then
$$\{\tfrac{1}{2}(a^{1/n}+b^{1/n})\}^n\to\sqrt{(ab)}.$$
[Take logarithms and use Ex. 16.]

18. Show that
$$1+\frac{1}{3}+\frac{1}{5}+\dots+\frac{1}{2n-1} = \tfrac{1}{2}\log n+\log 2+\tfrac{1}{2}\gamma+o(1),$$
where γ is Euler's constant (Ex. XC. 1).

19. Show that

$$1+\tfrac{1}{3}-\tfrac{1}{2}+\tfrac{1}{5}+\tfrac{1}{7}-\tfrac{1}{4}+\tfrac{1}{9}+\ldots=\tfrac{3}{2}\log 2,$$

the series being formed from the series $1-\frac{1}{2}+\frac{1}{3}-\ldots$ by taking alternately two positive terms and then one negative. [The sum of the first $3n$ terms is

$$1+\frac{1}{3}+\frac{1}{5}+\ldots+\frac{1}{4n-1}-\frac{1}{2}\left(1+\frac{1}{2}+\ldots+\frac{1}{n}\right)$$
$$=\tfrac{1}{2}\log 2n+\log 2+\tfrac{1}{2}\gamma+o(1)-\tfrac{1}{2}\{\log n+\gamma+o(1)\}.]$$

20. Prove that

$$\sum_1^n \frac{1}{\nu(36\nu^2-1)}=-3+3\,\Sigma_{3n+1}-\Sigma_n-S_n,$$

where $S_n=1+\dfrac{1}{2}+\ldots+\dfrac{1}{n}$, $\Sigma_n=1+\dfrac{1}{3}+\ldots+\dfrac{1}{2n-1}$. Hence prove that the sum of the series when continued to infinity is

$$-3+\tfrac{3}{2}\log 3+2\log 2. \qquad \textit{(Math. Trip. 1905)}$$

21. Prove that the sums of the four series

$$\sum_1^\infty \frac{1}{4n^2-1},\quad \sum_1^\infty \frac{(-1)^{n-1}}{4n^2-1},\quad \sum_1^\infty \frac{1}{(2n+1)^2-1},\quad \sum_1^\infty \frac{(-1)^{n-1}}{(2n+1)^2-1}$$

are $\frac{1}{2}$, $\frac{1}{4}\pi-\frac{1}{2}$, $\frac{1}{4}$, $\frac{1}{2}\log 2-\frac{1}{4}$ respectively.

22. Investigate the convergence or divergence of

$$\Sigma\left(1-\frac{x\log n}{n}\right)^n,\quad \Sigma(\log n)^{-x\log n},\quad \Sigma\left(\log 2-\sum_{n+1}^{2n}\frac{1}{\nu}\right)^x.$$

(*Math. Trip.* 1935)

23. Examine the convergence or divergence of

$$\Sigma n^{-a}e^{-b\sqrt{n}+cni}$$

for all real values of a, b, c. (*Math. Trip.* 1925)

24. The series Σu_n is rearranged in the form

$$u_1+u_2+u_4+u_3+u_5+u_7+u_9+u_6+u_8+\ldots+u_{20}+u_{11}+\ldots$$

(one term of odd rank, then two of even, then four of odd, then eight of even, ...). Examine the convergence or divergence of the rearranged series when

$$(1)\ u_n=\frac{(-1)^{n-1}}{n},\qquad (2)\ u_n=\frac{(-1)^{n-1}}{n\log(n+1)}.$$

(*Math. Trip.* 1930)

25. Prove that $n!(a/n)^n$ tends to 0 or to ∞ according as $a<e$ or $a>e$.

26. Prove that if $u_n = n!e^n n^{-n-\frac{1}{2}}$ then

$$\frac{u_n}{u_{n+1}} = 1 + O\left(\frac{1}{n^2}\right).$$

Deduce that, if a is fixed, and s is the integer nearest to $a\sqrt{n}$, then

$$\binom{2n}{n+s} \Big/ \binom{2n}{n} \to e^{-a^2}. \qquad \textit{(Math. Trip. 1928)}$$

27. If $u_n > 0$ and $\dfrac{u_{n+1}}{u_n} = 1 - \dfrac{a}{n} + O\left(\dfrac{1}{n^2}\right)$, then $u_n \sim Kn^{-a}$, where K is a constant. [For

$$\log \frac{u_{n+1}}{u_n} = -\frac{a}{n} + \rho_n,$$

where $\rho_n = O(n^{-2})$. Hence

$$\log \frac{u_n}{u_1} = -a \sum_1^{n-1} \frac{1}{\nu} + \sum_1^{n-1} \rho_\nu = -a(\log n + \gamma) + H + o(1),$$

where $H = \Sigma \rho_\nu$.]

28. Prove that

$$\frac{(a+1)(a+2)\dots(a+n)}{(b+1)(b+2)\dots(b+n)} \sim Kn^{a-b},$$

where K is a constant. [Use Ex. 27].

29. Prove that, in the notation of Ex. XC. 6, $\Sigma(u_n/s_n)$ converges or diverges with Σu_n. [The proof is the same in the case of convergence. If Σu_n is divergent, and $u_n < s_{n-1}$ from a certain value of n onwards, then $s_n < 2s_{n-1}$, and the divergence of $\Sigma(u_n/s_n)$ follows from that of $\Sigma(u_n/s_{n-1})$. If on the other hand $u_n \geqq s_{n-1}$ for an infinity of values of n, as might happen with a rapidly divergent series, then $u_n/s_n \geqq \frac{1}{2}$ for all these values of n.]

30. Prove that if $x > -1$ then

$$\frac{1}{(x+1)^2} = \frac{1}{(x+1)(x+2)} + \frac{1!}{(x+1)(x+2)(x+3)}$$
$$+ \frac{2!}{(x+1)(x+2)(x+3)(x+4)} + \dots. \qquad \textit{(Math. Trip. 1908)}$$

[The difference between $1/(x+1)^2$ and the sum of the first n terms of the series is

$$\frac{1}{(x+1)^2} \frac{n!}{(x+2)(x+3)\dots(x+n+1)}.]$$

31. Find the limit as $x \to \infty$ of

$$\left(\frac{a_0 + a_1 x + \dots + a_r x^r}{b_0 + b_1 x + \dots + b_r x^r}\right)^{\lambda_0 + \lambda_1 x},$$

distinguishing the different cases which may arise. *(Math. Trip. 1886)*

32. The general solution of $f(xy) = f(x)f(y)$, where f is a differentiable function, is x^a, where a is a constant: and that of

$$f(x+y)+f(x-y) = 2f(x)f(y)$$

is $\cosh ax$ or $\cos ax$, according as $f''(0)$ is positive or negative. [In proving the second result assume that f has derivatives of the first three orders. Then

$$2f(x)+y^2f''(x)+o(y^2) = 2f(x)\{f(0)+yf'(0)+\tfrac{1}{2}y^2f''(0)+o(y^2)\},$$

and therefore $f(0) = 1$, $f'(0) = 0$ and $f''(x) = f''(0)f(x)$.]

33. The equation $e^x = ax+b$ has one real root if $a<0$ or $a = 0$, $b>0$. If $a>0$ then it has two real roots or none, according as $a\log a>b-a$ or $a\log a<b-a$.

34. Show by graphical considerations that the equation

$$e^x = ax^2+2bx+c$$

has one, two, or three real roots if $a>0$, none, one, or two if $a<0$; and show how to distinguish between the different cases.

35. Prove that the equation $a^2e^x = x^2$ has three real roots if $a^2<4e^{-2}$, and that, when a is small, the small positive root is

$$a+\tfrac{1}{2}a^2+\tfrac{3}{8}a^3+\dots. \qquad (\textit{Math. Trip.}\ 1931)$$

36. Find the equation giving the values of x for which

$$y = Ae^{-x^2}+Be^{-(x-c)^2}$$

is stationary, and prove that the value of y corresponding to such a value x_1 of x is

$$\frac{Ac}{c-x_1}e^{-x_1^2}.$$

Show also that, when A, B, c are positive, the equation has just two roots, one greater than c and the other negative; and that they correspond to a minimum and a maximum respectively. (*Math. Trip.* 1923)

37. Trace the curve $y = \dfrac{1}{x}\log\left(\dfrac{e^x-1}{x}\right)$, showing that the point $(0, \tfrac{1}{2})$ is a centre of symmetry, and that as x increases through all real values, y steadily increases from 0 to 1. Deduce that the equation

$$\frac{1}{x}\log\left(\frac{e^x-1}{x}\right) = \alpha$$

has no real root unless $0<\alpha<1$, and then one, whose sign is the same as that of $\alpha-\tfrac{1}{2}$. [In the first place

$$y-\tfrac{1}{2} = \frac{1}{x}\left\{\log\left(\frac{e^x-1}{x}\right)-\log e^{\frac{1}{2}x}\right\} = \frac{1}{x}\log\left(\frac{\sinh\frac{1}{2}x}{\frac{1}{2}x}\right)$$

is clearly an odd function of x. Also

$$\frac{dy}{dx} = \frac{1}{x^2}\left\{\tfrac{1}{2}x\coth\tfrac{1}{2}x - 1 - \log\left(\frac{\sinh\frac{1}{2}x}{\frac{1}{2}x}\right)\right\}.$$

The function inside the large bracket tends to zero as $x \to 0$; and its derivative is

$$\frac{1}{x}\left\{1 - \left(\frac{\frac{1}{2}x}{\sinh\frac{1}{2}x}\right)^2\right\},$$

which has the sign of x. Hence $dy/dx > 0$ for all values of x.]

38. Trace the curve $y = e^{1/x}\sqrt{(x^2+2x)}$, and show that the equation

$$e^{1/x}\sqrt{(x^2+2x)} = \alpha$$

has no real roots if α is negative, one negative root if

$$0 < \alpha < a = e^{1/\sqrt{2}}\sqrt{(2+2\sqrt{2})},$$

and two positive roots and one negative if $\alpha > a$.

39. Show that the equation $f_n(x) = 1 + x + \dfrac{x^2}{2!} + \ldots + \dfrac{x^n}{n!} = 0$ has one real root if n is odd and none if n is even.

[Assume this proved for $n = 1, 2, \ldots, 2k$. Then $f_{2k+1}(x) = 0$ has at least one real root, since its degree is odd, and it cannot have more since, if it had, $f'_{2k+1}(x)$ or $f_{2k}(x)$ would have to vanish once at least. Hence $f_{2k+1}(x) = 0$ has just one root, and so $f_{2k+2}(x) = 0$ cannot have more than two. If it has two, say α and β, then $f'_{2k+2}(x)$ or $f_{2k+1}(x)$ must vanish once at least between α and β, say at γ; and

$$f_{2k+2}(\gamma) = f_{2k+1}(\gamma) + \frac{\gamma^{2k+2}}{(2k+2)!} > 0.$$

But $f_{2k+2}(x)$ is also positive when x is large (positively or negatively), and a glance at a figure will show that these results are contradictory. Hence $f_{2k+2}(x) = 0$ has no real roots.]

40. Prove that if a and b are positive and nearly equal then

$$\log\frac{a}{b} = \tfrac{1}{2}(a-b)\left(\frac{1}{a} + \frac{1}{b}\right),$$

approximately, the error being about $\frac{1}{6}(a-b)^3 a^{-3}$. [Use the logarithmic series. This formula is interesting historically as having been employed by Napier for the numerical calculation of logarithms.]

41. Prove by multiplication of series that if $-1 < x < 1$ then

$$\tfrac{1}{2}\{\log(1+x)\}^2 = \tfrac{1}{2}x^2 - \tfrac{1}{3}(1+\tfrac{1}{2})x^3 + \tfrac{1}{4}(1+\tfrac{1}{2}+\tfrac{1}{3})x^4 - \ldots,$$

$$\tfrac{1}{2}(\arctan x)^2 = \tfrac{1}{2}x^2 - \tfrac{1}{4}(1+\tfrac{1}{3})x^4 + \tfrac{1}{6}(1+\tfrac{1}{3}+\tfrac{1}{5})x^6 - \ldots.$$

42. The first $n+2$ terms in the expansion of $\log\left(1+x+\frac{x^2}{2!}+\ldots+\frac{x^n}{n!}\right)$ in powers of x are

$$x-\frac{x^{n+1}}{n!}\left\{\frac{1}{n+1}-\frac{x}{1!(n+2)}+\frac{x^2}{2!(n+3)}-\ldots+(-1)^n\frac{x^n}{n!(2n+1)}\right\}.$$

(*Math. Trip.* 1899)

43. Show that the expansion of $\exp\left(-x-\frac{x^2}{2}-\ldots-\frac{x^n}{n}\right)$ in powers of x begins with the terms

$$1-x+\frac{x^{n+1}}{n+1}-\sum_{s=1}^{n}\frac{x^{n+s+1}}{(n+s)(n+s+1)}.$$

(*Math. Trip.* 1909)

44. Use the identity

$$\log(1-x^3)=\log(1-x)+\log(1+x+x^2)$$

to prove that

$$\sum_{\frac{1}{2}k\leqq n\leqq k}\frac{(-1)^{n-1}(n-1)!}{(k-n)!(2n-k)!}$$

is k^{-1} if k is not a multiple of 3 and $-2k^{-1}$ if it is. (*Math. Trip.* 1932)

45. Prove that if x is small and y is the positive value of $(1+x+x^2)^{x^{-2}}$, then

$$y=e^{x^{-1}+\frac{1}{2}}\{1-\tfrac{2}{3}x+O(x^2)\}.$$

Find the limits of y and $\frac{dy}{dx}$ when $x\to 0$ by positive and by negative values, and sketch the graph of y near $x=0$. (*Math. Trip.* 1924)

46. Prove that $\int_0^\infty\frac{dx}{(x+a)(x+b)}=\frac{1}{a-b}\log\left(\frac{a}{b}\right)$ if $a>b>0$.

47. Prove that if α, β, γ are all positive, and $\beta^2>\alpha\gamma$, then

$$\int_0^\infty\frac{dx}{\alpha x^2+2\beta x+\gamma}=\frac{1}{\sqrt{(\beta^2-\alpha\gamma)}}\log\left\{\frac{\beta+\sqrt{(\beta^2-\alpha\gamma)}}{\sqrt{(\alpha\gamma)}}\right\};$$

and evaluate the integral when $\alpha>0$ and $\alpha\gamma>\beta^2$.

48. Prove that if $a>-1$ then

$$\int_1^\infty\frac{dx}{(x+a)\sqrt{(x^2-1)}}=\int_0^\infty\frac{dt}{\cosh t+a}=2\int_1^\infty\frac{du}{u^2+2au+1};$$

and deduce that the value of the integral is

$$\frac{2}{\sqrt{(1-a^2)}}\arctan\sqrt{\left(\frac{1-a}{1+a}\right)}$$

if $-1<a<1$, and

$$\frac{1}{\sqrt{(a^2-1)}}\log\frac{\sqrt{(a+1)}+\sqrt{(a-1)}}{\sqrt{(a+1)}-\sqrt{(a-1)}}=\frac{2}{\sqrt{(a^2-1)}}\operatorname{arg\,tanh}\sqrt{\left(\frac{a-1}{a+1}\right)}$$

if $a>1$. Discuss the case in which $a=1$.

49. If $0<\alpha<1$, $0<\beta<1$, then

$$\int_{-1}^{1}\frac{dx}{\sqrt{\{(1-2\alpha x+\alpha^2)(1-2\beta x+\beta^2)\}}}=\frac{1}{\sqrt{(\alpha\beta)}}\log\frac{1+\sqrt{(\alpha\beta)}}{1-\sqrt{(\alpha\beta)}}.$$

50. Prove that if $a>b>0$ then

$$\int_{-\infty}^{\infty}\frac{d\theta}{a\cosh\theta+b\sinh\theta}=\frac{\pi}{\sqrt{(a^2-b^2)}}.$$

51. Prove that

$$\int_0^1 x\log(1+\tfrac{1}{2}x)\,dx=\tfrac{3}{4}-\tfrac{3}{2}\log\tfrac{3}{2}<\tfrac{1}{2}\int_0^1 x^2\,dx=\tfrac{1}{6},$$

$$\int_1^{\infty}\frac{\log x}{x^n}\,dx=\frac{1}{(n-1)^2}\quad(n>1),\qquad \int_0^{\infty}\frac{dx}{\{x+\sqrt{(x^2+1)}\}^n}=\frac{n}{n^2-1}\quad(n>1),$$

$$\int_{\frac{1}{2}}^{1}\frac{dx}{x^4\sqrt{(1-x^2)}}=2\sqrt{3},\qquad \int_1^{\infty}\frac{dx}{(x+1)^2(x^2+1)}=\tfrac{1}{4}(1-\log 2),$$

$$\int_0^{\infty}\frac{dx}{(1+e^x)(1+e^{-x})}=1.$$

(*Math. Trip.* 1913, 1928, 1932, 1933, 1934)

52. Prove that

$$\int_0^1\frac{\log x}{1+x^2}\,dx=-\int_1^{\infty}\frac{\log x}{1+x^2}\,dx,\qquad \int_0^{\infty}\frac{\log x}{1+x^2}\,dx=0;$$

and deduce that if $a>0$ then

$$\int_0^{\infty}\frac{\log x}{a^2+x^2}\,dx=\frac{\pi}{2a}\log a.$$

[Use the substitutions $x=1/t$ and $x=au$.]

53. Prove that $\int_0^{\infty}\log\left(1+\frac{a^2}{x^2}\right)dx=\pi a$ if $a>0$. [Integrate by parts.]

54. Prove that

$$\lim_{t\to 1-0}(1-t)^{\frac{1}{2}}(t+t^4+t^9+t^{16}+\dots)=\int_0^{\infty}e^{-x^2}\,dx.$$

(*Math. Trip.* 1932)

[It follows from § 180 that

$$\int_h^{(n+1)h}e^{-x^2}\,dx<h\sum_{\nu=1}^{n}e^{-\nu^2h^2}<\int_0^{nh}e^{-x^2}\,dx.$$

Put $t=e^{-h^2}$ and make $n\to\infty$.]

CHAPTER X

THE GENERAL THEORY OF THE LOGARITHMIC, EXPONENTIAL, AND CIRCULAR FUNCTIONS

227. Functions of a complex variable. In Ch. III we defined the complex variable

$$z = x + iy,*$$

and we considered a few simple properties of some classes of expressions involving z, such as the polynomial $P(z)$. It is natural to describe such expressions as *functions* of z, and in fact we did describe the quotient $P(z)/Q(z)$, where $P(z)$ and $Q(z)$ are polynomials, as 'rational function'. We have however given no general definition of what is meant by a function of z.

It might seem natural to define a function of z in the same way as that in which we defined a function of the real variable x, i.e. to say that Z is a function of z if there is any relation between z and Z in virtue of which a value or values of Z corresponds to some or all values of z. But it will be found, on closer examination, that this defini ion is not one from which any profit can be derived. For if z is given, so are x and y, and conversely: to assign a value of z is just the same thing as to assign a pair of values of x and y. Thus a 'function of z', according to the definition suggested, is merely *a complex function*

$$f(x, y) + ig(x, y),$$

of the two real variables x and y. For example

$$x - iy, \quad xy, \quad |z| = \sqrt{(x^2 + y^2)}, \quad \operatorname{am} z = \arctan(y/x)$$

are 'functions of z'. The definition, although quite legitimate, is futile because it does not really define a new idea at all.

* In this chapter we shall generally find it convenient to write $x + iy$ rather than $x + yi$.

It is therefore more convenient to use the expression 'function of the complex variable z' in a more restricted sense, or in other words to pick out, from the general class of complex functions of the two real variables x and y, a special class to which the expression shall be restricted. If we were to explain how this selection is made, and what are the characteristic properties of the special class of functions selected, we should be led far beyond the limits of this book. We shall therefore not attempt to give any general definitions, but shall confine ourselves entirely to special functions defined directly.

228. We have already defined *polynomials* in z (§ 39), *rational functions* of z (§ 46), and *roots* of z (§ 47). There is no difficulty in extending to the complex variable the definitions of *algebraical functions*, explicit and implicit, which we gave (§§ 26–27) in the case of the real variable x. In all these cases we shall call the complex number z, the argument (§ 44) of the point z, the *argument* of the function $f(z)$ under consideration. The question which will occupy us in this chapter is that of defining, and determining the principal properties of, the logarithmic, exponential, and trigonometrical or circular functions of z. These functions are so far defined for real values of z only, the logarithm indeed for positive values only.

We shall begin with the logarithmic function. It is natural to attempt to define it by means of some extension of the definition

$$\log x = \int_1^x \frac{dt}{t} \qquad (x > 0);$$

and in order to do this we shall find it necessary to consider briefly some extensions of the notion of an integral.

229. Real and complex curvilinear integrals. Let AB be an arc C of a curve defined by the equations

$$x = \phi(t), \quad y = \psi(t),$$

where ϕ and ψ are functions of t with continuous differential coefficients ϕ' and ψ'; and suppose that, as t varies from t_0 to t_1, the point (x, y) moves along the curve, in the same direction, from A to B.

Then we define the *curvilinear integral*

$$\int_C \{g(x,y)\,dx + h(x,y)\,dy\} \quad\text{.................(1)},$$

where g and h are continuous functions of x and y, as the ordinary integral obtained by effecting the formal substitutions $x = \phi(t)$, $y = \psi(t)$, i.e. as

$$\int_{t_0}^{t_1} \{g(\phi,\psi)\,\phi' + h(\phi,\psi)\,\psi'\}\,dt.$$

We call C the *path of integration.*

Let us suppose now that

$$z = x + iy = \phi(t) + i\psi(t),$$

so that z describes the curve C in Argand's diagram as t varies. Further let us suppose that

$$f(z) = u + iv$$

is a polynomial in z or rational function of z. Then we define

$$\int_C f(z)\,dz \quad\text{..........................(2)}$$

as meaning $\int_C (u+iv)\,(dx + i\,dy)$,

which is itself defined as meaning

$$\int_C (u\,dx - v\,dy) + i\int_C (v\,dx + u\,dy).$$

230. The definition of $\operatorname{Log}\zeta$**.** Now let $\zeta = \xi + i\eta$ be any complex number. We define $\operatorname{Log}\zeta$, the general logarithm of ζ, by the equation

$$\operatorname{Log}\zeta = \int_C \frac{dz}{z},$$

where C is a curve which starts from 1 and ends at ζ and does not pass through the origin. Thus (Fig. 51) the paths (*a*), (*b*), (*c*) are paths such as are contemplated in the definition. The value of $\operatorname{Log}\zeta$ is thus defined when the particular path of integration has been chosen. But at present it is not clear how far the value of $\operatorname{Log}\zeta$ resulting from the definition depends upon what path is chosen. Suppose for example that ζ is real and positive, say

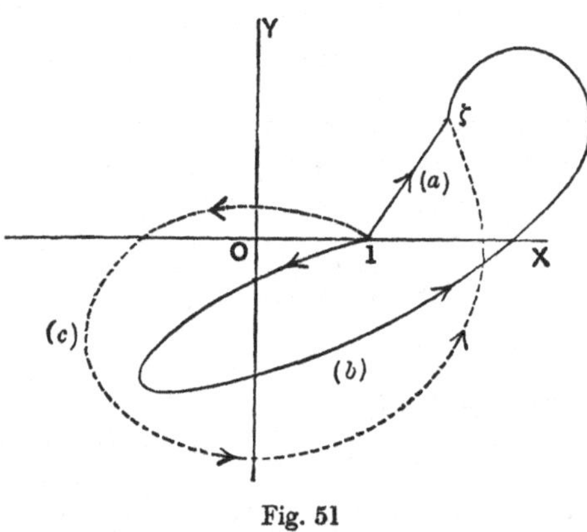

Fig. 51

equal to ξ. Then one possible path of integration is the straight line from 1 to ξ, a path which we may suppose to be defined by the equations $x = t$, $y = 0$. In this case, and with this particular choice of the path of integration, we have

$$\operatorname{Log}\xi = \int_1^\xi \frac{dt}{t},$$

so that $\operatorname{Log}\xi$ is equal to $\log\xi$, the logarithm of ξ according to the definition given in the last chapter. Thus one value at any rate of $\operatorname{Log}\xi$, when ξ is real and positive, is $\log\xi$. But in this case, as in the general case, the path of integration can be chosen in an infinite variety of different ways. There is nothing to show that *every* value of $\operatorname{Log}\xi$ is equal to $\log\xi$; and in fact we shall see that this is not so. This is why we have adopted the notation $\operatorname{Log}\zeta$, $\operatorname{Log}\xi$ instead of $\log\zeta$, $\log\xi$. $\operatorname{Log}\xi$ is (possibly at any rate) a many-valued function, and $\log\xi$ is only one of its values. And in the general case, so far as we can see at present, three alternatives are possible, viz. that

(1) we may always get the same value of $\operatorname{Log}\zeta$, by whatever path we go from 1 to ζ;

(2) we may get a different value corresponding to every different path;

(3) we may get a number of different values each of which corresponds to a whole class of paths:

and the truth or falsehood of any one of these alternatives is in no way implied by our definition.

231. The values of Log ζ. Let us suppose that the polar coordinates of the point $z = \zeta$ are ρ and ϕ, so that

$$\zeta = \rho(\cos\phi + i\sin\phi).$$

We suppose for the present that $-\pi < \phi < \pi$, while ρ may have any positive value. Thus ζ may have any value other than zero or a real negative value.

The coordinates (x, y) of any point on the path C are functions of t, and so also are its polar coordinates (r, θ). Also

$$\operatorname{Log}\zeta = \int_C \frac{dz}{z} = \int_C \frac{dx + i\,dy}{x + iy}$$

$$= \int_{t_0}^{t_1} \frac{1}{x+iy}\left(\frac{dx}{dt} + i\frac{dy}{dt}\right) dt,$$

in virtue of the definitions of § 229. But $x = r\cos\theta$, $y = r\sin\theta$, and

$$\frac{dx}{dt} + i\frac{dy}{dt} = \left(\cos\theta\frac{dr}{dt} - r\sin\theta\frac{d\theta}{dt}\right) + i\left(\sin\theta\frac{dr}{dt} + r\cos\theta\frac{d\theta}{dt}\right)$$

$$= (\cos\theta + i\sin\theta)\left(\frac{dr}{dt} + ir\frac{d\theta}{dt}\right);$$

so that $$\operatorname{Log}\zeta = \int_{t_0}^{t_1} \frac{1}{r}\frac{dr}{dt}\,dt + i\int_{t_0}^{t_1} \frac{d\theta}{dt}\,dt = [\log r] + i[\theta],$$

where $[\log r]$ denotes the difference between the values of $\log r$ at the points corresponding to $t = t_1$ and $t = t_0$, and $[\theta]$ has a similar meaning.

It is clear that

$$[\log r] = \log\rho - \log 1 = \log\rho;$$

but the value of $[\theta]$ requires a little more consideration. Let us suppose first that the path of integration is the straight line from

1 to ζ. The initial value of θ is the amplitude of 1, or rather one of the amplitudes of 1, viz. $2k\pi$, where k is any integer. Let us suppose that initially $\theta = 2k\pi$. It is evident from the figure that θ increases from $2k\pi$ to $2k\pi+\phi$ as t moves along the line. Thus

$$[\theta] = (2k\pi+\phi)-2k\pi = \phi,$$

and, when the path of integration is a straight line,

$$\operatorname{Log}\zeta = \log\rho+i\phi.$$

We shall call this particular value of $\operatorname{Log}\zeta$ the *principal value*. When ζ is real and positive, $\zeta = \rho$ and $\phi = 0$, so that the principal value of $\operatorname{Log}\zeta$ is the ordinary logarithm $\log\zeta$. Hence it

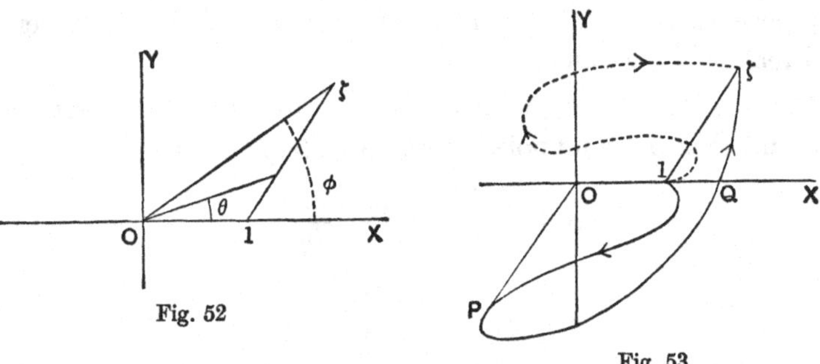

Fig. 52

Fig. 53

will be convenient generally to denote the principal value of $\operatorname{Log}\zeta$ by $\log\zeta$. Thus

$$\log\zeta = \log\rho+i\phi,$$

and the principal value is characterised by the fact that its imaginary part lies between $-\pi$ and π.

Next let us consider any path such that the area included between the path and the straight line from 1 to ζ does not include the origin: two such paths are shown in Fig. 53. It is easy to see that $[\theta]$ is still equal to ϕ. Along the curve shown in the figure by a continuous line, for example, θ, initially equal to $2k\pi$, first decreases to the value

$$2k\pi-XOP$$

and then increases again, being equal to $2k\pi$ at Q and finally to $2k\pi+\phi$. The dotted curve shows a similar but slightly more complicated case in which the straight line and the curve bound two areas, neither of which includes the origin. Thus *if the path of integration is such that the closed curve formed by it and the line from* 1 *to* ζ *does not include the origin, then*

$$\operatorname{Log} \zeta = \log \zeta = \log \rho + i\phi.$$

On the other hand it is easy to construct paths of integration such that $[\theta]$ is not equal to ϕ. Consider, for example, the curve indicated by a continuous line in Fig. 54. If θ is initially equal to $2k\pi$, it will have increased by 2π when we get to P and by 4π when we get to Q; and its final value will be $2k\pi + 4\pi + \phi$, so that $[\theta] = 4\pi + \phi$ and

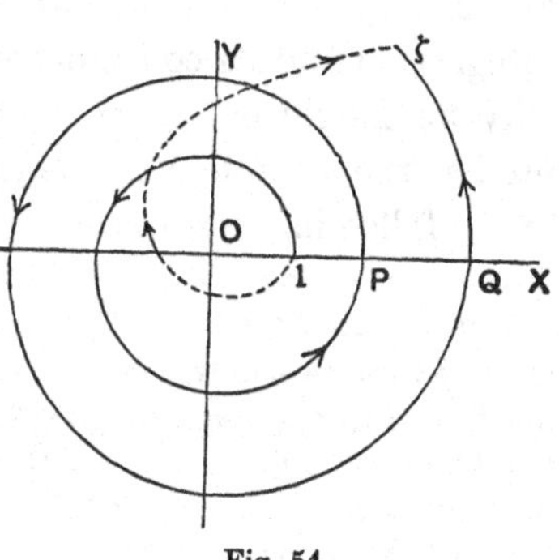

Fig. 54

$$\operatorname{Log} \zeta = \log \rho + i(4\pi + \phi).$$

In this case the path of integration winds twice round the origin in the positive sense. If we had taken a path winding k times round the origin we should have found in the same way that $[\theta] = 2k\pi + \phi$ and

$$\operatorname{Log} \zeta = \log \rho + i(2k\pi + \phi).$$

Here k is positive. By making the path wind round the origin in the opposite direction (as shown in the dotted path in Fig. 54), we obtain a similar series of values in which k is negative. Since $|\zeta| = \rho$, and the different angles $2k\pi + \phi$ are the different values of $\operatorname{am} \zeta$, we conclude that every value of $\log |\zeta| + i \operatorname{am} \zeta$ is a value of $\operatorname{Log} \zeta$; and it is clear from the preceding discussion that every value of $\operatorname{Log} \zeta$ must be of this form.

We may summarise our conclusions as follows: *the general value of* $\operatorname{Log} \zeta$ *is*

$$\log |\zeta| + i \operatorname{am} \zeta = \log \rho + i(2k\pi + \phi),$$

where k is any positive or negative integer. The value of k is *determined by the path of integration chosen. If this path is a straight line then $k = 0$ and*

$$\operatorname{Log}\zeta = \log\zeta = \log\rho + i\phi.$$

In what precedes we have used ζ to denote the argument of the function $\operatorname{Log}\zeta$, and (ξ, η) or (ρ, ϕ) to denote the coordinates of ζ; and z, (x, y), (r, θ) to denote an arbitrary point on the path of integration and its coordinates. There is however no reason now why we should not revert to the natural notation in which z is used as the argument of the function $\operatorname{Log} z$, and we shall do this in the following examples.

Examples XCIV. 1. We supposed above that $-\pi < \theta < \pi$, and so excluded the case in which z is *real and negative*. In this case the straight line from 1 to z passes through 0, and is therefore not admissible as a path of integration. Both π and $-\pi$ are values of $\operatorname{am} z$, and θ is equal to one or other of them; and $r = -z$. The values of $\operatorname{Log} z$ are still the values of $\log|z| + i \operatorname{am} z$, viz.

$$\log(-z) + (2k+1)\pi i,$$

where k is an integer. The values $\log(-z) + \pi i$ and $\log(-z) - \pi i$ correspond to paths from 1 to z lying respectively entirely above and entirely below the real axis. Either of them may be taken as the principal value of $\operatorname{Log} z$, as convenience dictates. We shall choose the value $\log(-z) + \pi i$ corresponding to the first path.

2. The real and imaginary parts of any value of $\operatorname{Log} z$ are both continuous functions of x and y, except for $x = 0$, $y = 0$.

3. **The functional equation satisfied by** $\operatorname{Log} z$. The function $\operatorname{Log} z$ satisfies the equation

$$\operatorname{Log} z_1 z_2 = \operatorname{Log} z_1 + \operatorname{Log} z_2 \quad \text{.....................(1)},$$

in the sense that *every* value of either side of this equation is *one* of the values of the other side. This follows at once by putting

$$z_1 = r_1(\cos\theta_1 + i\sin\theta_1), \quad z_2 = r_2(\cos\theta_2 + i\sin\theta_2),$$

and applying the formula of p. 453. It is however not true always that

$$\log z_1 z_2 = \log z_1 + \log z_2 \quad \text{.......................(2)}.$$

If, e.g.,

$$z_1 = z_2 = \tfrac{1}{2}(-1 + i\sqrt{3}) = \cos\tfrac{2}{3}\pi + i\sin\tfrac{2}{3}\pi,$$

then $\log z_1 = \log z_2 = \frac{2}{3}\pi i$, and $\log z_1 + \log z_2 = \frac{4}{3}\pi i$, which is one of the values of $\operatorname{Log} z_1 z_2$, but not the principal value. In fact $\log z_1 z_2 = -\frac{2}{3}\pi i$.

An equation such as (1), in which every value of either side is a value of the other, we shall call a *complete* equation, or an equation which is *completely true*.

4. The equation $\operatorname{Log} z^m = m \operatorname{Log} z$, where m is an integer, is not completely true: every value of the right-hand side is a value of the left-hand side, but the converse is not true.

5. The equation $\operatorname{Log}(1/z) = -\operatorname{Log} z$ is completely true. It is also true that $\log(1/z) = -\log z$, except when z is real and negative.

6. The equation

$$\log\left(\frac{z-a}{z-b}\right) = \log(z-a) - \log(z-b)$$

is true if z lies outside the region bounded by the line joining the points $z = a$, $z = b$, and lines through these points parallel to OX and extending to infinity in the negative direction.

7. The equation

$$\log\left(\frac{a-z}{b-z}\right) = \log\left(1-\frac{a}{z}\right) - \log\left(1-\frac{b}{z}\right)$$

is true if z lies outside the triangle formed by the three points 0, a, b.

8. Draw the graph of the function $\mathbf{I}(\operatorname{Log} x)$ of the real variable x. [The graph consists of the positive halves of the lines $y = 2k\pi$ and the negative halves of the lines $y = (2k+1)\pi$.]

9. The function $f(x)$ of the real variable x, defined by

$$\pi f(x) = p\pi + (q-p)\,\mathbf{I}(\log x),$$

is equal to p when x is positive and to q when x is negative.

10. The function $f(x)$ defined by

$$\pi f(x) = p\pi + (q-p)\,\mathbf{I}\{\log(x-1)\} + (r-q)\,\mathbf{I}(\log x)$$

is equal to p when $x > 1$, to q when $0 < x < 1$, and to r when $x < 0$.

11. For what values of z is (i) $\log z$ and (ii) any value of $\operatorname{Log} z$ (a) real or (b) purely imaginary?

12. If $z = x + iy$ then $\operatorname{Log}\operatorname{Log} z = \log R + i(\Theta + 2k'\pi)$, where

$$R^2 = (\log r)^2 + (\theta + 2k\pi)^2$$

and Θ is the least positive angle determined by the equations

$$\cos\Theta : \sin\Theta : 1 :: \log r : \theta + 2k\pi : \sqrt{\{(\log r)^2 + (\theta + 2k\pi)^2\}}.$$

Plot roughly the doubly infinite set of values of $\operatorname{Log}\operatorname{Log}(1+i\sqrt{3})$, indicating which of them are values of $\log\operatorname{Log}(1+i\sqrt{3})$ and which of $\operatorname{Log}\log(1+i\sqrt{3})$.

232. The exponential function. In Ch. IX we defined a function e^y of the real variable y as the inverse of the function $y = \log x$. It is naturally suggested that we should define a function of the complex variable z which is the inverse of the function $\operatorname{Log} z$.

DEFINITION. *If any value of* $\operatorname{Log} z$ *is equal to* ζ, *we call* z *the exponential of* ζ *and write*

$$z = \exp \zeta.$$

Thus $z = \exp \zeta$ if $\zeta = \operatorname{Log} z$. It is certain that to any given value of z correspond infinitely many different values of ζ. It would not be unnatural to suppose that, conversely, to any given value of ζ correspond infinitely many values of z, or in other words that $\exp \zeta$ is an infinitely many-valued function of ζ. This is however untrue, as is proved by the following theorem.

THEOREM. *The exponential function* $\exp \zeta$ *is a one-valued function of* ζ.

For suppose that

$$z_1 = r_1(\cos \theta_1 + i \sin \theta_1), \quad z_2 = r_2(\cos \theta_2 + i \sin \theta_2)$$

are both values of $\exp \zeta$. Then

$$\zeta = \operatorname{Log} z_1 = \operatorname{Log} z_2,$$

and so $$\log r_1 + i(\theta_1 + 2m\pi) = \log r_2 + i(\theta_2 + 2n\pi),$$

where m and n are integers. This involves

$$\log r_1 = \log r_2, \quad \theta_1 + 2m\pi = \theta_2 + 2n\pi.$$

Thus $r_1 = r_2$, and θ_1 and θ_2 differ by a multiple of 2π. Hence $z_1 = z_2$.

COROLLARY. *If* ζ *is real then* $\exp \zeta = e^\zeta$, *the real exponential function of* ζ *defined in Ch. IX.*

For if $z = e^\zeta$ then $\log z = \zeta$, i.e. one of the values of $\operatorname{Log} z$ is ζ. Hence $z = \exp \zeta$.

233. The value of $\exp \zeta$**.** Let $\zeta = \xi + i\eta$ and

$$z = \exp \zeta = r(\cos \theta + i \sin \theta).$$

Then $$\xi + i\eta = \operatorname{Log} z = \log r + i(\theta + 2m\pi),$$

where m is an integer. Hence $\xi = \log r$, $\eta = \theta + 2m\pi$, or

$$r = e^{\xi}, \quad \theta = \eta - 2m\pi;$$

and accordingly

$$\exp(\xi + i\eta) = e^{\xi}(\cos\eta + i\sin\eta).$$

If $\eta = 0$ then $\exp\xi = e^{\xi}$, as we have already inferred in § 232. It is clear that both the real and the imaginary parts of $\exp(\xi + i\eta)$ are continuous functions of ξ and η for all values of ξ and η.

234. The functional equation satisfied by $\exp\zeta$. Let $\zeta_1 = \xi_1 + i\eta_1$, $\zeta_2 = \xi_2 + i\eta_2$. Then

$$\begin{aligned}\exp\zeta_1 \times \exp\zeta_2 &= e^{\xi_1}(\cos\eta_1 + i\sin\eta_1) \times e^{\xi_2}(\cos\eta_2 + i\sin\eta_2)\\ &= e^{\xi_1+\xi_2}\{\cos(\eta_1+\eta_2) + i\sin(\eta_1+\eta_2)\}\\ &= \exp(\zeta_1+\zeta_2).\end{aligned}$$

The exponential function therefore satisfies the functional relation $f(\zeta_1+\zeta_2) = f(\zeta_1)f(\zeta_2)$, an equation which we have proved already (§ 213) to be true for real values of ζ_1 and ζ_2.

235. The general power a^{ζ}. It might seem natural, since $\exp\zeta = e^{\zeta}$ when ζ is real, to adopt the same notation when ζ is complex and to drop the notation $\exp\zeta$ altogether. We shall not follow this course because we shall have to give a more general definition of the meaning of the symbol e^{ζ}. We shall find then that e^{ζ} represents a function with infinitely many values of which $\exp\zeta$ is only one.

We have already defined the meaning of the symbol a^{ζ} in a considerable number of cases. It is defined in elementary algebra in the cases in which a is real and positive and ζ rational, or a real and negative and ζ a rational fraction whose denominator is odd. According to the definitions there given a^{ζ} has at most two values. In Ch. III we extended our definitions to cover the case in which a is any real or complex number and ζ any rational number p/q; and in Ch. IX we gave a new definition, expressed by the equation

$$a^{\zeta} = e^{\zeta\log a},$$

which applies whenever ζ is real and a real and positive.

Thus we have, in one way or another, attached a meaning to

$$3^{\frac{1}{2}},\quad (-1)^{\frac{1}{3}},\quad (\sqrt{3}+\tfrac{1}{2}i)^{-\frac{1}{2}},\quad (3{\cdot}5)^{1+\sqrt{2}};$$

but we have as yet given no definitions which enable us to attach any meaning to

$$(1+i)^{\sqrt{2}},\quad 2^{i},\quad (3+2i)^{2+3i}.$$

We shall now give a general definition of a^ζ which applies to all values of a and ζ, real or complex, with the one limitation that a must not be zero.

DEFINITION. *The function a^ζ is defined by the equation*

$$a^\zeta = \exp(\zeta \operatorname{Log} a),$$

where $\operatorname{Log} a$ *is any value of the logarithm of* a.

We must first satisfy ourselves that this definition is consistent with the previous definitions and includes them all as particular cases.

(1) If a is positive and ζ real, then one value of $\zeta \operatorname{Log} a$, viz. $\zeta \log a$, is real: and $\exp(\zeta \log a) = e^{\zeta \log a}$, which agrees with the definition adopted in Ch. IX. The definition of Ch. IX is, as we saw then, consistent with the definition given in elementary algebra; and so our new definition is so too.

(2) If $a = e^\tau(\cos\psi + i\sin\psi)$, then

$$\operatorname{Log} a = \tau + i(\psi + 2m\pi),$$

$$\exp\left(\frac{p}{q}\operatorname{Log} a\right) = e^{p\tau/q}\operatorname{Cis}\left\{\frac{p}{q}(\psi + 2m\pi)\right\},$$

where m may have any integral value. It is easy to see that if m assumes all possible integral values then this expression assumes q and only q different values, which are just the values of $a^{p/q}$ found in § 48. Hence our new definition is also consistent with that of Ch. III.

236. The general value of a^ζ. Let

$$\zeta = \xi + i\eta,\quad a = \sigma(\cos\psi + i\sin\psi),$$

where $-\pi < \psi \leqq \pi$, so that, in the notation of § 235, $\sigma = e^{\tau}$ or $\tau = \log\sigma$. Then

$$\zeta \operatorname{Log} a = (\xi + i\eta)\{\log\sigma + i(\psi + 2m\pi)\} = L + iM,$$

where

$$L = \xi\log\sigma - \eta(\psi + 2m\pi), \quad M = \eta\log\sigma + \xi(\psi + 2m\pi);$$

and $$a^{\zeta} = \exp(\zeta \operatorname{Log} a) = e^{L}(\cos M + i\sin M).$$

Thus the general value of a^{ζ} is

$$e^{\xi\log\sigma - \eta(\psi + 2m\pi)}[\cos\{\eta\log\sigma + \xi(\psi + 2m\pi)\} + i\sin\{\eta\log\sigma + \xi(\psi + 2m\pi)\}].$$

In general a^{ζ} is an infinitely many-valued function. For

$$|a^{\zeta}| = e^{\xi\log\sigma - \eta(\psi + 2m\pi)}$$

has a different value for every value of m, unless $\eta = 0$. If on the other hand $\eta = 0$, then the moduli of all the different values of a^{ζ} are the same. But any two values differ unless their amplitudes are the same or differ by a multiple of 2π. This requires that $\xi(\psi + 2m\pi)$ and $\xi(\psi + 2n\pi)$, where m and n are different integers, shall differ, if at all, by a multiple of 2π. But if

$$\xi(\psi + 2m\pi) - \xi(\psi + 2n\pi) = 2k\pi,$$

then $\xi = k/(m-n)$ is rational. We conclude that *a^{ζ} is infinitely many-valued unless ζ is real and rational*. On the other hand we have already seen that, when ζ is real and rational, a^{ζ} has but a finite number of values.

The *principal value* of $a^{\zeta} = \exp(\zeta \operatorname{Log} a)$ is obtained by giving $\operatorname{Log} a$ its principal value, i.e. by supposing $m = 0$ in the general formula. Thus the principal value of a^{ζ} is

$$e^{\xi\log\sigma - \eta\psi}\{\cos(\eta\log\sigma + \xi\psi) + i\sin(\eta\log\sigma + \xi\psi)\}.$$

Two particular cases are of especial interest. If a is real and positive and ζ real, then $\sigma = a$, $\psi = 0$, $\xi = \zeta$, $\eta = 0$, and the principal value of a^{ζ} is $e^{\zeta\log a}$, which is the value defined in Ch. IX. If $|a| = 1$ and ζ is real, then $\sigma = 1$, $\xi = \zeta$, $\eta = 0$, and the principal value of $(\cos\psi + i\sin\psi)^{\zeta}$ is $\cos\zeta\psi + i\sin\zeta\psi$. This is a further generalisation of de Moivre's theorem (§§ 45, 49).

Examples XCV. 1. Find all the values of i^i. [By definition

$$i^i = \exp(i \operatorname{Log} i).$$

But $\qquad i = \cos\frac{1}{2}\pi + i\sin\frac{1}{2}\pi, \quad \operatorname{Log} i = (2k+\frac{1}{2})\pi i,$

where k is any integer. Hence

$$i^i = \exp\{-(2k+\tfrac{1}{2})\pi\} = e^{-(2k+\frac{1}{2})\pi}.$$

All the values of i^i are therefore real and positive.]

2. The values of a^ζ, when plotted in the Argand diagram, are the vertices of an equiangular polygon inscribed in an equiangular spiral whose angle is independent of a. (*Math. Trip.* 1899)

[If $a^\zeta = r(\cos\theta + i\sin\theta)$ we have

$$r = e^{\xi\log\sigma - \eta(\psi+2m\pi)}, \quad \theta = \eta\log\sigma + \xi(\psi+2m\pi);$$

and all the points lie on the spiral $r = \sigma^{(\xi^2+\eta^2)/\xi}e^{-\eta\theta/\xi}$.]

3. **The function e^ζ.** If we write e for a in the general formula, so that $\log\sigma = 1$, $\psi = 0$, we obtain

$$e^\zeta = e^{\xi-2m\pi\eta}\{\cos(\eta+2m\pi\xi) + i\sin(\eta+2m\pi\xi)\}.$$

The principal value of e^ζ is $e^\xi(\cos\eta + i\sin\eta)$, which is equal to $\exp\zeta$ (§ 233). In particular, if ζ is real, so that $\eta = 0$, we obtain

$$e^\zeta(\cos 2m\pi\zeta + i\sin 2m\pi\zeta)$$

as the general and e^ζ as the principal value, e^ζ denoting here the positive value of the exponential defined in Ch. IX.

4. Show that $\operatorname{Log} e^\zeta = (1+2m\pi i)\zeta + 2n\pi i$, where m and n are any integers, and that in general $\operatorname{Log} a^\zeta$ has a double infinity of values.

5. The equation $1/a^\zeta = a^{-\zeta}$ is completely true (Ex. XCIV. 3): it is also true of the principal values.

6. The equation $a^\zeta \times b^\zeta = (ab)^\zeta$ is completely true but not always true of the principal values.

7. The equation $a^\zeta \times a^{\zeta'} = a^{\zeta+\zeta'}$ is not completely true, but is true of the principal values. [Every value of the right-hand side is a value of the left-hand side, but the general value of $a^\zeta \times a^{\zeta'}$, viz.

$$\exp\{\zeta(\log a + 2m\pi i) + \zeta'(\log a + 2n\pi i)\},$$

is not as a rule a value of $a^{\zeta+\zeta'}$ unless $m = n$.]

8. What are the corresponding results as regards the equations

$$\operatorname{Log} a^\zeta = \zeta \operatorname{Log} a, \quad (a^\zeta)^{\zeta'} = (a^{\zeta'})^\zeta = a^{\zeta\zeta'}?$$

9. A necessary and sufficient condition that all the values of a^ζ should be real is that 2ξ and $\{\eta\log|a| + \xi \operatorname{am} a\}/\pi$, where $\operatorname{am} a$ denotes any

value of the amplitude, should both be integral. What are the corresponding conditions that all the values should be of unit modulus?

10. The general value of $|x^i + x^{-i}|$, where $x > 0$, is

$$e^{-(m-n)\pi}\sqrt{[2\{\cosh 2(m+n)\pi + \cos(2\log x)\}]}.$$

11. Explain the fallacy in the following argument: since $e^{2m\pi i} = e^{2n\pi i} = 1$, where m and n are any integers, therefore, raising each side to the power i, we obtain $e^{-2m\pi} = e^{-2n\pi}$.

12. In what circumstances are any of the values of x^x, where x is real, themselves real? [If $x > 0$ then

$$x^x = \exp(x \operatorname{Log} x) = \exp(x \log x) \operatorname{Cis} 2m\pi x,$$

the first factor being real. The principal value, for which $m = 0$, is always real.

If x is a rational fraction $p/(2q+1)$, or is irrational, then there is no other real value. But if x is of the form $p/2q$, then there is one other real value, viz. $-\exp(x \log x)$, given by $m = q$.

If $x = -\xi < 0$ then

$$x^x = \exp\{-\xi \operatorname{Log}(-\xi)\} = \exp(-\xi \log \xi) \operatorname{Cis}\{-(2m+1)\pi\xi\}.$$

The only case in which any value is real is that in which $\xi = p/(2q+1)$, when $m = q$ gives the real value

$$\exp(-\xi \log \xi) \operatorname{Cis}(-p\pi) = (-1)^p \xi^{-\xi}.$$

The cases of reality are illustrated by the examples

$$(\tfrac{1}{3})^{\frac{1}{3}} = \sqrt[3]{\tfrac{1}{3}},\quad (\tfrac{1}{2})^{\frac{1}{2}} = \pm\sqrt{\tfrac{1}{2}},\quad (-\tfrac{2}{3})^{-\frac{2}{3}} = \sqrt[3]{\tfrac{9}{4}},\quad (-\tfrac{1}{3})^{-\frac{1}{3}} = -\sqrt[3]{3}.]$$

13. **Logarithms to any base.** We may define $\zeta = \operatorname{Log}_a z$ in two different ways. We may say (i) that $\zeta = \operatorname{Log}_a z$ if the *principal* value of a^ζ is equal to z; or we may say (ii) that $\zeta = \operatorname{Log}_a z$ if *any* value of a^ζ is equal to z.

Thus if $a = e$ then $\zeta = \operatorname{Log}_e z$, according to the first definition, if the principal value of e^ζ is equal to z, or if $\exp \zeta = z$; and so $\operatorname{Log}_e z$ is identical with $\operatorname{Log} z$. But, according to the second definition, $\zeta = \operatorname{Log}_e z$ if

$$e^\zeta = \exp(\zeta \operatorname{Log} e) = z, \quad \zeta \operatorname{Log} e = \operatorname{Log} z,$$

or $\zeta = (\operatorname{Log} z)/(\operatorname{Log} e)$, any values of the logarithms being taken. Thus

$$\zeta = \operatorname{Log}_e z = \frac{\log|z| + (\operatorname{am} z + 2m\pi)i}{1 + 2n\pi i},$$

so that ζ is a doubly infinitely many-valued function of z. And generally, according to this definition, $\operatorname{Log}_a z = (\operatorname{Log} z)/(\operatorname{Log} a)$.

14. $$\operatorname{Log}_e 1 = \frac{2m\pi i}{1 + 2n\pi i}, \quad \operatorname{Log}_e(-1) = \frac{(2m+1)\pi i}{1 + 2n\pi i},$$

where m and n are any integers.

237. The exponential values of the sine and cosine. From the formula

$$\exp(\xi + i\eta) = \exp\xi\,(\cos\eta + i\sin\eta),$$

we can deduce a number of very important subsidiary formulae. Taking $\xi = 0$, we obtain $\exp(i\eta) = \cos\eta + i\sin\eta$; and, changing the sign of η, $\exp(-i\eta) = \cos\eta - i\sin\eta$. Hence

$$\cos\eta = \tfrac{1}{2}\{\exp(i\eta) + \exp(-i\eta)\},$$

$$\sin\eta = -\tfrac{1}{2}i\{\exp(i\eta) - \exp(-i\eta)\}.$$

We can of course deduce expressions for any of the trigonometrical ratios of η in terms of $\exp(i\eta)$.

238. Definition of $\sin\zeta$ and $\cos\zeta$ for all values of ζ. We saw in the last section that, when ζ is real,

$$\cos\zeta = \tfrac{1}{2}\{\exp(i\zeta) + \exp(-i\zeta)\} \quad \text{............(1}a\text{)},$$

$$\sin\zeta = -\tfrac{1}{2}i\{\exp(i\zeta) - \exp(-i\zeta)\} \quad \text{............(1}b\text{)}.$$

The left-hand sides of these equations are defined, by the ordinary geometrical definitions adopted in elementary trigonometry, only for real values of ζ. The right-hand sides have, on the other hand, been defined for all values of ζ, real or complex. We are therefore naturally led to adopt the formulae (1) as the *definitions* of $\cos\zeta$ and $\sin\zeta$ for all values of ζ. These definitions agree, in virtue of the results of § 237, with the elementary definitions for real values of ζ.

Having defined $\cos\zeta$ and $\sin\zeta$, we define the other trigonometrical ratios by the equations

$$\tan\zeta = \frac{\sin\zeta}{\cos\zeta}, \quad \cot\zeta = \frac{\cos\zeta}{\sin\zeta}, \quad \sec\zeta = \frac{1}{\cos\zeta}, \quad \operatorname{cosec}\zeta = \frac{1}{\sin\zeta} \quad \text{...(2)}.$$

It is evident that $\cos\zeta$ and $\sec\zeta$ are even functions of ζ, and $\sin\zeta$, $\tan\zeta$, $\cot\zeta$, and $\operatorname{cosec}\zeta$ odd functions. Also, if $\exp(i\zeta) = t$, we have

$$\cos\zeta = \tfrac{1}{2}\{t + t^{-1}\}, \quad \sin\zeta = -\tfrac{1}{2}i\{t - t^{-1}\},$$

$$\cos^2\zeta + \sin^2\zeta = \tfrac{1}{4}\{(t + t^{-1})^2 - (t - t^{-1})^2\} = 1 \quad \text{......(3)}.$$

We can moreover express the trigonometrical functions of $\zeta + \zeta'$ in terms of those of ζ and ζ' by the same formulae as those which hold in elementary trigonometry. For if $\exp(i\zeta) = t$, $\exp(i\zeta') = t'$, we have

$$\cos(\zeta+\zeta') = \frac{1}{2}\left(tt'+\frac{1}{tt'}\right) = \frac{1}{4}\left\{\left(t+\frac{1}{t}\right)\left(t'+\frac{1}{t'}\right)+\left(t-\frac{1}{t}\right)\left(t'-\frac{1}{t'}\right)\right\}$$
$$= \cos\zeta\cos\zeta' - \sin\zeta\sin\zeta' \quad \text{......(4);}$$

and similarly we can prove that

$$\sin(\zeta+\zeta') = \sin\zeta\cos\zeta' + \cos\zeta\sin\zeta' \quad \text{.........(5).}$$

In particular

$$\cos(\zeta+\tfrac{1}{2}\pi) = -\sin\zeta, \quad \sin(\zeta+\tfrac{1}{2}\pi) = \cos\zeta \quad \text{......(6).}$$

All the ordinary formulae of elementary trigonometry are algebraical corollaries of the equations (2)–(6); and so all such relations hold also for the generalised trigonometrical functions defined in this section.

239. The generalised hyperbolic functions. In Ex. LXXXVIII. 20 we defined $\cosh\zeta$ and $\sinh\zeta$, for real values of ζ, by the equations

$$\cosh\zeta = \tfrac{1}{2}\{\exp\zeta + \exp(-\zeta)\}, \quad \sinh\zeta = \tfrac{1}{2}\{\exp\zeta - \exp(-\zeta)\} \quad \text{...(1)}$$

We can now extend this definition to complex values of the variable; i.e. we can agree that the equations (1) are to define $\cosh\zeta$ and $\sinh\zeta$ for all values of ζ real or complex. The reader will easily verify that

$$\cos i\zeta = \cosh\zeta, \quad \sin i\zeta = i\sinh\zeta, \quad \cosh i\zeta = \cos\zeta, \quad \sinh i\zeta = i\sin\zeta.$$

We have seen that any elementary trigonometrical formula, such as $\cos 2\zeta = \cos^2\zeta - \sin^2\zeta$ remains true when ζ is allowed to assume complex values. It therefore remains true if we write $\cos i\zeta$ for $\cos\zeta$, $\sin i\zeta$ for $\sin\zeta$ and $\cos 2i\zeta$ for $\cos 2\zeta$; or, in other words, if we write $\cosh\zeta$ for $\cos\zeta$, $i\sinh\zeta$ for $\sin\zeta$, and $\cosh 2\zeta$ for $\cos 2\zeta$. Hence

$$\cosh 2\zeta = \cosh^2\zeta + \sinh^2\zeta.$$

The same process of transformation may be applied to any trigonometrical identity. It is this which explains the correspondence noted in Ex. LXXXVIII. 22 between the formulae for the hyperbolic and those for the ordinary trigonometrical functions.

240. Formulae for $\cos(\xi+i\eta)$, $\sin(\xi+i\eta)$, etc. It follows from the addition formulae that

$$\cos(\xi+i\eta) = \cos\xi\cos i\eta - \sin\xi\sin i\eta = \cos\xi\cosh\eta - i\sin\xi\sinh\eta,$$
$$\sin(\xi+i\eta) = \sin\xi\cos i\eta + \cos\xi\sin i\eta = \sin\xi\cosh\eta + i\cos\xi\sinh\eta.$$

These formulae are true for all values of ξ and η. The interesting case is that in which ξ and η are real. They then give expressions for the real and imaginary parts of the cosine and sine of a complex number.

Examples XCVI. 1. Determine the values of ζ for which $\cos\zeta$ and $\sin\zeta$ are (i) real, (ii) purely imaginary. [For example $\cos\zeta$ is real when $\eta = 0$ or when ξ is any multiple of π.]

2. $$|\cos(\xi+i\eta)| = \sqrt{(\cos^2\xi+\sinh^2\eta)} = \sqrt{\{\tfrac{1}{2}(\cosh 2\eta+\cos 2\xi)\}},$$
$$|\sin(\xi+i\eta)| = \sqrt{(\sin^2\xi+\sinh^2\eta)} = \sqrt{\{\tfrac{1}{2}(\cosh 2\eta-\cos 2\xi)\}}.$$

[Use (e.g.) the equation
$$|\cos(\xi+i\eta)| = \sqrt{\{\cos(\xi+i\eta)\cos(\xi-i\eta)\}}.]$$

3. $$\tan(\xi+i\eta) = \frac{\sin 2\xi+i\sinh 2\eta}{\cosh 2\eta+\cos 2\xi},\quad \cot(\xi+i\eta) = \frac{\sin 2\xi-i\sinh 2\eta}{\cosh 2\eta-\cos 2\xi}.$$

[For example
$$\tan(\xi+i\eta) = \frac{\sin(\xi+i\eta)\cos(\xi-i\eta)}{\cos(\xi+i\eta)\cos(\xi-i\eta)} = \frac{\sin 2\xi+\sin 2i\eta}{\cos 2\xi+\cos 2i\eta},$$
which leads at once to the result given.]

4. $$\sec(\xi+i\eta) = \frac{\cos\xi\cosh\eta+i\sin\xi\sinh\eta}{\tfrac{1}{2}(\cosh 2\eta+\cos 2\xi)},$$
$$\operatorname{cosec}(\xi+i\eta) = \frac{\sin\xi\cosh\eta-i\cos\xi\sinh\eta}{\tfrac{1}{2}(\cosh 2\eta-\cos 2\xi)}.$$

5. If $|\cos(\xi+i\eta)| = 1$ then $\sin^2\xi = \sinh^2\eta$, and if $|\sin(\xi+i\eta)| = 1$ then $\cos^2\xi = \sinh^2\eta$.

6. If $|\cos(\xi+i\eta)| = 1$, then
$$\sin\{\operatorname{am}\cos(\xi+i\eta)\} = \pm\sin^2\xi = \pm\sinh^2\eta.$$

7. Prove that $\operatorname{Log}\cos(\xi+i\eta) = A+iB$, where
$$A = \tfrac{1}{2}\log\{\tfrac{1}{2}(\cosh 2\eta+\cos 2\xi)\}$$
and B is any angle such that
$$\frac{\cos B}{\cos\xi\cosh\eta} = -\frac{\sin B}{\sin\xi\sinh\eta} = \frac{1}{\sqrt{\{\tfrac{1}{2}(\cosh 2\eta+\cos 2\xi)\}}}.$$
Find a similar formula for $\operatorname{Log}\sin(\xi+i\eta)$.

8. **Solution of the equation $\cos\zeta = \alpha$, where α is real.** Putting $\zeta = \xi+i\eta$, and equating real and imaginary parts, we obtain
$$\cos\xi\cosh\eta = \alpha,\quad \sin\xi\sinh\eta = 0.$$
Hence either $\eta = 0$ or ξ is a multiple of π. If (i) $\eta = 0$ then $\cos\xi = \alpha$, which is impossible unless $-1 \leqq \alpha \leqq 1$. This hypothesis leads to the solution
$$\zeta = 2k\pi \pm \operatorname{arc}\cos\alpha,$$

where arc cos α lies between 0 and $\frac{1}{2}\pi$. If (ii) $\xi = m\pi$ then $\cosh\eta = (-1)^m\alpha$ so that either $\alpha \geqq 1$ and m is even, or $\alpha \leqq -1$ and m is odd. If $\alpha = \pm 1$ then $\eta = 0$, and we are led back to our first case. If $|\alpha| > 1$ then $\cosh\eta = |\alpha|$, and we are led to the solutions

$$\zeta = 2k\pi \pm i\log\{\alpha + \sqrt{(\alpha^2 - 1)}\} \quad (\alpha > 1),$$
$$\zeta = (2k+1)\pi \pm i\log\{-\alpha + \sqrt{(\alpha^2 - 1)}\} \quad (\alpha < -1).$$

For example, the general solution of $\cos\zeta = -\frac{5}{3}$ is $\zeta = (2k+1)\pi \pm i\log 3$.

Solve $\sin\zeta = \alpha$ similarly.

9. **Solution of** $\cos\zeta = \alpha + i\beta$, **where** $\beta \neq 0$. We may suppose $\beta > 0$, since the results when $\beta < 0$ may be deduced by merely changing the sign of i. In this case

$$\cos\xi\cosh\eta = \alpha, \quad \sin\xi\sinh\eta = -\beta \quad \text{...............(1)},$$

and

$$\frac{\alpha^2}{\cosh^2\eta} + \frac{\beta^2}{\sinh^2\eta} = 1.$$

If we put $\cosh^2\eta = x$ we find that

$$x^2 - (1 + \alpha^2 + \beta^2)x + \alpha^2 = 0$$

or $x = (A_1 \pm A_2)^2$, where

$$A_1 = \tfrac{1}{2}\sqrt{\{(\alpha+1)^2 + \beta^2\}}, \quad A_2 = \tfrac{1}{2}\sqrt{\{(\alpha-1)^2 + \beta^2\}}.$$

Suppose that $\alpha > 0$. Then $A_1 > A_2 > 0$ and $\cosh\eta = A_1 \pm A_2$. Also

$$\cos\xi = \frac{\alpha}{\cosh\eta} = A_1 \mp A_2,$$

and since $\cosh\eta > \cos\xi$ we must take

$$\cosh\eta = A_1 + A_2, \quad \cos\xi = A_1 - A_2.$$

The general solutions of these equations are

$$\xi = 2k\pi \pm \text{arc cos}\, M, \quad \eta = \pm\log\{L + \sqrt{(L^2 - 1)}\} \quad \text{.........(2)},$$

where $L = A_1 + A_2$, $M = A_1 - A_2$, and arc cos M lies between 0 and $\frac{1}{2}\pi$.

The values of η and ξ thus found above include, however, the solutions of the equations

$$\cos\xi\cosh\eta = \alpha, \quad \sin\xi\sinh\eta = \beta \quad \text{...............(3)},$$

as well as those of the equations (1), since we have only used the second of the latter equations after squaring it. To distinguish the two sets of solutions we observe that the sign of $\sin\xi$ is the same as the ambiguous sign in the first of the equations (2), and the sign of $\sinh\eta$ is the same as the ambiguous sign in the second. Since $\beta > 0$, these two signs must be different. Hence the general solution required is

$$\zeta = 2k\pi \pm [\text{arc cos}\, M - i\log\{L + \sqrt{(L^2 - 1)}\}].$$

Work out the cases in which $\alpha < 0$ and $\alpha = 0$ in the same way.

10. If $\beta = 0$ then $L = \frac{1}{2}|\alpha+1| + \frac{1}{2}|\alpha-1|$ and $M = \frac{1}{2}|\alpha+1| - \frac{1}{2}|\alpha-1|$. Verify that the results thus obtained agree with those of Ex. 8.

11. Show that if α and β are positive then the general solution of $\sin\zeta = \alpha + i\beta$ is

$$\zeta = k\pi + (-1)^k[\arcsin M + i\log\{L + \sqrt{(L^2-1)}\}],$$

where $\arcsin M$ lies between 0 and $\frac{1}{2}\pi$. Obtain the solution in the other possible cases.

12. Solve $\tan\zeta = \alpha$, where α is real. [All the roots are real.]

13. Show that the general solution of $\tan\zeta = \alpha + i\beta$, where $\beta \neq 0$, is

$$\zeta = k\pi + \tfrac{1}{2}\theta + \tfrac{1}{4}i\log\left\{\frac{\alpha^2+(1+\beta)^2}{\alpha^2+(1-\beta)^2}\right\},$$

where θ is the numerically least angle such that

$$\cos\theta : \sin\theta : 1 :: 1-\alpha^2-\beta^2 : 2\alpha : \sqrt{\{(1-\alpha^2-\beta^2)^2+4\alpha^2\}}.$$

14. Prove that

$$|\exp\exp(\xi+i\eta)| = \exp(\exp\xi\cos\eta),$$

$$\mathbf{R}\{\cos\cos(\xi+i\eta)\} = \cos(\cos\xi\cosh\eta)\cosh(\sin\xi\sinh\eta),$$

$$\mathbf{I}\{\sin\sin(\xi+i\eta)\} = \cos(\sin\xi\cosh\eta)\sinh(\cos\xi\sinh\eta).$$

15. Prove that $|\exp\zeta|$ tends to ∞ if ζ moves away towards infinity along any straight line through the origin making an angle less than $\frac{1}{2}\pi$ with OX, and to 0 if ζ moves away along a similar line making an angle between $\frac{1}{2}\pi$ and π with OX.

16. Prove that $|\cos\zeta|$ and $|\sin\zeta|$ tend to ∞ if ζ moves away towards infinity along any straight line through the origin other than either half of the real axis.

17. Prove that $\tan\zeta$ tends to $-i$ or to i if ζ moves away to infinity along the straight line of Ex. 16, to $-i$ if the line lies above the real axis and to i if it lies below.

241. The connection between the logarithmic and the inverse trigonometrical functions. We found in Ch. VI that the integral of a rational or algebraical function $\phi(x, \alpha, \beta, \ldots)$, where $\alpha, \beta, \ldots$ are constants, often assumes different forms according to the values of $\alpha, \beta, \ldots$; sometimes it can be expressed by means of logarithms, and sometimes by means of inverse trigonometrical functions. Thus, for example,

$$\int\frac{dx}{x^2+a} = \frac{1}{\sqrt{a}}\arctan\frac{x}{\sqrt{a}} \quad \ldots\ldots\ldots\ldots(1)$$

if $a > 0$, but

$$\int\frac{dx}{x^2+a} = \frac{1}{2\sqrt{(-a)}}\log\left|\frac{x-\sqrt{(-a)}}{x+\sqrt{(-a)}}\right| \quad \ldots\ldots\ldots\ldots(2)$$

if $a<0$. These formulae suggest the existence of some functional connection between the logarithmic and the inverse circular functions. That there is such a connection may also be inferred from the facts that we have expressed the circular functions of ζ in terms of $\exp i\zeta$, and that the logarithm is the inverse of the exponential function.

Let us consider more particularly the equation

$$\int \frac{dx}{x^2-\alpha^2} = \frac{1}{2\alpha}\log\left(\frac{x-\alpha}{x+\alpha}\right),$$

which holds when α is real and $(x-\alpha)/(x+\alpha)$ is positive. If we could write $i\alpha$ instead of α in this equation, we should be led to the formula

$$\operatorname{arc\,tan}\left(\frac{x}{\alpha}\right) = \frac{1}{2i}\log\left(\frac{x-i\alpha}{x+i\alpha}\right)+C \quad \text{.................(3)},$$

where C is a constant, and the question is suggested whether, now that we have defined the logarithm of a complex number, this equation will not be found to be true.

Now (§ 231)

$$\operatorname{Log}(x \pm i\alpha) = \tfrac{1}{2}\log(x^2+\alpha^2) \pm i(\phi+2k\pi),$$

where k is an integer and ϕ is the numerically least angle such that $\cos\phi : \sin\phi : 1 :: x : \alpha : \sqrt{(x^2+\alpha^2)}$. Thus

$$\frac{1}{2i}\operatorname{Log}\left(\frac{x-i\alpha}{x+i\alpha}\right) = -\phi - l\pi,$$

where l is an integer, and this in fact differs by a constant from any value of $\operatorname{arc\,tan}(x/\alpha)$.

The standard formula connecting the logarithmic and inverse circular functions is

$$\operatorname{arc\,tan} x = \frac{1}{2i}\operatorname{Log}\left(\frac{1+ix}{1-ix}\right) \quad \text{.....................(4)},$$

where x is real. It is most easily verified by putting $x = \tan y$, when the right-hand side reduces to

$$\frac{1}{2i}\operatorname{Log}\left(\frac{\cos y + i\sin y}{\cos y - i\sin y}\right) = \frac{1}{2i}\operatorname{Log}(\exp 2iy) = y + k\pi,$$

where k is any integer, so that the equation (4) is 'completely' true (Ex. XCIV. 3). The reader should also verify the formulae

$$\operatorname{arc\,cos} x = -i\operatorname{Log}\{x \pm i\sqrt{(1-x^2)}\}, \quad \operatorname{arc\,sin} x = -i\operatorname{Log}\{ix \pm \sqrt{(1-x^2)}\} \quad \text{......(5)},$$

where $-1 \leqq x \leqq 1$; each of these formulae also is 'completely' true.

Example. Solving the equation

$$\cos u = x = \tfrac{1}{2}(y+y^{-1}),$$

where $y = \exp(iu)$, with respect to y, we obtain $y = x \pm i\sqrt{(1-x^2)}$. Thus

$$u = -i \operatorname{Log} y = -i \operatorname{Log}\{x \pm i\sqrt{(1-x^2)}\},$$

which is equivalent to the first of the equations (5). Obtain the remaining equations (4) and (5) by similar reasoning.

242. The power series for $\exp z$*. We saw in § 219 that when z is real

$$\exp z = 1 + z + \frac{z^2}{2!} + \dots \quad \dots\dots\dots\dots\dots\dots (1).$$

We also saw in § 198 that the series on the right-hand side remains convergent (indeed absolutely convergent) when z is complex. It is naturally suggested that the equation (1) also remains true, and we shall now prove that this is so.

Let the sum of the series (1) be denoted by $F(z)$. The series being absolutely convergent, it follows by direct multiplication (as in Ex. LXXXI. 7) that $F(z)$ satisfies the functional equation

$$F(z+h) = F(z)\,F(h) \quad \dots\dots\dots\dots\dots\dots (2),$$

and in particular that

$$F(x+iy) = F(x)\,F(iy).$$

Now
$$F(x) = 1 + x + \frac{x^2}{2!} + \dots = e^x,$$

and
$$F(iy) = 1 - \frac{y^2}{2!} + \frac{y^4}{4!} - \dots + i\left(y - \frac{y^3}{3!} + \dots\right) = \cos y + i \sin y.$$

Hence
$$F(z) = e^x(\cos y + i \sin y) = \exp z$$

if $z = x + iy$.

There is an alternative proof which is interesting because it does not demand a knowledge of the power series for $\cos y$ and $\sin y$.

If $F(iy) = f(y)$ then $f(y+k) = f(y)f(k)$ and

$$\frac{f(y+k)-f(y)}{k} = f(y)\frac{f(k)-1}{k}$$

$$= if(y)\left\{1 + \frac{ik}{2!} + \frac{(ik)^2}{3!} + \dots\right\} = if(y)\,(1+\rho),$$

* It will be convenient now to use z instead of ζ as the argument of the exponential function.

where $$|\rho| \leqq \frac{|k|}{2!} + \frac{|k|^2}{3!} + \ldots \leqq (e-2)|k|$$

for small k, so that ρ tends to zero with k. Hence $f(y)$ is differentiable and
$$f'(y) = if(y).$$

It follows that
$$g(y) = f(y)(\cos y - i \sin y)$$
is differentiable.* Also
$$g'(y) = if(y)(\cos y - i \sin y) - f(y)(\sin y + i \cos y) = 0,$$
so that $g(y)$ is constant. Hence
$$g(y) = g(0) = 1$$
and $$f(y) = \frac{1}{\cos y - i \sin y} = \frac{\cos y + i \sin y}{\cos^2 y + \sin^2 y} = \cos y + i \sin y.$$
Finally $F(iy) = f(y) = \cos y + i \sin y$ and
$$F(x+iy) = F(x)\,F(iy) = e^x(\cos y + i \sin y).$$

243. The power series for $\cos z$ and $\sin z$. From the result of the last section and the equations (1) of § 238 it follows that
$$\cos z = 1 - \frac{z^2}{2!} + \frac{z^4}{4!} - \ldots, \quad \sin z = z - \frac{z^3}{3!} + \frac{z^5}{5!} - \ldots$$
for all values of z.

Examples XCVII. 1. Prove that
$$|\cos z| \leqq \cosh |z|, \quad |\sin z| \leqq \sinh |z|.$$

2. Prove that if $|z| < 1$ then $|\cos z| < 2$ and $|\sin z| < \frac{6}{5}|z|$.

3. Since $\sin 2z = 2 \sin z \cos z$ we have
$$(2z) - \frac{(2z)^3}{3!} + \frac{(2z)^5}{5!} - \ldots = 2\left(z - \frac{z^3}{3!} + \ldots\right)\left(1 - \frac{z^2}{2!} + \ldots\right).$$
Prove by multiplying the two series on the right-hand side (§ 202) and equating coefficients (§ 201) that
$$\binom{2n+1}{1} + \binom{2n+1}{3} + \ldots + \binom{2n+1}{2n+1} = 2^{2n}.$$
Verify the result by means of the binomial theorem. Derive similar identities from the equations
$$\cos^2 z + \sin^2 z = 1, \quad \cos 2z = 2\cos^2 z - 1 = 1 - 2\sin^2 z.$$

* The argument which followed in earlier editions contained a curious fallacy. That adopted here was suggested by Mr Love.

4. Show that $\exp\{(1+i)z\} = \sum_0^\infty 2^{\frac{1}{2}n} \exp(\tfrac{1}{4}n\pi i)\dfrac{z^n}{n!}$.

5. Expand $\cos z \cosh z$ in powers of z. [We have

$$\cos z \cosh z - i \sin z \sinh z = \cos\{(1+i)z\} = \tfrac{1}{2}[\exp\{(1+i)z\} + \exp\{-(1+i)z\}]$$

$$= \tfrac{1}{2}\sum_0^\infty 2^{\frac{1}{2}n}\{1+(-1)^n\}\exp(\tfrac{1}{4}n\pi i)\frac{z^n}{n!},$$

and similarly

$$\cos z \cosh z + i \sin z \sinh z = \cos(1-i)z$$

$$= \tfrac{1}{2}\sum_0^\infty 2^{\frac{1}{2}n}\{1+(-1)^n\}\exp(-\tfrac{1}{4}n\pi i)\frac{z^n}{n!}.$$

Hence

$$\cos z \cosh z = \tfrac{1}{2}\sum_0^\infty 2^{\frac{1}{2}n}\{1+(-1)^n\}\cos\tfrac{1}{4}n\pi\,\frac{z^n}{n!} = 1 - \frac{2^2z^4}{4!} + \frac{2^4z^8}{8!} - \dots.]$$

6. Expand $\sin^2 z$ and $\sin^3 z$ in powers of z. [Use the formulae

$$\sin^2 z = \tfrac{1}{2}(1-\cos 2z), \quad \sin^3 z = \tfrac{1}{4}(3\sin z - \sin 3z).$$

It is clear that the same method may be used to expand $\cos^n z$ and $\sin^n z$, where n is any integer.]

7. Sum the series

$$C = 1 + \frac{\cos z}{1!} + \frac{\cos 2z}{2!} + \frac{\cos 3z}{3!} + \dots, \quad S = \frac{\sin z}{1!} + \frac{\sin 2z}{2!} + \frac{\sin 3z}{3!} + \dots.$$

[Here $C + iS = 1 + \dfrac{\exp(iz)}{1!} + \dfrac{\exp(2iz)}{2!} + \dots = \exp\{\exp(iz)\}$

$$= \exp(\cos z)\{\cos(\sin z) + i\sin(\sin z)\},$$

and similarly

$$C - iS = \exp\{\exp(-iz)\} = \exp(\cos z)\{\cos(\sin z) - i\sin(\sin z)\}.$$

Hence $C = \exp(\cos z)\cos(\sin z)$, $S = \exp(\cos z)\sin(\sin z)$.]

8. Sum $1 + \dfrac{a\cos z}{1!} + \dfrac{a^2\cos 2z}{2!} + \dots$, $\dfrac{a\sin z}{1!} + \dfrac{a^2\sin 2z}{2!} + \dots$.

9. Sum $1 - \dfrac{\cos 2z}{2!} + \dfrac{\cos 4z}{4!} - \dots$, $\dfrac{\cos z}{1!} - \dfrac{\cos 3z}{3!} + \dots$

and the corresponding series involving sines.

10. Show that

$$1 + \frac{\cos 4z}{4!} + \frac{\cos 8z}{8!} + \dots = \tfrac{1}{2}\{\cos(\cos z)\cosh(\sin z) + \cos(\sin z)\cosh(\cos z)\}.$$

11. Show that the expansions of $\cos(x+h)$ and $\sin(x+h)$ in powers of h, found in § 152 (1), are valid for all values of x and h, real or complex.

244. The logarithmic series. We found in § 220 that

$$\log(1+z) = z - \tfrac{1}{2}z^2 + \tfrac{1}{3}z^3 - \dots \qquad \text{(1)}$$

when z is real and numerically less than unity. The series on the right-hand side is convergent, indeed absolutely convergent, when z has any complex value whose modulus is less than unity. It is naturally suggested that the equation (1) remains true for such complex values of z. That this is true may be proved by a modification of the argument of § 220. We shall in fact prove rather more than this, viz. that (1) is true for all values of z such that $|z| \leqq 1$, with the exception of the value -1.

It will be remembered that $\log(1+z)$ is the principal value of $\operatorname{Log}(1+z)$, and that

$$\log(1+z) = \int_C \frac{du}{u},$$

where C is the straight line joining the points 1 and $1+z$ in the plane of the complex variable u. We may suppose that z is not real, since the formula (1) has been proved already for real values of z.

If we put $\qquad z = r(\cos\theta + i\sin\theta) = \zeta r,$

so that $r \leqq 1$, and $\qquad u = 1 + \zeta t,$

then u will describe C as t increases from 0 to r. And

$$\begin{aligned}
\int_C \frac{du}{u} &= \int_0^r \frac{\zeta\, dt}{1+\zeta t} \\
&= \int_0^r \left\{\zeta - \zeta^2 t + \zeta^3 t^2 - \dots + (-1)^{m-1}\zeta^m t^{m-1} + \frac{(-1)^m \zeta^{m+1} t^m}{1+\zeta t}\right\} dt \\
&= \zeta r - \frac{(\zeta r)^2}{2} + \frac{(\zeta r)^3}{3} - \dots + (-1)^{m-1}\frac{(\zeta r)^m}{m} + R_m \\
&= z - \frac{z^2}{2} + \frac{z^3}{3} - \dots + (-1)^{m-1}\frac{z^m}{m} + R_m \qquad \text{(2)},
\end{aligned}$$

where $\qquad R_m = (-1)^m \zeta^{m+1} \displaystyle\int_0^r \frac{t^m\, dt}{1+\zeta t} \qquad \text{(3)}.$

It follows from (1) of § 170 that

$$|R_m| \leqq \int_0^r \frac{t^m\, dt}{|1+\zeta t|} \qquad \text{(4)}.$$

Now $|1+\zeta t|$ or $|u|$ is never less than ϖ, the perpendicular from O on to the line C*. Hence

$$|R_m| \leqq \frac{1}{\varpi}\int_0^r t^m\,dt = \frac{r^{m+1}}{(m+1)\varpi} \leqq \frac{1}{(m+1)\varpi},$$

and so $R_m \to 0$ when $m \to \infty$. It now follows from (2) that

$$\log(1+z) = z - \tfrac{1}{2}z^2 + \tfrac{1}{3}z^3 - \dots \quad \dots\dots\dots\dots\dots(5).$$

We have shown in the course of our proof that the series is convergent: this however has been proved already (Ex. LXXX. 4). The series is in fact absolutely convergent when $|z|<1$ and conditionally convergent when $|z| = 1$.

Changing z into $-z$ we obtain

$$\log\left(\frac{1}{1-z}\right) = -\log(1-z) = z + \tfrac{1}{2}z^2 + \tfrac{1}{3}z^3 + \dots \dots\dots(6),$$

for $|z| \leqq 1$, $z \neq 1$.

245. Now

$$\log(1+z) = \log\{(1+r\cos\theta) + ir\sin\theta\}$$
$$= \tfrac{1}{2}\log(1+2r\cos\theta+r^2) + i\,\text{arc}\tan\left(\frac{r\sin\theta}{1+r\cos\theta}\right).$$

That value of the inverse tangent must be taken which lies between $-\frac{1}{2}\pi$ and $\frac{1}{2}\pi$. For, since $1+z$ is the vector represented by the line from -1 to z, the principal value of am $(1+z)$ always lies between these limits when z lies within the circle $|z| = 1$†.

Since $z^m = r^m(\cos m\theta + i\sin m\theta)$, we obtain, on equating the real and imaginary parts in equation (5) of § 244,

$$\tfrac{1}{2}\log(1+2r\cos\theta+r^2) = r\cos\theta - \tfrac{1}{2}r^2\cos 2\theta + \tfrac{1}{3}r^3\cos 3\theta - \dots,$$

$$\text{arc}\tan\left(\frac{r\sin\theta}{1+r\cos\theta}\right) = r\sin\theta - \tfrac{1}{2}r^2\sin 2\theta + \tfrac{1}{3}r^3\sin 3\theta - \dots.$$

These equations hold when $0 \leqq r \leqq 1$, and for all values of θ, except that, when $r = 1$, θ must not be equal to an odd multiple of π. It is easy to see that they also hold when $-1 \leqq r \leqq 0$, except that, when $r = -1$, θ must not be equal to an even multiple of π.

* Since z is not real, C cannot pass through O when produced. The reader is recommended to draw a figure to illustrate the argument.

† See the preceding footnote.

A particularly interesting case is that in which $r = 1$. In this case we have

$$\log(1+z) = \log(1+\operatorname{Cis}\theta) = \tfrac{1}{2}\log(2+2\cos\theta) + i\arctan\left(\frac{\sin\theta}{1+\cos\theta}\right)$$

if $-\pi < \theta < \pi$, and so $\qquad = \tfrac{1}{2}\log(4\cos^2\tfrac{1}{2}\theta) + \tfrac{1}{2}i\theta,$

$$\cos\theta - \tfrac{1}{2}\cos 2\theta + \tfrac{1}{3}\cos 3\theta - \ldots = \tfrac{1}{2}\log(4\cos^2\tfrac{1}{2}\theta),$$
$$\sin\theta - \tfrac{1}{2}\sin 2\theta + \tfrac{1}{3}\sin 3\theta - \ldots = \tfrac{1}{2}\theta.$$

The sums of the series, for other values of θ, are easily found from the consideration that they are periodic functions of θ with the period 2π. Thus the sum of the cosine series is $\tfrac{1}{2}\log(4\cos^2\tfrac{1}{2}\theta)$ for all values of θ save odd multiples of π (for which values the series is divergent), while the sum of the sine series is $\tfrac{1}{2}(\theta - 2k\pi)$ if $(2k-1)\pi < \theta < (2k+1)\pi$, and zero if θ is an odd multiple of π. The graph of the function represented by the sine series is shown in Fig. 55. The function is discontinuous for $\theta = (2k+1)\pi$.

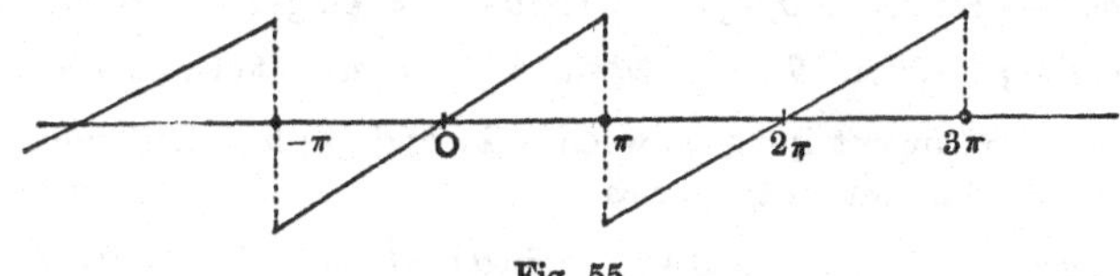

Fig. 55

If we write iz and $-iz$ for z in (5), and subtract, we obtain

$$\frac{1}{2i}\log\left(\frac{1+iz}{1-iz}\right) = z - \tfrac{1}{3}z^3 + \tfrac{1}{5}z^5 - \ldots.$$

If z is real and numerically less than unity, we are led, by the results of § 241, to the formula

$$\arctan z = z - \tfrac{1}{3}z^3 + \tfrac{1}{5}z^5 - \ldots,$$

already proved in a different manner in § 221.

Examples XCVIII. 1. Prove that, in any triangle in which $a > b$,

$$\log c = \log a - \frac{b}{a}\cos C - \frac{b^2}{2a^2}\cos 2C - \ldots.$$

(*Math. Trip.* 1915)

[Use the formula $\log c = \tfrac{1}{2}\log(a^2 + b^2 - 2ab\cos C)$.]

2. Prove that if $-1<r<1$ and $-\frac{1}{2}\pi<\theta<\frac{1}{2}\pi$ then

$$r\sin 2\theta-\tfrac{1}{2}r^2\sin 4\theta+\tfrac{1}{3}r^3\sin 6\theta-\ldots=\theta-\arctan\left\{\left(\frac{1-r}{1+r}\right)\tan\theta\right\},$$

the inverse tangent lying between $-\frac{1}{2}\pi$ and $\frac{1}{2}\pi$. Determine the sum of the series for all other values of θ.

3. Prove, by considering the expansions of $\log(1+iz)$ and $\log(1-iz)$ in powers of z, that if $-1<r<1$ then

$$r\sin\theta+\tfrac{1}{2}r^2\cos 2\theta-\tfrac{1}{3}r^3\sin 3\theta-\tfrac{1}{4}r^4\cos 4\theta+\ldots=\tfrac{1}{2}\log(1+2r\sin\theta+r^2),$$

$$r\cos\theta+\tfrac{1}{2}r^2\sin 2\theta-\tfrac{1}{3}r^3\cos 3\theta-\tfrac{1}{4}r^4\sin 4\theta+\ldots=\arctan\left(\frac{r\cos\theta}{1-r\sin\theta}\right),$$

$$r\sin\theta-\tfrac{1}{3}r^3\sin 3\theta+\ldots=\tfrac{1}{4}\log\left(\frac{1+2r\sin\theta+r^2}{1-2r\sin\theta+r^2}\right),$$

$$r\cos\theta-\tfrac{1}{3}r^3\cos 3\theta+\ldots=\tfrac{1}{2}\arctan\left(\frac{2r\cos\theta}{1-r^2}\right),$$

the inverse tangents lying between $-\frac{1}{2}\pi$ and $\frac{1}{2}\pi$.

4. Prove that

$$\cos\theta\cos\theta-\tfrac{1}{2}\cos 2\theta\cos^2\theta+\tfrac{1}{3}\cos 3\theta\cos^3\theta-\ldots=\tfrac{1}{2}\log(1+3\cos^2\theta),$$

$$\sin\theta\sin\theta-\tfrac{1}{2}\sin 2\theta\sin^2\theta+\tfrac{1}{3}\sin 3\theta\sin^3\theta-\ldots=\operatorname{arc\,cot}(1+\cot\theta+\cot^2\theta),$$

the inverse cotangent lying between $-\frac{1}{2}\pi$ and $\frac{1}{2}\pi$; and find similar expressions for the sums of the series

$$\cos\theta\sin\theta-\tfrac{1}{2}\cos 2\theta\sin^2\theta+\ldots,\quad \sin\theta\cos\theta-\tfrac{1}{2}\sin 2\theta\cos^2\theta+\ldots.$$

246. Some applications of the logarithmic series. The exponential limit. Let z be any complex number, and h a real number small enough to ensure that $|hz|<1$. Then

$$\log(1+hz)=hz-\tfrac{1}{2}(hz)^2+\tfrac{1}{3}(hz)^3-\ldots,$$

and so
$$\frac{\log(1+hz)}{h}=z+\phi(h,z),$$

where
$$\phi(h,z)=-\tfrac{1}{2}hz^2+\tfrac{1}{3}h^2z^3-\tfrac{1}{4}h^3z^4+\ldots,$$

$$|\phi(h,z)|\leqq|hz^2|(1+|hz|+|h^2z^2|+\ldots)=\frac{|hz^2|}{1-|hz|},$$

so that $\phi(h,z)\to 0$ as $h\to 0$. It follows that

$$\lim_{h\to 0}\frac{\log(1+hz)}{h}=z \quad\ldots\ldots\ldots\ldots\ldots\ldots\ldots(1).$$

If in particular we suppose $h = 1/n$, where n is a positive integer, we obtain

$$\lim_{n\to\infty} n\log\left(1+\frac{z}{n}\right) = z,$$

and so $$\lim_{n\to\infty}\left(1+\frac{z}{n}\right)^n = \lim_{n\to\infty}\exp\left\{n\log\left(1+\frac{z}{n}\right)\right\} = \exp z \quad \ldots(2).$$

This is a generalisation of the result proved in § 215 for real values of z.

From (1) we can deduce some other results which we shall require in the next section. If t and h are real, and h is sufficiently small, we have

$$\frac{\log(1+tz+hz)-\log(1+tz)}{h} = \frac{1}{h}\log\left(1+\frac{hz}{1+tz}\right),$$

which tends to the limit $z/(1+tz)$ when $h\to 0$. Hence

$$\frac{d}{dt}\{\log(1+tz)\} = \frac{z}{1+tz} \quad \ldots\ldots\ldots\ldots\ldots\ldots(3).$$

We shall also require a formula for the differentiation of $(1+tz)^m$, where m is any number real or complex, with respect to t. We observe first that, if $\phi(t) = \psi(t)+i\chi(t)$ is a complex function of t, whose real and imaginary parts $\phi(t)$ and $\chi(t)$ are differentiable, then

$$\begin{aligned}\frac{d}{dt}(\exp\phi) &= \frac{d}{dt}\{(\cos\chi+i\sin\chi)\exp\psi\}\\ &= \{(\cos\chi+i\sin\chi)\,\psi'+(-\sin\chi+i\cos\chi)\,\chi'\}\exp\psi\\ &= (\psi'+i\chi')(\cos\chi+i\sin\chi)\exp\psi\\ &= (\psi'+i\chi')\exp(\psi+i\chi) = \phi'\exp\phi,\end{aligned}$$

so that the rule for differentiating $\exp\phi$ is the same as when ϕ is real. This being so we have

$$\begin{aligned}\frac{d}{dt}(1+tz)^m &= \frac{d}{dt}\exp\{m\log(1+tz)\}\\ &= \frac{mz}{1+tz}\exp\{m\log(1+tz)\} = mz(1+tz)^{m-1} \quad \ldots(4).\end{aligned}$$

Here both $(1+tz)^m$ and $(1+tz)^{m-1}$ have their principal values.

247. The general form of the binomial theorem. We have proved already (§ 222) that the sum of the series

$$1+\binom{m}{1}z+\binom{m}{2}z^2+\dots$$

is $(1+z)^m = \exp\{m\log(1+z)\}$, for all real values of m and all real values of z between -1 and 1. If a_n is the coefficient of z^n then

$$\left|\frac{a_{n+1}}{a_n}\right| = \frac{|m-n|}{n+1} \to 1,$$

whether m is real or complex. Hence (Ex. LXXX. 3) the series is always convergent if the modulus of z is less than unity, and we shall now prove that its sum is still $\exp\{m\log(1+z)\}$, i.e. the principal value of $(1+z)^m$.

It follows from § 246 that if t is real then

$$\frac{d}{dt}(1+tz)^m = mz(1+tz)^{m-1},$$

z and m having any real or complex values and each side having its principal value. Hence, if $\phi(t) = (1+tz)^m$, we have

$$\phi^{(n)}(t) = m(m-1)\dots(m-n+1)\,z^n(1+tz)^{m-n}.$$

This formula still holds if $t=0$, so that

$$\frac{\phi^n(0)}{n!} = \binom{m}{n}z^n.$$

It follows from (1) and (2) of § 167 (if we remember the remark made at the end of § 170) that

$$\phi(1) = \phi(0)+\phi'(0)+\frac{\phi''(0)}{2!}+\dots+\frac{\phi^{(n-1)}(0)}{(n-1)!}+R_n,$$

where $$R_n = \frac{1}{(n-1)!}\int_0^1 (1-t)^{n-1}\phi^{(n)}(t)\,dt.$$

We write $$z = r(\cos\theta + i\sin\theta),\quad m=\mu+i\nu,$$

and determine an upper bound for R_n.

We have on the one hand

$$|1+tz|<2,$$

and on the other

$$|1+tz| = \sqrt{(1+2tr\cos\theta+t^2r^2)} \geqq 1-tr \geqq 1-r;$$

while $-\pi \leqq \operatorname{am}(1+tz) \leqq \pi$. Also

$$\begin{aligned}|(1+tz)^{m-1}| &= \exp\{(\mu-1)\log|1+tz| - \nu\operatorname{am}(1+tz)\}\\ &= |1+tz|^{\mu-1}e^{-\nu\operatorname{am}(1+tz)}.\end{aligned}$$

The first factor here does not exceed $2^{\mu-1}$ if $\mu \geqq 1$, or $(1-r)^{\mu-1}$ if $\mu < 1$; and the second does not exceed $e^{\pi|\nu|}$. Hence $|(1+tz)^{m-1}|$ has an upper bound K independent of t (and n); and therefore

$$\begin{aligned}|R_n| &= \frac{|m(m-1)\dots(m-n+1)|}{(n-1)!}|z|^n\\ &\quad\times\left|\int_0^1 (1+tz)^{m-1}\left(\frac{1-t}{1+tz}\right)^{n-1}dt\right|\\ &\leqq K\frac{|m(m-1)\dots(m-n+1)|}{(n-1)!}r^n\int_0^1\left(\frac{1-t}{1-tr}\right)^{n-1}dt.\end{aligned}$$

Finally $1-tr > 1-t$, so that

$$|R_n| < K\frac{|m(m-1)\dots(m-n+1)|}{(n-1)!}r^n = \rho_n,$$

say. But

$$\frac{\rho_{n+1}}{\rho_n} = \frac{|m-n|}{n}r \to r,$$

and so (Ex. XXVII. 6) $\rho_n \to 0$. Hence $R_n \to 0$, and we arrive at the following theorem.

THEOREM. *The sum of the binomial series*

$$1+\binom{m}{1}z+\binom{m}{2}z^2+\dots$$

is $\exp\{m\log(1+z)\}$, *where the logarithm has its principal value, for all values of m, real or complex, and all values of z such that* $|z|<1$.

A more complete discussion of the binomial series, including the more difficult case in which $|z| = 1$, will be found on pp. 287 *et seq.* of Bromwich's *Infinite series* (2nd edition).

Examples XCIX. 1. Suppose m real. Then since

$$\log(1+z) = \tfrac{1}{2}\log(1+2r\cos\theta+r^2)+i\arctan\left(\frac{r\sin\theta}{1+r\cos\theta}\right),$$

we obtain

$$\sum_0^\infty \binom{m}{n} z^n = \exp\{\tfrac{1}{2}m\log(1+2r\cos\theta+r^2)\}\operatorname{Cis}\left\{m\arctan\left(\frac{r\sin\theta}{1+r\cos\theta}\right)\right\}$$

$$= (1+2r\cos\theta+r^2)^{\frac{1}{2}m}\operatorname{Cis}\left\{m\arctan\left(\frac{r\sin\theta}{1+r\cos\theta}\right)\right\},$$

all the inverse tangents lying between $-\frac{1}{2}\pi$ and $\frac{1}{2}\pi$. In particular, if we suppose $\theta = \frac{1}{2}\pi$, $z = ir$, and equate the real and imaginary parts, we obtain

$$1-\binom{m}{2}r^2+\binom{m}{4}r^4-\dots = (1+r^2)^{\frac{1}{2}m}\cos(m\arctan r),$$

$$\binom{m}{1}r-\binom{m}{3}r^3+\binom{m}{5}r^5-\dots = (1+r^2)^{\frac{1}{2}m}\sin(m\arctan r).$$

2. Prove that if $0 \leqq r < 1$ then

$$1-\frac{1.3}{2.4}r^2+\frac{1.3.5.7}{2.4.6.8}r^4-\dots = \sqrt{\left\{\frac{\sqrt{(1+r^2)}+1}{2(1+r^2)}\right\}},$$

$$\frac{1}{2}r-\frac{1.3.5}{2.4.6}r^3+\frac{1.3.5.7.9}{2.4.6.8.10}r^5-\dots = \sqrt{\left\{\frac{\sqrt{(1+r^2)}-1}{2(1+r^2)}\right\}}.$$

[Take $m = -\frac{1}{2}$ in the last two formulae of Ex. 1.]

3. Prove that if $-\frac{1}{2}\pi < \theta < \frac{1}{2}\pi$ then

$$\cos m\theta = \cos^m\theta\left\{1-\binom{m}{2}\tan^2\theta+\binom{m}{4}\tan^4\theta-\dots\right\},$$

$$\sin m\theta = \cos^m\theta\left\{\binom{m}{1}\tan\theta-\binom{m}{3}\tan^3\theta+\dots\right\},$$

for all real values of m. [These results follow at once from the equations

$$\cos m\theta+i\sin m\theta = (\cos\theta+i\sin\theta)^m = \cos^m\theta(1+i\tan\theta)^m.]$$

4. We proved (Ex. LXXXI. 6), by direct multiplication of series, that $f(m,z) = \Sigma\binom{m}{n}z^n$, where $|z|<1$, satisfies the functional equation

$$f(m,z)f(m',z) = f(m+m',z).$$

Deduce, by an argument similar to that of § 223, and without assuming the general result of p. 477, that if m is real and rational then

$$f(m,z) = \exp\{m\log(1+z)\}.$$

5. If z and μ are real, and $-1<z<1$, then

$$\Sigma\binom{i\mu}{n}z^n = \cos\{\mu\log(1+z)\}+i\sin\{\mu\log(1+z)\}.$$

MISCELLANEOUS EXAMPLES ON CHAPTER X

1. Show that the real part of $i^{\log(1-i)}$ is

$$e^{\frac{1}{8}(4k+1)\pi^2}\cos\{\tfrac{1}{4}(4k+1)\pi\log 2\},$$

where k is any integer.

2. If $a\cos\theta+b\sin\theta+c=0$, where a, b, c are real and $c^2>a^2+b^2$, then

$$\theta = m\pi+\alpha\pm i\log\frac{|c|+\sqrt{(c^2-a^2-b^2)}}{\sqrt{(a^2+b^2)}},$$

where m is any odd or any even integer, according as c is positive or negative, and α is an angle whose cosine and sine are $a/\sqrt{(a^2+b^2)}$ and $b/\sqrt{(a^2+b^2)}$.

3. Prove that if $z=re^{i\theta}$, and $r<1$, then the imaginary part of

$$\log(1+iz)-\log(1-iz)$$

(where the logarithms are principal values) is that value of

$$\arctan\left(\frac{2r\cos\theta}{1-r^2}\right)$$

which lies between $-\frac{1}{2}\pi$ and $\frac{1}{2}\pi$. (*Math. Trip.* 1916)

4. Show that if x is real and $A=a+ib$, then

$$\frac{d}{dx}\exp Ax = A\exp Ax,\quad \int\exp Ax\,dx=\frac{\exp Ax}{A}.$$

Deduce the results of Ex. LXXXVIII. 5.

5. Show that if $a>0$ then $\int_0^\infty \exp\{-(a+ib)x\}\,dx=\frac{1}{a+ib}$, and deduce the results of Ex. LXXXVIII. 6.

6. Show that if $(x/a)^2+(y/b)^2=1$ is the equation of an ellipse, and $f(x,y)$ denotes the terms of highest degree in the equation of any other algebraic curve, then the sum of the eccentric angles of the points of intersection of the ellipse and the curve differs by a multiple of 2π from

$$-i\{\log f(a,ib)-\log f(a,-ib)\}.$$

[The eccentric angles are given by $f(a\cos\alpha, b\sin\alpha)+\dots=0$ or by

$$f\left\{\tfrac{1}{2}a\left(u+\frac{1}{u}\right),\ -\tfrac{1}{2}ib\left(u-\frac{1}{u}\right)\right\}+\dots=0,$$

where $u=\exp i\alpha$; and $\Sigma\alpha$ is equal to one of the values of $-i\operatorname{Log}P$, where P is the product of the roots of this equation.]

7. Determine the number and approximate positions of the roots of the equation $\tan z=az$, where a is real.

[We know already (Ex. XVII. 4) that the equation has infinitely many real roots. Now let $z = x+iy$, and equate real and imaginary parts. We obtain

$$\frac{\sin 2x}{\cos 2x+\cosh 2y} = ax, \quad \frac{\sinh 2y}{\cos 2x+\cosh 2y} = ay,$$

so that, unless x or y is zero, we have

$$\frac{\sin 2x}{2x} = \frac{\sinh 2y}{2y}.$$

This is impossible, the left-hand side being numerically less, and the right-hand side numerically greater than unity. Thus $x = 0$ or $y = 0$. If $y = 0$ we come back to the real roots of the equation. If $x = 0$ then $\tanh y = ay$. It is easy to see that this equation has no real root other than zero if $a \leqq 0$ or $a \geqq 1$, and two such roots if $0 < a < 1$. Thus there are two purely imaginary roots if $0 < a < 1$; otherwise all the roots are real.]

8. The equation $\tan z = az+b$, where a and b are real and b is not equal to zero, has no complex roots if $a \leqq 0$. If $a > 0$ then the real parts of all the complex roots are numerically greater than $|b/2a|$.

9. The equation $\tan z = a/z$, where a is real, has no complex roots, but has two purely imaginary roots if $a < 0$.

10. The equation $\tan z = a \tanh cz$, where a and c are real, has an infinity of real and of purely imaginary roots, but no complex roots.

11. Show that if x is real then

$$e^{ax}\cos bx = \sum_0^\infty \frac{x^n}{n!}\left\{a^n - \binom{n}{2}a^{n-2}b^2 + \binom{n}{4}a^{n-4}b^4 \ldots\ldots\right\},$$

where there are $\frac{1}{2}(n+1)$ or $\frac{1}{2}(n+2)$ terms inside the large brackets. Find a similar series for $e^{ax}\sin bx$.

12. If $n\phi(z,n) \to z$ as $n \to \infty$, then $\{1+\phi(z,n)\}^n \to \exp z$.

13. If $\phi(t)$ is a complex function of the real variable t, then

$$\frac{d}{dt}\log\phi(t) = \frac{\phi'(t)}{\phi(t)}.$$

[Use the formulae

$$\phi = \psi + i\chi, \quad \log\phi = \tfrac{1}{2}\log(\psi^2+\chi^2) + i\arctan(\chi/\psi).]$$

14. **Transformations.** In Ch. III (Exs. XXI. 21 *et seq.*, and Misc. Exs. 22 *et seq.*) we considered some simple examples of the geometrical relations between figures in the planes of two variables z, Z connected by a relation $z = f(Z)$. We shall now consider some cases in which the relation involves logarithmic, exponential, or circular functions.

Suppose first that

$$z = \exp\frac{\pi Z}{a}, \quad Z = \frac{a}{\pi}\operatorname{Log} z,$$

where a is positive. To one value of Z corresponds one of z, but to one of z infinitely many of Z. If x, y, r, θ are the coordinates of z, and X, Y, R, Θ those of Z, we have the relations

$$x = e^{\pi X/a}\cos\frac{\pi Y}{a}, \quad y = e^{\pi X/a}\sin\frac{\pi Y}{a},$$

$$X = \frac{a}{\pi}\log r, \qquad Y = \frac{a\theta}{\pi} + 2ka,$$

where k is any integer. If we suppose that $-\pi < \theta \leqq \pi$, and that $\operatorname{Log} z$ has its principal value $\log z$, then $k = 0$, and Z is confined to a strip of its plane parallel to the axis OX and extending to a distance a from it on each side, one point of this strip corresponding to one of the whole z-plane, and conversely. By taking a value of $\operatorname{Log} z$ other than the principal value we obtain a similar relation between the z-plane and another strip of breadth $2a$ in the Z-plane.

To the lines in the Z-plane for which X and Y are constant correspond the circles and radii vectores in the z-plane for which r and θ are constant. To one of the latter lines corresponds the whole of a parallel to OX, but to a circle for which r is constant corresponds only a part, of length $2a$, of a parallel to OY. To make Z describe the whole of the latter line we must make z move continually round and round the circle.

15. Show that to a straight line in the Z-plane corresponds an equiangular spiral in the z-plane.

16. Discuss similarly the transformation $z = c\cosh(\pi Z/a)$, showing in particular that the whole z-plane corresponds to any one of an infinite number of strips in the Z-plane, each parallel to the axis OX and of breadth $2a$. Show also that to the line $X = X_0$ corresponds the ellipse

$$\left\{\frac{x}{c\cosh(\pi X_0/a)}\right\}^2 + \left\{\frac{y}{c\sinh(\pi X_0/a)}\right\}^2 = 1,$$

and that these ellipses, for different values of X_0, form a confocal system; and that the lines $Y = Y_0$ correspond to the associated system of confocal hyperbolas. Trace the variation of z as Z describes the whole of a line $X = X_0$ or $Y = Y_0$. How does Z vary as z describes the degenerate ellipse and hyperbola formed by the segment between the foci of the confocal system and the remaining segments of the axis of x?

17. Verify that the results of Ex. 16 are in agreement with those of Ex. 14 and those of Ch. III, Misc. Ex. 26. [The transformation

$$z = c\cosh\frac{\pi Z}{a}$$

may be regarded as compounded from the transformations

$$z = cz_1, \quad z_1 = \frac{1}{2}\left(z_2 + \frac{1}{z_2}\right), \quad z_2 = \exp\frac{\pi Z}{a}.]$$

18. Discuss similarly the transformation $z = c\tanh(\pi Z/a)$, showing that to the lines $X = X_0$ correspond the coaxal circles

$$\{x - c\coth(2\pi X_0/a)\}^2 + y^2 = c^2 \operatorname{cosech}^2(2\pi X_0/a),$$

and to the lines $Y = Y_0$ the orthogonal system of coaxal circles.

19. **The stereographic and Mercator's projections.** The points of a unit sphere whose centre is the origin are projected from the south pole (whose coordinates are 0, 0, -1) on to the tangent plane at the north pole O. The coordinates of a point on the sphere are ξ, η, ζ, and Cartesian axes OX, OY are taken on the tangent plane, parallel to the axes of ξ and η. Show that the coordinates of the projection of the point are

$$x = \frac{2\xi}{1+\zeta}, \quad y = \frac{2\eta}{1+\zeta},$$

and that $x + iy = 2\tan\frac{1}{2}\theta \operatorname{Cis}\phi$, where ϕ is the longitude (measured from the plane $\eta = 0$) and θ the north polar distance of the point on the sphere.

This projection gives a map of the sphere on the tangent plane, generally known as the *stereographic projection*. If now we introduce a new complex variable

$$Z = X + iY = -i\log\tfrac{1}{2}z = -i\log\tfrac{1}{2}(x+iy),$$

so that $X = \phi$, $Y = \log\cot\frac{1}{2}\theta$, we obtain another map in the plane of Z, usually called *Mercator's projection*. In this map parallels of latitude and longitude are represented by straight lines parallel to the axes of X and Y respectively.

20. Discuss the transformation given by the equation $z = \operatorname{Log}\left(\dfrac{Z-a}{Z-b}\right)$, showing that the straight lines for which x and y are constant correspond to two orthogonal systems of coaxal circles in the Z-plane.

21. Discuss the transformation

$$z = \operatorname{Log}\left\{\frac{\sqrt{(Z-a)} + \sqrt{(Z-b)}}{\sqrt{(b-a)}}\right\},$$

showing that the straight lines for which x and y are constant correspond to sets of confocal ellipses and hyperbolas whose foci are the points $Z = a$ and $Z = b$.

[We have
$$\sqrt{(Z-a)} + \sqrt{(Z-b)} = \sqrt{(b-a)}\exp(x+iy),$$
$$\sqrt{(Z-a)} - \sqrt{(Z-b)} = \sqrt{(b-a)}\exp(-x-iy);$$

and it will be found that

$$|Z-a| + |Z-b| = |b-a|\cosh 2x, \quad |Z-a| - |Z-b| = |b-a|\cos 2y.]$$

22. The transformation $z = Z^i$. If $z = Z^i$, where the imaginary power has its principal value, we have

$$\exp(\log r + i\theta) = z = \exp(i \log Z) = \exp(i \log R - \Theta),$$

so that $\log r = -\Theta$, $\theta = \log R + 2k\pi$, where k is an integer. Since all values of k give the same point z, we may suppose that $k = 0$, when

$$\log r = -\Theta, \quad \theta = \log R \qquad (1).$$

The whole plane of Z is covered when R varies through all positive values of Θ from $-\pi$ to π: then r has the range $\exp(-\pi)$ to $\exp \pi$ and θ ranges through all real values. Thus the Z-plane corresponds to the ring bounded by the circles $r = \exp(-\pi)$, $r = \exp \pi$; but this ring is covered infinitely often. If however θ is allowed to vary only between $-\pi$ and π, so that the ring is covered only once, then R can vary only from $\exp(-\pi)$ to $\exp \pi$, so that the variation of Z is restricted to a ring similar in all respects to that within which z varies. Each ring, moreover, must be regarded as having a barrier along the negative real axis which z (or Z) must not cross, since its amplitude must not transgress the limits $-\pi$ and π.

We thus obtain a correspondence between two rings, given by the pair of equations

$$z = Z^i, \quad Z = z^{-i},$$

where each power has its principal value. To circles whose centre is the origin in one plane correspond straight lines through the origin in the other.

23. Trace the variation of z when Z, starting at the point $\exp \pi$, moves round the larger circle in the positive direction to the point $-\exp \pi$, along the barrier, round the smaller circle in the negative direction, back along the barrier, and round the remainder of the larger circle to its original position.

24. If $z = Z^i$, any value of the power being taken, and Z moves along an equiangular spiral whose pole is the origin in its plane, then z moves along an equiangular spiral whose pole is the origin in its plane.

25. How does $Z = z^{ai}$, where a is real, behave as z approaches the origin along the real axis? [Z moves round and round a circle whose centre is the origin (the unit circle if z^{ai} has its principal value), and the real and imaginary parts of Z both oscillate finitely.]

26. Show that the region of convergence of a series of the type $\sum_{-\infty}^{\infty} a_n z^{nai}$, where a is real, is an angle, i.e. a region bounded by inequalities of the type $\theta_0 < \operatorname{am} z < \theta_1$. [The angle may reduce to a line, or cover the whole plane.]

27. **Level curves.** If $f(z)$ is a function of the complex variable z, we call the curves for which $|f(z)|$ is constant the *level curves* of $f(z)$. Sketch the forms of the level curves of

$z-a$ (*concentric circles*), $(z-a)(z-b)$ (*Cartesian ovals*),

$(z-a)/(z-b)$ (*coaxal circles*), $\exp z$ (*straight lines*).

28. Sketch the forms of the level curves of $(z-a)(z-b)(z-c)$.

29. Sketch the forms of the level curves of (i) $z\exp z$, (ii) $\sin z$. [See Fig. 56*, which represents the level curves of $\sin z$. The curves marked I–VIII correspond to $k = \cdot 35, \cdot 50, \cdot 71, 1\cdot 00, 1\cdot 41, 2\cdot 00, 2\cdot 83, 4\cdot 00$.]

30. Sketch the forms of the level curves of $\exp z - c$, where c is a real constant. [Fig. 57 shows the level curves of $|\exp z - 1|$, the curves I–VII

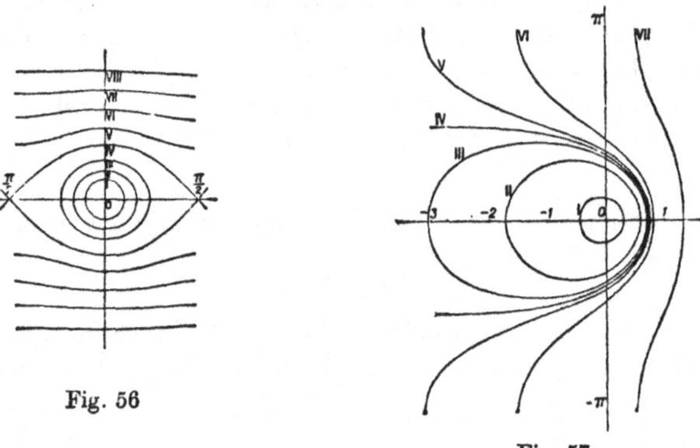

Fig. 56

Fig. 57

corresponding to the values of k given by $\log k = -1\cdot 00, -\cdot 20, -\cdot 05, 0\cdot 00, \cdot 05, \cdot 20, 1\cdot 00$.]

31. The level curves of $\sin z - c$, where c is a positive constant, are sketched in Figs. 58, 59. [The nature of the curves differs in the two cases $c<1$ and $c>1$. In Fig. 58 we have taken $c = \cdot 5$, and the curves I–VIII correspond to $k = \cdot 29, \cdot 37, \cdot 50, \cdot 87, 1\cdot 50, 2\cdot 60, 4\cdot 50, 7\cdot 79$. In Fig. 59 we have taken $c = 2$, and the curves I–VII correspond to $k = \cdot 58, 1\cdot 00, 1\cdot 73, 3\cdot 00, 5\cdot 20, 9\cdot 00, 15\cdot 59$. If $c = 1$ then the curves are the same as those of Fig. 56, except that the origin and scale are different.]

32. Prove that if $0<\theta<\pi$ then

$$\cos\theta + \tfrac{1}{3}\cos 3\theta + \tfrac{1}{5}\cos 5\theta + \dots = \tfrac{1}{4}\log\cot^2\tfrac{1}{2}\theta,$$

$$\sin\theta + \tfrac{1}{3}\sin 3\theta + \tfrac{1}{5}\sin 5\theta + \dots = \tfrac{1}{4}\pi,$$

* The figures were drawn for me by Mr (now Prof.) E. H. Neville when an undergraduate.

and determine the sums of the series for all other values of θ for which they are convergent. [Use the equation

$$z+\tfrac{1}{3}z^3+\tfrac{1}{5}z^5+\ldots = \tfrac{1}{2}\log\left(\frac{1+z}{1-z}\right),$$

where $z = \cos\theta + i\sin\theta$. When θ is increased by π the sum of each series changes its sign. It follows that the first formula holds for all values of θ save multiples of π (for which the series diverges), while the sum of the second series is $\frac{1}{4}\pi$ if

$$2k\pi < \theta < (2k+1)\pi,$$

$-\frac{1}{4}\pi$ if $$(2k+1)\pi < \theta < (2k+2)\pi,$$

and 0 if θ is a multiple of π.]

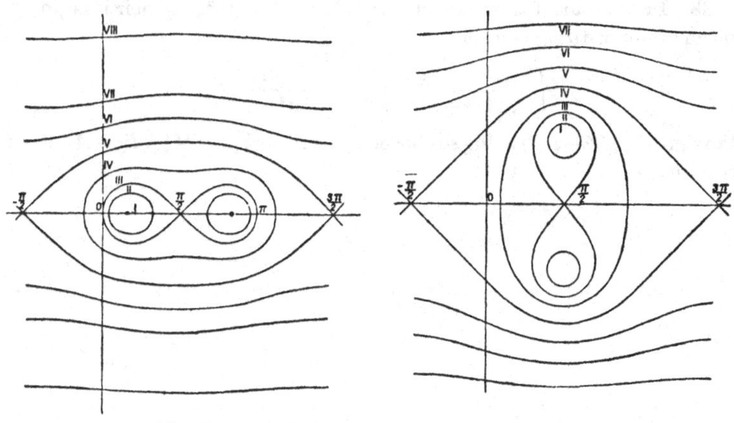

Fig. 58 Fig. 59

33. Prove that $$\sum_1^\infty \frac{\sin n\theta}{n} = \pi\left(\frac{1}{2}-\frac{\theta}{2\pi}+\left[\frac{\theta}{2\pi}\right]\right)$$

for all real θ. (*Math. Trip.* 1932)

34. Prove that if $0<\theta<\frac{1}{2}\pi$ then

$$\cos\theta - \tfrac{1}{3}\cos 3\theta + \tfrac{1}{5}\cos 5\theta - \ldots = \tfrac{1}{4}\pi,$$
$$\sin\theta - \tfrac{1}{3}\sin 3\theta + \tfrac{1}{5}\sin 5\theta - \ldots = \tfrac{1}{4}\log(\sec\theta+\tan\theta)^2;$$

and determine the sums of the series for all other values of θ for which they are convergent.

35. Prove that

$$\cos\theta\cos\alpha + \tfrac{1}{2}\cos 2\theta\cos 2\alpha + \tfrac{1}{3}\cos 3\theta\cos 3\alpha + \ldots = -\tfrac{1}{4}\log\{4(\cos\theta-\cos\alpha)^2\},$$

unless $\theta-\alpha$ or $\theta+\alpha$ is a multiple of 2π.

36. Prove that if neither a nor b is real then

$$\int_0^\infty \frac{dx}{(x-a)(x-b)} = -\frac{\log(-a)-\log(-b)}{a-b},$$

each logarithm having its principal value. Verify the result when $a = ci$, $b = -ci$, where c is positive. Discuss also the cases in which a or b or both are real and negative.

37. Prove that if α and β are real, and $\beta > 0$, then

$$\int_0^\infty \frac{dx}{x^2-(\alpha+i\beta)^2} = \frac{\pi i}{2(\alpha+i\beta)}.$$

What is the value of the integral when $\beta < 0$?

38. Prove that, if the roots of $Ax^2+2Bx+C = 0$ have their imaginary parts of opposite signs, then

$$\int_{-\infty}^\infty \frac{dx}{Ax^2+2Bx+C} = \frac{\pi i}{\sqrt{(B^2-AC)}},$$

the sign of $\sqrt{(B^2-AC)}$ being so chosen that the real part of $\{\sqrt{(B^2-AC)}\}/Ai$ is positive.

APPENDIX I

The inequalities of Hölder and Minkowski

There are three inequalities which are particularly important in analysis, the theorem of the arithmetic and geometric means and Hölder's and Minkowski's inequalities. The first of these is needed in a more general form than that of p. 34: the other two can then be deduced from it.

In what follows all letters denote (strictly) positive numbers. We can prove as on p. 34* that

$$\frac{a_1+a_2+\ldots+a_n}{n} > (a_1 a_2 \ldots a_n)^{1/n} \qquad \ldots\ldots\ldots\ldots\ldots\ldots(1)$$

unless all the a are equal (in which case the two means are equal). If we suppose that they fall into m groups of equal numbers, p_1 being equal to a_1, p_2 to a_2, and so on, so that

$$p_1+p_2+\ldots+p_m = n,$$

then (1) becomes

$$q_1 a_1 + q_2 a_2 + \ldots + q_m a_m > a_1^{q_1} a_2^{q_2} \ldots a_m^{q_m} \qquad \ldots\ldots\ldots\ldots\ldots(2),$$

where

$$q_\nu = \frac{p_\nu}{p_1+p_2+\ldots+p_m} \qquad \ldots\ldots\ldots\ldots\ldots\ldots\ldots\ldots(3),$$

so that

$$q_1+q_2+\ldots+q_m = 1 \qquad \ldots\ldots\ldots\ldots\ldots\ldots\ldots\ldots(4).$$

Here again the inequality reduces to an equality when all the a are equal.

Conversely, if $q_1, q_2, \ldots, q_m$ are any positive rational numbers whose sum is 1, we can reduce them to a common denominator and write them in the form (3), when (2) reduces to a case of (1).

We shall now prove that (2) is true (unless all the a are equal) for all real q whose sum is 1. In other words, we shall remove the restriction that the q are *rational*. We shall call this theorem the 'general theorem of the means', and refer to it as G_m, or simply as G. The proof will be independent of what precedes.

We can reduce the proof to a proof of the special case G_2. For suppose that $m>2$, and that G_k has been proved for $k = 2, 3, \ldots, m-1$. Let

$$q_1+q_2+\ldots+q_{m-1} = q,$$

so that

$$q+q_m = 1;$$

and write

$$q_1' = q_1/q, \ \ldots, \ q_{m-1}' = q_{m-1}/q,$$

so that

$$q_1'+q_2'+\ldots+q_{m-1}' = 1.$$

* We did not actually prove quite so much, but the argument needs only slight changes. It is unnecessary to state these in detail, since (1) is included in (2) and we give an independent proof of (2).

Then
$$a_1^{q_1}\dots a_{m-1}^{q_{m-1}}a_m^{q_m} = (a_1^{q_1'}\dots a_{m-1}^{q'_{m-1}})^q a_m^{q_m}$$
$$\leqq q(a_1^{q_1'}\dots a_{m-1}^{q'_{m-1}}) + q_m a_m$$
$$\leqq q(q_1' a_1 + \dots + q'_{m-1}a_{m-1}) + q_m a_m$$
$$= q_1 a_1 + q_2 a_2 + \dots + q_m a_m,$$

by G_2 and G_{m-1}. There is inequality in the second line unless

$$a_1^{q_1'}\dots a_{m-1}^{q'_{m-1}} = a_m,$$

and in the third unless $a_1 = a_2 = \dots = a_{m-1}$; and therefore there is inequality somewhere unless $a_1 = a_2 = \dots = a_{m-1} = a_m$. Hence G_k is true for $k = m$, and therefore it is true generally.

It remains to prove G_2. Changing the notation, we can write G_2 as

$$a^\alpha b^{1-\alpha} < \alpha a + (1-\alpha)b \quad (0<\alpha<1) \quad \dots\dots\dots(5)$$

(unless $a = b$). We may plainly suppose, without loss of generality, that $b > a$. Then (5) is

$$b^{1-\alpha} - a^{1-\alpha} < (1-\alpha)(b-a)a^{-\alpha} \dots\dots\dots(6).$$

But, by the Mean Value Theorem (p. 242),

$$b^{1-\alpha} - a^{1-\alpha} = (1-\alpha)(b-a)\xi^{-\alpha},$$

where $a < \xi < b$, and this gives (6), since $-\alpha < 0$ and so $\xi^{-\alpha} < a^{-\alpha}$. This proves G_2 and so G_m.

We can also write the general inequality G_m in a form like (5), viz.

$$a^\alpha b^\beta \dots l^\lambda < \alpha a + \beta b + \dots + \lambda l \quad \dots\dots\dots(7),$$

where $\alpha + \beta + \dots + \lambda = 1$.

One question may occur to the reader: can we not deduce the general theorem, by a limiting process, from the special case in which the q are rational? We can approximate to each q_ν by a sequence of rationals $q_\nu^{(r)}$ in such a way that

$$q_1^{(r)} + q_2^{(r)} + \dots + q_m^{(r)} = 1,$$

for each r, and $q_\nu^{(r)} \to q_\nu$, for each ν, when $r \to \infty$. Then

$$q_1^{(r)}a_1 + q_2^{(r)}a_2 + \dots + q_m^{(r)}a_m > a_1^{q_1^{(r)}} a_2^{q_2^{(r)}} \dots a_m^{q_m^{(r)}} \quad \dots\dots\dots(8)$$

for each r, and the two sides of (8) tend to those of (2) when $r \to \infty$.

This argument would suffice if we were content to prove (2) in the less sharp form in which '$>$' is replaced by '$\geqq$'. But '$>$' degenerates into '$\geqq$' when $r \to \infty$: $x^{(r)} \to x$, $y^{(r)} \to y$, and $x^{(r)} > y^{(r)}$ imply only $x \geqq y$ and not necessarily $x > y$. The difficulty can be overcome (see *Inequalities*, p. 18), but at the cost of a little ingenuity, and we prefer to follow a more direct course.

The inequality (6) is one of those proved, in § 74, with the restriction that α is rational. The reader will find it instructive to prove that all the inequalities of § 74 are true for general, not necessarily rational, indices. This would plainly have been impossible in § 74, since x^α is not defined, for irrational α, until § 214.

There is another interesting way of looking at G_2. Since

$$\frac{d^2}{dx^2}\log x = -\frac{1}{x^2} < 0,$$

the function $\log x$ is *concave* (i.e. its graph has everywhere negative curvature), and all chords of the curve $y = \log x$ lie below the curve. If P is $(a, \log a)$ and Q is $(b, \log b)$, then the point R which divides PQ so that

$$\alpha \,.\, PR = (1-\alpha)\, RQ$$

has abscissa $\alpha a + (1-\alpha)b$ and ordinate $\alpha \log a + (1-\alpha)\log b$. Hence

$$\alpha \log a + (1-\alpha)\log b < \log\{\alpha a + (1-\alpha) b\},$$

which is (5).

Hölder's inequality (*H*)

If $k > 1$ *and* $k' = k/(k-1)$, *so that* $k' > 1$ *and*

$$\frac{1}{k} + \frac{1}{k'} = 1 \quad \text{.................................(9)};$$

and $a_1, a_2, \ldots, a_n$ *and* $b_1, b_2, \ldots, b_n$ *are two sets of positive numbers; then*

$$\sum_{m=1}^{n} a_m b_m \leqq \left(\sum_{m=1}^{n} a_m^k\right)^{1/k} \left(\sum_{m=1}^{n} b_m^{k'}\right)^{1/k'} \quad \text{...............(10)}.$$

There is inequality unless the sets (*a*) *and* (*b*) *are proportional,* i.e. *unless* a_m/b_m *is independent of* m.

This is a corollary of (5). Since each side of (10) is homogeneous (of degree 1) in both the a and the b, we may suppose without loss of generality that

$$\Sigma a = 1, \quad \Sigma b = 1 \quad \text{........................(11)}.$$

If also we write α for $1/k$ and β for $1/k'$, so that $\alpha + \beta = 1$, and replace a and b by a^α and b^β, then (10) becomes

$$\Sigma a^\alpha b^\beta \leqq (\Sigma a)^\alpha (\Sigma b)^\beta \quad \text{........................(12)}.$$

But, by (5),

$$\Sigma a^\alpha b^\beta \leqq \Sigma(\alpha a + \beta b) = \alpha + \beta = 1 = (\Sigma a)^\alpha (\Sigma b)^\beta.$$

There is inequality unless $a_m = b_m$ for every m; and therefore, when we drop the conditions (11), unless a_m/b_m is independent of m.

More generally

$$\Sigma a^\alpha b^\beta \ldots l^\lambda < (\Sigma a)^\alpha (\Sigma b)^\beta \ldots (\Sigma l)^\lambda \text{.................(13)}$$

if

$$\alpha + \beta + \ldots + \lambda = 1 \quad \text{.......................(14)},$$

unless the sets (a), (b), ..., (l) are all proportional. This may be deduced from (7) as we deduced (12) from (5), or by induction from (12) itself.

Minkowski's inequality (M)

If $k>1$ and $a_1, a_2, \ldots, a_n$ and $b_1, b_2, \ldots, b_n$ are two sets of positive numbers, then

$$\left\{\sum_{m=1}^{n}(a_m+b_m)^k\right\}^{1/k} \leqq \left(\sum_{m=1}^{n}a_m^k\right)^{1/k}+\left(\sum_{m=1}^{n}b_m^k\right)^{1/k} \quad\text{.........(15).}$$

There is inequality unless the sets (a) and (b) are proportional.

This may be deduced from (10). We write

$$S=\left\{\sum_{m=1}^{n}(a_m+b_m)^k\right\}^{1/k}=\{\Sigma(a+b)^k\}^{1/k}$$

(dropping the suffixes). Then

$$S=\Sigma a(a+b)^{k-1}+\Sigma b(a+b)^{k-1}.$$

Applying (10) to each term on the right, and observing that $(k-1)k'=k$, we obtain

$$\begin{aligned} S &\leqq (\Sigma a^k)^{1/k}\{\Sigma(a+b)^k\}^{1/k'}+(\Sigma b^k)^{1/k}\{\Sigma(a+b)^k\}^{1/k'} \\ &= \{(\Sigma a^k)^{1/k}+(\Sigma b^k)^{1/k}\}\,S^{1/k'}, \end{aligned}$$

and (15) follows when we divide by $S^{1/k'}$. There is inequality unless each of (a) and (b) is proportional to $(a+b)$, i.e. unless (a) and (b) are proportional.

There is a useful companion inequality (in the opposite direction). Suppose that $a+b=1$. Then $a<1$, $b<1$ and so (since $k>1$) $a^k<a$, $b^k<b$ and

$$a^k+b^k<a+b=1=(a+b)^k \quad\text{.................(16).}$$

Since both sides of the final inequality are homogeneous (of degree k), it is true generally (without the restriction $a+b=1$). It follows that

$$\Sigma(a+b)^k>\Sigma a^k+\Sigma b^k \quad\text{.......................(17).}$$

There are no cases of equality when (as we have assumed throughout) the a and b are strictly positive.

Miscellaneous remarks on the inequalities

When $k=2$, $k'=2$, and H reduces to

$$(\Sigma ab)^2<\Sigma a^2\Sigma b^2,$$

Cauchy's inequality (p. 34). If we suppose that $k=2$ and $n=3$ in M, and take the sets (a) and (b) to be x_1, y_1, z_1 and x_2, y_2, z_2, it becomes

$$\sqrt{\{(x_1+x_2)^2+(y_1+y_2)^2+(z_1+z_2)^2\}}<\sqrt{(x_1^2+y_1^2+z_1^2)}+\sqrt{(x_2^2+y_2^2+z_2^2)}.$$

This expresses the fact that one side of the triangle whose vertices are $(0,0,0)$, (x_1,y_1,z_1), $(-x_2,-y_2,-z_2)$ is less than the sum of the other two. The inequality reduces to equality when x_1, y_1, z_1 are proportional to x_2, y_2, z_2, that is to say when the triangle degenerates. In general, M

extends the 'triangle inequality' to a space of n dimensions in which the distance between two points P_1, P_2 is defined as

$$(|x_1-x_2|^k+|y_1-y_2|^k+|z_1-z_2|^k+\dots)^{1/k}.$$

The inequality (7) of p. 33 is a corollary of H, since

$$(\Sigma a)^k = (\Sigma a\,.\,1)^k < \Sigma a^k(\Sigma 1)^{k/k'} = n^{k-1}\Sigma a^k.$$

But (6) of p. 33 is not deducible from any of the inequalities proved here, and is in fact a case of an inequality of a different type, Tchebychef's inequality: see *Inequalities*, p. 43.

When k is rational, H and M are *algebraical* theorems, and it is desirable that the proofs should also be algebraical, i.e. that they should not depend on limit processes of any kind. Such proofs will be found in Ch. II of *Inequalities* (where many analogues and extensions of the theorems are also discussed). If k is irrational, x^k is not an algebraical function; and then there can be no question of an algebraical proof. In this book, for example, x^k was defined as $\exp(k\log x)$, and it is natural that the proofs which we have given should depend on the processes of the calculus and the theory of the logarithmic and exponential functions.

APPENDIX II

The proof that every equation has a root

The theorem that 'every algebraical equation has a root' is usually called 'the fundamental theorem of algebra', but it belongs more properly to analysis, since it cannot be proved without appeal somewhere to considerations of continuity. It may be worth while to give a sketch of the two most familiar proofs.

(A) The first is a natural development of the ideas of Chs. III and X. Let

$$Z = f(z) = a_0 z^n + a_1 z^{n-1} + \dots + a_n$$

be a polynomial in z with real or complex coefficients. We may assume that $a_0 \neq 0$.

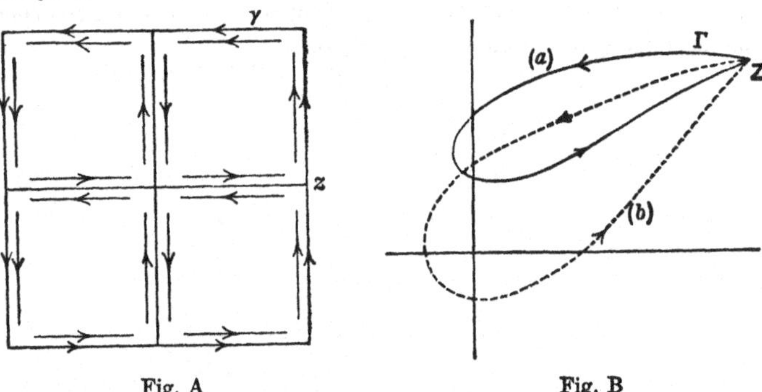

Fig. A Fig. B

Suppose that z describes a closed path γ in the z-plane: actually, γ will always be a square, with sides parallel to the axes, described in the positive direction. Then Z describes a closed path Γ in the Z-plane. We may assume, for our present purpose, that Γ does not pass through the origin, since to suppose that it did, at any stage of the argument, would be to admit the truth of the theorem.

To any value of Z correspond an infinity of values of am Z, differing by multiples of 2π, and each of these values varies continuously as Z describes Γ*. We choose a particular value of am Z (say the value for which $-\pi < \text{am}\, Z \leqq \pi$) corresponding to the initial value of Z, and follow its variation along Γ. We thus define a value of am Z (which we call simply am Z) corresponding to each Z on Γ.

* It is here that we use the assumption that Γ does not pass through the origin.

When Z returns to its original position, am Z may be unchanged or may differ from its original value by a multiple of 2π. Thus if Γ does not enclose the origin, like (a) in Fig. B, am Z will be unchanged; but if Γ winds once round the origin in the positive direction, like (b), am Z will have increased by 2π. We denote the increment of am Z when z describes γ by $\Delta(\gamma)$.

Suppose first that γ is the square S, of side $2R$, defined by the lines $x = \pm R$, $y = \pm R$. Then $|z| \geqq R$ on S. We can choose R so large that

$$\frac{|a_1|}{|a_0|R}+\frac{|a_2|}{|a_0|R^2}+\dots+\frac{|a_n|}{|a_0|R^n}<\tfrac{1}{2},$$

and then
$$Z = a_0 z^n\left(1+\frac{a_1}{a_0 z}+\dots+\frac{a_n}{a_0 z^n}\right) = a_0 z^n(1+\eta),$$

where $|\eta| < \frac{1}{2}$ at all points of S. The amplitude of $1+\eta$ is plainly unchanged when z describes S, and that of z^n is increased by $2n\pi$. Hence that of Z is increased by $2n\pi$, i.e. $\Delta(S) = 2n\pi$. All that we shall actually need to know is that $\Delta(S) \neq 0$.

The square S is divided into four equal squares $S^{(1)}$, $S^{(2)}$, $S^{(3)}$, $S^{(4)}$, of side R, by the coordinate axes. We can take any one of these for γ, and assume again that the corresponding Γ does not pass through the origin. Then

$$\Delta(S) = \Delta(S_1^{(1)})+\Delta(S_1^{(2)})+\Delta(S_1^{(3)})+\Delta(S_1^{(4)}) \quad \text{...........(1)}.$$

For if z describes each of $S_1^{(1)}, \dots$ in turn, it will (see Fig. A) have described each side of S once, and each side l of a smaller square which is not part of a side of S twice in opposite directions, and the two contributions of l to the sum (1) will cancel one another. Since $\Delta(S) \neq 0$, one at least of $\Delta(S_1^{(1)}), \dots$ is not zero; we choose the first which is not, and call the square thus chosen S_1. Then $\Delta(S_1) \neq 0$.

We now divide S_1 into four equal squares by parallels to the axes, and repeat the argument, thus obtaining a square S_2, of side $\frac{1}{2}R$, for which $\Delta(S_2) \neq 0$. Continuing the process, we obtain a series of squares $S, S_1, S_2, \dots, S_n, \dots$, of sides $2R, R, \frac{1}{2}R, \dots, 2^{-n+1}R$, each lying inside the one which precedes, and with $\Delta(S_n) \neq 0$ for every n.

If the south-west and north-east corners of S_n are (x_n, y_n) and (x_n', y_n'), so that $x_n'-x_n = y_n'-y_n = 2^{-n+1}R$, then (x_n) and (y_n) are increasing sequences and (x_n') and (y_n') decreasing sequences; x_n and x_n' tend to a common limit x_0, and y_n and y_n' to a common limit y_0. The point (x_0, y_0), or P, lies in, or on the boundary of, every S_n*. Given any positive δ, we can choose n so that the distance of every point of S_n from P is less than δ. Hence P has the property that, however small be δ, there is a square S_n

* There is nothing in the argument so far to show that it does not lie on the boundary of S, though it will appear later that it cannot.

containing P, and with all its points at a distance from P less than δ, for which $\Delta(S_n) \neq 0$.

We can now prove that

$$f(z_0) = f(x_0 + iy_0) = 0.$$

For suppose that $f(z_0) = c$, where $|c| = \rho > 0$. Since $f(x_0 + iy_0)$ is a continuous function of x_0 and y_0, we can choose n large enough to make

$$|f(z) - f(z_0)| < \tfrac{1}{2}\rho$$

at all points of S_n. Then

$$Z = f(z) = c + \sigma = c(1 + \eta),$$

where $|\sigma| < \frac{1}{2}\rho$, $|\eta| < \frac{1}{2}$, at all points of S_n. It follows that am Z is unchanged when z describes S_n; a contradiction. Hence $f(z_0) = 0$*.

(B) Our second proof depends on an extension to functions of several variables of the results of §§ 103 *et seq.*

We define, as in § 103, the upper and lower bounds of a function $F(x, y)$ in the domain D bounded by a square such as S. We can prove (much as in the last paragraph of § 105†) that a continuous function attains its upper and lower bounds in any such domain D.

Let

$$F(x, y) = |f(x + iy)| = |f(z)| = |Z|.$$

Then $F(x, y)$ is continuous and non-negative, and has a non-negative lower bound m in D, which is attained at some point z_0 of D. It is easy to see that, if R is large, z_0 lies inside D‡.

Suppose that $m > 0$. If we put $z = z_0 + \zeta$, and rearrange $f(z)$ in powers of ζ, we obtain

$$f(z) = f(z_0) + A_1\zeta + A_2\zeta^2 + \dots + A_n\zeta^n,$$

where $A_1, A_2, \dots, A_n$ are independent of ζ. Let A_k be the first of these coefficients which is not zero, and write

$$f(z_0) = me^{i\mu}, \quad A_k = ae^{i\alpha}, \quad \zeta = \rho e^{i\phi}.$$

We can suppose ρ small enough to ensure that $a\rho^k < m$ and

$$|A_{k+1}\zeta^{k+1} + \dots + A_n\zeta^n| < \tfrac{1}{2}a\rho^k.$$

Then

$$f(z) = me^{i\mu} + a\rho^k e^{i(\alpha + k\phi)} + g,$$

* And so, incidentally, z_0 is not on S, since Z is large at all points of S.

† The first proof of § 105 depends on a Dedekind section, and has no analogue in two dimensions.

‡ For suppose (putting in evidence the dependence of S, D, m_0 and z_0 on R) that $m_0(R)$ and $z_0(R)$ correspond to $S(R)$ and $D(R)$: then $z_0(R)$ might (so far as its definition shows) lie on $S(R)$. However, given R_1, we can choose R_2 so that $|Z|$ is greater than $\frac{1}{2}|a_0|R_2^n$, and so certainly greater than $m(R_1)$, at all points on and outside $S(R_2)$; and then $z_0(R)$ lies inside $S(R_2)$, and so certainly inside $S(R)$, for $R \geqq R_2$. Actually, $m(R)$ and $z_0(R)$ are, from a certain R onwards, independent of R.

where $|g| < \frac{1}{2}a\rho^k$. We choose ϕ so that

$$\alpha + k\phi = \mu + \pi \quad \text{.............................(2)},$$

and then
$$f(z) = e^{i\mu}\{m - a\rho^k + ge^{-i\mu}\},$$
$$|f(z)| = |m - a\rho^k + ge^{i\mu}| \leqq m - a\rho^k + |g| < m - \tfrac{1}{2}a\rho^k < m,$$

a contradiction. It follows that $m = 0$, i.e. $f(z_0) = 0$.

When we choose ϕ so as to satisfy (1), we are in effect solving the equation

$$\zeta^k = -\rho^k e^{i(\mu-\alpha)}.$$

In other words, we are using the fact that an equation of the special form

$$z^n - c = 0 \quad \text{.................................(3)}$$

has always a root, i.e. that the 'fundamental theorem' is true for *binomial* equations. We know of course, after § 48 (and our later rigorous accounts of the circular and exponential functions), that (3) has in fact n roots.

There is, however, some logical interest in finding a proof of the theorem independent of the theory of the circular functions. Our argument would give such a proof if we already possessed one for the special equation (3); and Littlewood, in a note in vol. 16 of the *Journal of the London Mathematical Society*, has shown how we can complete this proof by applying the 'lower bound' argument to the special function

$$f(z) = z^n - c \quad \text{.............................(4)},$$

where $c = a + ib \neq 0$.

We know (p. 94, Ex. 14) that any quadratic equation, and in particular the equation $z^2 = c$, has roots. They are in fact

$$\pm\sqrt{[\tfrac{1}{2}\{\sqrt{(a^2+b^2)}+a\}]} \pm i\sqrt{[\tfrac{1}{2}\{\sqrt{(a^2+b^2)}-a\}]},$$

the signs being like if $b > 0$ and unlike if $b < 0$. Hence, if $n = 2^\nu N$, where N is odd, we can, by the solution of ν quadratics, reduce the solution of (3) to that of an equation $z^N - d = 0$. We may therefore suppose n *odd*.

We now argue as before with the special function (4). There are two possibilities: either $z_0 \neq 0$ or $z_0 = 0$. If $z_0 \neq 0$, then

$$f(z_0 + \zeta) = f(z_0) + nz_0^{n-1}\zeta + \ldots = f(z_0) + A_1\zeta + \ldots,$$

where $A_1 \neq 0$, so that $k = 1$. The completion of the proof then involves only the solution of a *linear* equation. If on the other hand $z_0 = 0$, then

$$f(z) = f(\zeta) = \zeta^n - c.$$

If we give ζ the four values $\pm\rho$, $\pm i\rho$, with a small ρ, then (since n is odd) $f(\zeta)$ takes the four values

$$-c \pm \delta^n, \quad -c \pm i\delta^n.$$

In other words, if P is the point $f(z_0)$ or $-c$ in the Argand diagram, the four points representing $f(z)$ in the four cases are obtained from P by small displacements in each of the four possible directions parallel to the axes.

At least one of these brings P nearer to the origin* and, if ζ has the appropriate value, $|f(z)|<|f(z_0)|$. We thus obtain the contradiction required to complete the proof. The main ideas of the proof can be found in Cauchy's *Exercises de mathématiques*, t. 4, pp. 65–128 (though in a less concise and accurate form). There is an account of this proof in ch. 2 of Todhunter's *Theory of equations*.

Of the many proofs which have been given of the 'fundamental theorem' perhaps the one most satisfactory to an algebraist is 'Gauss's second proof' (in one of the simplified forms given by later writers). See Gauss, *Werke*, vol. iii, pp. 33–56 or Perron, *Algebra*, vol. i, pp. 258–266. These proofs are however much longer.

EXAMPLES ON APPENDIX II

1. Show that the number of roots of $f(z) = 0$ lying within a closed contour which does not pass through any root is equal to the increment of

$$\frac{1}{2\pi i}\log f(z)$$

when z describes the contour.

2. Show that if R is any number such that

$$\frac{|a_1|}{R}+\frac{|a_2|}{R^2}+\dots+\frac{|a_n|}{R^n}<1,$$

then all the roots of $z^n+a_1z^{n-1}+\dots+a_n = 0$ are in absolute value less than R. In particular show that all the roots of $z^5-13z-7 = 0$ are in absolute value less than $2\frac{1}{67}$.

3. Determine the numbers of the roots of the equation $z^{2p}+az+b=0$, where a and b are real and p odd, which have their real parts positive and negative. Show that if $a>0$, $b>0$ then the numbers are $p-1$ and $p+1$; if $a<0$, $b>0$ they are $p+1$ and $p-1$; and if $b<0$ they are p and p. Discuss the particular cases in which $a = 0$ or $b = 0$. Verify the results when $p = 1$.

[Trace the variation of $\operatorname{am}(z^{2p}+az+b)$ as z describes the contour formed by a large semi-circle whose centre is the origin and whose radius is R, and the part of the imaginary axis intercepted by the semi-circle.]

4. Consider similarly the equations

$$z^{4q}+az+b = 0,\quad z^{4q-1}+az+b = 0,\quad z^{4q+1}+az+b = 0.$$

* Leaving the ordinate unaltered and decreasing the absolute value of the abscissa, or *vice versa*.

5. Show that if α and β are real then the numbers of the roots of the equation $z^{2n}+\alpha^2 z^{2n-1}+\beta^2=0$ which have their real parts positive and negative are $n-1$ and $n+1$, or n and n, according as n is odd or even. (*Math. Trip.* 1891)

6. The points z_1, z_2, z_3 form a triangle in the complex plane, the interior of the triangle lying to the left of the side from z_1 to z_2. Show that, when z moves along the straight line joining the points $z=z_1$, $z=z_2$, from a point near z_1 to a point near z_2, the increment of

$$\operatorname{am}\left(\frac{1}{z-z_1}+\frac{1}{z-z_2}+\frac{1}{z-z_3}\right)$$

is nearly equal to π.

7. A contour enclosing the three points $z=z_1$, $z=z_2$, $z=z_3$ is defined by parts of the sides of the triangle formed by z_1, z_2, z_3, and the parts exterior to the triangle of three small circles with their centres at those points. Show that when z describes the contour the increment of

$$\operatorname{am}\left(\frac{1}{z-z_1}+\frac{1}{z-z_2}+\frac{1}{z-z_3}\right)$$

is equal to -2π.

8. Prove that a closed oval path which surrounds all the roots of a cubic equation $f(z)=0$ also surrounds those of the derived equation $f'(z)=0$. [Use the equation

$$f'(z)=f(z)\left(\frac{1}{z-z_1}+\frac{1}{z-z_2}+\frac{1}{z-z_3}\right),$$

where z_1, z_2, z_3 are the roots of $f(z)=0$, and the result of Ex. 7.]

9. Show that the roots of $f'(z)=0$ are the foci of the ellipse which touches the sides of the triangle (z_1, z_2, z_3) at their middle points. [For a proof see Cesàro's *Elementares Lehrbuch der algebraischen Analysis*, p. 352.]

10. Extend the result of Ex. 8 to equations of any degree.

11. If $f(z)$ and $\phi(z)$ are two polynomials in z, γ is a contour which does not pass through any root of $f(z)$, and $|\phi(z)|<|f(z)|$ at all points on γ, then the numbers of the roots of the equations

$$f(z)=0, \quad f(z)+\phi(z)=0$$

which lie inside γ are the same.

12. Show that the equations

$$e^z=az, \quad e^z=az^2, \quad e^z=az^3,$$

where $a>e$, have respectively (i) one positive root, (ii) one positive and one negative root, and (iii) one positive and two complex roots within the circle $|z|=1$. (*Math. Trip.* 1910)

APPENDIX III

A note on double limit problems

In Chapters IX and X we came into contact with some special cases of a general problem of great importance in analysis.

In § 220 we proved that

$$\log(1+x) = x - \tfrac{1}{2}x^2 + \tfrac{1}{3}x^3 - \dots,$$

where $-1 < x \leqslant 1$, by integrating the equation

$$\frac{1}{1+t} = 1 - t + t^2 - \dots$$

between the limits 0 and x. What we proved amounted to this, that

$$\int_0^x \frac{dt}{1+t} = \int_0^x dt - \int_0^x t\,dt + \int_0^x t^2\,dt - \dots;$$

or in other words that *the integral of the sum of the infinite series*

$$1 - t + t^2 - \dots,$$

taken between the limits 0 *and* x, *is equal to the sum of the integrals of its terms taken between the same limits*. Another way of expressing this is to say that the operations of summation from 0 to ∞, and of integration from 0 to x, are *commutative* when applied to the function $(-1)^n t^n$, i.e. that it does not matter in what order they are performed on the function.

Again, in § 223, we proved that the differential coefficient of the exponential function

$$\exp x = 1 + x + \frac{x^2}{2!} + \dots$$

is itself equal to $\exp x$, or that

$$D_x\left(1 + x + \frac{x^2}{2!} + \dots\right) = D_x 1 + D_x x + D_x \frac{x^2}{2!} + \dots;$$

that is to say that *the differential coefficient of the sum of the series is equal to the sum of the differential coefficients of its terms*, or that the operations of summation from 0 to ∞ and of differentiation with respect to x are commutative when applied to $x^n/n!$.

Incidentally we proved in the same section that the function $\exp x$ is a continuous function of x, or in other words that

$$\lim_{x\to\xi}\left(1+x+\frac{x^2}{2!}+\dots\right)=1+\xi+\frac{\xi^2}{2!}+\dots=\lim_{x\to\xi}1+\lim_{x\to\xi}x+\lim_{x\to\xi}\frac{x^2}{2!}+\dots;$$

i.e. that the limit of the sum of the series is equal to the sum of the limits of the terms, or that the sum of the series is continuous for $x=\xi$, or that the operations of summation from 0 to ∞ and of making x tend to ξ are commutative when applied to $x^n/n!$.

In each of these cases we gave a special proof of the correctness of the result. We have not proved any general theorem from which the truth of any one of them could be inferred immediately. In Ex. XXXVII. 1 we saw that the sum of a finite number of continuous terms is itself continuous, and in § 114 that the differential coefficient of the sum of a finite number of terms is equal to the sum of their differential coefficients; and in § 165 we stated the corresponding theorem for definite integrals. Thus we have proved that in certain circumstances the operations symbolised by

$$\lim_{x\to\xi}\dots,\quad D_k\dots,\quad \int_a^b\dots dx$$

are commutative with respect to the operation of summation of a *finite* number of terms. And it is natural to suppose that, in certain circumstances which it should be possible to define precisely, they should be commutative also with respect to the operation of summation of an *infinite* number. It is natural to suppose so; but that is all that we have a right to say at present.

A few further instances of commutative and non-commutative operations may help to elucidate these points.

(1) Multiplication by 2 and multiplication by 3 are always commutative, since

$$2\times 3\times x = 3\times 2\times x$$

for all values of x.

(2) The operation of taking the real part of z is never commutative with that of multiplication by i, except when $z=0$; for

$$i\times \mathbf{R}(x+iy)=ix,\quad \mathbf{R}\{i\times(x+iy)\}=-y.$$

(3) The operations of proceeding to the limit zero with each of two variables x and y may or may not be commutative when applied to a function $f(x,y)$. Thus

$$\lim_{x\to 0}\{\lim_{y\to 0}(x+y)\}=\lim_{x\to 0}x=0,\quad \lim_{y\to 0}\{\lim_{x\to 0}(x+y)\}=\lim_{y\to 0}y=0;$$

but on the other hand

$$\lim_{x\to 0}\left(\lim_{y\to 0}\frac{x-y}{x+y}\right)=\lim_{x\to 0}\frac{x}{x}=\lim_{x\to 0}1=1,$$

$$\lim_{y\to 0}\left(\lim_{x\to 0}\frac{x-y}{x+y}\right)=\lim_{y\to 0}\frac{-y}{y}=\lim_{y\to 0}(-1)=-1.$$

(4) The operations $\sum\limits_1^\infty \ldots$, $\lim\limits_{x\to 1}\ldots$ may or may not be commutative. Thus if $x\to 1$ through values less than 1 then

$$\lim_{x\to 1}\left\{\sum_1^\infty \frac{(-1)^{n-1}}{n}x^n\right\}=\lim_{x\to 1}\log(1+x)=\log 2,$$

$$\sum_1^\infty\left\{\lim_{x\to 1}\frac{(-1)^{n-1}}{n}x^n\right\}=\sum_1^\infty\frac{(-1)^{n-1}}{n}=\log 2;$$

but on the other hand

$$\lim_{x\to 1}\left\{\sum_1^\infty (x^{n-1}-x^n)\right\}=\lim_{x\to 1}\{(1-x)+(x-x^2)+\ldots\}=\lim_{x\to 1}1=1,$$

$$\sum_1^\infty\left\{\lim_{x\to 1}(x^{n-1}-x^n)\right\}=\sum_1^\infty(1-1)=0+0+0+\ldots=0.$$

The preceding examples suggest that there are three possibilities with respect to the commutation of two given operations, viz.: (1) the operations may always be commutative; (2) they may never be commutative, except in very special circumstances; (3) they may be commutative in most of the cases which occur commonly in analysis.

The really important case is (as is suggested by the instances which we quoted from Ch. IX) that in which each operation is one which involves a passage to the limit, such as a differentiation or the summation of an infinite series: such operations are called *limit operations*. The problem of deciding whether two given limit operations are commutative is one of the most important in mathematics; but to attempt to deal with problems of this character by means of general theorems would carry us far beyond the scope of this volume.

We may however remark that the answer to the general question is on the lines suggested by the examples above. If L and L' are two limit operations then the numbers $LL'z$ and $L'Lz$ are not *generally* equal, in the strict sense of the word 'general'. We can always, by the exercise of a little ingenuity, find z so that $LL'z$ and $L'Lz$ shall differ from one another.

But they are equal generally if we use the word in a more 'practical' sense, viz. as meaning 'in the great majority of such cases as are likely to occur naturally'. In practice, a result obtained by assuming that two limit operations are commutative is *probably* true; at any rate it gives a valuable suggestion of the answer to the problem under consideration. But an answer thus obtained must, in default of a further study of the general question, or a special investigation of the particular problem such as we gave in § 220, be regarded as suggested only and not proved.

APPENDIX IV

The infinite in analysis and geometry

Some, though not all, systems of analytical geometry contain 'infinite' elements, the line at infinity, the circular points at infinity, and so on. The object of this brief note is to point out that these concepts are in no way dependent upon the analytical doctrine of limits.

In what may be called 'common Cartesian geometry' a *point* is a *pair of real numbers* (x, y). A *line* is the class of points which satisfy a linear relation $ax+by+c=0$, in which a and b are not both zero. There are no infinite elements, and two lines may have no point in common.

In a system of real homogeneous geometry a point is *a class of triads of real numbers* (x, y, z), not all zero, triads being classed together when their constituents are proportional. A line is a class of points which satisfy a linear relation $ax+by+cz=0$, where a, b, c are not all zero. In some systems one point or line is on exactly the same footing as another. In others certain 'special' points and lines are regarded as peculiarly distinguished, and it is on the relations of other elements to these special elements that emphasis is laid. Thus, in what may be called 'real homogeneous Cartesian geometry', those points are special for which $z=0$, and there is one special line, viz. the line $z=0$. This special line is called 'the line at infinity'.

This is not a treatise on geometry, and there is no occasion to develop the matter in detail. The point of importance is this. The infinite of analysis is a 'limiting' and not an 'actual' infinite. The symbol '∞' has, throughout this book, been regarded as an 'incomplete symbol', a symbol to which no independent meaning has been attached, though one has been attached to certain phrases containing it. But *the infinite of geometry is an actual and not a limiting infinite.* The 'line at infinity' is a line in precisely the same sense in which other lines are lines.

It is possible to set up a correlation between 'homogeneous' and 'common' Cartesian geometry in which all elements of the first system, *the special elements excepted*, have correlates in the second. The line

$$ax+by+cz=0,$$

for example, corresponds to the line

$$ax+by+c=0.$$

Every point of the first line has a correlate on the second, except one, viz. the point for which $z = 0$. When (x, y, z) varies on the first line, in such a manner as to tend in the limit to the special point for which $z = 0$, the corresponding point on the second line varies so that its distance from the origin tends to infinity. This correlation is historically important, since it is from it that the vocabulary of the subject has been derived, and it is often useful for purposes of illustration. It is however no more than an illustration, and no rational account of the geometrical infinite can be based upon it. The confusion about these matters so prevalent among students arises from the fact that, in the commonly used text-books of analytical geometry, the illustration is sometimes taken for the reality.

Readers interested in the relations between analysis and geometry may consult

D. Hilbert, *Grundlagen der Geometrie* (English translation *Foundations of geometry*, Chicago, 1938);

C. W. O'Hara and D. R. Ward, *An introduction to projective geometry*, Oxford, 1937;

G. de B. Robinson, *The foundations of geometry*, Toronto, 1940;

O. Veblen and J. W. Young, *Projective geometry*, vol. 1, New York, 1910;

and an article by the author 'What is geometry?' in the *Mathematical Gazette*, vol. 12, 1925, pp. 309–316.

INDEX

Printed in the United States
By Bookmasters